FUNDAMENTALS OF OIL & GAS ACCOUNTING

5TH EDITION

FUNDAMENTALS OF OIL & GAS ACCOUNTING

5TH EDITION

Charlotte J. Wright, PhD, CPA

and

Rebecca A. Gallun, PhD

> **Disclaimer**
>
> The recommendations, advice, descriptions, and the methods in this book are presented solely for educational purposes. The author and publisher assume no liability whatsoever for any loss or damage that results from the use of any of the material in this book. Use of the material in this book is solely at the risk of the user.

Copyright© 2008 by
PennWell Corporation
1421 South Sheridan Road
Tulsa, Oklahoma 74112-6600 USA

800.752.9764
+1.918.831.9421
sales@pennwell.com
www.pennwellbooks.com
www.pennwell.com

Marketing Manager: Julie Simmons
National Account Executive: Barbara McGee

Director: Mary McGee
Managing Editor: Marla Patterson
Production Manager: Sheila Brock
Production Editor: Tony Quinn
Cover Designer: Karla Pfeifer
Book Layout: Andrew Hicks

Library of Congress Cataloging-in-Publication Data

Wright, Charlotte J.
  Fundamentals of oil and gas accounting / Charlotte J. Wright and Rebecca A. Gallun. -- 5th ed.
    p. cm.
  Rev. ed. of: Fundamentals of oil & gas accounting / Rebecca A. Gallun ... [et al.]. 4th ed.
  Includes bibliographical references and index.
  ISBN 978-1-59370-137-6
  1. Petroleum industry and trade--Accounting. 2. Gas industry--Accounting. I. Gallun, Rebecca A. II. Fundamentals of oil & gas accounting. III. Title.
  HF5686.P3G3 2008
  657'.862--dc22

                                                                        2007041136

All rights reserved. No part of this book may be reproduced, stored in a retrieval system, or transcribed in any form or by any means, electronic or mechanical, including photocopying and recording, without the prior written permission of the publisher.

Printed in the United States of America

9 10   16 15 14

This work is dedicated to our sons Chad Gallun, Todd Wright, and Tate Wright, to our precious grandchildren Grant, Tyler, Hannah, Alec, Ethan, and Olivia, and to our daughters-in-law Susanna, Dawn, and Ashley.

Acknowledgments

The authors wish to acknowledge and thank Caleb Van Dolah for his hard work and efforts in regard to problem material and the solution manual that accompanies this book.

Contents

1 Upstream Oil and Gas Operations 1
    Brief history of the U.S. oil and gas industry 2
    Origin of petroleum 5
    Exploration methods and procedures 8
    Acquisition of mineral interests in property 11
        Mineral rights 11
        Mineral interests 12
        Retained ORI 14
        Carved-out production payment 15
        Carved-out net profits interest created from working interest 16
        Net profits interest creatd from mineral interest 16
    Lease provisions 16
    Drilling operations 23
    Recovery processes 27
    Production and sales 28
    Common state and federal regulations 28
    What does the future hold? 30
    Problems 32
    References 36

2 Introduction to Oil and Gas Accounting 37
    Historical cost accounting methods 44
    Historical development of accounting methods and current status 48
    Introduction to successful efforts accounting 51
    Chart of accounts for a successful efforts company 55
    Introduction to full cost accounting 63
    Chart of accounts for a full cost company 66
    Problems 70

3 Nondrilling Exploration Costs—Successful Efforts 75
    G&G costs 77
    Carrying and retaining costs 80
    Test-well contributions 82
    Support equipment and facilities 86
    Offshore and international operations 87
    Problems 87
    References 91

4 Acquisition Costs of Unproved Property—Successful Efforts 93
Purchase in fee 98
Internal costs 99
Options to lease 100
Delinquent taxes and mortgage payments. 103
Top leasing 105
Disposition of capitalized costs—impairment of unproved properties 106
Disposition of capitalized costs—surrender or abandonment of property 112
Post-balance sheet events 115
Disposition of capitalized costs—reclassification of an unproved property 116
Land department 127
Problems. 128
References 136

5 Drilling and Development Costs—Successful Efforts 137
Income tax accounting for drilling costs 137
Financial accounting for drilling and development costs 141
Well classification 142
Exploratory drilling costs 144
Development drilling costs 151
Stratigraphic test wells 154
AFE's and drilling contracts 155
Special drilling operations and problems 156
    Workovers 156
    Damaged or lost equipment and materials 158
    Fishing and sidetracking 159
Additional development costs 160
Support equipment and facilities 162
Drilling and development seismic 162
Post-balance sheet events 163
Accounting for suspended well costs 164
Interest capitalization 165
Offshore and international operations 167
Problems 167
References 178

6 Proved Property Cost Disposition—Successful Efforts 179
Cost disposition through amortization 181
    Reserves owned or entitled to 181
    DD&A calculation 183
    DD&A on a field-wide basis 188

  DD&A when oil and gas reserves are produced jointly 191
  Estimated future dismantlement, site restoration, and abandonment costs . . . 194
  Exclusion of costs or reserves 195
  Depreciation of support equipment and facilities 197
  Cost disposition—nonworking interests 199
  Revision of DD&A rates 202
 Cost disposition through abandonment or retirement of proved property 206
 Successful efforts impairment 216
 Problems 216
 References 226

7 Full Cost Accounting 227

 Disposition of capitalized costs 234
  Inclusion of estimated future development expenditures 237
  Inclusion of estimated future decommissioning costs 237
  Exclusion of costs 239
  Impairment of unproved properties costs 242
  Abandonment of properties 244
  Reclassification of properties 245
  Support equipment and facilities 253
  DD&A under successful efforts versus full cost 253
 Reserves in place—purchase 256
 Interest capitalization 257
 Limitation on capitalized costs—a ceiling 258
  Asset retirement obligations 260
  Deferred taxes 261
  Income tax effects 261
  Assessment of the ceiling test 263
  SFAS No. 144 265
  Post-balance sheet events and the ceiling test 265
 Problems 266

8 Accounting for Production Activities 281

 Accounting treatment 282
  Cost of production versus inventory 283
  Recognition of inventories 284
  Lower-of-cost-or-market valuation 285
 Accumulation and allocation of costs 286
 Individual production costs 288
  Secondary and tertiary recovery 290
  Gathering systems 291

Saltwater disposal systems. 291
   Tubular goods 291
   Severance taxes 292
  Production costs statements. 292
  Joint interest operations 293
  Decision to complete a well 293
  Project analysis and investment decision making 296
   Payback method. 296
   Accounting rate of return 298
   Net present value method 299
   Internal rate of return 301
   Profitability index. 301
  Problems. 305
  References. 312

9 Accounting for asset retirement obligations and asset impairment. . 313
  Accounting for asset retirement obligations 314
   Scope of SFAS No. 143. 314
   Legally enforceable obligations 315
   Obligating event. 316
   Asset recognition 317
   Initial measurement—fair value 317
   Subsequent recognition and measurement 322
   Funding and assurance provisions 328
   Conditional AROs. 329
  Accounting for the impairment and disposal of long-lived assets 331
   Scope 331
   Asset groups. 332
   Long-lived assets to be held and used. 332
   Long-lived assets to be disposed of. 337
   Impairment for full cost companies 338
  Problems. 338
  References. 348

10 Accounting for Revenue from Oil and Gas Sales. 349
  Definitions 349
  Measurement and sale of oil and natural gas 351
   Crude oil measurement 351
   Run ticket calculation 354
   Crude oil sales 360
   Natural gas measurement 361
   Natural gas sales 363

Standard division order 364
Determination of revenue 366
Unitizations 370
Oil and gas revenues 372
Recording oil revenue 373
Crude oil exchanges 379
Recording gas revenue 380
    Vented or flared gas 385
    Nonprocessed natural gas 385
    Natural gas processing 386
    Stored natural gas 386
    Take-or-pay provisions 386
Timing of revenue recognition 388
    Revenue from crude oil 388
    Revenue from natural gas 389
Revenue reporting to interest owners 391
Additional topics 391
    Gas imbalances 391
    Producer gas imbalances 391
    Pipeline gas imbalances 396
    Allocation of oil and gas 398
    Minimum royalty—an advance revenue to royalty owners 402
Problems 403
References 413

11 Basic Oil and Gas Tax Accounting 415

Lessee's transactions 416
    Nondrilling costs 416
    Acquisition costs 419
    Drilling operations 422
    Equipment costs 423
    Production operations 428
    Losses from unproductive property 432
    Percentage depletion 446
    Property 448
    Recapture of IDC and depletion 449
Lessor's transactions 452
    Acquisition costs 452
    Revenue 453
Problems 454
References 564

xii FUNDAMENTALS OF OIL & GAS ACCOUNTING

12 Joint Interest Accounting 465
    Joint operations 465
        Joint venture contracts. 466
    The joint operating agreement 467
    The accounting procedure 491
        General provisions 516
        Direct charges 517
        Overhead 521
        Pricing of joint account material purchases, transfers, and dispositions. 523
        Inventories 524
    Joint interest accounting. 524
        Booking charges to the joint account: accumulation of joint costs in
            operator's regular accounts. 524
        Booking charges to the joint account: distribution of joint costs as incurred. . 526
        Nonconsent operations 527
        Accounting for materials. 530
    Offshore operations 536
    Joint interest audits. 536
    Problems 538
    References 543

13 Conveyances 545
    Mineral interests 545
        Types of interest 546
    Conveyances: general rules 549
    Conveyances: exchanges and poolings 551
        Farm-ins/farm-outs 553
        Farm-ins/farm-outs with a reversionary working interest 555
        Free wells 560
        Carried interests or sole risk 563
        Joint venture operations 568
        Poolings and unitizations. 570
        Unitizations 571
    Conveyances: sales 573
        Unproved property sales 574
        Proved property sales. 581
    Conveyances: production payments. 589
        Retained production payments 589
        Carved-out production payments 595
    Conveyances—full cost 604
    Problems 607
    References 616

14 Oil and Gas Disclosures 617
Required disclosures..618
Illustrative example...619
Proved reserve quantity information 622
    Reserve definitions..623
    Use of end-of-year prices 624
    Reserve quantity disclosure....................................624
Capitalized costs relating to oil and gas producing activities 627
Costs incurred for property acquisition, exploration, and development activities..631
Results of operations for oil and gas producing activities 633
Standardized measure of discounted future net cash flows relating
    to proved oil and gas reserve quantities 637
Changes in the standardized measure of discounted future net cash flows relating
    to proved oil and gas reserve quantities 649
    Analysis of reasons for changes in value of standardized measure 12/31/XB..652
Conclusion 666
Problems..667
References 675

15 Accounting for International Petroleum Operations................ 677
Petroleum fiscal systems...678
Concessionary systems..679
Concessionary agreements with government participation 680
Contractual systems..681
    Government involvement in operations 681
Production sharing contracts...682
    Signature and production bonuses 682
    Royalties...683
    Government participation.....................................684
    Cost recovery 684
    Profit oil 685
    Other terms and fiscal incentives...............................688
Service contracts...689
Joint operating agreements..691
    Recoverable and non-recoverable costs..........................691
Financial accounting issues 692
    Financial accounting versus contract accounting 692
    Disclosure of proved reserves—SFAS No. 69.....................693
International accounting standards....................................695
Problems..695
References 699

16 Analysis of Oil and Gas Companies' Financial Statements 701

- Source of data. 702
- Comparing financial reports 703
- Reserve ratios 704
  - Reserve replacement ratio 704
  - Reserve life ration 707
  - Net wells to gross wells ratio 708
  - Average reserves per well ratio. 709
  - Average daily production per well. 710
- Reserve cost ratios 711
  - Finding costs ratios 711
  - Lifting costs per BOE 716
  - DD&A per BOE 718
- Reserve value ratios 719
  - Value of proved reserve additions per BOE 719
  - Value added ratio. 722
- Financial ratios. 724
- Problems 726
- References. 729

Appendix A: Authorization for Expenditure 731

Appendix B: *Regulation SX 4-10* 735

Index 749

1
UPSTREAM OIL AND GAS OPERATIONS

Traditionally, oil and gas operations have been classified as being either upstream or downstream. **Upstream** activities include exploration, acquisition, drilling, developing, and producing oil and gas. In other words, upstream activities generally include all of the activities involved in finding and producing oil and gas up to the initial point that the oil or gas is capable of being sold or used. Upstream activities are frequently referred to as exploration and production activities, or E&P activities. **Downstream** activities generally include refining, processing, marketing, and distribution.

Classification of oil and gas activities as being upstream versus downstream may seem straightforward initially. However, modern petroleum industry operations often are conducted in exotic locations and have become increasingly complex, making classification of operations into any single category difficult. As a consequence, today some activities that have characteristics of both upstream and downstream activities are actually referred to as **midstream**.

An **integrated oil and gas company** is one involved in E&P activities as well as at least one downstream activity. An **independent oil and gas company** is one involved primarily in only E&P activities.

Upstream E&P activities are uniquely characterized by the following:

- A high level of risk
- A long time span before a return on investment is received
- A lack of correlation between the magnitude of expenditures and the value of any resulting reserves
- A high level of regulation
- Complex tax rules
- Unique cost-sharing agreements

For accounting, the classification of oil and gas activities as being upstream versus midstream or downstream is of special significance. This is due to the fact that a specialized set of accounting rules and standards apply to the financial accounting for and reporting of upstream oil and gas operations. These rules, which are complex and relate to companies ranging in size from mega-majors to smaller U.S. domestic producers, are the primary focus of this book.

The procedures and steps involved in locating and acquiring mineral interests, drilling and completing oil and gas wells, and producing and selling petroleum products are briefly reviewed in this chapter. These procedures and steps are essentially the same for any type or size of oil and gas company. A more detailed discussion of these procedures may be found in publications such as *Introduction to Oil and Gas Production*, *A Primer of Oilwell Drilling*, or *The Petroleum Industry: A Nontechnical Guide*.[1-3] A basic knowledge of these procedures is necessary in order to understand their accounting implications, which are discussed in the following chapters.

This book provides the fundamentals of a broad range of accounting and reporting issues that relate to upstream E&P operations conducted in the United States. It also provides an overview for oil and gas operations conducted outside the United States. A more detailed discussion of the accounting issues and challenges encountered in upstream oil and gas operations conducted outside the United States is given in *International Petroleum Accounting*.[4]

Brief History of the U.S. Oil and Gas Industry

The modern history of the U.S. oil and gas industry began in the latter half of the 19th century with the first commercial oil drilling venture in Pennsylvania. The product demand at that time was for kerosene, which was used as lamp fuel. The petroleum industry expanded greatly in the 20th century as new uses for oil were developed. The invention of the automobile was responsible for much of the industry's growth.

Early in the 20th century, World War I strained the industry's ability to supply fuel, prompting the decade of the 1920s to be one of great U.S. oil discovery. In that decade, many U.S. companies also began exploring for foreign oil in the Middle East, South America, Africa, and the Far East. The onset of World War II in 1937 once again tested the industry's ability to supply oil. During World War II, drilling began in U.S. waters on offshore structures that resembled the offshore drilling platforms of today. During and following World War II, U.S. companies also increased foreign exploration, especially in the Persian Gulf area.

Following World War II, natural gas was established as a major fuel for industry and home heating. It was also during the postwar period that natural gas transmission pipelines were constructed, linking remote producing areas to large population centers.

The United States increased its reliance on foreign oil during the 1950s and 1960s. In 1960, the Organization of Petroleum Exporting Countries (OPEC) was formed. Although OPEC was ineffective during its first decade of existence, its oil embargo of 1973–74, coupled with an increase in the price of crude oil, created an energy crisis in the United States. The resulting shortages of petroleum products caused the nation for the first time to become acutely aware of the myriad of problems and issues resulting from its dependence on foreign oil.

This energy crisis prompted the passage of the Energy Policy and Conservation Act of 1975, the purpose of which was to encourage energy conservation, reduce reliance on foreign oil, and encourage the development of alternative energy sources. Unfortunately, with the easing of the energy crisis, the nation's attention turned away from energy matters. By 1977, U.S. dependence on foreign sources of petroleum reached a high of 47%. A second round of price increases occurred from 1979 through 1981. The U.S. dependence on foreign sources of petroleum fell to 27% by 1985; however, according to the U.S. Department of Energy, the U.S. reliance on imported oil increased by 142% between 1984 and 2004.

Historically, both the supply and demand of crude oil have been inelastic in the short run. As a consequence, political events that affect supply tend to cause volatility in short-run oil prices, with disruptions in supply resulting in increases in crude oil prices. Higher prices have, as expected, resulted in E&P companies expanding their exploration efforts and the U.S. public conserving energy. Both efforts have historically relieved the supply and demand pressure, resulting in downward price movement in the years following a spike in prices. Recent political instability in the Middle East, along with increased Asian demand for oil, have had a substantial impact on world oil and gas prices. Whether or not this trend will persist in the future is yet to be seen.

Figure 1–1 reports crude oil prices in U.S. dollar/barrel (bbl) from 1861 through 2005 in both "money of the day," i.e., nominal dollars, as well as prices adjusted to 2005 U.S. dollars. This chart illustrates the volatility of crude oil prices whether denominated in nominal U.S. dollars or adjusted for inflation. Figure 1–2 shows the upward trend in U.S. natural gas prices over a 30-year period measured in nominal U.S. dollars (not adjusted for inflation).

4 Fundamentals of Oil & Gas Accounting

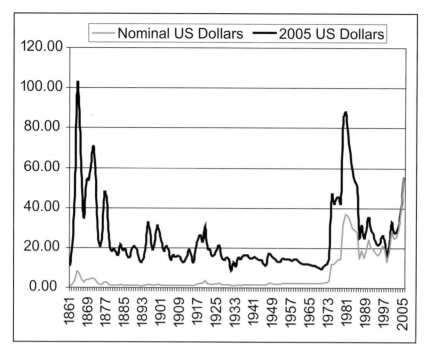

Fig. 1–1. Historical world oil prices
Source: BP Statistical Review of World Energy. June 2006.

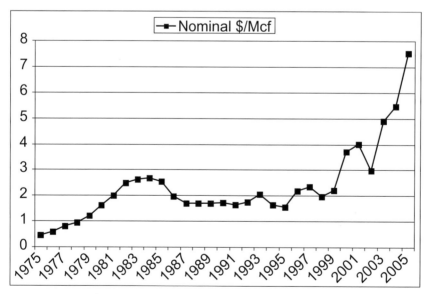

Fig. 1–2. Natural gas historical prices—average U.S. wellhead price by year
Source: Energy Information Administration. 2006. National Energy Review. December (www.eia.doe.gov).

Origin of Petroleum

The most widely accepted hypothesis of the origin of oil and gas is the organic theory. This theory holds that petroleum (hydrocarbons) is formed from organic material including marine plants and animals that lived millions of years ago in low-lying areas—normally in the oceans of the world. These plant and animal remains were deposited throughout the years, along with layer after layer of eroded particles of igneous rock. The weight and pressure of the overlying layers caused the eroded rock particles to form **sedimentary rock**. The weight and pressure of overlying layers—and other not fully understood factors, such as chemical and bacterial processes—changed, and still change, the organic material into oil and gas.

After formation, oil and gas move upward through the layers of sedimentary rock due to pressure and the natural tendency of oil and gas to rise through water. (Salt water is often contained in the pore space of sedimentary rock.) The petroleum migrates upward through porous and permeable rock formations until it becomes trapped by an impervious layer of rock. The impervious rock that prevents further movement of the oil and gas is called a **trap** (Fig. 1–3).

Some of the common types of traps are as follows:

- **Fault trap.** A trap formed when the movement of the earth's crust causes different rock strata to offset or shear off. In a fault trap, a nonporous rock formation that has shifted stops the movement of oil or gas within an offsetting formation that allows petroleum to migrate.
- **Anticline.** A trap formed by the folding of the earth's crust into a dome. This upward folding is caused by pressures developed from the earth's molten core. An impervious or nonporous layer of rock overlying the anticline traps the oil or gas in the anticline structure. Most of the earth's oil and gas reserves are found in anticlines.
- **Salt dome.** With substantial heat and pressure, salt buried deep within the Earth may slowly begin to move upward. As it does so, the rock layers are cracked, bent, and folded. Often oil and gas are trapped in these cracked and folded rock beds.

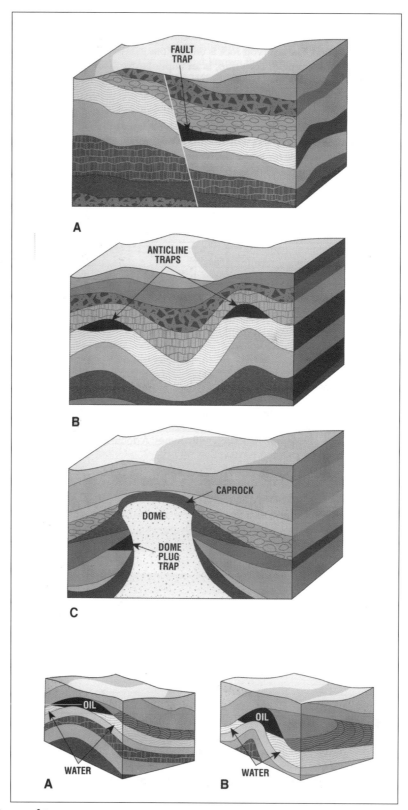

Fig. 1–3. Types of traps
Source: Van Dyke, Kate. 1997. Fundamentals of Petroleum. 4th ed. Austin, TX: Petroleum Extension Service. p. 17.

In order for oil and gas production to be feasible, there must be a sufficient quantity of hydrocarbons to justify the cost of producing it. Such an accumulation is commonly referred to as a **reservoir**. In order for an oil or gas reservoir to have been formed, four conditions must have been present:

- A source of petroleum, i.e., remains of land and sea life
- Conditions such as heat and pressure resulting in the transformation of the organic material into petroleum
- Porous and permeable rock through which the petroleum was able to migrate after formation
- An impervious rock formation that acts as a trap or cap rock, permitting petroleum to accumulate in substantial quantities

Assuming these four conditions are present, the reservoir rock itself must have porosity and permeability. **Porosity** is the measure of the pore space, i.e., openings in a rock in which petroleum can exist. The greater the porosity, the more fluid (petroleum) the rock can hold. **Permeability** measures the "connectability" of the pores, which determines the ability of the petroleum to flow through the rock from one pore space to another. High porosity is often accompanied by high permeability (Fig. 1–4 and Fig. 1–5).

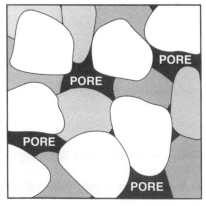

Fig. 1–4. Porosity
Source: Van Dyke, Kate. 1997. Fundamentals of Petroleum. 4th ed. Austin, TX: Petroleum Extension Service. p. 14.

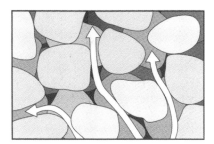

Fig. 1–5. Permeability
Source: Van Dyke, Kate. 1997. Fundamentals of Petroleum. 4th ed. Austin, TX: Petroleum Extension Service. p. 15.

To be commercially productive, a petroleum reservoir must have adequate porosity and permeability, and also must have a sufficient physical area of rock that contains hydrocarbons (oil and natural gas). In other words, the reservoir must contain enough oil and natural gas to, when produced and sold, cover attendant costs.

If an oil and gas reservoir has low permeability, procedures exist to increase the flow of oil or gas through the formation to the wellbore. One method used to increase the flow from a "tight" formation is **fracturing**. This method usually involves introducing sand mixed with water or oil into the formation under high pressure to open or clean channels between the pores. Another common method used to increase the permeability of the formation is **acidizing**. Acidizing usually involves introducing hydrochloric acid into the formation to enlarge or reopen the channels between the pores. While these techniques to increase permeability are commonly used, they are effective only for a small area of the formation around the wellbore.

Exploration Methods and Procedures

Oil and gas exploration involves the work of geoscientists using a variety of geological and geophysical (G&G) techniques to identify areas far beneath the earth's surface that may contain petroleum reserves. Geological methods rely on the identification of rocks and minerals on or near the surface and the understanding of the environments in which they were formed. Geological studies, which involve surface studies, involve any number of methods, depending on the size of the area being examined. These methods may include, for example, the use of aerial photography, satellite imaging, imaging radar, and topographical and geological mapping. Such methods are aimed at gathering data about surface features that can be used to make inferences regarding the potential existence of petroleum-bearing subsurface formations.

Geophysical methods, which involve subsurface studies, are aimed at locating and detecting the presence of subsurface structures and the determination of their size, shape, depth, and physical properties in order to identify the presence of certain physical characteristics that are indicative of oil and gas reservoirs. Geophysical methods include gravitational studies, magnetic and electromagnetic evaluation, and seismic studies.

Seismology is one of the most important tools in oil and gas exploration today. These studies provide detailed information about subsurface structures by recording the reflection of sound waves on subsurface formations. Innovations in seismology, such as 3-D seismic studies, have significantly increased drilling success rates. The use of seismic studies has extended beyond exploration, with seismic studies now being used extensively in field development and production planning. Time-lapse, or 4-D seismic, involves repeating a series of 3-D seismic surveys over time in order to monitor how certain reservoir properties (i.e., movement of fluids, temperature, and pressure) change in response to production. In this way the movement of oil, gas, and/or water can be anticipated before it affects production.

Today, the search for oil and gas is taking E&P companies to increasingly exotic and challenging locations and involves the application of increasingly complex, cutting-edge technologies. For example, advances in offshore technology, supercomputing, and 3-D imaging techniques contributed to recent exploration successes in ultra deepwater drilling in the Gulf of Mexico. Here wells were drilled through 10,000 feet of water and more than five miles of rock to locations that would have been completely inaccessible just a few years ago.

By examining all the G&G data collected—including the data collected by seismic studies—a map can be made to indicate formations favorable to the accumulation of petroleum, as well as interest areas that warrant further investigation. A **reconnaissance survey** is a G&G study covering a large or broad area. A **detailed survey** is a G&G study covering a smaller area, called an **area of interest**. Due to the high cost of seismic studies, they are usually performed after general reconnaissance studies have indicated a formation where high potential for oil and gas accumulation exists. If an area of interest is identified by a detailed survey, the area may be leased, if available and if not already leased.

Even the best G&G techniques cannot normally guarantee that oil or gas exists in economically producible quantities. Often the only definite way to determine whether an economically viable petroleum reservoir exists is to drill wells into the formation. In the past, with traditional G&G technology, approximately only 10% of wells drilled in new, unproved areas (areas with no known reservoirs) were successful. With the advent of 3-D and 4-D seismic technology, success rates in unproved areas have increased to approximately 60%. With current technology, more than 75% of all wells drilled (including in both unproved areas and areas with proved reservoirs) are successful.

A successful well is a well that finds reserves in economically producible quantities. However, even though a well is classified as a success, it may nevertheless be unprofitable. (See chapter 8.)

The following flowchart (Fig. 1–6) outlines the procedure in exploring for oil and gas.

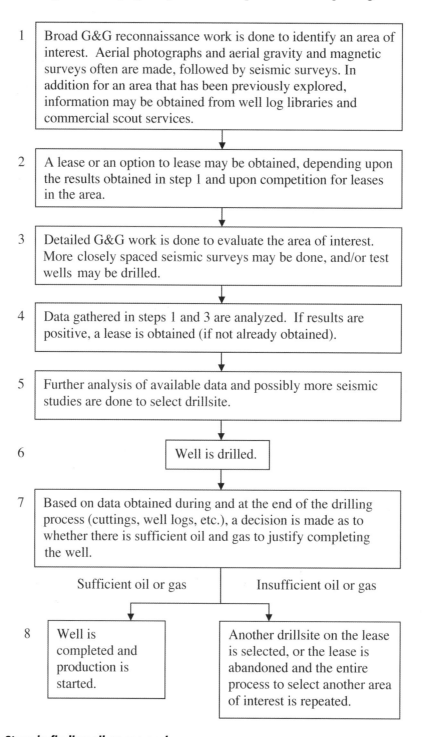

Fig. 1–6. Steps in finding oil or gas onshore

ACQUISITION OF MINERAL INTERESTS IN PROPERTY

After an E&P company has identified an area with potential, the company will seek to acquire the right to explore, develop, and produce any minerals that might exist beneath the property, unless it already holds this right. This right, along with the right to simply share in proceeds from the sale of any minerals produced, is referred to as a **mineral interest** or an **economic interest**. The specific type of mineral interest that is owned largely determines how costs and revenues are shared.

The United States is one of the few countries in the world where individual ownership of mineral rights is allowed. In the United States, most mineral rights are owned by individuals; however, in some instances (e.g., offshore), the U.S. government or a state government owns the mineral rights. Therefore, oil and gas companies wishing to obtain a mineral interest in the United States must typically do so by executing lease agreements with individuals. In most locations outside the United States, ownership of mineral rights resides with a governmental entity (e.g., a ministry of petroleum). Therefore, oil and gas companies seeking to obtain mineral interests outside the United States, in most cases, must contract with the relevant governmental authority in the country in which the property is located.

The following section discusses the most common types of mineral interests and the typical methods of acquiring them. The discussion relates to the United States, where mineral interests are typically acquired via leasing. Contracting in locations outside the United States is discussed in detail in chapter 15.

Mineral rights

U.S. law assumes that for ownership purposes, the surface of a piece of property can be separated from minerals existing underneath the surface. When a piece of land is purchased, one may acquire ownership of the surface rights only, the mineral rights only, or both rights. An ownership of both the surface and mineral rights is called a **fee interest**.

The **surface owner** has the right to use the surface in any legal way that the owner deems appropriate. For example, the surface owner may use the land for farming, ranching, building a residence, building apartments, etc. **Mineral rights (MR)** refer to the ownership, conveyed by deed, of any mineral beneath the surface. If the mineral rights are owned by one party and the surface is owned by another, the surface owner must allow the mineral rights owner, or his lessee, access to the surface area that is required to conduct exploration and production operations. The surface owner is entitled to compensation for any damages that may result from exploration and production operations.

> **Mr. A owns mineral rights (MR) in property**

Mineral interests

A **mineral interest (MI)** is an economic interest or ownership of minerals-in-place, giving the owner the right to a share of the minerals produced either in-kind or in the proceeds from the sale of the minerals. (**Sharing in-kind** means the company or individual has elected to receive the oil or gas itself rather than the proceeds from the sale of the minerals.) When the owner of the mineral rights enters into a lease agreement or contract, two types of mineral interests are created—a working interest and a royalty interest. Descriptions of the basic types of mineral interests are given in the following paragraphs. Examples and additional types of mineral interests are provided in chapter 13.

Royalty interest (RI). This type of mineral interest is created by leasing. The royalty interest is retained by the owner of the mineral rights when that owner enters into a lease agreement with another party. The royalty interest typically receives a specified portion of the minerals produced or a specified portion of the gross revenue from selling the production, free and clear of any costs of exploring, developing, or operating the property. The royalty interest owner is responsible for any severance or production taxes assessed on his or her share of production from the property. While the amount of the royalty is subject to negotiation, a 1/8 royalty is common in the United States on nongovernment leases.

Because the royalty owner is typically not responsible for exploration, development, and production costs, in most instances, the working interest owner bears all of the costs related to drilling, completion, testing, and other similar costs. However, most leases provide that the royalty owner bear a proportionate share of postproduction costs. Postproduction costs typically include costs related to the transportation of the saleable product as well as costs necessary to get the product into marketable condition, such as dehydration, compression, and liquid extraction costs. As stated above, the royalty owner is also responsible for his or her share of production or severance taxes.

Since the royalty interest owner is not responsible for the exploration, development, or production of the property, the interest is referred to as a **nonoperating interest** or **nonworking interest**.

Working interest or operating interest (WI). This interest is created via leasing and is responsible for the exploration, development, and operation of a property. The working interest owner or owners are responsible for paying all (100%) of the cost of exploring, drilling, developing, and producing the property. The working interest's share of revenue, however, is the amount that remains after deducting the share of the royalty interest and other nonworking interests. (Other types of nonworking interests are defined in following paragraphs.)

Mr. A owns MR	Mr. A leases → → → → → → Property to Co. B	Mr. A = RI = 1/8
		Co. B = WI = 100% (7/8 revenue interest)

In the previous diagrammed illustration, the working interest is 100%. Specifically, the working interest owner would be responsible for paying 100% of the cost of exploration, development, and production and, after paying the 1/8 royalty, would retain 7/8 of the gross revenues. The royalty interest owner would receive 1/8 of the gross revenues and pay none of the exploration, development, and production costs. Note that the working interest is stated in terms of how the costs are to be shared, while the royalty interest is stated in terms of its share of gross revenue.

A working interest can be either an undivided interest or a divided interest. An undivided mineral interest exists when multiple owners share and share alike, according to their proportion of ownership in any minerals severed from the ground. In contrast, a divided interest exists when specific parties own specific acreage, minerals, or equipment.

For example, assume that Company A enters into an agreement to lease 640 acres from Rachael Brown, who receives a 1/8 royalty. Company A owns a 100% working interest, and Rachael Brown owns a 1/8 royalty interest. Now, Company A sells 50% of the working interest to Company B. Thus both Company A and Company B have an **undivided** 50% working interest in the entire 640 acres. This means that they will each pay 50% of the cost of exploring, developing, and producing the property. Additionally, they will each be entitled to 50% of the revenue net of the royalty (i.e., 50% of 7/8 of gross sales). Now, instead of selling 50% of the working interest in 640 acres, assume that Company A sells 100% of the working interest in 320 acres to Company B. This transaction results in a **divided** working interest. Afterwards, each company owns 100% of the working interest in 320 acres or one-half of the original acreage. Company A and Company B would each be responsible for paying 100% of the cost of exploring, developing, and producing their 320 acres, and each would be entitled to 100% of 7/8 of the gross revenue generated from their own 320 acres. In addition, Company A and Company B would each be responsible for paying Rachael Brown her royalty of 1/8 of the gross revenue from each respective 320 acre property.

Joint working interest. A joint working interest is an undivided working interest owned by two or more parties. Sharing the working interest is common in the oil and gas industry, since it provides a means for companies to share the costs and risks of operations. Further, since less money is invested in any one property, companies are able to invest in more properties, thus spreading their risk over many properties.

In a joint working interest, one of the parties is designated as the **operator** of the property and all the other working interest owners are called **nonoperators**. The operator manages the day-to-day operations of the property, with the nonoperators each being responsible for paying their proportionate share of any costs incurred. These arrangements are often referred to as **E&P joint ventures**. In a typical E&P joint venture operation, each working interest owner accounts for its own share of costs. This practice is referred to as **proportionate consolidation**. Partnership accounting is typically not used to account for joint oil and gas operations.

Most E&P joint operations are not set up as separate legal entities; rather, the joint ventures involve undivided interests in jointly controlled assets. In the event that the joint venture is incorporated or otherwise established as a separate legal entity, consolidation or use of the equity method of accounting may be required.

Overriding royalty interest (ORI). An ORI is a nonworking interest created from the working interest. The ORI's share of revenue is a stated percentage of the share of revenue belonging to the working interest from which it was created. Like a royalty interest, the owner of an ORI does not pay any of the exploration, development, or operating costs but is responsible for its share of any severance or production taxes. An overriding royalty interest may be created either by being retained by the working interest owner when the working interest is sold or otherwise transferred, or by being carved out. A carved-out ORI is created when the working interest owner sells or transfers the ORI and retains the working interest.

Retained ORI

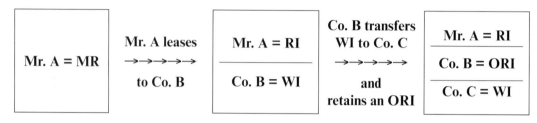

In the example above, if Mr. A leases his property to Co. B, retaining a 1/8 royalty interest, and Co. B subsequently conveys its interest to Co. C, retaining a 1/7 ORI, the costs and revenues would be shared as follows:

	Costs	Share of Gross Revenues
Mr. A	0%	1/8
Co. B	0%	1/7 × 7/8
Co. C	100%	6/7 × 7/8

Production payment interest (PPI). This interest is a nonworking interest created out of a working interest and is similar to an ORI, except that a production payment interest is limited to a specified amount of oil or gas, money, or time, after which it reverts back to the interest from which it was created and ceases to exist. If a production payment is payable with money, the payment is typically stated as a percentage of the working interest's share of revenue. If it is payable in product (i.e., oil, gas, etc.), payment is typically stated as a percentage of the working interest's share of current production. Like ORIs, production payments are created by carve-out or by retention. The following diagram illustrates a carved-out production payment.

Carved-out production payment

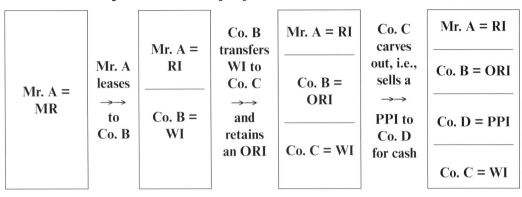

Both overriding royalty interests and production payments are created out of the working interest. These interests, as well as other interests discussed in chapter 13, are often created by the working interest owner in order to obtain financing or assistance in exploring and developing a property and to spread the risk involved.

Net profits interest (NPI). This nonworking interest is created on onshore property typically from the working interest. Offshore, a net profits interest is the type of interest that the government, as the mineral rights owner, often retains when leasing an offshore block to a petroleum company. This type of interest is similar to a royalty interest or an ORI except that the amount to be received is a specified percentage of net profit from the property versus a percentage of the gross revenues from the property. The allowed deductions from gross revenues to calculate the net profit are usually specified in the lease agreement. While net profits interest owners are entitled to a percentage of the profits, they are not responsible for any portion of losses incurred in property development and operations. These losses, however, may be recovered by the working interest owner from future profits.

When created from the working interest, a net profits interest may be created either by being retained by the working interest owner when the working interest is sold or otherwise transferred, or by being carved out. A carved-out net profits interest is created when the working interest owner sells or transfers the net profits interest and retains the working interest. The first of the following diagrams illustrates a net profits interest created onshore from the working interest by being carved out. The second diagram illustrates a net profits interest being created offshore when the property is leased by the government to an oil company.

Carved-out net profits interest created from working interest

Net profits interest created from mineral interest

Government owns MR	Government leases to Co. B →→→→→→	Government = NPI
		Co. B = WI = 100%

Pooled or unitized working interest. This type of interest is created when the working interests as well as the nonworking interests in two or more properties are combined. Each interest owner now owns the same type of interest (but a smaller percentage) in the total combined property as they held previously in the separate property. The properties are operated as one unit, resulting in a more efficient, economical operation.

The terms **pooling** and **unitization** are often used interchangeably. These terms refer to the combining of small or irregular tracts of land into a unit large enough to meet state spacing regulations for drilling, or the combining of tracts in a field or reservoir in order to facilitate enhanced recovery projects. (Enhanced recovery projects are more complicated and expensive production techniques used after the first stages of production of a reservoir.) However, the most common usage of the term **pooling** is the combining of undrilled acreage to form a drilling unit. In contrast, the term **unitization** is commonly used to refer to a larger combination involving an entire producing field or reservoir for purposes of enhanced oil and gas recovery. In most states, pooling or unitization can be forced to facilitate maximum recovery and more efficient production.

LEASE PROVISIONS

Petroleum companies must obtain the rights to explore, drill, and produce subsurface minerals before conducting those activities. While these rights may be acquired through an outright purchase of a fee interest or the mineral rights, they are usually acquired through an oil and gas lease agreement (a copy of a typical lease contract is provided in Fig. 1–7).

Fig. 1–7. Oil, gas, and mineral lease

Source: Pound Printing Company. (Most of the forms sold by Pound Printing Company are copyrighted and, as such, are protected under penalty of law. The photocopying, scanning, or reproduction in any manner of the copyrighted forms is strictly prohibited without the express written consent of Pound Printing Company.)

NOTICE OF CONFIDENTIALITY RIGHTS: IF YOU ARE A NATURAL PERSON, YOU MAY REMOVE OR STRIKE ANY OR ALL OF THE FOLLOWING INFORMATION FROM THIS INSTRUMENT BEFORE IT IS FILED FOR RECORD IN THE PUBLIC RECORDS: YOUR SOCIAL SECURITY NUMBER OR YOUR DRIVER'S LICENSE NUMBER.

Producers 88 (4/76) Revised - Rental
With 640 Acres Pooling Provision

©POUND PRINTING COMPANY
P.O. Box 683046, Houston, Texas 77268-3046, (713) 552-9797

OIL, GAS AND MINERAL LEASE

THIS AGREEMENT made this   day of   ,   , between

Lessor (whether one or more), whose address is:

and
Lessee, WITNESSETH:

1. Lessor in consideration of   Dollars
($), in hand paid, of the royalties herein provided, and of the agreements of Lessee herein contained, hereby grants, leases and lets exclusively unto Lessee for the purpose of investigating, exploring, prospecting, drilling and mining for and producing oil, gas and all other minerals, conducting exploration, geologic and geophysical surveys by seismograph, core test, gravity and magnetic methods, injecting gas, water and other fluids, and air into subsurface strata, laying pipe lines, building roads, tanks, power stations, telephone lines and other structures thereon and on, over and across lands owned or claimed by Lessor adjacent and contiguous thereto, to produce, save, take care of, treat, transport and own said products, and housing its employees, the following described land in   County, Texas, to-wit:

This lease also covers and includes all land owned or claimed by Lessor adjacent or contiguous to the land particularly described above, whether the same be in said survey or surveys or in adjacent surveys, although not included within the boundaries of the land particularly described above. For the purpose of calculating the rental payments hereinafter provided for, said land is estimated to comprise   acres, whether it actually comprises more or less.

2. Subject to the other provisions herein contained, this lease shall be for a term of   years from this date (called "primary term") and as long thereafter as oil, gas or other mineral is produced from said land or land with which said land is pooled hereunder.

3. The royalties to be paid by Lessee are: (**a**) on oil, one-eighth of that produced and saved from said land, the same to be delivered at the wells or to the credit of Lessor into the pipelines to which the wells may be connected; Lessee may from time to time purchase any royalty oil in its possession, paying the market price therefor prevailing for the field where produced on the date of purchase; (**b**) to pay Lessor on gas and casinghead gas produced from said land (1) when sold by Lessee, one-eighth of the amount realized by Lessee, computed at the mouth of the well, or (2) when used by Lessee off said land or in the manufacture of gasoline or other products, one-eighth of the amount realized from the sale of gasoline or other products extracted therefrom and one-eighth of the amount realized from the sale of residue gas after deducting the amount used for plant fuel and/or compression; while there is a gas well on this lease or on acreage pooled therewith but gas is not being sold or used, Lessee may pay as royalty, on or before ninety (90) days after the date on which (1) said well is shut in, or (2) the land covered hereby or any portion thereof is included in a pooled unit on which a well is located, or (3) this lease ceases to be otherwise maintained as provided herein, whichever is the later date, and thereafter at annual intervals on or before the anniversary of the date the first payment is made, a sum equal to the amount of the annual rental payable in lieu of drilling operations during the primary term on the number of acres subject to this lease at the time such payment is made, and if such payment is made or tendered, this lease shall not terminate, and it will be considered that gas is being produced from this lease in paying quantities; and (**c**) on all other minerals mined and marketed, one-tenth either in kind or value at the well or mine, at Lessee's election, except that on sulphur mined and marketed the royalty shall be fifty cents (50¢) per long ton. Lessee shall have free use of oil, gas, coal, and water from said land, except water from Lessor's wells, for all operations hereunder, and the royalty on oil, gas and coal shall be computed after deducting any so used.

Fig. 1–7. (Continued)

4. Lessee, at its option. is hereby given the right and power to pool or combine the acreage covered by this lease or any portion thereof as to oil and gas, or either of them, with any other land covered by this lease, and/or with any other land, lease or leases in the immediate vicinity thereof to the extent hereinafter stipulated, when in Lessee's judgment it is necessary or advisable to do so in order properly to explore, or to develop and operate said leased premises in compliance with the spacing rules of the Railroad Commission of Texas, or other lawful authority, or when to do so would, in the judgment of Lessee, promote the conservation of oil and gas in and under and that may be produced from said premises. Units pooled for oil hereunder shall not substantially exceed 40 acres each in area, and units pooled for gas hereunder shall not substantially exceed in area 640 acres each plus a tolerance of ten percent (10%) thereof, provided that should governmental authority having jurisdiction prescribe or permit the creation of units larger than those specified, for the drilling or operation of a well at a regular location or for obtaining maximum allowable from any well to be drilled, drilling or already drilled, units thereafter created may conform substantially in size with those prescribed or permitted by governmental regulations. Lessee under the provisions hereof may pool or combine acreage covered by this lease or any portion thereof as above provided as to oil in any one or more strata and as to gas in any one or more strata. The units formed by pooling as to any stratum or strata need not conform in size or area with the unit or units into which the lease is pooled or combined as to any other stratum or strata, and oil units need not conform as to area with gas units. The pooling in one or more instances shall not exhaust the rights of the Lessee hereunder to pool this lease or portions thereof into other units. Lessee shall file for record in the appropriate records of the county in which the leased premises are situated an instrument describing and designating the pooled acreage as a pooled unit; and upon such recordation the unit shall be effective as to all parties hereto, their heirs, successors, and assigns, irrespective of whether or not the unit is likewise effective as to all other owners of surface, mineral, royalty, or other rights in land included in such unit. Lessee may at its election exercise its pooling option before or after commencing operations for or completing an oil or gas well on the leased premises, and the pooled unit may include, but it is not required to include, land or leases upon which a well capable of producing oil or gas in paying quantities has theretofore been completed or upon which operations for the drilling of a well for oil or gas have theretofore been commenced. In the event of operations for drilling on or production of oil or gas from any part of a pooled unit which includes all or a portion of the land covered by this lease, regardless of whether such operations for drilling were commenced or such production was secured before or after the execution of this instrument or the instrument designating the pooled unit, such operations shall be considered as operations for drilling on or production of oil and gas from land covered by this lease whether or not the well or wells be located on the premises covered by this lease and in such event operations for drilling shall be deemed to have been commenced on said land within the meaning of paragraph 5 of this lease; and the entire acreage constituting such unit or units, as to oil and gas, or either of them, as herein provided, shall be treated for all purposes, except the payment of royalties on production from the pooled unit, as if the same were included in this lease. For the purpose of computing the royalties to which owners of royalties and payments out of production and each of them shall be entitled on production of oil and gas, or either of them, from the pooled unit, there shall be allocated to the land covered by this lease and included in said unit (or each separate tract within the unit if this lease covers separate tracts within the unit) a pro rata portion of the oil and gas, or either of them, produced from the pooled unit after deducting that used for operations on the pooled unit. Such allocation shall be on an acreage basis - that is to say, there shall be allocated to the acreage covered by this lease and included in the pooled unit (or to each separate tract within the unit if this lease covers separate tracts within the unit) that pro rata portion of the oil and gas, or either of them, produced from the pooled unit which the number of surface acres covered by this lease (or in each such separate tract) and included in the pooled unit bears to the total number of surface acres included in the pooled unit. Royalties hereunder shall be computed on the portion of such production, whether it be oil and gas, or either of them, so allocated to the land covered by this lease and included in the unit just as though such production were from such land. The production from an oil well will be considered as production from the lease or oil pooled unit which it is producing and not as production from a gas pooled unit; and production from a gas well will be considered as production from the lease or gas pooled unit from which it is producing and not from an oil pooled unit. The formation of any unit hereunder shall not have the effect of changing the ownership of any delay rental or shut-in production royalty which may become payable under this lease. If this lease now or hereafter covers separate tracts, no pooling or unitization of royalty interest as between any such separate tracts is intended or shall be implied or result merely from the inclusion of such separate tracts within this lease but Lessee shall nevertheless have the right to pool as provided above with consequent allocation of production as above provided. As used in this paragraph 4, the words "separate tract" mean any tract with royalty ownership differing, now or hereafter, either as to parties or amounts, from that as to any other part of the leased premises.

5. If operations for drilling are not commenced on said land or on acreage pooled therewith as above provided on or before one year from this date, the lease shall then terminate as to both parties, unless on or before such anniversary date Lessee shall pay or tender (or shall make a bona fide attempt to pay or tender, as hereinafter stated) to Lessor or to the credit of Lessor in Bank at ,

(which bank and its successors are Lessor's agent and shall continue as the depository for all rentals payable hereunder regardless of changes in ownership of said land or the

rentals) the sum of Dollars ($), (herein called rentals),

which shall cover the privilege of deferring commencement of drilling operations for a period of twelve (12) months. In like manner and upon like payments or tenders annually, the commencement of drilling operations may be further deferred for successive periods of twelve (12) months each during the primary term. The payment or tender of rental under this paragraph and of royalty under paragraph 3 on any gas well from which gas is not being sold or used may be made by the check or draft of Lessee mailed or delivered

Fig. 1–7. (Continued)

to the parties entitled thereto or to said bank on or before the date of payment. If such bank (or any successor bank) should fail, liquidate or be succeeded by another bank, or for any reason fail or refuse to accept rental, Lessee shall not be held in default for failure to make such payment or tender or rental until thirty (30) days after Lessor shall deliver to Lessee a proper recordable instrument naming another bank as agent to receive such payments or tenders. If Lessee shall, on or before any anniversary date, make a bona fide attempt to pay or deposit rental to a Lessor entitled thereto according to Lessee's records or to a Lessor, who, prior to such attempted payment or deposit, has given Lessee notice, in accordance with subsequent provisions of this lease, of his right to receive rental, and if such payment or deposit shall be ineffective or erroneous in any regard, Lessee shall be unconditionally obligated to pay to such Lessor the rental properly payable for the rental period involved, and this lease shall not terminate but shall be maintained in the same manner as if such erroneous or ineffective rental payment or deposit had been properly made, provided that the erroneous or ineffective rental payment or deposit be corrected within 30 days after receipt by Lessee of written notice from such Lessor of such error accompanied by such instruments as are necessary to enable Lessee to make proper payment. The down cash payment is consideration for this lease according to its terms and shall not be allocated as a mere rental for a period. Lessee may at any time or times execute and deliver to Lessor or to the depository above named or place of record a release or releases of this lease as to all or any part of the above-described premises, or of any mineral or horizon under all or any part thereof, and thereby be relieved of all obligations as to the released land or interest. If this lease is released as to all minerals and horizon under a portion of the land covered by this lease, the rentals and other payments computed in accordance therewith shall thereupon be reduced in the proportion that the number of surface acres within such released portion bears to the total number of surface acres which was covered by this lease immediately prior to such release.

6. If prior to discovery and production of oil, gas or other mineral on said land or on acreage pooled therewith, Lessee should drill a dry hole or holes thereon, or if after discovery and production of oil, gas or other mineral, the production thereof should cease from any cause, this lease shall not terminate if Lessee commences operations for drilling or reworking within sixty (60) days thereafter or if it be within the primary term, commences or resumes the payment or tender of rentals or commences operations for drilling or reworking on or before the rental paying date next ensuing after the expiration of sixty (60) days from the date of completion of dry hole or cessation of production. If at any time subsequent to sixty (60) days prior to the beginning of the last year of the primary term and prior to the discovery of oil, gas or other mineral on said land, or on the acreage pooled therewith, Lessee should drill a dry hole thereon, no rental payment or operations are necessary in order to keep the lease in force during the remainder of the primary term. If at the expiration of the primary term, oil, gas or other mineral is not being produced on said land, or on acreage pooled therewith, but Lessee is then engaged in drilling or reworking operations thereon or shall have completed a dry hole thereon within sixty (60) days prior to the end of the primary term, the lease shall remain in force so long as operation on said well or for drilling or reworking of any additional well are prosecuted with no cessation of more than sixty (60) consecutive days, and if they result in the production of oil, gas or other mineral, so long thereafter as oil, gas or other mineral is produced from said land or acreage pooled therewith. Any pooled unit designated by Lessee in accordance with the terms hereof may be dissolved by Lessee by instrument filed for record in the appropriate records of the county in which the leased premises are situated at any time after the completion of a dry hole or the cessation of production on said unit. In the event a well or wells producing oil or gas in paying quantities shall be brought in on adjacent land and within three hundred thirty (330) feet of and draining the leased premises, or acreage pooled therewith, Lessee agrees to drill such offset wells as a reasonably prudent operator would drill under the same or similar circumstances.

7. Lessee shall have the right at any time during or after the expiration of this lease to remove all property and fixtures placed by Lessee on said land, including the right to draw and remove all casing. When required by Lessor, Lessee will bury all pipe lines below ordinary plow depth, and no well shall be drilled within two hundred (200) feet of any residence or barn now on said land without Lessor's consent.

8. The rights of either party hereunder may be assigned in whole or in part, and the provisions hereof shall extend to their heirs, successors and assigns; but no change or division in ownership of the land, rentals or royalties, however accomplished, shall operate to enlarge the obligations or diminish the rights of Lessee; and no change or division in such ownership shall be binding on Lessee until thirty (30) days after Lessee shall have been furnished by registered U.S. mail at Lessee's principal place of business with a certified copy of recorded instrument or instruments evidencing same. In the event of assignment hereof in whole or in part, liability for breach of any obligation hereunder shall rest exclusively upon the owner of this lease or of a portion thereof who commits such breach. In the event of the death of any person entitled to rentals hereunder, Lessee may pay or tender such rentals to the credit of the deceased or the estate of the deceased until such time as Lessee is furnished with proper evidence of the appointment and qualification of an executor or administrator of the estate, or if there be none, then until Lessee is furnished with evidence satisfactory to it as to the heirs or devisees of the deceased and that all debts of the estate have been paid. If at any time two or more persons be entitled to participate in the rental payable hereunder, Lessee may pay or tender said rental jointly to such persons or to their joint credit in the depository named herein; or, at Lessee's election, the proportionate part of said rentals to which each participant is entitled may be paid or tendered to him separately or to his separate credit in said depository; and payment or tender to any participant of his portion of the rentals hereunder shall maintain this lease as to such participant. In event of assignment of this lease as to a segregated portion of said land, the rentals payable hereunder shall be apportionable as between the several leasehold owners ratably according to the surface area of each, and default in rental payment by one shall not affect the rights of other leasehold owners hereunder. If six or more parties become entitled to royalty hereunder, Lessee may withhold payment thereof unless and until furnished with a recordable instrument executed by all such parties designating an agent to receive payment for all.

Fig. 1–7. (Continued)

9. The breach by Lessee of any obligation arising hereunder shall not work a forfeiture or termination of this lease nor cause a termination or reversion of the estate created hereby nor be grounds for cancellation hereof in whole or in part. In the event Lessor considers that operations are not at any time being conducted in compliance with this lease, Lessor shall notify Lessee in writing of the facts relied upon as constituting a breach hereof, and Lessee, if in default, shall have sixty days after receipt of such notice in which to commence the compliance with the obligations imposed by virtue of this instrument. After the discovery of oil, gas or other mineral in paying quantities on said premises, Lessee shall develop the acreage retained hereunder as a reasonably prudent operator, but in discharging this obligation it shall in no event be required to drill more than one well per forty (40) acres of the area retained hereunder and capable of producing oil in paying quantities and one well per 640 acres plus an acreage tolerance not to exceed 10% of 640 acres of the area retained hereunder and capable of producing gas or other mineral in paying quantities.

10. Lessor hereby warrants and agrees to defend the title to said land and agrees that Lessee at its option may discharge any tax, mortgage or other lien upon said land, either in whole or in part, and in event Lessee does so, it shall be subrogated to such lien with right to enforce same and apply rentals and royalties accruing hereunder toward satisfying same. Without impairment of Lessee's rights under the warranty in event of failure of title, it is agreed that if this lease covers a less interest in the oil, gas, sulphur, or other minerals in all or any part of said land than the entire and undivided fee simple estate (whether Lessor's interest is herein specified or not), or no interest therein, then the royalties, delay rental, and other monies accruing from any part as to which this lease covers less than such full interest, shall be paid only in the proportion which the interest therein, if any, covered by this lease, bears to the whole and undivided fee simple estate therein. All royalty interest covered by this lease (whether or not owned by Lessor) shall be paid out of the royalty herein provided. Should any one or more of the parties named above as Lessors fail to execute this lease, it shall nevertheless be binding upon the party or parties executing the same. Failure of Lessee to reduce rental paid hereunder shall not impair the right of Lessee to reduce royalties.

11. Should Lessee be prevented from complying with any express or implied covenant of this lease, from conducting drilling or reworking operations thereon or from producing oil or gas therefrom by reason of scarcity of or inability to obtain or to use equipment or material, or by operation of force majeure, any Federal or state law or any order, rule or regulation of governmental authority, then while so prevented, Lessee's obligation to comply with such covenant shall be suspended, and Lessee shall not be liable in damages for failure to comply therewith; and this lease shall be extended while and so long as Lessee is prevented by any such cause from conducting drilling or reworking operations on or from producing oil or gas from the leased premises; and the time while Lessee is so prevented shall not be counted against Lessee, anything in this lease to the contrary notwithstanding.

IN WITNESS WHEREOF, this instrument is executed on the date first above written.

LESSOR   SS. OR TAX I.D. NO.   LESSOR   SS. OR TAX I.D. NO.

STATE OF
COUNTY OF   ACKNOWLEDGMENT

This instrument was acknowledged before me on the   day of   , by

Notary Public, State of Texas
Notary's name (printed):
Notary's commission expires:

STATE OF
COUNTY OF   ACKNOWLEDGMENT

This instrument was acknowledged before me on the   day of   , by

Notary Public, State of Texas
Notary's name (printed):
Notary's commission expires:

Producers 88 (4/76) Revised - Rental
With 640 Acres Pooling Provision

©POUND PRINTING COMPANY

In the United States, oil and gas leases are typically obtained through the use of a **landman**, an individual who specializes in searching for and obtaining leases. A landman in many cases acts as an agent for an undisclosed principal in trying to obtain a lease at the lowest possible price.

An oil and gas lease embodies the legal rights, privileges, and duties pertaining to the lessor and lessee. The **lessor** is the mineral rights owner who leases the property to another party and retains a royalty interest. The **lessee**, the party leasing the property, receives a working interest. The lessee's working interest provides for investigating, exploring, prospecting, drilling, and mining for oil, gas, and other minerals, as well as for conducting G&G surveys, installing production equipment, and producing said products.

Most lease contracts contain the following provisions:

- **Lease bonus.** Initial amount paid to the mineral rights owner in return for the rights to explore, drill, and produce. Lease bonus payments usually are a dollar amount per acre. The amount of the lease bonus and other payments made to the mineral rights owner is a result of negotiations or bargaining between the mineral rights owner and the oil company representative.

- **Royalty provision.** Specified fraction of the oil and gas produced free and clear of any costs (except severance taxes and certain costs to market the product) to which the royalty interest owner is entitled. Royalty payments are provided in many different formats. In the United States, the most common method is to provide a share of the proceeds from the sale of production. This amount varies greatly, e.g., 1/8, 1/4, etc., of production. On offshore federal leases, complicated formulas may be used for determining royalty payments. In some instances, the mineral rights owner will receive a portion of the net profits as royalty, i.e., net profits leasing. Some offshore leases also provide for a sliding scale royalty, whereby the amount of royalty paid is determined by the amount of oil and gas produced.

- **Primary term.** Initial term of the lease. The primary term is the maximum time that the lessee has to begin drilling or commence production from the property. Technically, during this term, the lessee must begin drilling within one year from the signing of the lease. However, in the absence of drilling or production, the lessee can keep the lease in effect during this term by making an annual payment called a **delay rental payment**.

- **Delay rental payment.** Yearly payment made during the primary term in the absence of drilling operations (or production) in order to retain the lease. In other words, a delay rental payment is an annual payment made to allow the lessee to delay drilling operations for one additional year. Delay rental payments are typically based upon a dollar amount per acre. They must be paid on or before one year from the date the lease was signed and each year following during the primary term if drilling operations (or production) have not commenced. Generally, the lease will require the payment to be made to a specified bank. Some short-term leases (two- or three-year primary term), called **paid up leases**, require the lessee to pay the delay rentals at the inception of the lease. After the primary term, the lease can be held only by drilling or production, i.e., after the primary term, a delay rental payment can no longer keep the lease from terminating. However, once production begins, the revenue from production from the lease keeps the lease in effect whether during or after the primary term. In the case where production commences before the expiration of the primary term, no additional delay rental payments are necessary: the royalty provides compensation to the royalty interest owner.

- **Shut-in payments.** If the well is capable of producing oil or gas in paying quantities but is **shut in** (not producing), the lessee may hold the lease by making shut-in payments. Shut-in payments are usually made in natural gas situations where access to a pipeline is not available or an oversupply of gas exists. Shut-in payments are generally not recoverable from future royalty payments.
- **Right to assign interest.** The rights of each party may be assigned in whole or in part without the approval of the other party. For example, the working interest owner may carve out a production payment interest or ORI from the working interest without notifying the royalty interest owner.
- **Rights to free use of resources for lease operations.** The operator usually has the right to use, without cost, any oil or gas produced on the lease to carry out operations on that lease.
- **Option payment.** Payment made to obtain a preleasing agreement that gives the oil company (the lessee) a specified period of time to obtain a lease from the entity receiving the payment. In addition to specifying the period of time within which the lessee may lease the property, the option contract will also typically specify the lease form, royalty interest, bonus to be paid, etc.
- **Offset clause.** If a producing well is drilled on Lease B close to the property line of Lease A (Fig. 1–8), within a distance specified in the lease contract, the offset clause requires the lessee of Lease A to drill an offset well on Lease A, within a specified period of time, in order to prevent the well on Lease B from draining the reservoir. If, however, the leases are within a state with forced pooling or unitization, the leases can be forced to be pooled and operated as one, and the offset clause becomes irrelevant. If the state in question does not have forced pooling or unitization, then the only recourse for the interest owners of Lease A is for the lessee to assume the burden of offset drilling.

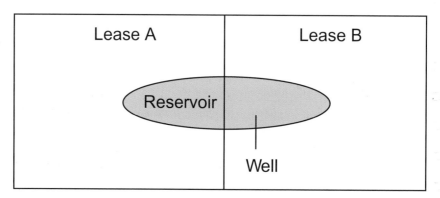

Fig. 1–8. Offset well situation

- **Minimum royalty.** A minimum royalty clause provides for the payment of a stipulated amount to the lessor regardless of production. A minimum royalty is similar to a shut-in royalty except that it is commonly recoverable from future royalty payments.
- **Pooling provisions.** Modern lease forms provide that if the working interest owner forms a pool or unit with other leases, the royalty interest and other nonworking interest owners may also be forced to combine their interests with the nonworking interest owners of the other leases forming the unit.

DRILLING OPERATIONS

Onshore drilling operations include building access roads to the drillsite, preparing the site for the drilling rig, transporting the rig to the site, and drilling the wellbore. Most drilling operations in the United States are performed by drilling contractors. A **drilling contract** is an agreement between the lessee (working interest owners) and a drilling contractor for the drilling of a well. The contract details the rights and obligations of the lessee and the contractor. Drilling contracts generally provide payment on a **day rate** (payment based on the number of days drilled), a **footage rate** (payment based on the number of feet drilled), or a **turnkey basis** (payment of a fixed sum of money based on drilling to a certain depth or stage of completion).

The first step in drilling an oil and gas well is selecting the actual drillsite. Seismic studies, particularly 3-D seismic studies, are usually performed, and the results of the studies are examined to determine the optimal site for the well. Once the site has been selected and a drilling contract signed, site preparations can begin. The well site is normally surveyed and staked, then access roads are built, and the site is graded and leveled. Reserve and waste pits are also prepared, and a water supply is obtained.

After the site is prepared, often the initial 20 to 100 feet of the well will be drilled with a small truck-mounted rig. The drilling rig and related equipment are next moved in and set up, a process called **rigging up**. The well is then ready to be **spudded in**. The **spud date** is the date the rotary drilling bit touches ground.

Routine rotary drilling consists of rotating a drill bit downwards through the formations towards target depth, cutting away pieces of the formation called **cuttings**. During the drilling process, drilling fluid, i.e., **mud**, is constantly circulated down the wellbore. Drilling mud serves several purposes. It raises the cuttings to the surface, lubricates the drilling bit, and keeps formation fluids from entering the wellbore.

Approximately every 30 feet as the hole is deepened, a joint of drill pipe is added, a process called making a **mousehole connection**. Periodically, when the drill bit becomes worn or damaged, all of the drill pipe has to be removed from the hole in a process called **tripping out**. After a new drill bit is attached, the pipe is lowered back into the hole, called **tripping in**. Normally the drill pipe is removed and lowered three joints, i.e., a **stand**, at a time, depending upon the height of the derrick or mast. A derrick or mast is a four-legged, load-bearing structure that is part of the drilling rig. The height of the derrick correlates with the depth of the well, since the derrick must support the weight of the drill string, i.e., the drill pipe, etc., suspended downhole.

Tripping out is also necessary when casing must be set. **Casing** is steel pipe that is **set**, i.e., cemented, into the wellbore. Functions of casing include preventing the caving in of the hole, protecting fresh water sands, excluding water from the producing formations, confining production to the wellbore, and controlling formation pressure.

Although most onshore wellbores are drilled vertically, some wells are drilled at an angle. These wells can be either directional or horizontal wells. **Directional wells** are wells that are normally drilled straight to a predetermined depth and then curved or angled so that the bottom of the wellbore is at the desired location. **Horizontal wells** are also initially drilled straight down but then are gradually curved until the hole runs parallel to the earth's surface, with drilling actually achieving a horizontal direction through the formation. Figure 1–9 shows a conventional well, a directional well, and a horizontal well.

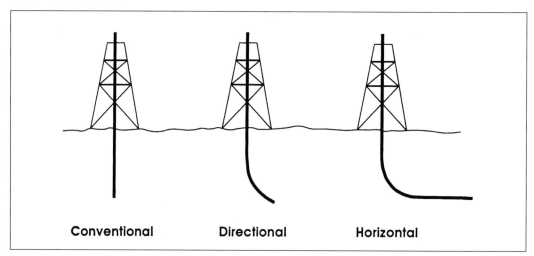

Fig. 1–9. Types of wells

Directional drilling is used in situations where the drilling objective cannot be achieved with a vertical wellbore. For example, directional drilling may be necessary in urban locations where limitations exist regarding well location or to sidetrack around an obstruction. Offshore, directional drilling is normally necessary so that multiple wells can be drilled from a single offshore platform.

Horizontal drilling is a subset of directional drilling. Unlike a directional well that is drilled to position a reservoir entry point, a horizontal well is commonly defined as any well in which the lower part of the wellbore parallels the pay zone. The angle of the wellbore does not have to reach 90° for the well to be considered a horizontal well.

The objective of horizontal drilling is to expose more reservoir rock to the wellbore than would have been possible with a conventional vertical well. Most oil and gas reservoirs have greater horizontal dimensions than vertical thickness. By drilling a portion of a well parallel to the reservoir, the well is capable of accessing oil and gas that would otherwise not be accessible. Horizontal wells have become a preferred method of recovering hydrocarbons from reservoirs in which the oil- and gas-bearing zones are more or less horizontal. These wells are drilled approximately vertically down to a depth just above the target. At that point, the wellbore bears off on an arc to intersect the reservoir and then continues on horizontally. The cost of drilling horizontally directed wells may be 2 or 3 times that of drilling conventional vertical wells. However, the production factor can be enhanced as much as 15 or 20 times, making horizontal drilling very attractive to producers.

During drilling operations, the petroleum engineer or geologist examines data from a number of sources in order to decide whether there is sufficient oil or gas to justify the cost of completing and producing the well. As the well is drilled, mud is circulated in the hole. The mud and the cuttings are analyzed to identify evidence of hydrocarbons and to gain insight into possible fluid content and rock structure. Periodically, a larger piece of formation rock may be obtained by placing a core barrel on the end of the drill stem and cutting a cylindrical core approximately 4 inches in diameter and up to 60 feet in length from the formation. This piece of formation rock, called a **core sample**, is then analyzed to determine formation rock characteristics, sequence of rock layers in the earth, and fluid content of the formation. Seismic data, especially 3-D seismic, is also often used to aid this analysis.

After total depth has been reached, the well is **logged** by lowering a device to the bottom of the well and then pulling it back up to the surface. As the device passes up the hole, it measures and records properties of the formations and the fluids residing in them. Based on the results of the analyses and the test methods discussed above, as well as other evaluation methods, a decision is made as to whether to complete the well. If the well is judged incapable of providing oil or gas in commercial quantities, it is plugged and abandoned. If the well is judged capable of producing oil or gas in commercial quantities, the well is completed.

Activities incident to completing a well and placing it on production include the following:

- Obtaining and installing production casing (steel pipe lining the wellbore)
- Installing tubing (steel pipe suspended in the well through which the oil and gas are produced)
- **Perforating** (setting off charges to create holes in the casing and cement so formation fluids can flow from the formation into the wellbore)
- Installing the **Christmas tree** (valves and fittings controlling production at the wellhead that somewhat resemble a Christmas tree)
- Constructing **production facilities** (separators, heater-treaters, tanks, etc.) and installing flow lines

Activities incident to plugging and abandoning a well would include removal of any equipment possible and cementing the wellbore to seal the hole.

Some wells penetrate more than one zone containing oil or gas in commercial quantities. In these cases, the wells may be completed either to produce from only one zone or from multiple zones. In a **multiple completion**, the well is capable of simultaneous production from multiple zones containing oil or gas.

Determination of whether or not to complete a well entails a comparison of the incremental costs to complete the well with the net cash flows expected to be received from the sale of petroleum products from the well. A well is judged capable of commercial production if the expected net proceeds from production exceed the cost of completing the well. The costs incurred in drilling prior to completion—past costs or sunk costs—are not considered in making the decision to plug and abandon or complete the well.

EXAMPLE

Completion Decision

Tyler Company incurred $300,000 in drilling costs up to target depth. The estimated costs to complete the well (installing casing, Christmas tree, etc.) are $200,000. Tyler's estimated discounted future net cash flow from the sale of the oil and gas from this well is expected to be $350,000. Should the well be completed or abandoned?

Answer:

The well should be completed because the estimated discounted future net cash flow is $350,000 and the incremental costs (after drilling) are $200,000. If the well is not completed, Tyler will have lost $300,000. If the well is completed, Tyler's loss is only $150,000. (See chapter 8 for further discussion concerning the decision to complete a well.)

The activities involved in offshore drilling are somewhat different than in onshore operations. In territorial waters offshore the United States, operators may acquire mineral leases from state governments or from the federal government. In contrast to most onshore federal leases, offshore federal leases are obtained through a system of closed competitive bidding on available offshore tracts. Normally, the federal government keeps a 1/6 royalty on the tracts. Competitive bidding is also common for state leases.

Drilling operations are much more expensive offshore. For example, some offshore drilling contracts today are as high as $800,000 per day. Consequently, most offshore drilling is done in the form of a joint venture or joint interest operation, i.e., several companies pool capital resources to explore, develop, and produce an offshore tract. While joint interest operations are also common onshore, they are generally the rule offshore.

In some offshore areas, very little may be known about the types and depths of the subsurface formations. In those areas, a **stratigraphic test well**, a well drilled for information only, may be drilled. Often, such a well will be drilled prior to the bidding process and paid for by multiple companies that agree to share the information. Stratigraphic test wells are often drilled offshore to determine the existence and quantity of proved reserves and also to determine the location for the permanent development drilling and production platform. Advances in seismic technology have significantly reduced the need to drill stratigraphic test wells.

Exploratory drilling offshore is almost always done from mobile rigs. (Exploratory wells are wells drilled in an area not known to contain oil or gas.) The choice of a rig to be used is based on the specific locations. These temporary platforms can take several forms:

- **Drilling barges and ships.** These are towed to location.
- **Jack-up drilling platforms.** The platforms are towed to location, and then the legs are lowered to the ocean floor. The body of the structure is then jacked up high enough above the water level so as to clear the waves and provide a stable platform for drilling.

- **Submersible and semisubmersible drilling platforms.** The platforms are towed to location, and then the pontoon-like legs are flooded with water for extra stability in the open ocean.

Development wells, which produce a reservoir that has been discovered with exploratory drilling, are often drilled from fixed platforms containing production and well maintenance facilities. These fixed platforms, which are very expensive, are usually made from steel or concrete, depending on their location.

The drilling operations of an offshore rig are similar to those of onshore rigs, with the exception of specialized technical adaptations that have been made to deal with the hostile marine environment. In addition, directional drilling is commonly used offshore, since that technique can reach thousands of feet away from the platform. This allows the drilling of multiple wells (as many as 40 or more) from the same development platform.

In addition to production from fixed platforms, offshore production may also be achieved via subsea completions. Subsea completions are subsea satellite wells that are situated on the ocean floor (as opposed to being located on a production platform). The production from these wells is moved directly to platforms, floating production/storage/offloading vessels (FPSOs), or to the shore, where it is processed and stored pending sale. Subsea completions are commonly used in areas where it is not feasible to access the production location from a production platform and/or in situations where the economics do not support the cost of a platform.

Many industry experts predict that the future of the offshore industry is in ultra-deepwater areas. **Ultra deepwater** is defined as outer continental areas where the water depths are 1,500 meters or greater. The cost of drilling in these areas is extremely high. In September 2006, Chevron Corp. announced a major discovery in ultra deepwater 175 miles off the coast of Louisiana. Oil analysts and company executives indicate that the discovery could yield total recoverable oil in excess of 3 billion barrels, and perhaps as much as 15 billion barrels of oil. This single discovery has the potential of increasing U.S. reserves by 50% and is the biggest discovery in the United States since the Prudhoe Bay field in Alaska was discovered in the 1960s.

RECOVERY PROCESSES

Several types of production processes may be employed in order to move the oil or gas from the reservoir to the well. These production processes are commonly divided into three types of recovery methods: primary, secondary, and tertiary.

The initial or **primary recovery** of oil and gas is either by natural reservoir drive or by pumping. Natural drive occurs when sufficient water or gas exists in the reservoir under high pressure. The reservoir pressure provides the natural energy needed to drive the oil to the wellbore. If insufficient natural drive exists, the oil may be pumped to the surface using a beam pumping unit.

When the maximum amount of oil and gas has been recovered by primary recovery methods and the reservoir pressure has been largely depleted, secondary recovery methods may be instituted. **Secondary recovery** consists of inducing an artificial drive into the formation to replace the natural drive. The most common method is waterflooding, which involves injecting water under pressure into the formation to drive the oil to the wellbore.

Following the second attempt to recover oil, a third attempt, called **tertiary recovery**, may be instituted. (The distinction between secondary and tertiary recovery methods may actually be obscure. Some secondary recovery methods are also considered as tertiary recovery methods, and vice versa.) Tertiary recovery involves the use of enhanced recovery methods to produce oil. Tertiary recovery methods include injection of chemicals, gas, or heat into the well to modify the fluid properties and thereby enhance the movement of the oil through the formation. A newer form of tertiary recovery uses microwave technology, which introduces microwaves into reservoirs in northern climates to warm the oil. Tertiary methods may be very expensive, and many methods are still in the developmental stage.

Even with the best recovery methods, a large amount of oil remains locked in the formation. Some experts have estimated that 50% or more of the oil cannot be recovered with current technology. The amount of oil and gas to be recovered in the future from old wells is partially determined by oil and gas prices. Many more secondary and tertiary projects will be attempted when petroleum product pricing makes such recovery methods economically feasible.

PRODUCTION AND SALES

Fluids produced from a well normally will contain a combination of crude oil and natural gas, as well as basic sediment and water (BS&W). Before the oil and gas are sold, the well fluid must be separated, treated, and measured. Flow lines take well fluid from individual wells to a central gathering point. At this location, separators, heater-treaters, and other equipment are used to perform the initial separation of the liquids and gases and to remove much of the BS&W and other impurities. From this point, crude oil generally goes to stock tanks for storage until it is delivered to a buyer. When the oil is sold, it is measured as it is transferred from the storage tanks to either truck or ship transport or an oil pipeline. Gas, which is not stored on site, is measured as it is gathered, processed, and transferred to a gas pipeline.

The amount of oil transferred from the storage tanks is recorded on a document called a **run ticket**. The amount of payment for the oil is based upon information contained in the run ticket (chapter 10). The **gas settlement statement** is used to record similar information for the production and sale of gas.

COMMON STATE AND FEDERAL REGULATIONS

In the 1920s and the 1930s, overdrilling was commonplace. In that era, overdrilling was so extreme that the legs of one drilling rig might be inside the legs of another rig. Since those days of frantic drilling, many states have established agencies to oversee oil and gas drilling, and those agencies have passed regulations to eliminate waste and uneconomical methods of producing oil and gas. For example, the state agency that oversees oil and gas drilling in Texas is the Texas Railroad Commission. In Oklahoma, it is the Oklahoma Corporation Commission. Exploration of federal offshore areas as well as federal onshore lands is regulated by the Department of the Interior.

One common regulation is related to well spacing. The fact that mineral rights in the United States can be owned by individual parties results in numerous leases that often cover a relatively small acreage. Unless the lease is large, it is unlikely that an oil and gas reservoir would be within a single lease. In the lower-48 United States, there are likely to be numerous leases associated with any given reservoir. Without regulations, many more wells than necessary (and wells that are too closely spaced) would be drilled on the typically multiple leases associated with given reservoirs. Figure 1–10 illustrates how four leases of varying sizes may be associated with a single reservoir.

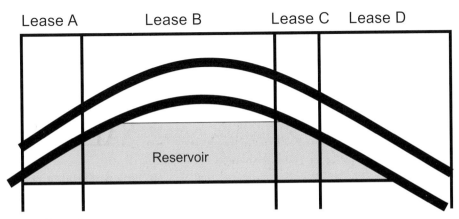

Fig. 1–10. Well spacing

Today the various states as well as the U.S. government regulate the number of wells drilled into a reservoir through the use of spacing and density regulations in order to prevent economic waste and to maximize recovery from the reservoir. Economic waste occurs when too many wells are drilled into a reservoir, since an increased number of wells drilled does not necessarily increase oil and gas recovery. Further, the reservoir's natural drive may be rapidly depleted if more wells are drilled in one area than in another, causing premature water or gas encroachment. While exceptions are possible, a common spacing requirement in the lower-48 United States is no more than one oil well per 40 acres, and no more than one gas well per 640 acres.

Another common regulation relates to drilling permits. Prior to starting the drilling process, whether on private or public land, a drilling permit is generally required from the state, or if on federal lands or water, from the federal government. An application for a permit is made to the appropriate governmental agency outlining the location, depth, etc., of the well. Generally, the drilling permit will not be granted unless the well spacing requirements are met or an exception is granted. In addition to obtaining a permit to drill, various types of reports must be filed, e.g., protection of water-bearing strata, various completion reports, and plug-and-abandon reports. For offshore leases, the process of obtaining a drilling permit is much more complicated and time-consuming. Permits may be required from both the state and federal governments. The application for a federal offshore permit requires more information than is requested for an onshore permit. Information to be provided includes a description of the vessels, platforms, and other required structures, targeted locations for each well, and forms of protection against environmental contamination.

Another important state regulation deals with restrictions of production. If demand is adequate, states will typically allow all wells and leases to produce at the **maximum efficiency rate (MER)**. (The MER is the maximum rate at which oil or gas can be produced without damaging the reservoir's natural energy.) In past years, demand for oil and gas was sometimes lower than U.S. production capacity. In those periods of low demand for oil and gas, states frequently restricted production by a proration process. A state agency (regulatory commission) decided the amount to be produced within the state for a given period of time, usually a month, and then prorated this amount among the state's producing fields (field allowable). The field allowable was then prorated to the various leases and wells. The leases and wells would then have an allowable for the month, i.e., the maximum number of barrels of oil or cubic feet of gas that could be produced from the lease and wells during the month.

Pooling or unitization, as discussed earlier, involves combining two or more leases and operating them as a single property. The basic purpose is to recover the maximum amount of oil and gas possible in the most efficient and economical manner. The federal and state governments encourage pools and units, with most states imposing mandatory pooling and unitization. Typically, units (especially mandatory units) must be approved by the appropriate state regulatory agencies, or by the federal government if on federal lands.

In Figure 1–10, in order to effectively manage the reservoir and conform to spacing regulations, the leases normally would be pooled together and operated as a single property. If the spacing requirements resulted in wells being drilled on some leases, such as A, B, and D, but not Lease C, the latter would nevertheless share in the production. In other words, the spacing requirements do not penalize any given lease in a unit. The ownership interests would be recomputed for the entire unit, and the production would be prorated to the various leases based on the amount of reservoir "pay sand" or acreage contributed by each lease.

What Does the Future Hold?

The U.S. oil and gas industry is currently facing a challenge to survive and prosper in the future. In 2006, estimates indicated that the United States has less than 3% of the world's reserves of oil in contrast to the Middle East, which is estimated to have 65% of world reserves. (See Fig. 1–11.) Further, the average Middle East well outproduces the average U.S. well by more than 1,500 to 1. These statistics do not paint a bright picture for the U.S. domestic industry. The situation for gas is not as bleak. (See Fig. 1–12.)

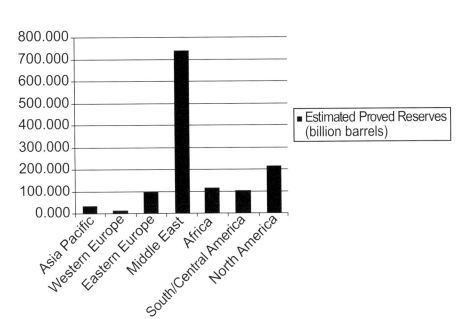

Fig. 1–11. World oil reserves (12/2006)
Source: PennWell Corporation. 2006. ***Oil and Gas Journal***. Vol. 104, no. 47 (Dec. 18).

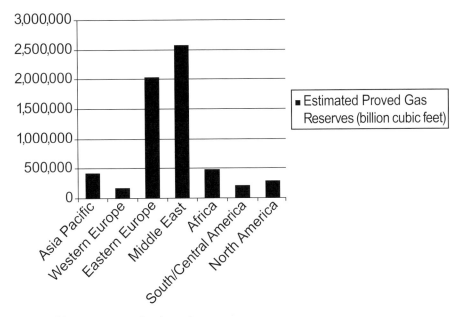

Fig. 1–12. World gas reserves (12/2006)
Source: PennWell Corporation. 2006. ***Oil and Gas Journal***. Vol. 104, no. 47 (Dec. 18).

Technology is the key if oil production is to have a future in the United States. Many reserves remain in the ground in wells that have been plugged because remaining reserves could not be economically produced with current recovery methods. Enhanced recovery methods are currently being researched that may make it economically feasible to produce many of those reserves.

For example, the method of horizontal drilling is now being used to produce from areas that were once abandoned as not economically feasible. The development of horizontal drilling in the oil industry has been compared to the invention of the transistor in the electronics industry. Further, increases in the price of oil in the future will make exploration in increasingly hostile areas as well as expensive enhanced recovery methods economically feasible.

Environmental issues will be at the forefront as a concern for the industry for many years to come. The industry has a long tradition of acting responsibly to develop environmental safeguards in order to protect the environment. Hopefully, a balance can be reached between environmentalists and the industry that will allow for exploration activities to continue to increase in both onshore and offshore U.S. areas.

• • •

This profile of petroleum operations provides the framework for the remaining chapters of this text. Accounting applications of the various activities described in this chapter are investigated from the three oil and gas accounting methodologies:

- Successful efforts
- Full cost
- Tax accounting

In addition, the final two chapters present an overview of international oil and gas accounting and the analysis of oil and gas companies' financial statements. Also included in the book are appendices containing an authorization for expenditure (AFE) and authoritative literature by the Securities and Exchange Commission.

PROBLEMS

1. Describe the organic theory of the origin of oil and gas.

2. Terms

   a. Define the following:

      fault trap

      anticline

      salt dome

      porosity

      permeability

   b. Define the following:

      day-rate contract

      footage-rate contract

      turnkey contract

      horizontal drilling

c. Explain the following:

   petroleum reservoir

   primary recovery

   secondary recovery

   tertiary recovery

3. List the steps in finding oil and gas.

4. What is the difference between an operating (working) interest and a nonoperating (nonworking) interest?

5. Define the following:

   economic interest in oil and gas

   mineral rights

   mineral interest

   royalty interest

   working interest

   overriding royalty interest

   production payment interest

6. Define and discuss the important provisions of the typical oil and gas lease.

7. What are the drilling operations that give rise to accounting implications?

8. Which of the following would not be a mineral interest?

   a. production payment interest

   b. working interest

   c. overriding royalty interest

   d. surface rights interest

   e. net profits interest

   f. royalty interest

   g. joint working interest

9. Celsius Oil Company signed a lease contract on January 1, 2016. The primary term specified in the contract was a four-year term.

   a. On what date is the first delay rental payment due?

   b. What is the maximum number of delay rental payments that may be made?

   c. By what date must drilling be commenced in order to keep the lease from terminating?

   d. Assume Celsius Oil begins drilling a well on January 2, 2017.

      1) Would the first delay rental be necessary to keep the lease from terminating?

2) If the well is still in process 14 months later, would the second delay rental be necessary?

3) If instead, the well was completed and production begun by October 3, 2017, would the second delay rental be necessary?

4) If production ceased by December 25, 2018, would the third delay rental payment be necessary?

10. Mr. Zeman owns the mineral rights in a property in Grant County, Oklahoma. He leases the property to Force Petroleum, reserving a 1/5 royalty. Force drills a successful well and begins producing oil. Revenue from the first year of operations totaled $20,000 and costs of development and operation totaled $150,000. How much revenue will each party receive? How much of the costs will each party pay?

11. Pressure Oil Corporation owns a working interest in an oil and gas lease. Lacking the funds to develop the lease, Pressure assigns the working interest to Tritium Oil Company, reserving 1/32 of 6/7 of production. What kind of interest has Tritium acquired? What kind of interest has Pressure retained?

12. Dwight Energy owns the working interest in a tract of land in Texas. Lacking the funds to develop the property, Dwight assigns Bartz Oil 30,000 barrels of oil to be paid out of 1/7 of the working interest's share of production in exchange for $600,000 in cash. What type of interest has Bartz acquired?

13. Aggie Company obtained a lease with a three-year primary term on August 1, 2016.

    a. Drilling operations were commenced on June 1, 2017, and continued until October 15, 2017, when the well was determined to be dry.

       1) Would the first delay rental payment be required?

       2) How many more delay rentals would be necessary to hold the lease without further drilling?

    b. Drilling operations were started on May 1, 2019, and the well was completed on October 12, 2019, as a producer.

       1) Did the lease terminate on August 1, 2019? Explain.

       2) How many years will the lease continue, assuming production in commercial quantities?

14. Cowboy Oil Corporation incurred $275,000 in drilling costs prior to deciding whether to complete the well. Estimated completion costs are $175,000. The expected net cash flows from the sale of the oil and gas from this well are $300,000. Should the well be completed?

15. Answer the following questions related to horizontal drilling:

    a. Under what conditions would horizontal drilling operations be considered?

    b. Would horizontal drilling operations be more difficult and expensive than the regular vertical drilling process? Explain.

    c. Would horizontal drilling operations be appropriate for most producing formations? Explain.

16. Discuss the following:

    well spacing

    proration

    field and well allowable

    drilling permit

17. Discuss the requirements generally necessary to exist for a petroleum reservoir to be commercially productive.

18. Explain the following terms:

    fracturing

    acidizing

    tripping in/out

    well casing

19. Describe the primary types of geological and geophysical studies.

20. Explain the role of a landman in oil and gas operations.

References

1. American Petroleum Institute. 1983. *Introduction to Oil and Gas Production.* Washington, D.C.: American Petroleum Institute.

2. Petex. 1996. *A Primer of Oilwell Drilling.* 5th ed., rev. Austin, TX: Petroleum Extension Service.

3. Conaway, Charles. 1999. *The Petroleum Industry: A Nontechnical Guide.* Tulsa, OK: PennWell.

4. Wright, Charlotte, and Rebecca Gallun. 2005. *International Petroleum Accounting.* Tulsa, OK: PennWell.

2

INTRODUCTION TO OIL AND GAS ACCOUNTING

Oil and gas accounting, as discussed in this book, relates to accounting for the four basic costs incurred by companies with oil and gas exploration and producing activities. These four basic types of costs are as follows:

- **Acquisition costs.** Costs incurred in acquiring property, i.e., costs incurred in acquiring the rights to explore, drill, and produce oil and natural gas. In the United States, these rights are normally acquired by obtaining an oil, gas, and mineral lease as described in chapter 1. A variety of contracts are used in countries outside the United States. These contracts are described in chapter 15.

- **Exploration costs.** Costs incurred in exploring property. Exploration involves identifying areas that may warrant examination and examining specific areas, including drilling exploratory wells.

- **Development costs.** Costs incurred in preparing proved reserves for production, i.e., costs incurred to obtain access to proved reserves and to provide facilities for extracting, treating, gathering, and storing oil and gas.

- **Production costs.** Costs incurred in lifting the oil and gas to the surface and in gathering, treating, and storing the oil and gas.

To account for these costs, knowledge of the industry terms and procedures discussed in chapter 1 is imperative. Additional and important terms are defined in the following glossary as they are used in accounting. The terms *proved reserves*, *proved developed reserves*, and *proved undeveloped reserves* are especially important in oil and gas accounting.

Glossary of Common Terms

The Securities and Exchange Commission (SEC) *Reg. S-X 4-10* includes the following definitions:

Reservoir. *A porous and permeable underground formation containing a natural accumulation of producible oil and/or gas that is confined by impermeable rock or water barriers and is individual and separate from other reservoirs.*

Proved reserves. *Proved oil and gas reserves are the estimated quantities of crude oil, natural gas, and natural gas liquids which geological and engineering data demonstrate with reasonable certainty to be recoverable in future years from known reservoirs under existing economic and operating conditions, i.e., prices and costs as of the date the estimate is made. Prices include consideration of changes in existing prices provided only by contractual arrangements, but not on escalations based upon future conditions.*

(i) Reservoirs are considered proved if economic producibility is supported by either actual production or conclusive formation test. The area of a reservoir considered proved includes

  (A) that portion delineated by drilling and defined by gas-oil and/or oil-water contacts, if any; and

  (B) the immediately adjoining portions not yet drilled, but which can be reasonably judged as economically productive on the basis of available geological and engineering data. In the absence of information on fluid contacts, the lowest known structural occurrence of hydrocarbons controls the lower proved limit of the reservoir.

(ii) Reserves which can be produced economically through application of improved recovery techniques (such as fluid injection) are included in the "proved" classification when successful testing by a pilot project, or the operation of an installed program in the reservoir, provides support for the engineering analysis on which the project or program was based.

(iii) Estimates of proved reserves do not include the following:

  (A) oil that may become available from known reservoirs but is classified separately as "indicated additional reserves";

  (B) crude oil, natural gas, and natural gas liquids, the recovery of which is subject to reasonable doubt because of uncertainty as to geology, reservoir characteristics, or economic factors;

  (C) crude oil, natural gas, and natural gas liquids, that may occur in undrilled prospects; and

  (D) crude oil, natural gas, and natural gas liquids, that may be recovered from oil shales, coal, gilsonite and other such sources.

Proved developed reserves. *Proved developed oil and gas reserves are reserves that can be expected to be recovered through existing wells with existing equipment and operating methods. Additional oil and gas expected to be obtained through the application of fluid injection or other* *improved recovery techniques for supplementing the natural forces and mechanisms of primary recovery should be included as "proved developed reserves" only after testing by a pilot project or after the operation of an installed program has confirmed through production response that increased recovery will be achieved.*

Proved undeveloped reserves. *Proved undeveloped oil and gas reserves are reserves that are expected to be recovered from new wells on undrilled acreage, or from existing wells where a relatively major expenditure is required for recompletion. Reserves on undrilled acreage shall be limited to those drilling units offsetting productive units that are reasonably certain of production when drilled. Proved reserves for other undrilled units can be claimed only where it can be demonstrated with certainty that there is continuity of production from the existing productive formation. Under no circumstances should estimates for proved undeveloped reserves be attributable to any acreage for which an application of fluid injection or other improved recovery technique is contemplated, unless such techniques have been proved effective by actual tests in the area and in the same reservoir.*

Field. *An area consisting of a single reservoir or multiple reservoirs all grouped on or related to the same individual geological structural feature and/or stratigraphic condition. There may be two or more reservoirs in a field which are separated vertically by intervening impervious strata, or laterally by local geologic barriers, or by both. Reservoirs that are associated by being in overlapping or adjacent fields may be treated as a single or common operational field. The geological terms "structural feature" and "stratigraphic condition" are intended to identify localized geological features as opposed to the broader terms of basins, trends, provinces, plays, areas-of-interest, etc.*

Proved area. *The portion of a property at a certain depth to which proved reserves have been specifically attributed.*

Exploratory well. *A well drilled to find and produce oil or gas in an unproved area, to find a new reservoir in a field previously found to be productive of oil or gas in another reservoir, or to extend a known reservoir. Generally, an exploratory well is any well that is not a development well, a service well, or a stratigraphic test well as those items are defined below.*

Development well. *A well drilled within the proved area of an oil or gas reservoir to the depth of an horizon known to be productive.*

Service well. *A well drilled or completed for the purpose of supporting production in an existing field. Specific purposes of service wells include gas injection, water injection, steam injection, air injection, salt-water disposal, water supply for injection, observation, or injection for in-situ combustion.*

Stratigraphic test well. *A drilling effort, geologically directed, to obtain information pertaining to a specific geologic condition. Such wells customarily are drilled without the intention of being completed for hydrocarbon production. This classification also includes tests identified as core tests and all types of expendable holes related to hydrocarbon exploration. Stratigraphic test wells are classified as (i) "exploratory type," if not drilled in a proved area, or (ii) "development type," if drilled in a proved area.*

Oil and gas producing activities.

(i) Such activities include:

(A) The search for crude oil, including condensate and natural gas liquids, or natural gas ("oil and gas") in their natural states and original locations.

(B) The acquisition of property rights or properties for the purpose of further exploration and/or for the purpose of removing the oil or gas from existing reservoirs on those properties.

(C) The construction, drilling and production activities necessary to retrieve oil and gas from its natural reservoirs, and the acquisition, construction, installation, and maintenance of field gathering and storage systems—including lifting the oil and gas to the surface and gathering, treating, field processing (as in the case of processing gas to extract liquid hydrocarbons) and field storage. For purposes of this section, the oil and gas production function shall normally be regarded as terminating at the outlet valve on the lease or field storage tank; if unusual physical or operational circumstances exist, it may be appropriate to regard the production functions as terminating at the first point at which oil, gas, or gas liquids are delivered to a main pipeline, a common carrier, a refinery, or a marine terminal.

(ii) Oil and gas producing activities do not include:

(A) The transporting, refining, and marketing of oil and gas.

(B) Activities relating to the production of natural resources other than oil and gas.

(C) The production of geothermal steam or the extraction of hydrocarbons as a by-product of the production of geothermal steam or associated geothermal resources as defined in the Geothermal Steam Act of 1970.

(D) The extraction of hydrocarbons from shale, tar sands, or coal.

Proved properties. *Properties with proved reserves.*

Unproved properties. *Properties with no proved reserves.*

Source: Reg. S-X 4-10

The reserve categories defined in the glossary are summarized in Figure 2–1:

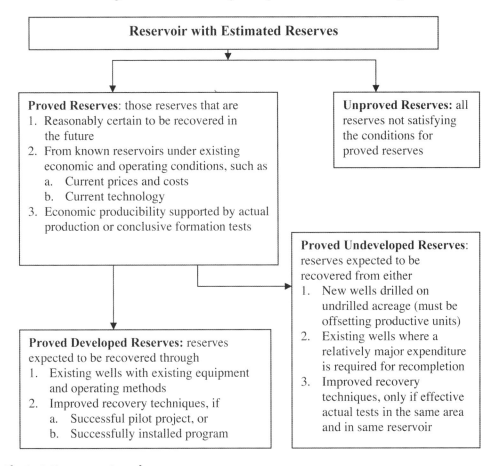

Fig. 2–1. Reserve categories

Figure 2–2 illustrates the difference between an exploratory well and a development well. As shown, a well drilled in an unproved area is an **exploratory well**. If an exploratory well finds proved reserves, an area around the well at the depth of the proved reservoir is designated a proved area. If additional wells are drilled within the proved area, to the proved depth, the wells are considered **development wells**. Note that a well drilled within a proved area with respect to the surface acreage, but to shallower depth, is not being drilled in a proved area, and therefore is considered an exploratory well.

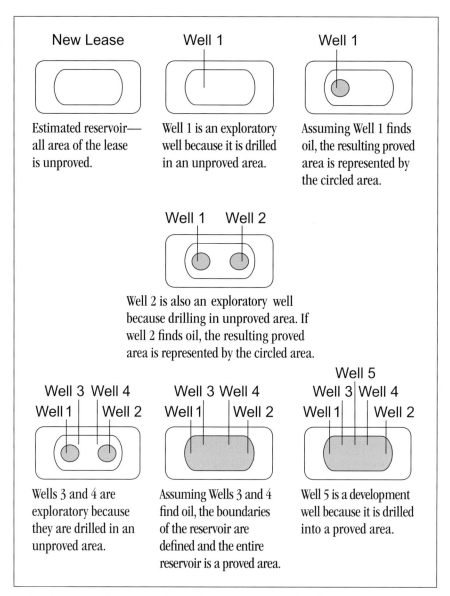

Fig. 2–2. Examples of exploratory and development wells. All wells are assumed to find oil at the same depth or horizon.

The number of wells that must be drilled in order to classify a reservoir as proved depends on many factors. Such factors include the size of the reservoir, the extent to which exploration has occurred in the area, and whether developed reservoirs with similar characteristics exist in the area.

When proved reserves have been confirmed, the property is reclassified from an unproved property to a proved property. However, although the first successful well or wells on a property prove the property, they will prove only a designated area at a specified depth. The well or wells will not prove the entire area of the property. Therefore, additional exploratory wells may still be drilled and are likely to be drilled on a property, even if that property has been proved as a result of a previous successful well.

The classification of a well as either an exploratory well or a development well depends upon engineering and geological data and is typically made by engineers and geologists, not accountants. In making classifications that will be used for financial reporting, the engineers and geologists must follow the definitions of *proved reserves* and *proved developed reserves* given in *Reg. S-X 4-10*.

Even though *Reg. S-X 4-10* provides definitions for exploratory and development wells, in practice, classification of wells is often difficult. For example, it is possible for a well to be partially exploratory and partially development. Consider the possibility that a well is being drilled to a depth of a horizon known to be productive. However, since the marginal cost for drilling additional footage for the purpose of testing a lower unproved formation is relatively low, the decision is made to drill beyond the producing formation to test for the presence of oil and gas in a previously unexplored horizon. If the well is drilled partially for the purpose of producing known proved reserves and partially for the purpose of searching for new reserves, the well should be classified as being partially development and partially exploratory. When this occurs, drilling costs must be allocated between exploratory well cost and development well cost.

Well classification is also complicated by the fact that there are a variety of terms used to refer to wells, and these terms are not always used consistently from one firm to another. For example, the terms **new field wildcat well**, **delineation well**, and **extension well** are commonly used in the oil and gas industry. Typically **new field wildcat well** is used to refer to the first well drilled in a location where there has been no previous drilling or production. Obviously, since new field wildcat wells are drilled in new, unproved areas, new field wildcat wells should be classified as exploratory wells.

Classification of delineation and extension wells is more difficult. While the terminology may vary from one company to another, commonly the term **delineation well** is used to refer to a well drilled along what the engineers believe to be the outer perimeter of the reservoir. If the well is drilled in an area where the reserves are unproved, it should be classified as an exploratory well. If, on the other hand, the well is drilled in a proved area for the purpose of developing proved undeveloped reserves, the well should be treated as a development well.

The term **extension well** often refers to a well that is drilled to test whether a known, proved reservoir actually extends beyond what engineers had previously believed to be the outer reservoir perimeter. Since the exploratory well definition includes wells drilled to extend a known reservoir, extension wells should typically be classified as exploratory wells.

A distinguishing characteristic of an exploratory well is whether it is drilled with the *original* intention of *adding* new proved reserves. Wells that are drilled for the purpose of discovering new proved reserves are exploratory wells, while wells that are drilled in a proved area for the purpose of producing proved undeveloped reserves are development wells.

Occasionally a company may decide that it is appropriate to increase its estimates of proved reserves as a consequence of information that was obtained through the drilling of a development well. Such an occurrence does not change the original classification of the well. If the original purpose for drilling the well was to produce proved reserves, i.e., a well drilled in a proved area, the well is classified as development even if the drilling results in an upward revision in estimated proved reserves.

Historical Cost Accounting Methods

The four basic types of costs incurred by oil and gas companies in exploration and production activities must be accounted for using one of two generally accepted historical cost methods: the **successful efforts method** or the **full cost method**. In connection with the four basic costs, the fundamental accounting issue is whether to capitalize or expense the incurred costs. If capitalized, the costs may be expensed as expiration takes place either through abandonment, impairment, or depletion as reserves are produced. If expensed as incurred, the costs are treated as period expenses and charged against revenue in the current period. The primary difference between successful efforts and full cost is in whether a cost is capitalized or expensed when incurred. In other words, the primary difference between the two methods is in the timing of the expense or loss charge against revenue.

The other basic difference between the two accounting methods is the size of the cost center over which costs are accumulated and amortized. For successful efforts, the cost center is a lease, field, or reservoir. In contrast, the cost center under full cost is a country. The cost center size has implications in computing depreciation, depletion, and amortization (DD&A) and also in computing ceiling and impairment writedowns (chapters 6, 7, and 9).

The successful efforts method adopted by the SEC (*Reg. S-X 4-10*) is essentially the same as prescribed by *SFAS No. 19*, "Financial Accounting and Reporting by Oil and Gas Producing Companies." The successful efforts method is generally consistent with financial accounting theory. Paragraph 143 of *SFAS No. 19* states:

> *In the presently accepted financial accounting framework, an asset is an economic resource that is expected to provide future benefits, and nonmonetary assets generally are accounted for at the cost to acquire or construct them. Costs that do not relate directly to specific assets having identifiable future benefits normally are not capitalized—no matter how vital those costs may be to the ongoing operations of the enterprise. If costs do not give rise to an asset with identifiable future benefits, they are charged to expense or recognized as a loss.*

Under successful efforts, a direct relationship is thus required between costs incurred and reserves discovered. Consequently, under successful efforts, only exploratory drilling costs that are successful, i.e., directly result in the discovery of proved reserves are considered to be part of the cost of finding oil or gas and thus are capitalized. Unsuccessful exploratory drilling costs do not result in an asset with future economic benefit and are therefore expensed.

In contrast, because there is no known way to avoid unsuccessful costs in searching for oil or gas, full cost considers both successful and unsuccessful costs incurred in the search for reserves as a necessary part of the cost of finding oil or gas. A direct relationship between costs incurred and reserves discovered is not required under full cost. Hence, both successful and unsuccessful costs are capitalized, even though the unsuccessful costs have no future economic benefit.

Specifically, successful efforts treats exploration costs that do not directly find oil or gas as period expenses, and successful exploration costs as capital expenditures. Under full cost, all exploration costs are capitalized. Under both methods, acquisition and development costs are capitalized and production costs are expensed. Although development costs could include an unsuccessful development well, all development costs are capitalized under successful efforts because the purpose of development activities is considered to be building a producing system of wells and related equipment and facilities, rather than searching for oil and gas. Table 2–1 shows the accounting treatment of these costs under successful efforts compared to full cost.

Table 2–1. Successful Efforts vs. Full Cost

Item	Successful Efforts	Full Cost
Acquisition costs	Capital	Capital
G&G costs	Expense	Capital
Exploratory dry hole	Expense	Capital
Exploratory well, successful	Capital	Capital
Development dry hole	Capital	Capital
Development well, successful	Capital	Capital
Production costs	Expense	Expense
Amortization cost center	Property, field, or reservoir	Country

The divergent accounting treatment of unsuccessful exploratory drilling costs under successful efforts versus full cost can have a substantial impact on the income statements of E&P companies. A company with a large exploratory drilling program and a normal unsuccessful drilling rate would, under successful efforts, have a significant amount of dry-hole expense. Those dry-hole costs would adversely affect the net income of a successful efforts company. On the other hand, a full cost company would capitalize exploratory dry-hole costs, and therefore, these costs would typically have no immediate effect on net income. They would, however, reduce net income through future amortization. The adverse effect on net income of expensing exploratory dry-hole costs under successful efforts may be especially significant for smaller companies.

The following example illustrates the impact of the full cost (FC) and successful efforts (SE) accounting methods on the financial statements of Tyler Company.

EXAMPLE

Financial Statements

Tyler Oil Company began operations on March 3, 2010, with the acquisition of a lease in Texas. During the first year, the following costs were incurred, DD&A (depreciation, depletion, and amortization) recognized, and the following revenue was earned:

G&G costs	$ 60,000		
Acquisition costs	100,000		
Exploratory dry holes	1,400,000		
Exploratory wells, successful	800,000		
Development costs	500,000		
Production costs	50,000		
DD&A expense	40,000 (SE);	90,000	(FC)
Revenue	250,000		

Income Statements

	Successful Efforts	Full Cost
Revenue	$ 250,000	$ 250,000
Expenses:		
G&G	$ 60,000	$ 0
Exploratory dry holes	1,400,000	0
Production costs	50,000	50,000
DD&A	40,000	90,000
Total expenses	1,550,000	140,000
Net income	$(1,300,000)	$ 110,000

Partial Balance Sheets

	Successful Efforts	Full Cost
G&G costs		$ 60,000
Acquisition costs	$ 100,000	100,000
Exploratory dry holes		1,400,000
Exploratory wells, successful	800,000	800,000
Development costs	500,000	500,000
Total assets	1,400,000	2,860,000
Less: Accumulated DD&A	(40,000)	(90,000)
Net assets	$ 1,360,000	$2,770,000

As shown in the example, the successful efforts method results in a much lower net "income" than the full cost method: a loss of $1,300,000 compared with a profit of $110,000. The majority of the difference is caused by the different treatment of the G&G costs and exploratory dry-hole costs. Under the successful efforts method, the costs were expensed, while under the full cost method, the costs were capitalized. Another difference is the amount of amortization (DD&A) recognized under each method. Under the full cost method, more costs were capitalized, resulting in a greater annual amortization expense.

Another major impact can be made on the income statements of successful efforts companies (but not full cost companies) by the order in which successful versus unsuccessful wells are drilled.

EXAMPLE

Order of Drilling

Tyler Oil, a successful efforts company, acquires a lease that has an oil reservoir at 10,000 feet. The reservoir has an unknown fault trap that contains no oil, located at the center of the reservoir. In attempting to locate, define, and develop the reservoir, Tyler drills a total of five wells as shown below at a cost of $300,000 each. Figure 2–3 depicts the drilling of five wells labeled A–E.

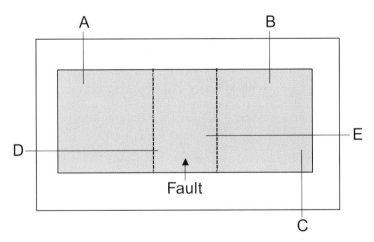

Fig. 2–3. Order of drilling

If the successful wells, A, B, and C, are drilled first and are considered to have delineated the reservoir (shaded area) shown in the figure, then the unsuccessful wells, D and E, would be classified as development wells. In this case, Tyler would have no dry-hole expense on its income statement relating to the five wells. If, however, wells D and E are drilled first, they would be classified as exploratory wells and expensed as dry holes. Wells A, B, and C would also be classified as exploratory but would be capitalized because they found proved reserves. Thus, in the situation described, merely changing the order in which the wells were drilled would result in a difference of $600,000 on Tyler's income statement.

If Tyler Company had been a full cost company, the order in which the wells were drilled would have had no effect on the income statement.

Historical Development of Accounting Methods and Current Status

Accounting for oil and gas producing activities poses many technical and theoretical problems and has been subject to much controversy. Reasons for the controversy surrounding the accounting procedures used by oil and gas exploration and production companies relate to the characteristics of the oil and gas industry. These characteristics include the following:

- High risk
- High cost of investment
- Lack of correlation between the size of expenditure and the value of any resulting reserves
- Long time span from when costs are first incurred until benefits are received

In addition, taxation, regulation, and the expansive use of joint operations (i.e., joint working interests) have added substantially to the complexity of companies' accounting and reporting. The federal taxation statutes contain many specialized rules applying only to oil and gas operations (e.g., intangible drilling costs and depletion rules). Further, federal and state governments have a long history of regulating the production and pricing of oil and natural gas. Finally, companies typically undertake oil and gas exploration and produce any resulting reserves as joint venture operations. In addition to placing extensive accounting responsibilities on the operator, joint venture accounting requires each of the parties to account for its proportionate share of costs and revenues and to separately report their income to the state and federal taxing authorities. For all the above reasons, accounting for oil and gas operations has complex and specialized accounting rules and procedures.

Prior to the release of *SFAS No. 19* in December 1977, the principal methods of financial accounting for oil and gas producing activities were successful efforts and full cost accounting. Both methods, and variations of the methods, were widely used, and the relative merits of each debated for many years.

In 1969, the American Institute of Certified Public Accountants (AICPA) issued *Accounting Research Study No. 11*, "Financial Reporting in the Extractive Industries." This study supported the successful efforts method of accounting. Despite this study, the Accounting Principles Board chose not to issue an opinion. As a result, its successor, the Financial Accounting Standards Board (FASB), was required to deal with the complex and politically sensitive issues surrounding an appropriate method of accounting for oil and gas exploration, development, and production costs.

In 1975, in response to OPEC's embargo and the consequent oil and gas shortage in the United States, the Federal Energy and Conservation Act was passed. This act required that the SEC must either prescribe oil and gas accounting rules or approve oil and gas accounting rules developed by the FASB. In response to a request from the SEC, the FASB began work on the oil and gas financial accounting and reporting problem and in 1977, issued *SFAS No. 19*. This statement, which prescribed the successful efforts method, was to have become effective for fiscal years after December 1978. However, in August 1978, the SEC issued *Accounting Series Release (ASR) 253*. This rule stated that oil and gas producing companies could use the full cost method according to *ASR 258*, or they could use the successful efforts method according to *ASR 257* or its equivalent, *SFAS No. 19*. *ASR 257* and *SFAS No. 19* are virtually identical, with the exception of certain reserve definitions. *SFAS No. 25*, "Suspension of Certain Accounting Requirements for Oil and Gas Producing Companies," was later issued that redefined the terms to agree with those in *ASR 257*. *SFAS No. 25*, as a result of the SEC's actions, also essentially suspended *SFAS No. 19*'s requirement that the successful efforts method of accounting be used.

When issuing *ASR 257* and *ASR 258*, the SEC stated that it believed neither full cost nor successful efforts provides sufficient information concerning the financial position or operating results of oil and gas companies. The primary deficiency, as perceived by the SEC, was the failure to include in the primary financial statements the most valuable asset an oil and gas company has, namely, oil and gas reserves. The SEC stated that a reserve valuation should be included in the primary financial statements, i.e., the balance sheet and income statement. For this reason, the SEC proposed a new method of accounting under which revenue would be recognized when reserves were discovered versus when they were produced and sold. Assets would be a valuation of the estimated future production of proved oil and gas reserves in place, discounted at a rate of 10%. The method was called *reserve recognition accounting (RRA)*. It was intended by the SEC to replace full cost and successful efforts as the basis for the primary financial statements after a trial period in which RRA statements would be presented as supplemental information.

In 1981, the SEC decided that RRA was not the answer and again called upon the FASB to provide a solution to the problem of oil and gas accounting. After much discussion, the FASB issued *SFAS No. 69*, "Disclosures about Oil and Gas Producing Activities," which established required disclosures for oil and gas producing companies.

SFAS No. 69 requires publicly traded companies with significant oil and gas producing activities to disclose supplementary information in their annual financial statements related to the following items:

Historical based:

- Proved reserve quantity information
- Capitalized costs relating to oil and gas producing activities
- Costs incurred for property acquisition, exploration, and development activities
- Results of operations for oil and gas producing activities

Value based:

- A standardized measure of discounted future net cash flows relating to proved oil and gas reserve quantities
- Changes in the standardized measure of discounted cash flows relating to proved oil and gas reserve quantities

Public and nonpublic companies are required to disclose two informational items:

1. Accounting method used in accounting for oil and gas producing activities
2. Manner of disposing of capitalized costs

In 1996, the SEC issued *Reg. S-X 4-10* establishing the current regulatory framework for accounting for E&P operations. According to *Reg. S-X 4-10*, successful efforts companies are to follow *SFAS No. 19* as amended, while full cost companies are to follow the SEC rules provided in *Reg. S-X 4-10*. In addition, whether a company uses the successful efforts or the full cost method, disclosures prepared according to *SFAS No. 69* must be presented as supplemental information to the financial statements. *Reg. S-X 4-10* is reproduced in appendix B. FASB standards and other pronouncements are available on the FASB website (www.FASB.org).

SFAS No. 19 and *Reg. S-X 4-10* provide authoritative guidance for financial accounting and reporting for *all* types of E&P operations. However, joint interest operations must also be accounted for according to the specific rules contained in the various contracts and agreements into which a company has entered. Accounting for oil and gas joint interest operations has been greatly influenced by the Council of Petroleum Accountants Societies (COPAS). COPAS was formed in 1961, with its activities and projects directed primarily toward issues and problems encountered in joint interest operations. COPAS issues "pronouncements" in the form of procedures, guidelines, and interpretations.

Perhaps the most widely used COPAS pronouncements are accounting procedures. When companies enter into joint operations (i.e., shared working interests), they must specify how costs are going to be shared and how the property is going to be managed. Typically this is done by executing a contract referred to as a **joint operating agreement**. Most, if not all, joint operating agreements include an accounting procedure. Accounting procedures address issues such as which costs are directly shared by the parties to the contract, determination of overhead rates, acquisition and transfer of materials, and conduct of audits and inventories. If the procedure is a part of the joint operating agreement contract, the parties are legally bound to follow the terms of the procedure.

Additionally, COPAS issues **Model Form Interpretations (MFIs)** to clarify the specific provisions of various accounting procedures. The purpose of a MFI relating to specific accounting procedures is to provide guidance and interpretation in implementing the procedures. COPAS also issues **Accounting Guidelines (AGs)**. These guidelines cover such topics as accounting for unitizations, imbalances, and conducting revenue audits. AGs provide advice regarding a variety of accounting issues that may arise in the industry. In addition, COPAS sponsors research projects on emerging issues facing the oil and gas industry and supports various educational efforts within the industry. COPAS accounting procedures are discussed in detail in chapter 12.

Introduction to Successful Efforts Accounting

Figure 2–4 presents a flow chart overview of the treatment of the four basic costs under successful efforts. An example follows that gives a brief illustration of typical costs incurred by oil and gas companies and their treatment under successful efforts. As shown in the flowchart and in the example, gross acquisition costs are capitalized as unproved property until either proved reserves are found or until the property is abandoned or impaired. If proved reserves are found, the property is reclassified from unproved property to proved property. Exploration costs are recorded in two different ways, depending upon the type of incurred costs. If the costs are nondrilling, as defined in chapter 3, they are expensed as incurred. If the exploration costs are drilling costs, they are capitalized temporarily as wells-in-progress until a determination is made whether proved reserves have been found. If proved reserves are found, the drilling costs are transferred to wells and equipment and then are charged to expense, specifically DD&A expense, as production occurs. If proved reserves are not found, i.e., a dry hole, the drilling costs are expensed. Development costs, which include the costs of drilling development wells, are capitalized regardless of whether or not proved reserves are found. All production costs are expensed as incurred.

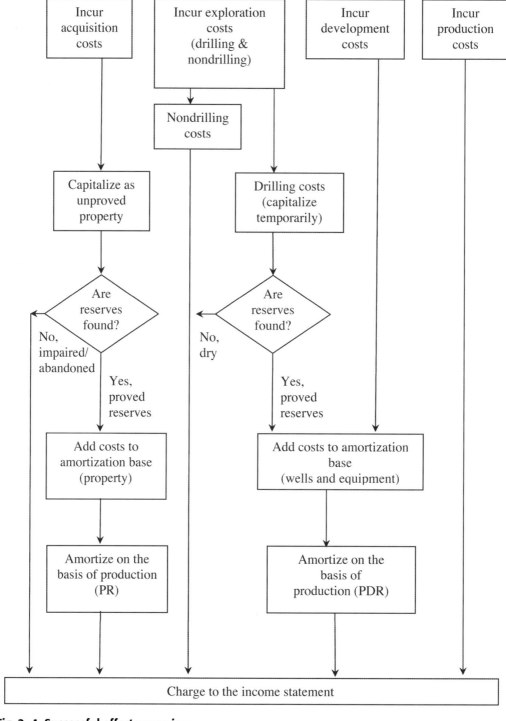

Fig. 2–4. Successful efforts overview

EXAMPLE

Overview of Entries—Successful Efforts

a. On January 1, Tyler Company spends $900,000 on G&G activities to locate and explore an oil prospect. (This is an exploration activity that cannot directly find oil or gas and so cannot be termed successful. Only by drilling a well can oil or gas normally be found.)

    Entry
G&G expense	900,000	
Cash		900,000

b. On January 15, Tyler Company acquires a 100-acre lease, paying a $500-per-acre bonus (acquisition cost).

    Entry
Unproved property (100 × $500)	50,000	
Cash		50,000

c. On February 20, Tyler Company drills a dry exploratory well at a cost of $700,000 (unsuccessful or nonproductive exploration cost).

    Entry
Dry-hole expense	700,000	
Cash		700,000

d. On March 29, Tyler Company drills a successful exploratory well at a cost of $825,000 (successful exploration cost).

    Entry*
Wells and equipment	825,000	
Cash		825,000

As a result of the successful exploratory well, Tyler must also reclassify the property.

Entry

Proved property	50,000	
Unproved property		50,000

e. On April 10, Tyler Company spends $850,000 on production facilities such as flow lines. (This cost is incurred in preparing proved reserves for production and, therefore, is a development cost.)

Entry

Wells and equipment	850,000	
Cash		850,000

f. On June 3, Tyler Company incurs $50,000 in production costs (production cost).

Entry

Production expense	50,000	
Cash		50,000

---

*The same entry would have been made if the well had been a successful development well or a dry development well.

A chart of accounts for a company using the successful efforts method of accounting is presented on the following pages. Each successful efforts company will have its own chart of accounts. The accounts shown in the following pages are illustrative of the accounts that a typical successful efforts company might have.

Chart of Accounts for a Successful Efforts Company

00000—00499 **CASH ACCOUNTS**
00001-000   Cash—Houston National Bank
00054-000   Cash—Chase NY

01000—01999 **ACCOUNTS RECEIVABLES**
01010-000   A/R—Lease Revenue
01410-000   A/R—Joint Venture
01411-000   A/R—Joint Venture Cash Calls
01901-000   A/R—Allowance for Doubtful Accounts

02500—02599 **MATERIAL & SUPPLIES**
02501-000   Materials & Supplies (M&S) Inventory
02501-020   Material & Supplies—Material Transfers, Pipe
02501-021   Material & Supplies—Material Transfers, Equipment

03000—03499 **PREPAIDS**
03001-000   Prepaid Insurance
03101-000   Miscellaneous Prepaids

03500—03999 **ACCRUALS**
03511-000   Accrued Gas Underdeliveries
03521-000   Accrued Pipeline Imbalance

10000—19999 **UNEVALUATED LEASEHOLDS—UNPROVED PROPERTY**
10101-000   Unevaluated Leaseholds
10101-904   Unevaluated Leaseholds, Overhead Allocation
10101-905   Unevaluated Leaseholds, Capitalized Interest
10101-907   Unevaluated Leaseholds, Allowance
10105-000   Accumulated Unevaluated Leasehold Improvements
10200-000   Unproved Nonworking Interests

12000—12299 **PRODUCING LEASEHOLDS—PROVED PROPERTY**
12011-000   Producing Leaseholds
12011-904   Producing Leaseholds—Overhead Allocation
12019-000   Producing Leaseholds—Capitalized Interest
12031-000   Producing Royalties
12041-000   Producing Property Retirements

12050—12310	**WELLS AND EQUIPMENT**
12051-000	Intangible Drilling Cost
12071-000	Lease & Well Equipment
12310-000—12310-999	**OFFSHORE DEVELOPMENT DRILLING (OFFDEVDRL) (WELLS-IN-PROGRESS)**
12310-040	Offdevdrl—Title Examination
12310-043	Offdevdrl—Survey, Location, Damages
12310-046	Offdevdrl—Mobile and Demobile
12310-049	Offdevdrl—Contract Drilling
12310-052	Offdevdrl—Turnkey Drilling
12310-055	Offdevdrl—Completion Rig
12310-058	Offdevdrl—Directional Drill, Well Survey
12310-061	Offdevdrl—Water
12310-063	Offdevdrl—Lost/Dam Rental Equipment
12310-064	Offdevdrl—Equipment Rental & Service
12310-067	Offdevdrl—Mud & Chemicals
12310-068	Offdevdrl—Completion W/O Fluids
12310-070	Offdevdrl—Drilling Bits
12310-071	Offdevdrl—Perforating
12310-074	Offdevdrl—Cement & Cement Service
12310-077	Offdevdrl—Well Log/Open Hole
12310-078	Offdevdrl—Well Log/Cased Hole
12310-080	Offdevdrl—Mud Logging
12310-082	Offdevdrl—Fishing Tools & Service
12310-083	Offdevdrl—Wireline Services
12310-086	Offdevdrl—Well Testing
12310-088	Offdevdrl—Tubular & Equipment Repair
12310-089	Offdevdrl—Tubular Inspection Service
12310-091	Offdevdrl—Stimulation
12310-092	Offdevdrl—Sand Control
12310-094	Offdevdrl—Reamers & Stabilizer
12310-095	Offdevdrl—Trucking, Land Transportation
12310-096	Offdevdrl—Boat, Water Transportation
12310-097	Offdevdrl—Helicopter, Air Transportation
12310-098	Offdevdrl—Abandonment Expense
12310-101	Offdevdrl—Company Labor
12310-103	Offdevdrl—Payroll Burden
12310-105	Offdevdrl—Contract Labor
12310-107	Offdevdrl—Consultants
12310-108	Offdevdrl—Meals/Entertainment
12310-109	Offdevdrl—Other Employee Expense
12310-123	Offdevdrl—Fuel & Electric

12310-150	Offdevdrl—Pollution Control
12310-179	Offdevdrl—Insurance
12310-180	Offdevdrl—Legal
12310-189	Offdevdrl—Federal Fuel Use Sales Tax
12310-190	Offdevdrl—State Fuel Use Sales Tax
12310-200	Offdevdrl—Overhead
12310-207	Offdevdrl—Misc., Unclassified
12310-299	Offdevdrl—Intangible J O Share
12310-300	Offdevdrl—Casing
12310-301	Offdevdrl—Tubing
12310-302	Offdevdrl—Wellhead Equipment
12310-303	Offdevdrl—Subsurface Equipment
12310-330	Offdevdrl—Miscellaneous Nonoperating Equipment
12310-399	Offdevdrl—Tangible J O Share
12310-902	Offdevdrl—Account Transfers
12311-000—12311-999	**OFFSHORE FACILITIES (WELLS-IN-PROGRESS)**
	(Many of the individual accounts are similar to accounts for offshore development drilling.)
12312-000—12312-999	**OFFSHORE WORKOVER (WELLS-IN-PROGRESS)**
	(Individual accounts are similar to accounts for offshore development drilling.)
12313-000—12313-999	**OFFSHORE RECOMPLETION (WELLS-IN-PROGRESS)**
	(Individual accounts are similar to accounts for offshore development drilling.)
12314-000—12314-999	**OFFSHORE DEVELOPMENT WELL PLUG AND ABANDONMENT (OFFDEVPA) (WELLS-IN-PROGRESS)**
12314-043	Offdevpa—Survey, Road, Location, Damage
12314-046	Offdevpa—Mobile and Demobile
12314-055	Offdevpa—Completion Rig
12314-061	Offdevpa—Water
12314-063	Offdevpa—Lost/Dam Rental Equipment
12314-064	Offdevpa—Equipment Rental & Service
12314-067	Offdevpa—Mud & Chemicals
12314-068	Offdevpa—Completion without Fluids
12314-070	Offdevpa—Drilling Bits
12314-071	Offdevpa—Perforation
12314-074	Offdevpa—Cement & Cement Service
12314-078	Offdevpa—Well Log/Cased Hole
12314-082	Offdevpa—Fishing Tools & Service
12314-083	Offdevpa—Wireline Services
12314-088	Offdevpa—Tubular & Equipment Repair

12314-089	Offdevpa—Tubular Inspection Services
12314-091	Offdevpa—Stimulation
12314-095	Offdevpa—Trucking, Land Transportation
12314-096	Offdevpa—Boat, Water Transportation
12314-097	Offdevpa—Helicopter, Air Transportation
12314-101	Offdevpa—Company Labor
12314-103	Offdevpa—Payroll Burden
12314-105	Offdevpa—Contract Labor
12314-107	Offdevpa—Consultants
12314-108	Offdevpa—Meals/Entertainment
12314-109	Offdevpa—Other Employee Expense
12314-123	Offdevpa—Fuel & Electric
12314-150	Offdevpa—Pollution Control
12314-179	Offdevpa—Insurance
12314-189	Offdevpa—Federal Fuel Use Sales Tax
12314-190	Offdevpa—State Fuel Use Sales Tax
12314-200	Offdevpa—Overhead
12314-207	Offdevpa—Miscellaneous Unclassified
12314-299	Offdevpa—Intangible Joint Operations Share
12314-300	Offdevpa—Casing
12314-301	Offdevpa—Tubing
12314-302	Offdevpa—Wellhead Equipment
12314-303	Offdevpa—Subsurface Equipment
12314-330	Offdevpa—Miscellaneous Nonoperating Equipment
12314-399	Offdevpa—Tangible Joint Operations Share
12317-000—12317-999	**OFFSHORE EXPLORATORY DRILLING (WELLS-IN-PROGRESS)**
12320-000—12320-999	**ONSHORE DEVELOPMENT DRILLING (WELLS-IN-PROGRESS)**
12321-000—12321-999	**ONSHORE FACILITIES (WELLS-IN-PROGRESS)**
12322-000—12322-999	**ONSHORE WORKOVER (WELLS-IN-PROGRESS)**
12323-000—12323-999	**ONSHORE RECOMPLETIONS (WELLS-IN-PROGRESS)**
12324-000—12324-999	**ONSHORE DEVELOPMENT WELL PLUG AND ABANDON (WELLS-IN-PROGRESS)**
12327-000—12327-999	**ONSHORE EXPLORATORY DRILLING (WELLS-IN-PROGRESS)**
12330-000—12330-999	**OFFSHORE LEASE ACQUISITION—UNPROVED PROPERTY**

12330-400	Offshore Lease Acquisition—General AFE Cost
12330-404	Offshore Lease Acquisition—Lease Bonus
12330-408	Offshore Lease Acquisition—Brokerage
12330-411	Offshore Lease Acquisition—Recording Fees
12330-415	Offshore Lease Acquisition—Unitization
12330-418	Offshore Lease Acquisition—Legal
12330-420	Offshore Lease Acquisition—Division Order Title Exam
14000—14131	**MISCELLANEOUS ASSETS**
13501-000	Plant Equipment & Facilities
14011-000	Transportation—Vehicles
14031-000	Marine Equipment
14051-000	Work Equipment
14071-000	Furniture & Fixtures
14091-000	Computer Equipment—Hardware
14101-000	Computer Equipment—Software
14111-000	Leasehold Improvements
14131-000	Telephone/Communication Equipment
16000—16999	**ACCUMULATED DD&A AND WRITEDOWNS**
18101-000	Long-Term Receivable
18201-000	Note Receivable—Branch
19001-000	Preacquisition Expenses
19021—19091	**DEFERRED FEDERAL INCOME TAX**
20000—20999	**ACCOUNTS PAYABLE**
21000—21999	**SHORT-TERM NOTES PAYABLE**
24000—24999	**ACCRUED STATE AND FEDERAL TAXES**
29000—29999	**ACCRUED LIABILITIES**
29001-000	Accrued Miscellaneous Liabilities
29006-000	Accrued Gas Marketing Liabilities
29007-000	Accrued Oil Marketing Liabilities
29010-000	Accrued Rent
29011-000	Accrued Accounting Fees
29111-000	Accrued Gas Overdeliveries
29501-000	Accrued Vacation Pay
30000—31999	**LONG-TERM NOTES PAYABLE**
40000—49999	Stockholders Equity and Related Accounts
40001-000	Common Stock
41001-000	Contributed Capital

42001-000	Retained Earnings
42005-000	Dividends Declared

53000—59999	**REVENUES AND GAINS**
53001-000	Revenue—Oil
53500-000	Revenue—Gas
53600-000	Royalty Revenue
54001-000	Plant Revenue
55001-000	Pipeline Transportation Revenue
59001-000	Miscellaneous Operating Income
59501-000	Gas Balance Settlements/Adjustments
59601-000	Gain on Sale/Retirement of Prop

60000—60999	**LEASE OPERATING EXPENSE (LOE)**
60001-061	LOE—Water
60001-096	LOE—Trucking & Land Transportation
60001-096	LOE—Boats, Water Transportation
60001-097	LOE—Helicopter, Air Transportation
60001-101	LOE—Company Labor
60001-103	LOE—Payroll Burden
60001-104	LOE—Employee Pension
60001-105	LOE—Contract Labor
60001-107	LOE—Consultants & Professional Service
60001-108	LOE—Meals/Entertainment
60001-109	LOE—Other Employee Expense
60001-112	LOE—Chemicals & Treating
60001-114	LOE—Supplies & Tools
60001-116	LOE—Gathering
60001-119	LOE—Transportation
60001-123	LOE—Fuel & Electric
60001-131	LOE—Saltwater Disposal
60001-135	LOE—Groceries & Food
60001-137	LOE—Other Rents
60001-139	LOE—Measurement & Testing
60001-143	LOE—Gas Marketing Service & Repair
60001-147	LOE—Safety & Supplies
60001-150	LOE—Pollution Control
60001-154	LOE—Compressor Parts
60001-158	LOE—Surface Repairs & Maintenance
60001-160	LOE—Communications
60001-162	LOE—Road & Location Maintenance
60001-164	LOE—Controllable Equipment

60001-166	LOE—Subsurface Repairs & Maintenance
60001-170	LOE—Well Workover
60001-172	LOE—Well Services
60001-176	LOE—Abandonment
60001-180	LOE—Legal
60001-182	LOE—Equipment Rental & Service
60001-183	LOE—Compressor Rentals
60001-185	LOE—Ad Valorem Tax
60001-188	LOE—Production/Severance Tax
60001-189	LOE—Federal Fuel Use Sales Tax
60001-190	LOE—State Fuel Use Sales Tax
60001-192	LOE—Prod./Severance Tax (20% prop.)
60001-194	LOE—Field Expense
60001-196	LOE—Facility Operations
60001-198	LOE—Net Profits Interest
60001-200	LOE—Overhead
60001-201	LOE—District Expense
60001-204	LOE—Miscellaneous
60001-205	LOE—Nonoperating Joint Cost
60001-250	LOE—Advances to Operators
60001-995	LOE—Cutback on Insurance
60001-999	LOE—Charges to Joint Owners
61000—61999	**MARKETING AND TRANSPORTATION EXPENSE**
66000—66999	**GENERAL AND ADMINISTRATIVE EXPENSE**
66001-501	G&A—Salaries & Wages
66001-505	G&A—Bonus Compensation
66001-509	G&A—Auto Allowance
66001-514	G&A—Company Vehicle
66001-519	G&A—Parking
66001-525	G&A—Remote Parking
66001-529	G&A—Relocation
66001-533	G&A—Education Reimbursement
66001-538	G&A—Personal Development
66001-542	G&A—Subscriptions/Publications
66001-543	G&A—Professional Dues
66001-545	G&A—Membership Dues
66001-547	G&A—Entertainment
66001-552	G&A—Meals
66001-556	G&A—Lodging
66001-560	G&A—Transportation
66001-564	G&A—Auto/Taxi

66001-568	G&A—Overtime Meal Allowance
66001-572	G&A—Other Employee Expense
66001-576	G&A—Group Insurance
66001-580	G&A—Workers Compensation
66001-584	G&A—Employer FICA
66001-588	G&A—Federal/State Unemployment
66001-592	G&A—Thrift Contributions
66001-594	G&A—Other Employee Benefits
66001-595	G&A—Employee Pension
66001-596	G&A—Office Contract Labor
66001-599	G&A—Contract Labor Benefits
66001-603	G&A—Accounting Fees
66001-608	G&A—Accounting—Tax Fees
66001-612	G&A—Legal Fees, Litigation
66001-616	G&A—Legal Fees, Other
66001-620	G&A—Data Processing Fees
66001-624	G&A—Reserve Engineering Fees
66001-628	G&A—Other Professional Fees
66001-630	G&A—Office Rent
66001-632	G&A—Communications
66001-636	G&A—Reproduction
66001-640	G&A—Postage
66001-644	G&A—Courier/Messenger Services
66001-648	G&A—Office Equip Rentals & Services
66001-652	G&A—Maintenance & Repairs
66001-656	G&A—Moving & Storage
66001-660	G&A—Office Supplies
66001-661	G&A—Data Processing Supplies
66001-664	G&A—Refreshments
66001-668	G&A—Contributions
66001-672	G&A—General Liability Insurance
66001-676	G&A—Property & Casualty Insurance
66001-680	G&A—Miscellaneous Taxes
66001-685	G&A—Ad Valorem Taxes
66001-686	G&A—Outside Service, Other
66001-688	G&A—Other G&A
68000—68999	**DD&A EXPENSE**

70000—70999	**NONPRODUCTIVE EXPENSES**
70000-100	Nonproductive Exploratory Drilling
70000-200	Dry-Hole and Bottom-Hole Contributions
70000-300	Delay Rentals
70000-400	G&G Services
70000-500	Carrying and Maintenance Costs
70000-600	Abandoned Leases
70000-700	Impairment
79600—85400	**MISCELLANEOUS EXPENSES**
79601-000	Bad Debt Expense
80001-000	Federal Income Tax—Current
81001-000	Deferred Federal Income Tax—Capitalized Interest

INTRODUCTION TO FULL COST ACCOUNTING

Figure 2–5 presents a flow chart overview of the treatment of the four basic costs under full cost. The example used to illustrate successful efforts accounting is also used to illustrate full cost accounting. As shown in the flow chart and in the example, acquisition, exploration, and development costs are capitalized under the full cost method, regardless of whether the costs result in a discovery of reserves. As in successful efforts accounting, gross acquisition costs are placed in an unproved property account and are moved to a proved property account if proved reserves are found. If the property is abandoned or impaired, the costs continue to be capitalized but are transferred to an abandoned or impaired cost account. All acquisition, exploration, and development costs incurred in each country are capitalized into one pool, so that theoretically, individual properties lose their identities. However, companies maintain subsidiary records on individual properties for tax, regulatory, and management purposes. All production costs are expensed as incurred.

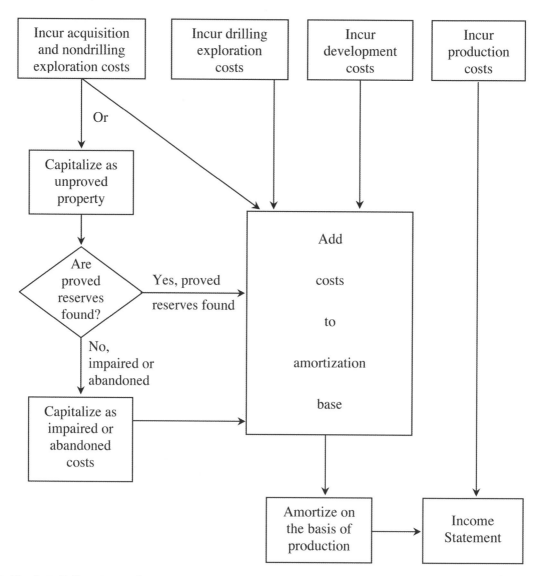

Fig. 2–5. Full cost overview

EXAMPLE

Overview of Entries—Full Cost

a. On January 1, Tyler Company spends $900,000 on G&G activities to locate and explore an oil prospect (exploration cost).

   Entry
   G&G costs   900,000
      Cash   900,000

b. On January 15, Tyler Company acquires a 100-acre lease, paying a $500-per-acre bonus (acquisition cost).

   Entry
   Unproved property—acquisition   50,000
      Cash   50,000

c. On February 20, Tyler Company drills a dry exploratory well at a cost of $700,000 (exploration cost).

   Entry
   Exploratory dry hole   700,000
      Cash   700,000

d. On March 29, Tyler Company drills a successful exploratory well at a cost of $825,000 (exploration cost).

   Entry
   Wells and equipment   825,000
      Cash   825,000

   As a result of the successful exploratory well, Tyler must also reclassify the property.

   Entry
   Proved property—acquisition   50,000
      Unproved property—acquisition   50,000

e. On April 10, Tyler Company spends $850,000 on production facilities such as flow lines (development cost).

Entry

Wells and equipment	850,000	
Cash		850,000

f. On June 3, Tyler Company incurs $50,000 in production costs (production cost).

Entry

Production expense	50,000	
Cash		50,000

A full cost chart of accounts is presented on the following pages. A full cost company, in contrast to a successful efforts company, does not have expense accounts for G&G costs, delay rentals, exploratory dry holes, etc. Instead, these costs are capitalized under the full cost method of accounting. Impairment and abandoned leasehold costs are also capitalized for full cost companies. Other than these differences, the chart of accounts shown on the following pages is essentially the same as the successful efforts chart of accounts shown earlier.

CHART OF ACCOUNTS FOR A FULL COST COMPANY

00000—00499	**CASH ACCOUNTS** (For detailed subaccounts, see the successful efforts chart of accounts.)
01000—01999	**ACCOUNTS RECEIVABLES** (For detailed subaccounts, see the successful efforts chart of accounts.)
02500—02599	**MATERIAL & SUPPLIES** (For detailed subaccounts, see the successful efforts chart of accounts.)
03000—03499	**PREPAIDS** (For detailed subaccounts, see the successful efforts chart of accounts.)
03500—03999	**ACCRUALS** (For detailed subaccounts, see the successful efforts chart of accounts.)

10000—10999	**UNEVALUATED LEASEHOLDS—UNPROVED PROPERTY**
	(For detailed subaccounts, see the successful efforts chart of accounts.)
11000—11999	**NONPRODUCTIVE COSTS**
11001-000	Capitalized Nonproductive Drilling
11011-000	Capitalized Dry-Hole and Bottom-Hole Contributions
11101-000	Capitalized Delay Rentals
11111-000	Capitalized Abandoned Leases
11131-000	Capitalized Impairment
11201-000	Capitalized G&G Services
12000—12299	**PRODUCING LEASEHOLDS—PROVED PROPERTY**
	(For detailed subaccounts, see the successful efforts chart of accounts.)
12050—12310	**WELLS AND EQUIPMENT**
12051-000	Intangible Drilling Cost
12071-000	Lease & Well Equipment
12310-000—12310-999	**OFFSHORE DEVELOPMENT DRILLING (OFFDEVDRL) (WELLS-IN-PROGRESS)**
	(For detailed subaccounts, see the successful efforts chart of accounts.)
12311-000—12311-999	**OFFSHORE FACILITIES (WELLS-IN-PROGRESS)**
	(Many of the individual accounts are similar to accounts for offshore development drilling.)
12312-000—12312-999	**OFFSHORE WORKOVER (WELLS-IN-PROGRESS)**
	(Individual accounts are similar to accounts for offshore development drilling.)
12313-000—12313-999	**OFFSHORE RECOMPLETION (WELLS-IN-PROGRESS)**
	(Individual accounts are similar to accounts for offshore development drilling.)
12314-000—12314-999	**OFFSHORE DEVELOPMENT WELL PLUG AND ABANDONMENT (WELLS-IN-PROGRESS)**
	(For detailed subaccounts, see the successful efforts chart of accounts.)
12317-000—12317-999	**OFFSHORE EXPLORATORY DRILLING (WELLS-IN-PROGRESS)**
12320-000—12320-999	**ONSHORE DEVELOPMENT DRILLING (WELLS-IN-PROGRESS)**

12321-000—12321-999	**ONSHORE FACILITIES (WELLS-IN-PROGRESS)**
12322-000—12322-999	**ONSHORE WORKOVER (WELLS-IN-PROGRESS)**
12323-000—12323-999	**ONSHORE RECOMPLETIONS (WELLS-IN-PROGRESS)**
12324-000—12324-999	**ONSHORE DEVELOPMENT WELL PLUG AND ABANDON (WELLS-IN-PROGRESS)**
12327-000—12327-999	**ONSHORE EXPLORATORY DRILLING (WELLS-IN-PROGRESS)**
12330-000—12330-999	**OFFSHORE LEASE ACQUISITION—UNPROVED PROPERTY**
	(For detailed subaccounts, see the successful efforts chart of accounts.)
12340-000—12340-999	**OFFSHORE G&G (WELLS-IN-PROGRESS)**
12340-400	Offshore G&G—General AFE Cost
12340-420	Offshore G&G—Acquisition
12340-424	Offshore G&G—Processing
12340-429	Offshore G&G—Permits
12345-000—12345-999	**ONSHORE G&G (WELLS-IN-PROGRESS)**
12345-400	Onshore G&G—General AFE Cost
12345-410	Onshore G&G—General Subcontract Cost
12345-420	Onshore G&G—Acquisition
12345-421	Onshore G&G—Seismic Cost
12345-422	Onshore G&G—Mapping Cost
12345-424	Onshore G&G—Processing
12345-429	Onshore G&G—Permits
12345-430	Onshore G&G—Consultants
12345-432	Onshore G&G—Technical Services
12345-434	Onshore G&G—Bulldozing
12345-436	Onshore G&G—Operations Equipment
12345-438	Onshore G&G—Communication Cost
12345-440	Onshore G&G—Environmental Costs
12345-441	Onshore G&G—Quality Control Costs
12345-442	Onshore G&G—Damages Costs
14000—14131	**MISCELLANEOUS ASSETS**
	(For detailed subaccounts, see the successful efforts chart of accounts.)
16000—16999	**ACCUMULATED DD&A AND WRITEDOWNS**

18000—18999	**LONG-TERM RECEIVABLES**
18101-000	Long-Term Receivable
18201-000	Note Receivable—Branch
19021—19091	**DEFERRED FEDERAL INCOME TAX**
20000—20999	**ACCOUNTS PAYABLE**
21000—21999	**SHORT-TERM NOTES PAYABLE**
24000—24999	**ACCRUED STATE AND FEDERAL TAXES**
29000—29999	**ACCRUED LIABILITIES** (For detailed subaccounts, see the successful efforts chart of accounts.)
30000—31999	Long-Term Notes Payable
40000—49999	**STOCKHOLDERS EQUITY AND RELATED ACCOUNTS** (For detailed subaccounts, see the successful efforts chart of accounts.)
53000—59999	**REVENUES AND GAINS** (For detailed subaccounts, see the successful efforts chart of accounts.)
60000—60999	**LEASE OPERATING EXPENSE** (For detailed subaccounts, see the successful efforts chart of accounts.)
61000—61999	**MARKETING AND TRANSPORTATION EXPENSE**
66000—66999	**GENERAL AND ADMINISTRATIVE EXPENSE** (For detailed subaccounts, see the successful efforts chart of accounts.)
68000—68999	**DD&A EXPENSE**
79600—85400	**MISCELLANEOUS EXPENSES** (For detailed subaccounts, see the successful efforts chart of accounts.)

The numbers in the above examples and in the other examples and homework problems throughout this book are not intended to be realistic and are intended for illustrative purposes and ease of computation. In addition, the detailed record keeping used in actual practice is not illustrated in the examples and problems. Subsidiary accounts and cost records would in actual practice be prepared for many of the entries given in the book. For example, if three leases were obtained, subsidiary records would be maintained for each lease, each well, and each owner's interest.

・・・

Chapters 3 through 5 detail the accounting for exploration, acquisition, and drilling activities according to successful efforts. Chapter 6 presents the capitalized cost disposition under successful efforts by both DD&A and abandonment of property. Chapter 7 provides a description of full cost accounting. Chapters 8 and 10 deal with production and revenue accounting, respectively. Chapter 9 details current rules for accounting for asset retirement obligations and asset impairment. Chapter 11 presents tax accounting methodology. Joint interest accounting is presented in chapter 12. Accounting for conveyances under successful efforts and full cost accounting is described in chapter 13. Chapter 14 explains the disclosures required by *SFAS No. 69* for companies with oil and gas producing activities. Chapter 15 gives an overview of the international dimensions of the industry. Chapter 16 presents financial statement analysis.

PROBLEMS

1. List the costs that are treated the same (i.e., capitalized or expensed) under successful efforts and full cost accounting. List the costs that are treated differently.

2. What accounting treatment is given to the following costs under successful efforts: acquisition costs, exploration costs, development costs, and production costs?

3. The successful efforts method of accounting for oil and gas operations is considered to follow generally accepted accounting principles as pronounced by the FASB. Explain.

4. Define the following:

    reserves

    proved reserves

    proved developed reserves

    proved undeveloped reserves

    proved area

    field

5. Define the following:

    exploratory well

    development well

    delineation well

    new field wildcat well

    extension well

service well

stratigraphic test well (exploratory and development)

6. When is a delineation well classified as an exploratory well versus a development well?

7. Lease A has a known productive horizon at 15,000 feet. A well is drilled to 10,000 feet. Would the well be classified as an exploratory or development well?

8. Lease B has a producing formation at 8,000 feet. A well is drilled to 11,000 feet and is classified as a development well. Comment.

9. Sauer Petroleum, a successful efforts company, drilled an exploratory well offshore at a cost of $1 million. The well was dry, but Sauer Petroleum felt that the G&G data obtained from the well was promising and drilled another well close to the first one. Should the first well be expensed or capitalized?

10. McGavin Oil Company incurred the following costs during calendar year 2016:

February 1	Cost of G&G activities to locate an oil prospect, $100,000
March 2	Acquisition costs for a 400-acre lease: lease bonus $50/acre; other costs incurred in acquiring the property, $1,000
May 30	Dry-hole costs of an exploratory well, $315,000
June 28	Successful exploratory well costs, $405,000
August 15	Cost of production facilities such as flow lines and separators, $225,000
September 1	Production costs, $50,000

Prepare journal entries for the above transactions using the successful efforts method of accounting.

11. Given the following costs for Lease A, all incurred during 2018, prepare income statements and unclassified partial balance sheets for a successful efforts (SE) company and a full cost (FC) company.

Acquisition costs	$ 30,000	
G&G costs	80,000	
Exploratory dry holes	1,500,000	
Successful exploratory holes	350,000	
Development wells, dry	200,000	
Development wells, successful	475,000	
Cost of production facilities	250,000	
Production costs	60,000	
DD&A*	55,000	(SE); 125,000 (FC)
Accumulated DD&A	150,000	(SE); 360,000 (FC)
Revenue from sale of oil	225,000	

*Depreciation, depletion, and amortization

12. Reida Oil Corporation incurred the following costs during the fiscal year ending May 31, 2015.

June 1, 2014	G&G costs	$ 50,000
August 10, 2014	Lease bonus on a 1,000 acre lease	80/acre
	Other acquisition costs	2,000
December 15, 2014	Dry-hole costs of an exploratory well. . . .	400,000
January 18, 2015	Successful well costs	600,000
April 10, 2015	Cost of production facilities.	500,000
April 30, 2015	Production costs	40,000

Prepare journal entries for the above transaction, assuming that Reida Oil Corporation uses the full cost method of accounting.

13. Revenue and costs for Optimistic Company for the year 2015 are presented below:

Revenue.	$ 600,000
G&G Costs.	600,000
Acquisition costs.	2,000,000
Exploratory dry holes	4,000,000
Successful exploratory wells	3,000,000
Development wells, dry	1,000,000
Development wells, successful.	800,000
Production facilities	700,000
Production costs	100,000

	Successful Efforts	Full Cost
Amortization for 2015	$200,000	$ 400,000
Accumulated DD&A	500,000	700,000

Prepare income statements and unclassified balance sheets for a successful efforts (SE) company and a full cost (FC) company and explain the difference in net income.

14. Indicate whether the following costs should be expensed (E) or capitalized (C) depending on whether the company uses successful efforts or the full cost method of accounting.

	Successful Efforts		Full Cost	
	Expense	Capital	Expense	Capital
Acquisition costs				
G&G costs				
Exploratory dry holes				
Successful exploratory wells				
Development wells, dry				
Development wells, successful				
Production facility costs				
Production costs				

15. Elizabeth Corporation incurred the following costs during 2015:

January 12	G&G Costs	$ 75,000
February 15	Lease acquisition costs	150,000
March 16	Exploratory dry-hole costs	350,000
April 29	Successful exploratory well costs	500,000
June 15	Developmental well costs	200,000
August 4	Production facility costs	100,000
September 20	Production costs	55,000

Elizabeth uses the successful efforts method of accounting. Prepare the journal entries for the year-end December 31, 2015.

3

NONDRILLING EXPLORATION COSTS—SUCCESSFUL EFFORTS

Exploration involves identifying and examining areas that may contain oil and gas reserves. Exploration costs are defined in *Reg. S-X 4-10*, par. 16 as follows:

> *Costs incurred in identifying areas that may warrant examination and in examining specific areas that are considered to have prospects of containing oil and gas reserves, including costs of drilling exploratory wells and exploratory-type stratigraphic test wells. Exploration costs may be incurred both before acquiring the related property (sometimes referred to in part as prospecting costs) and after acquiring the property. Principal types of exploration costs, which include depreciation and applicable operating costs of support equipment and facilities and other costs of exploration activities, are:*
>
> *(1) Costs of topographical, geographical and geophysical studies, rights of access to properties to conduct those studies, and salaries and other expenses of geologists, geophysical crews, and others conducting those studies. Collectively, these are sometimes referred to as geological and geophysical or "G&G" costs.*
>
> *(2) Costs of carrying and retaining undeveloped properties, such as delay rentals, ad valorem taxes on properties, legal costs for title defense, and the maintenance of land and lease records.*

(3) Dry hole contributions and bottom hole contributions.

(4) Costs of drilling and equipping exploratory wells.

(5) Costs of drilling exploratory-type stratigraphic test wells.

The first three types of costs are often incurred before any drilling activities begin and are nondrilling in nature. In order to distinguish them from type 4 and 5 costs, which are drilling costs, type 1, 2, and 3 costs are referred to in this book as **nondrilling exploration costs**. *SFAS No. 19* specifies that nondrilling exploration costs are to be charged to expense as incurred, as shown in Figure 3–1.

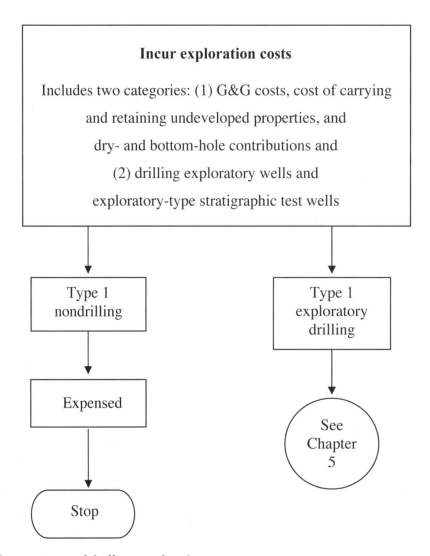

Fig. 3–1. Successful efforts, exploration costs

G&G Costs

The purpose of geological and geophysical (G&G) exploration is to locate or identify areas with the potential of producing oil and/or gas in commercial quantities. As discussed in chapter 1, both surface and subsurface G&G techniques are used to locate these areas. Surface techniques are used to evaluate the surface for evidence of subsurface formations with characteristics favorable to the accumulation of oil or gas. Subsurface techniques identify formations capable of containing oil or gas by utilizing the fact that all types of rocks have different characteristics and respond differently to stimuli such as sound or magnetic waves.

A **reconnaissance survey** is a G&G study covering a large or broad area, while a **detailed survey** is a G&G study covering a smaller area, one possibly identified as a result of the reconnaissance survey. By studying the data yielded by the G&G studies, geologists and geophysicists can identify interest areas that may warrant the acquisition of a working interest.

G&G studies may be conducted either before or after a mineral interest in a property has been acquired. In the United States, if a company wants to explore an onshore area prior to obtaining a mineral lease on the property, rights to access the property, called **shooting rights**, must first be obtained from the property owner. The E&P company pays a fee for the shooting rights, and typically, the rights are coupled with an option to lease. The option to lease guarantees that the E&P company has the opportunity to lease the property if the G&G data are promising.

If shooting rights coupled with an option to lease are obtained and if the surface owner is different from the mineral rights owner, the agreement must be signed with the mineral rights owner. In that case, the surface owner would be compensated for rent and any damages done to the surface (e.g., damage to crops, livestock, buildings, etc.). If shooting rights only are obtained and if the surface owner is different from the mineral rights owner, the E&P company may sign a contract with either the mineral rights owner or the surface rights owner.

When offshore U.S. tracts are involved, a permit, not usually requiring the payment of a fee, is normally obtained from the U.S. Geological Survey (USGS). If the offshore tract is in state waters, then a permit must be obtained from the state.

As discussed in chapter 1, in operations conducted outside the United States, the E&P company must obtain permission from the local government in order to come into the country and conduct G&G operations. Contracting for a mineral interest typically involves negotiations with a governmental agency, e.g., the ministry of petroleum. In these situations, G&G operations are typically undertaken only after a contract has been executed. Oil and gas exploration and production contracts may take many forms, the most common being concessionary agreements and production sharing contracts. These contracts are discussed in detail in chapter 15. Regardless of the location of the operations, if the E&P company is using the successful efforts method in compliance with U.S. generally accepted accounting principles (GAAP), it must apply the capitalization rules contained in *SFAS No. 19* and provide the disclosures mandated by *SFAS No. 69*.

G&G costs include all the costs related to conducting G&G studies and the cost of access rights to properties to conduct those studies, including any damages or rent paid to the surface interest owner. These G&G costs must be expensed as incurred regardless of whether they were incurred before or after acquisition of a working interest in the property. G&G costs are similar to research costs because they are incurred in the process of obtaining information. At the time the exploratory G&G studies are being done, there are no proved reserves and correlation of G&G costs with specific discoveries that might not occur for months or even years is very difficult or impossible. Therefore, the FASB concluded that G&G costs should be expensed as incurred.

EXAMPLE

G&G Costs

a. Tyler Company obtained shooting rights—access rights to a property so G&G studies may be conducted—to 10,000 acres, paying $1.00 per acre.

Entry

G&G expense (10,000 × $1.00)	10,000	
Cash		10,000

b. Tyler Company then hired ABC Company to conduct the G&G work and paid the company $250,000.

Entry

G&G expense	250,000	
Cash		250,000

As described above, broad G&G surveys determine interest areas that warrant leasing or further exploration. Detailed surveys are typically made to further delineate the more promising areas for possible leasing. Although not required by the successful efforts method, both broad and detailed survey costs are, in many cases, allocated to the leases acquired. However, even if allocated to individual leases by companies using successful efforts, G&G costs should still be expensed.

Since *SFAS No. 19* was written in 1977, significant advances have been made in 3-D and 4-D seismic technology. In 1977, seismic studies were used almost exclusively for exploration purposes. Today, 3-D and 4-D technology are used not only to explore for new reservoirs, they are also used to determine optimal well location and in the planning and execution of development operations. One question that has arisen is if all seismic must be written off as G&G exploration expense or whether well-related seismic should be accounted for as a drilling cost and development-related seismic should be accounted for as development costs. Recently the SEC has questioned the capitalization of well-related and development-related seismic. While no definitive guidance has been provided by the SEC, at the present time, they have not indicated that capitalization in such cases is improper. According to the *2001 PricewaterhouseCoopers Survey of U.S. Petroleum Accounting Practices,* 66.7% of the successful efforts companies responding to the survey indicate that they capitalize the cost of 3-D seismic studies of existing producing reservoirs as a development cost.[1] G&G studies performed to determine the location of a specific drillsite are part of the drilling process and theoretically should be accounted for as a drilling cost rather than a nondrilling exploration cost. These costs are accounted for as such throughout this book.

G&G studies may also be conducted on a property owned by another party in exchange for an interest in the property if proved reserves are found—if not found, the G&G costs incurred are reimbursed. In this situation, the G&G costs should be recorded as a receivable when incurred and, if proved reserves are found, should be transferred to become part of the cost of the proved property acquired (*SFAS No. 19,* par. 20). If proved reserves are not found, reimbursement is received.

EXAMPLE

G&G Studies Exchanged for Interest in Property

During 2012, Tyler Company paid for G&G studies to be performed on two leases owned by other parties. The agreements provided that Tyler Company would receive 25% of each working interest if proved reserves were found and would be reimbursed for G&G costs incurred if proved reserves were not found. The G&G costs incurred by Tyler were as follows:

Lease Alpha $40,000
Lease Beta $60,000

Drilling activity on the leases in 2012 was as follows:

Lease Alpha   Drilling resulted in a dry hole and the lease was abandoned. The owner of Lease Alpha reimbursed Tyler for the G&G costs.

Lease Beta   Drilling resulted in discovering proved reserves and, as per the agreement, Tyler received 25% of the working interest.

Entries
Lease Alpha:
Receivable—Lease Alpha   40,000
   Cash   40,000

Cash   40,000
   Receivable—Lease Alpha   40,000

Lease Beta:

Receivable—Lease Beta	60,000	
Cash		60,000
Proved property—Lease Beta	60,000	
Receivable—Lease Beta		60,000

CARRYING AND RETAINING COSTS

Carrying and retaining costs are incurred in U.S. operations primarily to maintain the lessee's economic interests, not to acquire those rights. Common carrying and retaining costs include:

- Delay rentals
- Property taxes
- Legal costs for title defense
- Clerical and record-keeping costs

Delay rentals are costs paid on or before the anniversary date of the lease during the primary term in order to delay drilling operations for a year. Ad valorem taxes or property taxes are assessed on the economic interest owned by the working interest and are levied by a governmental agency, i.e., city, county, school district, etc. (These property taxes are incurred to maintain the property, unlike the delinquent property taxes discussed in chapter 4 that may be incurred at the time of acquiring the property.) Legal costs for title defense paid by the lessee include attorney's fees, court costs, etc., incurred when the royalty interest owner is involved in a legal dispute regarding claims to title to the property. (The lessee may be willing to pay these fees to ensure the title problems are resolved satisfactorily.) Lease record maintenance costs are incurred by the land department in maintaining, evaluating, and updating the company's lease records. Employee salaries, materials, and supplies comprise the bulk of such maintenance costs.

Carrying costs do not increase the potential recoverable amount of oil and gas and do not enhance future benefits to be derived from the acquired properties. Instead, carrying and retaining costs are incurred primarily for the purpose of keeping the mineral interest and keeping a clear title. For these reasons, carrying and retaining costs are expensed as incurred.

The carrying and retaining costs under discussion are related to unproved properties and are usually insignificant in terms of cost. For instance, delay rentals are typically a nominal amount—as little as $1 or $2 per acre. Lease records maintenance cost is generally relatively small in dollar amount per lease. Ad valorem taxes on unproved properties will, if they are assessed, also be a small amount because the existence of minerals in economic quantities is not yet known. Legal costs, in contrast, may be relatively significant.

EXAMPLE

Carrying and Retaining Costs

a. Tyler Company acquired a 1,500-acre lease in Oklahoma. During the first year, the company did not begin drilling operations. Therefore to retain the lease, Tyler Company paid a delay rental payment of $3,000.

Entry

Delay rental expense	3,000	
Cash		3,000

b. Tyler Company paid $2,200 in ad valorem taxes, i.e., property taxes assessed on Tyler's economic interest in the lease.

Entry

Ad valorem tax expense	2,200	
Cash		2,200

c. The land department incurred allocable costs of $5,000 in maintaining land and lease records and allocated $750 of that cost to the property.

Entry

Record maintenance expense	750	
Cash		750

The carrying and retaining costs discussed in this chapter are associated with undeveloped properties and are classified as exploration costs. If ad valorem taxes or lease record maintenance costs are associated with proved properties, they are classified as production costs, not exploration costs (chapter 8).

TEST-WELL CONTRIBUTIONS

Test-well contributions, diagramed in Figure 3–2, result when one company drills a well and another company, which owns the working interest in nearby acreage, agrees to pay the drilling company in return for certain G&G information. For example, per a prearranged agreement, Company B drills a well and provides specified G&G information to Company A (an unrelated company with a working interest in acreage nearby). Company A receives the information and, in return, may have to pay a test-well contribution to Company B. From Company A's perspective, they are paying cash for G&G information, therefore the payment is treated as a G&G cost, and expensed as incurred. From Company B's perspective, they are receiving cash in return for information received as a consequence of drilling the well and providing the information to Company A. According to *SFAS No. 19*, the recipient of a test-well contribution (Company B) treats the amount received as a reduction of intangible drilling costs incurred as the well is drilled (chapter 5).

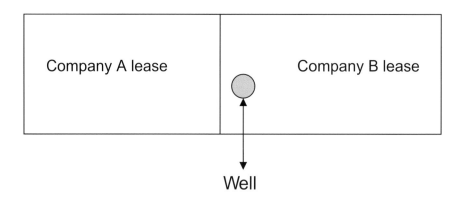

Fig. 3–2. Well drilled by Company B. Company A receives G&G information from Company B's well.

A test-well contribution may be one of two basic types:

- **Dry-hole contribution.** Payment is made only if the well is dry or not commercially producible.
- **Bottom-hole contribution.** Payment is made when an agreed-upon depth is reached, regardless of the outcome of the well.

EXAMPLE

Test-Well Contributions

Several wells were being drilled on leases in close proximity to an undeveloped lease owned by Tyler Company. In order to obtain G&G information from the wells, Tyler Company entered into the following agreements with the other companies that own the working interests on the various leases:

> Well 1: dry-hole contribution of $30,000
> Well 2: dry-hole contribution of $50,000
> Well 3: bottom-hole contribution of $20,000 to be paid if and when the well reaches a minimum depth of 5,000 feet
> Well 4: bottom-hole contribution of $40,000 to be paid if and when the well reaches a minimum depth of 8,000 feet

The following results were obtained and the entries made by Tyler Company:

Well 1—dry

Entry (dry-hole contribution)

Test-well contribution expense...............	30,000	
Cash		30,000

Well 2—completed as a producer (no payment is made, since the agreement is a dry-hole contribution agreement and the well is successful)

Entry (dry-hole contribution)
None

Well 3—drilled to agreed depth and determined to be dry

Entry (bottom-hole contribution)

Test-well contribution expense...............	20,000	
Cash		20,000

Well 4—well abandoned after drilling to 4,500 feet (no payment is made, since the well did not reach the agreed-upon depth)

Entry (bottom-hole contribution)
None

Note: The companies receiving test-well contributions should record the payment as a reduction in intangible drilling costs (chapter 5).

A dry-hole contribution is paid *only* if the well does not find proved reserves. However, whether or not proved reserves are found, the party drilling the well is obligated to furnish G&G information to the other company or companies involved in the dry-hole contribution agreement. A bottom-hole contribution is paid when the drilling operations reach the predetermined contract depth, even if proved reserves are not found. The bottom-hole contribution will not be paid if the drilling operations cease before reaching the predetermined depth. Depending on contract terms, the bottom-hole contributor may receive G&G information even if the well is not drilled to contract depth and no payment is made.

Test-well contributions provide benefits to both the drilling party and the test-well contributor. The test-well contributor receives valuable information (well logs, drillstem tests, etc.) without having to pay the cost and take the risk associated with drilling a well. On the other hand, the drilling party potentially receives money to partially offset the cost of drilling the well. A dry-hole contribution is entered into if the drilling party seeks to offset the cost of drilling the well only if the well is dry—the assumption being if the well finds proved reserves, the reimbursement is not necessary. In a bottom-hole contribution situation, the drilling party is seeking to minimize its cost regardless of the outcome.

The following example illustrates the accounting for different types of nondrilling exploration costs.

COMPREHENSIVE EXAMPLE

a. Tyler Company was interested in a large tract of land in western Montana. Tyler obtained shooting rights to 5,000 acres for $2.50 per acre (G&G cost).

   Entry
   G&G expense (5,000 × $2.50)   12,500
      Cash 12,500

b. Tyler paid a geological firm $500,000 to conduct a reconnaissance survey on the area (G&G cost).

   Entry
   G&G expense   500,000
      Cash 500,000

c. Based on the results of that study, Tyler acquired one 620-acre lease and immediately commissioned the same geological firm to conduct detailed G&G studies on the lease at a cost of $1,500,000 (G&G cost).

Entry

G&G expense	1,500,000	
Cash		1,500,000

d. During the first year, Tyler had to pay $5,000 in ad valorem taxes and $50,000 for title defense in connection with the property (carrying and retaining costs).

Entries

Ad valorem tax expense	5,000	
Cash		5,000
Legal expense—exploration	50,000	
Cash		50,000

e. By the end of the first year, Tyler had not yet begun any drilling efforts. Wanting to retain the lease, Tyler paid the first delay rental of $6,200 (carrying and retaining cost).

Entry

Delay rental expense	6,200	
Cash		6,200

f. Early in the second year, another company began drilling a well on a nearby property. Tyler entered into a bottom-hole contribution agreement with that company in order to obtain the G&G information from the well. The depth specified in the agreement was reached two months later, and Tyler paid $200,000 as per agreement (test-well contribution).

Entry

Test-well contribution expense	200,000	
Cash		200,000

Support Equipment and Facilities

Support equipment and facilities may not be directly related to one field or reservoir, but are necessary for efficient exploration, development, and production activities. Included as support equipment and facilities are seismic, drilling, construction, grading, or other equipment, warehouses and division or field offices, repair shops, and vehicles. Support equipment may be used in a single oil and gas producing activity, i.e., exploration, development, or production. The equipment and facilities also may be used to serve two or more of those activities or other activities, such as marketing or refining.

The acquisition cost of support equipment and facilities should be capitalized. Any depreciation or operating costs of support equipment and facilities should be classified as an exploration, development, or production cost to the extent the equipment or facility is used for that activity. In the case that equipment and facilities are used in more than one activity, the operating and depreciation costs should be allocated to the appropriate activities based on some measure of usage, such as hours utilized in each activity. Costs of support equipment and facilities used in nondrilling exploration activities should be expensed as incurred. Additional discussion is provided in chapter 6.

EXAMPLE

Support Equipment and Facilities

Depreciation of the seismic equipment used by Tyler Oil Company in West Texas was $100,000 for 2014. Operating costs were $91,000.

Entry to record depreciation
G&G expense—depreciation	100,000	
Accumulated depreciation		100,000

Entry to record operating costs
G&G expense—operating costs	91,000	
Cash		91,000

Offshore and International Operations

In offshore and in most international locations, nondrilling exploration costs, such as shooting rights and legal costs for title defense, are not applicable because a government entity owns the mineral rights. However, the government entity may impose other types of fees and taxes. Offshore G&G surveys are conducted differently than U.S. onshore G&G and may be much more expensive. In international operations, G&G surveys may be conducted either onshore or offshore, depending on the location of the contract area. In many situations, these costs are borne entirely by the foreign E&P company, even when a local, government-owned oil company shares in the working interest. Despite these differences, the costs of G&G and other nondrilling exploration activities incurred offshore and in international locations are accounted for in the same manner as for U.S. onshore operations. They are charged to expense as incurred by successful efforts companies.

Problems

1. Diane Oil Company obtained shooting rights on 10,000 acres at $2/acre and then hired an independent geological firm to conduct the initial G&G work for $60,000. As a result of the G&G work, Diane Company decided to lease 500 acres (ignore acquisition costs) and hired the same company to perform detailed G&G studies on the 500 acres at a cost of $15,000. Give the entries to record these transactions.

2. Gusher Petroleum obtained a lease on March 1, 2015. Being short of funds, Gusher Petroleum did not begin drilling operations during the first year of the primary term and on March 1, 2016, made a delay rental payment of $8,000. On May 12, 2016, the company paid a bottom-hole contribution of $30,000. The information obtained from this well was so encouraging that Gusher Petroleum decided to begin drilling operations. However, there were some title problems, and drilling was delayed. Legal costs incurred for title defense were $50,000. Give the entries.

3. During 2016, the exploration department of Black Gold Oil Corporation incurred the following costs in exploring the Oklahoma Anadarko Basin. Give the entries.

Shooting rights.	$ 12,000
Bottom-hole contribution	40,000
Supplies for exploration (G&G) activities	8,000
Salaries for exploration (G&G) activities	100,000
Mapping costs for exploration (G&G) activities	15,000
Depreciation of exploration (G&G) equipment	20,000
Transportation for seismic crew	5,000
Operating costs for exploration (G&G) equipment . .	3,000

4. Gamma Oil Company obtained the rights to shoot 25,000 acres at a cost of $0.20/acre on May 3, 2017. Gamma contracted and paid $80,000 for a reconnaissance survey during 2017. As a result of this broad exploration study, Lease A and Lease B were leased on January 9, 2018. (Ignore acquisition costs.) The two properties totaled 1,500 acres, and each had a delay rental clause requiring a payment of $2 per acre if drilling was not commenced by the end of each full year during the primary term. Detailed surveys costing a total of $30,000 were done during January and February on the leases.

During July, Gamma entered into two test-well contribution agreements: a bottom-hole contribution agreement for $15,000, with a specified depth of 10,000 feet, and a dry-hole contribution of $20,000, also with a specified depth of 10,000 feet. In November both wells were drilled to 10,000 feet. The well with the bottom-hole contribution was successful, but the well with the dry-hole contribution was dry.

The cost for maintaining land and lease records allocated to these two properties for 2018 was $2,000. Ad valorem taxes were assessed on Gamma's economic interest in both properties, amounting to $2,500 for 2018. After preparing their financial statements for 2018, Gamma decided to delay drilling on these properties until some time in 2020.

On April 15, 2020, enough money was left after paying taxes for a well to be drilled on Lease B. Before drilling the well, costs of $7,000 were incurred to successfully defend a title suit concerning Lease B.

Give all entries necessary to record these transactions. Assume any necessary delay rental payments were made.

5. Discuss why each party to a test-well contribution situation would enter into the transaction.

6. Pistol Oil Company purchased seismic equipment on March 1, 2017, costing $100,000. The seismic equipment was used in G&G operations for the remainder of the calendar year (2017). Compute straight-line depreciation for 2017, assuming a 10-year life and no salvage value, and prepare the entries to record the purchase and depreciation of the equipment.

7. Prospect Petroleum had the following transactions in 2017 concerning test-well contributions:

   a. Contracted with Alan Energy Corporation, agreeing to pay $50,000 if a well was drilled on Alan's lease to a depth of 10,000 feet.

   b. Contracted to pay Varsity Oil Company $40,000 if a well being drilled on Varsity's property was dry.

   c. Agreed to pay Richards Oil Company $100,000 if a well being drilled reached a depth of 7,500 feet.

   Results from the above transactions were the following:

   a. Because of mechanical difficulty, the Alan well was abandoned at 9,500 feet.

   b. The Varsity well was dry.

   c. The Richards well was completed as a producer at 12,000 feet.

Prepare entries for the above transactions, assuming Prospect Petroleum fulfilled its contractual obligations.

8. Basic Oil Company conducted G&G activities on leases owned by Artificial Oil Company and Universal Oil Company. Each agreement provides for Basic Oil Company to receive one-fourth of each working interest if proved reserves are found and to be reimbursed if proved reserves are not found. Basic Oil Company incurred the following G&G costs on Artificial's and Universal's leases:

Artificial $50,000

Universal $40,000

The well drilled on the Artificial Oil lease was successful, and one-fourth of the working interest was assigned. Drilling on the Universal lease resulted in a dry hole, and Basic was reimbursed for the G&G costs incurred. Prepare entries for the above transactions.

9. Foundation Energy obtained a three-year lease on 1,000 acres on May 1, 2014 that contained a $3 per acre delay rental clause. Drilling operations were started on June 15, 2015 and completed on October 16, 2015. The well, determined to be dry, was plugged and abandoned. No further drilling operations were started during the primary term. All required delay rentals were paid. Give all entries relating to the delay rental requirement.

10. Frank Energy Company entered into two test-well contribution agreements as follows:

    a. On May 17, 2014, a bottom-hole agreement was obtained requiring a payment of $45,000 when the contract depth of 10,000 feet was reached. The contract depth was reached on September 21, 2014, and the required payment was made.

    b. On September 30, 2014, a dry-hole test-well contribution was entered into, requiring payment of $50,000 if the well was dry but no payment if the well was successful.

        1) Assume the well is successful.

        2) Assume the well is dry.

    Prepare necessary entries for the above transactions.

11. Jayhawk Oil Company obtained seismic equipment on January 1, 2014, at a cost of $100,000. The equipment was used in G&G operations for calendar year 2014. The equipment has an estimated life of 10 years with a salvage value of $20,000. The company uses the straight-line method in computing depreciation. Record the depreciation for the year 2014.

12. During 2015, Contender Oil Company obtained the following leases:

Lease	Acres
A	3,000
B	4,000
C	5,000

After obtaining these leases, Contender Oil Company incurred shooting rights on Leases A and C at $0.50 an acre and incurred the following costs:

Salaries for exploration activities	$50,000
Mapping costs for exploration activities	20,000
Minor repairs of G&G exploration equipment . .	1,000

Give entries for the above transactions.

13. Define the following:

    shooting rights

    G&G costs

    carrying costs

    dry-hole contribution

    bottom-hole contribution

14. Clarence Oil and Gas Corporation agreed to conduct G&G studies and other exploration activities on a lease owned by Charles Energy Company in exchange for an interest in the property if proved reserves are found. If proved reserves are not found, Clarence will be reimbursed for costs incurred.

    a. Clarence incurs $200,000 of exploration costs.

    b. Assume proved reserves are found.

    c. Assume instead that proved reserves are not found.

    Give any entries required.

15. Discuss accounting for seismic costs. Be sure to include in your answer an explanation of the situations in which companies may opt to capitalize seismic costs and whether or not this practice is in compliance with U.S. GAAP.

References

1. Institute of Petroleum Accounting. 2001. *2001 PricewaterhouseCoopers Survey of U.S. Petroleum Accounting Practices.* Denton, TX: Institute of Petroleum Accounting.

4

ACQUISITION COSTS OF UNPROVED PROPERTY—SUCCESSFUL EFFORTS

In the United States, interest in a property may be acquired through the purchase of the mineral rights or the purchase of the fee interest. Land purchased *in fee* means that both the mineral rights and the surface rights are acquired rather than just the mineral rights. However, as discussed in chapter 1, leasing is the typical method of acquiring property.

Acquiring rights in mineral interests through leasing is complicated when, as is often the case, the mineral rights on a property are not owned by the same party that owns the surface. In that case, the E&P company seeking to lease the property must secure rights from the mineral owner first, and then in certain states obtain a separate agreement from the surface owner. The surface owner must permit the E&P company to utilize the surface area, either in the short-term or long-term, for construction of drilling and production facilities. If this occurs, the surface owner would be compensated. The surface owner is also entitled to receive payment for damages to the surface (e.g., crops, livestock, and structures) resulting from the E&P company's activities. In no circumstances can the surface owner deny the E&P company, by rights granted under the lease, access to the surface acreage above the leased minerals.

When operating in locations outside the United States, it is typically necessary for E&P companies to acquire mineral interests by contracting with the local government or government-owned oil company. The terms and conditions of international petroleum contracts differ dramatically from those in domestic U.S. leases. Nevertheless, if an E&P company is using the successful efforts method mandated by U.S. GAAP, its accounting must conform to *SFAS No. 19* no matter where its operations are located. *SFAS No. 19* requirements for accounting for the acquisition and impairment of unproved properties are discussed in this chapter.

An interesting characteristic of *SFAS No. 19* is its focus on the legal and operational environment that is found in the United States. As a consequence, accountants working in locations outside the United States may find it necessary to first become familiar with contracting in the United States in order to fully understand and apply the provisions of *SFAS No. 19* to their specific location. The discussion below as well as the discussions appearing in chapters 1 and 2 should aid in this process.

Acquisition costs are costs incurred in acquiring an economic interest in the mineral rights whether through leasing or purchase. The principal types of acquisition costs are, as specified in *Reg. S-X 4-10*, par. 14:

> *Costs incurred to purchase, lease or otherwise acquire a property, including costs of lease bonuses and options to purchase or lease properties, the portion of costs applicable to minerals when land including mineral rights is purchased in fee, brokers' fees, recording fees, legal costs, and other costs incurred in acquiring properties.*

A lease or signature bonus is paid by the lessee to the mineral rights owner at the time the working interest is acquired through a lease arrangement. This payment is typically the most significant unproved property acquisition cost. Certain incidental acquisition costs such as brokers' fees, legal costs, and options to lease may also be incurred. Brokers' fees are fees that are paid to real estate agents or to landmen for services in connection with leasing or purchasing property. These services include finding available property, negotiating, and consummating a lease or purchase. Also, brokers generally file the legal papers that result from purchase or lease transactions. Legal fees are paid to attorneys for preparation of legal documents (deeds, leases, etc.) and for title examinations, which involve determining if the seller or lessor has a good marketable title. A good marketable title is determined by tracing the chain of title back to the original land grant. The legal document resulting from the tracing of title is called an *abstract of title*. An option to purchase or lease a property is a contract that gives the prospective lessee a stated period of time in which to decide whether to purchase or lease the property.

In international operations, negotiation of petroleum contracts is often an intense and time-consuming process that involves a number of individuals. Such negotiating teams may include people from a number of disciplines (e.g., geologists, engineers, economists, attorneys, and accountants). The cost of time, travel, etc. of negotiation teams may sum to a sizeable amount. If material in amount, these costs should be accounted for along with other unproved property acquisition costs.

According to *SFAS No. 19*, the cost of acquiring unproved properties should be capitalized when incurred. See Figure 4–1 and Figure 4–2 for an overview of the accounting for acquisition costs.

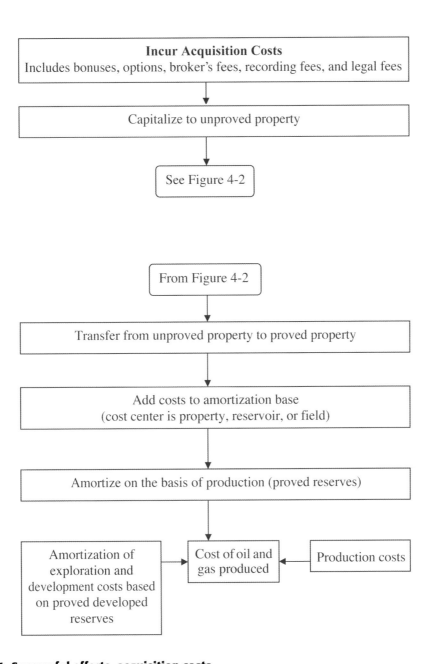

Fig. 4–1. Successful efforts, acquisition costs

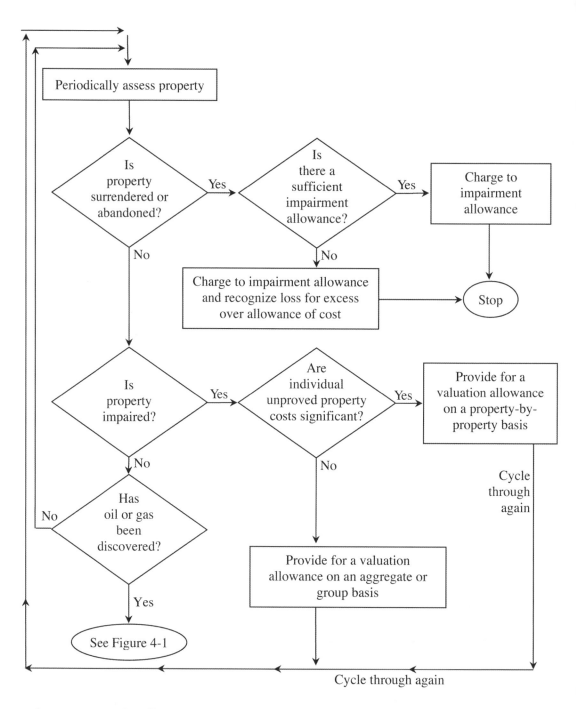

Fig. 4–2. Successful efforts, acquisition costs

EXAMPLE

Acquisition Costs

Tyler Company acquired a 620-acre unproved property. Acquisition costs included a lease bonus of $100 per acre and recording fees of $1,000.

Entry

Unproved property (620 × $100 + $1,000)	63,000	
Cash		63,000

Note: A proved property may also be acquired. In that case, proved property instead of unproved property would be debited. Acquisition costs of proved properties would include the costs discussed above and would be capitalized when incurred. When proved properties are acquired it may be necessary to allocate the purchase price between the cost of the mineral interest and the cost of the equipment and facilities installed on the property. This complicated process typically involves allocating to equipment and facilities an amount equal to their fair market value, and capitalizing the remainder of the acquisition cost to the proved property account. Accounting for the acquisition of proved properties is discussed in chapter 13.

When a company acquires a lease, it is necessary to set the property up in a property database. The database would include such information as location and legal description of the property, whether it is proved or unproved, names of the parties to the agreement, and a listing of all royalty owners, working interest owners, and any other parties with an economic interest in the minerals.

In addition to providing critical information about the property, the property database is accessed by both the joint interest billing system and the revenue system in allocating the appropriate amount of costs and revenues to the parties. For example, once oil or gas is discovered, before it can be sold, a division order must be prepared and signed by all interest owners. In most companies, the task of executing the division order is the responsibility of the division order group or department. The division order group must also monitor changes in ownership on an ongoing basis so that ownership and division of revenue records are always up-to-date.

The majority of acquisition costs incurred are simply capitalized as unproved property acquisition costs when incurred, as illustrated in the previous example. However, the treatment of some acquisition costs is not always so straightforward. Acquisition costs needing further discussion are the following:

- Purchase in fee
- Internal costs
- Options to purchase or lease
- Delinquent taxes and mortgage payments
- Top leasing

PURCHASE IN FEE

When land is purchased in fee, the purchase price must be allocated between the mineral and surface rights acquired. While both types of costs are initially capitalized, if production is ultimately achieved from the property, the cost allocated to the mineral rights will be subject to amortization. However, the cost assigned to the surface rights is not subject to amortization. Typically the allocation of the total purchase price between mineral and surface rights is based on the relative fair market values (FMVs) of the rights, if known.

EXAMPLE

Purchase in Fee, Both FMVs Known

Tyler Company purchased land in fee for $90,000. A qualified appraiser made the following estimate of the fair market values of the surface and mineral rights:

Surface rights	$ 60,000
Mineral rights	40,000
	$100,000

The $90,000 acquisition cost would be allocated as follows:

Land: (surface rights): $\dfrac{\$60,000}{\$100,000} \times \$90,000 = \$54,000$

Unproved property (mineral rights): $\dfrac{\$40,000}{\$100,000} \times \$90,000 = \underline{36,000}$

$\underline{\underline{\$90,000}}$

Entry

Land	54,000	
Unproved property	36,000	
Cash		90,000

If the fair market value of only one interest is available, then that interest should be allocated an amount equal to its fair market value and the remainder of the payment allocated to the other interest.

EXAMPLE

Purchase in Fee, One FMV Known

Tyler Company purchased land in fee for $90,000. A qualified appraiser estimated the fair market value of the surface rights to be $55,000. A reasonable estimate of the fair market value of the mineral rights was not possible. The $90,000 acquisition cost would be allocated as follows:

Land (surface rights):	$55,000
Unproved property (mineral rights):	$90,000 – $55,000 = $35,000

Entry

Land	55,000	
Unproved property	35,000	
Cash		90,000

INTERNAL COSTS

A number of E&P company employees (e.g., attorneys, economists, accountants, and engineers) may be involved in a variety of activities related to the acquisition of unproved properties. Such activities may include acquisition of mineral interests, examination of records before drilling is commenced, and negotiation and arrangement of joint ventures. Theoretically, the portion of the salaries of these employees along with other internal costs related to the acquisition of unproved properties should be capitalized. Directly allocating these types of costs to specific leases may be impractical if the costs are relatively low. However, if the costs are material, they should be allocated to the individual properties. Two reasonable allocation bases follow:

a. Capitalize the portion of the costs relating to acquisition activities and allocate to the individual leases acquired, based on total acreage acquired.

b. Allocate the portion of the costs relating to acquisition activities on an acreage basis to all prospects investigated, capitalizing the portion of the costs allocated to prospects acquired and expensing the portion of costs allocated to prospects not acquired.

SFAS No. 19 does not address the proper treatment of these types of costs. However, option b appears to be theoretically consistent with *SFAS No. 19*, because only those successful costs that result in lease acquisition are capitalized. In situations where these costs are deemed to be immaterial, expensing is appropriate.

EXAMPLE

Internal Costs

Tyler Company acquired an 80 acre undeveloped lease paying a $100 per acre bonus. Legal costs and recording fees were $2,000. The salary of a company attorney working solely on lease acquisition was $100,000. The attorney's salary is allocated to all leases acquired, based on relative acreage acquired. During the accounting period, leases totaling 800 acres were acquired.

Entry

Unproved property (80 × $100 + $2,000 + 80/800 × $100,000)   20,000
    Cash   20,000

OPTIONS TO LEASE

Instead of leasing a property before conducting G&G studies, a company may obtain shooting rights to a property so that certain G&G studies can be performed before a decision to lease is made. Generally, the shooting rights will be coupled with an option to lease. The option protects the lessee by keeping the right to lease the property open for a period of time and guarantees the company the right to lease the property at a specified price during that time period.

The cost of an option to lease is an acquisition cost and should be capitalized temporarily until a decision is made about whether to lease the property. If the property is leased, the option cost should be capitalized as part of unproved property; if the property is not leased, the option cost should be expensed. Shooting rights, in contrast to an option to lease, are a G&G exploration cost and should be expensed as incurred.

Two situations must be addressed when dealing with shooting rights coupled with an option to lease:

a. **The payment represents both an acquisition cost and a G&G exploration cost. If the acreage is leased, should the entire payment be capitalized or expensed or, if it can be determined, should the acquisition portion of the payment be capitalized and the exploration portion expensed?**

   Theoretically if the property is leased, the payment should be apportioned and treated as both an acquisition cost and a G&G exploration cost. In practice, however, the entire payment is generally capitalized as an acquisition cost of the property.

b. **If none or only a portion of the explored acreage is leased, how should the amount allocated to the option be handled?**

If none of the acreage is leased, then the option amount should be expensed. If only a portion of the acreage explored is leased, theoretically only a proportional amount of the option cost based on the number of acres leased relative to the total acreage explored should be capitalized as an acquisition cost of the property. In practice, sometimes the entire option amount is capitalized.

A flow diagram of the accounting for options to lease coupled with shooting rights is presented in Figure 4–3.

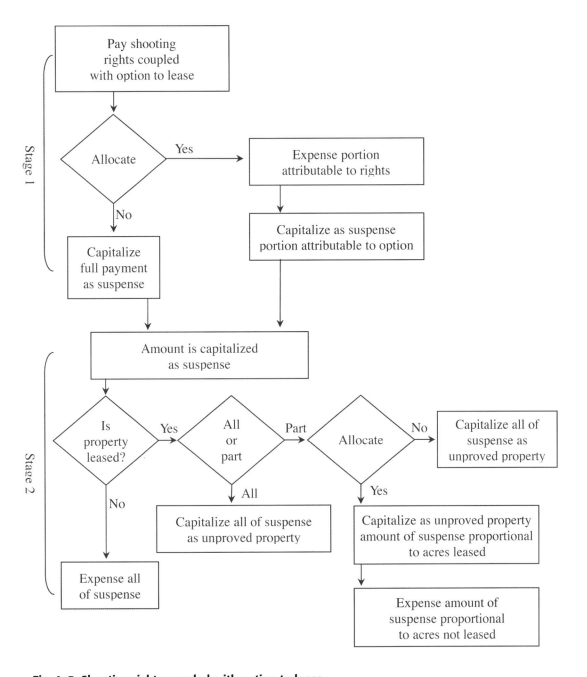

Fig. 4–3. Shooting rights coupled with option to lease

EXAMPLE

Option to Lease

Tyler Company obtained the following rights:

a. Shooting rights only for $5 per acre on 3,000 acres owned by Abby Dorr.

b. Shooting rights coupled with an option to lease for $45,000 on 4,000 acres owned by Alec Smith. These 4,000 acres are located adjacent to the 3,000 acres owned by Ms. Dorr.

c. Shooting rights coupled with an option to lease for $30,000 on 1,020 acres owned by Mark Cameron located in Timbuktu. Tyler Company has no reasonable basis for valuing the shooting rights versus the option to lease.

Entries

Abby Dorr's Acreage

G&G expense ($5 × 3,000)	15,000	
Cash		15,000

Alec Smith's Acreage—The cost of the shooting rights on Abby Dorr's acreage ($5 per acre) can be used to determine the value of the shooting rights versus the value of the option on Mr. Smith's acreage.

G&G expense ($5 × 4,000)	20,000	
Property purchase suspense (remainder)	25,000	
Cash		45,000

This entry is theoretically correct; however, the entire payment normally is capitalized as shown below.

Property purchase suspense	45,000	
Cash		45,000

Mark Cameron's Acreage—There exists no reasonable basis for apportionment, so the entire payment must be capitalized.

Property purchase suspense	30,000	
Cash		30,000

As a result of the G&G work performed, Tyler Company made the following decisions:

a. Not to lease any of Abby Dorr's acreage.
b. To lease 1,000 acres from Alec Smith, paid a $20,000 bonus.
c. To lease 680 acres from Mark Cameron, paid a $15,000 bonus.

Entries

Abby Dorr's Acreage—No entry.

Alec Smith's Acreage—Tyler decides to capitalize only the portion of the suspense that applies to acres leased.

Surrendered lease expense (3,000/4,000 × $25,000)	18,750	
Unproved property (1,000/4,000 × $25,000 + $20,000) . . .	26,250	
Property purchase suspense (close out)		25,000
Cash		20,000

Mark Cameron's Acreage—Tyler decides to capitalize only the portion of the suspense that applies to acres leased.

Surrendered lease expense (340/1,020 × $30,000)	10,000	
Unproved property (680/1,020 × $30,000 + $15,000)	35,000	
Property purchase suspense (close out)		30,000
Cash		15,000

Note that Tyler Company instead could have capitalized the entire amount in the suspense accounts relating to both Alec Smith's and Mark Cameron's acreages. In that case, the entry for Alec Smith's acreage would have been the following:

Unproved property ($25,000 + $20,000).	45,000	
Property purchase suspense (close out)		25,000
Cash		20,000

DELINQUENT TAXES AND MORTGAGE PAYMENTS

During the initial process of obtaining a lease, the lessee may agree to pay such items as the lessor's delinquent taxes or mortgage payments in order to obtain a lease on land with a clear title. Such payments may or may not be recoverable from future delay rental payments or from future royalty payments. Accounting treatment depends upon whether the payments are recoverable.

1. If nonrecoverable, the payments are additional costs of acquiring the property and should be capitalized as an additional leasehold cost, i.e., an acquisition cost.

2. If recoverable, the payments should be held in a receivable or suspense account until recovered. If not recovered, the payments should be charged to an expense account, i.e., surrendered lease expense or surrender and abandonment expense. (Nonrecovery of the payments would normally occur because the lease had been surrendered.)

In some cases, payments such as these may be made during the lease term by the lessee to protect the lessee's interest in the lease. This type of payment would not be an acquisition cost but a cost of maintaining the property, i.e., a nondrilling exploration cost (chapter 3). Consequently, if these payments are nonrecoverable, they should be expensed.

EXAMPLE

Delinquent Taxes and Mortgage Payments

Tyler Company paid delinquent property taxes of $350 on an unproved lease as called for in the lease agreement.

a. Assume these taxes were not recoverable.

Entry

Unproved property	350	
Cash		350

Since the taxes were part of the lease agreement and are nonrecoverable, the payment would be an acquisition cost.

b. Assume instead that the taxes were recoverable out of future delay rental payments or royalty payments.

Entry

Receivable from lessor	350	
Cash		350

c. Two months later, a delay rental of $500 was due. Since the $350 delinquent property tax paid by Tyler Company is recoverable from the delay rental, a net payment of $150 is remitted to the lessor.

Entry

Delay rental expense	500	
Receivable from lessor		350
Cash		150

d. Assume that instead of making the delay rental payment, Tyler Company had abandoned the lease without recovering the delinquent property taxes.

Entry

Surrendered lease expense................	350	
Receivable from lessor................		350

Payment of delinquent property taxes by the lessee for the lessor is different from the payment of property taxes discussed in chapter 3. The property taxes discussed in chapter 3 are property taxes assessed by a governmental body on the economic interest owned by the lessee. The delinquent property taxes discussed in this chapter are unpaid taxes paid by the lessee but assessed on the interest owned by the lessor.

TOP LEASING

A **top lease**, which occurs infrequently, is a new lease executed before expiration or termination of the existing lease. A top lease is usually sought when the current working interest owner wants to retain a property on which no drilling was done during the primary term. This situation is a two-party top lease. Sometimes, a different working interest owner will negotiate a new lease prior to expiration of the primary term on an existing lease. This is called a three-party top lease.

If the existing lease is still in effect when the new lease (same working interest owner) is executed, the book value of the old lease plus the acquisition costs of the new lease will be capitalized as acquisition costs. However, if the existing lease terminates prior to the date on which a new lease is obtained (not a top lease), the book value of the old lease will be written off, and only the acquisition costs of the new lease will be capitalized.

In the case where a different working interest owner signs a lease agreement with the mineral rights owner before the end of the primary term of an existing lease, the situation is no different from the acquisition of a new lease. Acquisition costs paid by the new lessee are capitalized as described earlier in the chapter.

EXAMPLE

Top Leasing

Tyler Company has two unproved leases with the following costs:

Lease A (320 acres)	$30,000
Lease B (640 acres)	$40,000

a. A new lease was executed on Lease A by Tyler, the original lessee, prior to the termination of the old lease. The bonus paid for the new lease was $65 per acre.

Entry

Unproved property—Lease A	50,800	
Unproved property—Lease A.		30,000
Cash ($65 × 320).		20,800

b. Lease B terminated on November 14, 2014, and a new lease was executed by Tyler to become effective on November 15. The bonus paid was $80 per acre.

Entries

November 14:

Allowance for impairment*	40,000	
Unproved property—Lease B.		40,000

November 15:

Unproved property—Lease B.	51,200	
Cash ($80 × 640).		51,200

---

* The account debited in this journal entry depends upon how and if impairment has been recognized relating to the lease. This topic is discussed in the following sections.

Disposition of Capitalized Costs—Impairment of Unproved Property

Each year E&P companies typically acquire the working interest in a number of unproved properties. At the time the properties are acquired, all of the properties have potential. However, only after wells are drilled is it possible to know with certainty which properties have oil and gas reserves and which do not. Ultimately, successful wells will be drilled on some of the leases, with those leases being proved, while on other leases, drilling efforts will prove the leases to be worthless. Those worthless leases, as well as leases whose primary terms expired without any drilling, must be written off. The accounting for unproved properties in the interim before they are either proved successful or are written off is discussed below.

At the time a lease is executed, the mineral interest acquired has probable future economic benefit, and the acquisition cost is capitalized and reported as an asset of the company.

However, unproved properties must be assessed periodically in order to determine whether they have been impaired (*SFAS No. 19*, par. 28).

Fundamentally, **impairment** occurs when there is an indication that a property's value has declined to an amount that is less than the purchase price. Estimating impairment is difficult and subjective. However, examining certain information may be helpful in determining whether a property has been impaired and, if impaired, in determining the extent of the impairment. The following questions may be asked to help determine whether impairment has occurred:

a. Have there been any dry holes drilled on the lease or on nearby leases, or has any additional negative G&G information been obtained?
b. How close is the expiration of the primary lease term?
c. Are there any firm plans for drilling?
d. Is the passage of time an indication that the company has reassessed its interest in the property?

Dry holes and negative G&G information would normally indicate that the property has become impaired. If neither drilling nor production is in progress at the end of the primary term, the lease terminates, and the leasehold costs must be written off. Therefore, a property may also be impaired if the end of the primary term is near and no firm plans for drilling have been established.

At any given time, it is possible that a company may have hundreds of unproved properties. This is especially true for operations in the United States, where individual leases typically cover a relatively small area. Assessing each and every lease for impairment annually would require a substantial commitment of time and resources. Accordingly the FASB permits companies to classify their unproved properties into one of two categories:

1. Those properties whose cost is individually significant
2. Those properties whose cost is not individually significant

According to *SFAS No. 19* par. 28:

> *Impairment of individual unproved properties whose acquisition costs are relatively significant shall be assessed on a property-by-property basis, and an indicated loss shall be recognized by providing a valuation allowance. When an enterprise has a relatively large number of unproved properties whose acquisition costs are not individually significant, it may not be practical to assess impairment on a property-by-property basis, in which case the amount of loss to be recognized and the amount of the valuation allowance needed to provide for impairment of those properties shall be determined by amortizing those properties, either in the aggregate or by groups, on the basis of the experience of the enterprise in similar situations and other information about such factors as the primary lease terms of those properties, the average holding period of unproved properties, and the relative proportion of such properties on which proved reserves have been found in the past.*

In international operations, contract areas tend to be quite large and are frequently expensive to obtain, while domestic operations are characterized by a large number of relatively less expensive leases. Companies may treat each international contract area as individually significant, since the costs are large and there are fewer properties. Domestically, the sheer number of properties may make the cost of each property in itself insignificant.

SFAS No.19 does not provide any guidance for companies in identifying individually significant properties. According to the *2001 PricewaterhouseCoopers Survey of U.S. Petroleum Accounting Practices*, a common approach is to base the classification on the actual cost of the property.[1] For example, properties with an individual cost of $1,000,000 and above might be classified as individually significant, and all properties with a cost below $1,000,000 might be treated as not being individually significant. The SEC has issued significance guidelines for full cost companies but not for successful efforts companies. Under the SEC's full cost guidelines, a property is significant if its costs exceed 10% of the net capitalized costs of the cost center, which for full cost is a country (chapter 7).

If the results of the assessment indicate that the property has been impaired, a loss should be recognized by providing a valuation allowance. The exact approach used depends upon whether the acquisition costs of the property under consideration are significant.

For individually significant properties, each lease is examined individually to determine whether it has been impaired. In accounting, determining impairment typically involves comparing cost or carrying value to fair value or net realizable value. *Statement of Financial Accounting Concepts (SFAC) No. 7*, "Using Cash Flow Information and Present Value in Accounting Measurements," calls for the use of future cash flows in determining the fair value of assets. Unproved property assessment in the oil and gas industry, however, is especially problematic. Since the properties are unproved, there are no reserves and thus no future cash flows to assess. Companies therefore must evaluate such variables as dry holes drilled, plans for drilling, and the nearness of the end of the lease term in determining impairment for each significant lease.

EXAMPLE

Impairment of Individually Significant Unproved Properties

On January 1, 2015, Tyler Company acquired an unproved lease at a total cost of $1,000,000. During the year, Tyler Company drilled two dry holes on the property. As a result of drilling these dry holes, Tyler Company decided on December 31 that the lease was 75% impaired, i.e., the impaired value was 75% less than cost.

Entry

Lease impairment expense ($1,000,000 × 0.75)	750,000	
Allowance for impairment		750,000

The allowance for impairment account is a contra asset account. The pertinent part of a balance sheet prepared following this entry would be as follows:

Assets: Unproved property	$1,000,000
Less: Allowance for impairment	(750,000)
	$ 250,000

If acquisition costs are not individually significant, *SFAS No. 19* indicates that property-by-property impairment is not necessary. In this case impairment may be assessed on the aggregate unproved properties or by grouping unproved properties based on some common feature. For example, in practice it is common for companies to treat U.S. onshore, lower-48 properties in one group, and U.S. offshore properties in another group. Accounting for impairment on a group basis is similar to providing an allowance for bad debts. The company estimates the impairment based on the historical experience of the company and factors such as the primary terms of the involved properties, the average holding period of individually insignificant unproved properties, and the historical percentage of such properties that have been proved.

An approach commonly used in practice involves using the company's actual success rate for proving properties in the group to estimate the total cost of properties that will ultimately be surrendered. This cost is then charged to expense over the average holding period, i.e., the amount of time that it normally takes the company to assess properties in this group. The following example illustrates this technique.

EXAMPLE

Impairment of Individually Insignificant (Group) Properties

Tyler Company groups individually insignificant unproved properties by year of acquisition. During 2012, Tyler acquired individually insignificant unproved property costing a total of $100,000. Past experience indicates that 40% of all unproved properties are eventually abandoned without ever being proved and that it takes Tyler an average of four years to determine whether a property will be proved or surrendered.

Calculation of Annual Impairment for Properties Acquired in 2012

$100,000 × 40% × 1/4 = $10,000

Entry

Lease impairment expense	10,000	
Allowance for impairment, group basis		10,000

An alternative approach used in practice is to determine the annual impairment charge based on the balance desired in the impairment allowance account. First, the desired balance is determined by applying the estimated impairment percentage to the total balance of the group of individually insignificant unproved properties. The amount of impairment recognized is the difference between the current balance in the impairment allowance account before impairment versus the desired balance. This approach emphasizes the net realizable value of the properties and focuses on the statement of financial position.

EXAMPLE

Impairment of Individually Insignificant (Group) Properties

On December 31, 2012, the unproved property account containing leases that are not individually significant had a $100,000 balance, and the allowance for impairment account had a $10,000 balance. Past experience indicates that 40% of all unproved properties in this group are eventually abandoned without ever being proved. Therefore, Tyler Company has a policy of providing at year-end an allowance equal to 40% of the cost of individually insignificant (group) unproved properties.

Entry

Lease impairment expense	30,000	
Allowance for impairment, group basis		30,000

Calculations

$100,000 × 40% = $40,000

Allowance	
10,000	← have
?	← 30,000
40,000	← desired balance

All the individually insignificant properties of a company may be amortized together in one group or the properties may be grouped into different groups, with each group being amortized separately. Individually insignificant properties are typically referred to as **group properties**. Groupings commonly used in practice include country by country, onshore versus offshore, or by year of acquisition. In the following example, Tyler Company operates onshore and offshore. Because its average success rate and average holding period are different between the two areas, Tyler decided to compute impairment for insignificant onshore properties separately from insignificant offshore properties.

EXAMPLE

Impairment of Multiple Groups of Properties

Tyler Company maintains two groups of individually insignificant unproved properties, onshore and offshore. Tyler has as success rate of 64% for proving individually insignificant offshore unproved properties and a success rate of only 48% for individually insignificant onshore properties. The average time that the properties are held for evaluation is three years for offshore and four years for onshore. During 2012 a total of 5 new offshore leases are acquired at a total cost of $2,000,000, and a total of 30 new onshore leases are acquired for $450,000. Tyler computes impairment by using its success rate for proving properties and the average holding period of the companies in the group.

$$\text{Offshore:} \quad \$2,000,000 \times 36\% \times 1/3 = \$240,000$$

$$\text{Onshore:} \quad \$450,000 \times 52\% \times 1/4 = \$58,500$$

Entry

Lease impairment expense, offshore	240,000	
Lease impairment expense, onshore	58,500	
Allowance for impairment, offshore group.		240,000
Allowance for impairment, onshore group		58,500

If a property is jointly owned, each working interest owner makes a separate decision as to whether the property is significant and how or whether to impair the property. Each company could legitimately compute a different amount of impairment based on their future plans for the property and their own financial accounting policies and procedures. For example, one company's geologist may, after examining data concerning the property, decide on full impairment. Another company's geologist may feel the lease has potential and recommend no impairment.

Unproved property should be assessed individually or in the aggregate at least once a year. Once impairment has been recognized on a property, the property should not subsequently be written back up.

In 2001, the FASB issued *SFAS No. 144*, "Accounting for the Impairment or Disposal of Long-Lived Assets." This standard calls for the use of net realizable value and fair value in the measurement of impairment for long-lived assets. (See chapter 9.) E&P companies that use the successful efforts method must apply the provisions of *SFAS No. 144* to all of their long-lived assets, including the capitalized cost of proved oil and gas properties. However, unproved properties are exempted from the impairment requirements of *SFAS No. 144* (par. 5), so long as the impairment provisions of *SFAS No. 19*, described above, are applied.

Disposition of Capitalized Costs—Surrender or Abandonment of Property

The actual abandonment of an unproved property may be accomplished in one of several ways. Failure to pay a delay rental payment when due or failure to begin drilling or production operations before the end of the primary term automatically terminates the lease.

In some situations, property may be abandoned before the date of the next delay rental payment or before the end of the primary term. In international operations, contracts often have specific provisions that permit early termination of the agreement if early stage exploration does not yield positive results. In either event, abandonment is required to establish worthlessness for tax purposes (chapter 11). Formal legal documents must be filed in this situation to justify the income tax deduction and establish the appropriate tax year.

The specific accounting entries required to record the surrender and abandonment of a property depend on whether the property was originally classified as being individually significant or not. When an individually significant unproved property is surrendered or abandoned, its capitalized acquisition costs should be charged against the related allowance for impairment account. If the allowance account is inadequate, the excess should be charged to surrendered lease expense (*SFAS No. 19*, par. 40). For properties assessed individually, abandonment results in the net carrying value (acquisition cost minus impairment allowance) of the abandoned properties being expensed.

EXAMPLE

Abandonment of Individually Significant Properties

Tyler Company abandoned the following unproved properties, all of which are treated as being individually significant for impairment purposes.

Lease	Acquisition Cost	Allowance for Impairment
A	$5,000,000	$5,000,000
B	2,800,000	1,800,000
C	1,700,000	0

Entries

LEASE A

Allowance for impairment	5,000,000	
Unproved property		5,000,000

LEASE B

Surrendered lease expense.	1,000,000	
Allowance for impairment	1,800,000	
Unproved property		2,800,000

LEASE C

Surrendered lease expense.	1,700,000	
Unproved property		1,700,000

When an individually insignificant unproved property assessed on a group basis is abandoned, the cost of the unproved property being abandoned should be charged against the allowance for impairment account, group basis. The allowance account should be adequate. However, in the event it is inadequate, the excess of the abandoned property cost over the allowance balance results in a debit balance in the allowance account. To avoid reporting complications at the end of an interim period, this debit balance should be charged to surrendered lease expense and an appropriate accrual made so that a sufficient allowance balance is established. An insufficient balance in the allowance account may indicate a need for the company to reevaluate its procedure for impairment estimation.

EXAMPLE

Abandonment of Individually Insignificant (Group) Properties

During December, Tyler Company abandoned two unproved leases that had been assessed on a group basis. The impairment allowance for this group of individually insignificant unproved properties had a $200,000 balance. Data for the abandoned leases are as follows:

Lease	Acquisition Cost
D	$50,000
E	75,000

Entries

LEASE D

Allowance for impairment, group basis	50,000	
Unproved property		50,000

LEASE E

Allowance for impairment, group basis	75,000	
Unproved property		75,000

Now assume that the impairment allowance for this group of individually insignificant unproved properties only had a balance of $100,000. The following entries would be made:

LEASE D

Allowance for impairment, group basis	50,000	
Unproved property		50,000

LEASE E

Allowance for impairment, group basis	75,000	
Unproved property		75,000

The cost of the abandoned leases exceeds the balance in the allowance account by $25,000, resulting in a debit balance in the allowance account. Since the allowance account should never have a debit balance, the following entry is required:

Entry

Surrendered lease expense.	25,000	
Allowance for impairment, group basis		25,000

After the above entry is made, the allowance balance is zero. So long as the company has other unproved properties in the group, the balance in the allowance account is obviously insufficient. Another entry is required in order to assure that the balance in the allowance account is adequate. The following additional entry is called for with the amount equal to the balance that should be in the allowance account at the present time.

Surrendered lease expense.	xx,xxx	
Allowance for impairment, group basis		xx,xxx

Partial abandonment may occur when a company surrenders part of a property but retains the remaining acreage. This situation occurs frequently in international operations where a contract is likely to cover a large geographical area. The agreement with the government frequently includes provisions requiring, over the exploration period, any acreage that is no longer of interest be relinquished from the contract area. The following example illustrates a partial abandonment.

EXAMPLE

Partial Abandonment

Tyler Company entered into an agreement with the government of Angola covering a 10,000 acre offshore block. The contract stipulated that Tyler Company must relinquish 20% of the acreage at the end of three years. Tyler has a balance of $7,000,000 in the unproved property account and treats the property as being individually significant. At the end of the third year, Tyler surrendered 2,000 acres and continued to evaluate the remaining 8,000 acres.

Entry

Surrendered lease expense ($7,000,000 × 20%)........	1,400,000	
Unproved property		1,400,000

Surrender or abandonment of proved property is discussed in chapter 6.

POST–BALANCE SHEET EVENTS

The time between the end of the fiscal year and the date on which the financial statements are actually issued is referred to as the post–balance sheet period. It is important that any events that would have a material effect on an investor's assessment of a firm's financial position be reflected in the most recent financial statements. Accordingly the FASB indicated that information becoming known during this period relating to impairment of unproved properties, dry exploratory wells, and other events that could have a material negative effect on the balance sheet is to be reflected in the year-end accounts. This treatment is required by *SFAS No. 19*, par. 39, even though the information was not known at the balance sheet date:

> Information that becomes available after the end of the period covered by the financial statements but before those financial statements are issued shall be taken into account in evaluating conditions that existed at the balance sheet date, for example, in assessing unproved properties (paragraph 28) and in determining whether an exploratory well or exploratory-type stratigraphic test well had found proved reserves.

Accordingly, a dry hole completed during the interim period (after the balance sheet date but before the financial statements are issued) may indicate that the lease was impaired at year-end. (The condition, i.e., no reserves, existed at balance sheet date, but was unknown at that time.) Any impairment amount should be reflected in the financial statements at fiscal year-end by making an adjusting entry recognizing impairment.

EXAMPLE

Post–Balance Sheet Events

Tyler Company has an individually significant unproved property on the books at a cost of $1,200,000 as of December 31, 2013. On January 18, 2014, prior to the issuance of the audited financial statements, a well on an adjacent lease owned by another party was determined to be dry. Tyler had not taken any impairment on Lease A prior to December 31, 2013. Due to the dry hole on the adjacent property, management now estimates that the lease was impaired 60% at December 31, 2013. The adjusting entry necessary for the year ending December 31, 2013, would be as follows:

Entry

Impairment expense ($1,200,000 × 60%).	720,000	
Allowance for impairment		720,000

DISPOSITION OF CAPITALIZED COSTS—RECLASSIFICATION OF AN UNPROVED PROPERTY

When proved reserves are discovered on an unproved or undeveloped property, the property should be reclassified from an unproved property to a proved property (*SFAS No. 19*, par. 29). For a property that has been assessed for impairment individually, the net carrying value should be transferred to proved property.

EXAMPLE

Reclassification of Individually Significant Properties

Tyler Company discovered proved reserves on the following unproved properties, both assessed individually.

Lease	Acquisition Costs	Impairment Allowance
A	$750,000	$ 0
B	500,000	300,000

Entries

LEASE A

Proved property	750,000	
Unproved property		750,000

LEASE B

Proved property	200,000	
Allowance for impairment	300,000	
Unproved property		500,000

For a property assessed on a group basis, the gross acquisition cost of the property should be transferred when reclassifying the property. The gross acquisition cost must be used due to the fact that a net carrying value cannot be determined on a separate property basis for properties assessed on a group basis.

EXAMPLE

Reclassification of Individually Insignificant (Group) Properties

Tyler Company discovered proved reserves on a lease that had been assessed on a group basis. The acquisition cost of the property was $45,000, and the impairment allowance for this group of properties had a balance of $200,000.

Entry

Proved property	45,000	
Unproved property		45,000

For a property to be proved, it is not necessary that the entire property be classified as a proved area. It is only necessary that proved reserves are found on the property. Thus, only a very small portion of a proved property may actually be a proved area. (Note that as a result, it is still possible to drill an exploratory well on a proved property.)

Occasionally, a property is so large that only the portion of the property to which the proved reserves relate should be reclassified from unproved to proved. *SFAS No. 19* gives as an example a foreign concession covering a large geographical area. Multiple reservoirs may actually exist under the acreage. As a company explores and drills wells in the contract area, some acreage may become proved, other acreage may be retained for future evaluation, and still other acreage may be relinquished. This situation may also occur in other instances, for example in the exploration of certain offshore properties whose evaluation requires multiple wells.

EXAMPLE

Partial Reclassification

Tyler Company has an offshore lease in the Gulf of Mexico that has a total capitalized cost of $20,000,000. This lease covers four tracts of 1,000 acres each. Proved reserves are found on one tract. Management decides to reclassify only one tract.

Entry

Proved property ($20,000,000 × 1/4)	5,000,000	
Unproved property		5,000,000

To reinforce and tie together the material presented in this chapter, a comprehensive example is presented next.

COMPREHENSIVE EXAMPLE #1

Tyler Company has the following unproved leases as of December 31, 2011.

Individually significant:

Lease	Acquisition Costs	Impairment Allowance
A	$2,000,000	$ 50,000
B	7,500,000	3,000,000
C	1,300,000	0

Individually insignificant (group):

Lease	Acquisition Costs	Impairment Allowance
D	$ 10,000	
E	25,000	(Not assessed
F	15,000	individually)
G	30,000	
Total	$ 80,000	$20,000 (credit balance)

a. At December 31, 2011, it is determined that Lease A should be impaired an additional 30% of acquisition cost.

   Entry

   Lease impairment expense ($2,000,000 × 0.30) 600,000
       Allowance for impairment 600,000

b. At December 31, 2011, it is determined that Lease C should be impaired 40% of acquisition cost.

   Entry

   Lease impairment expense ($1,300,000 × 0.40) 520,000
       Allowance for impairment 520,000

c. At December, 31, 2011, Tyler Company impairs the individually insignificant (group) unproved properties. The company's policy is to provide at year end an impairment allowance equal to 40% of the gross acquisition cost of the group of properties.

   Entry

   Lease impairment expense 12,000
       Allowance for impairment, group basis 12,000

Calculations

$80,000 \times 40\% = \$32,000$

```
             Allowance
         |
         |  20,000    ← have
         |    ?       ←12,000
         |  32,000    ← balance needed
```

d. On March 1, 2012, Tyler Company surrendered Lease A.

   Entry

Surrendered lease expense.	1,350,000	
Allowance for impairment ($50,000 + $600,000)	650,000	
Unproved property		2,000,000

e. On March 20, 2012, Tyler Company acquired an unproved lease, Lease H, paying a lease bonus of $25,000 and legal costs of $1,000. The property is not individually significant.

   Entry

Unproved property	26,000	
Cash		26,000

f. On April 13, 2012, Tyler Company discovered proved reserves on Lease C.

   Entry

Proved property	780,000	
Allowance for impairment	520,000	
Unproved property		1,300,000

g. On April 29, 2012, Tyler Company abandoned Lease E.

   Entry

Allowance for impairment, group basis	25,000	
Unproved property		25,000

h. On July 15, 2012, Tyler Company discovered proved reserves on Lease F. (The allowance for impairment, group basis account should not be prorated.)

   Entry

Proved property	15,000	
Unproved property		15,000

Tyler Company has the following unproved leases as of December 31, 2012, as a result of the previous transactions.

Individually significant leases:

Lease	Acquisition Costs	Impairment Allowance
B	$7,500,000	$3,000,000

Individually insignificant leases:

Lease	Acquisition Costs	Impairment Allowance
D	$10,000	
G	30,000	(Not assessed
H	26,000	individually)
Total	$66,000	$7,000

i. At December 31, 2012, it is determined that Lease B should be impaired an additional 30% of acquisition cost.

Entry

Lease impairment expense ($7,500,000 × 0.30)	2,250,000	
Allowance for impairment		2,250,000

j. At December 31, 2012, Tyler Company impairs the individually insignificant unproved properties. Tyler Company retains its policy of providing at year-end a 40% impairment allowance for individually insignificant unproved properties.

Entry

Lease impairment expense	19,400	
Allowance for impairment, group basis		19,400

Calculations

$66,000 × 40% = $26,400

```
                    Allowance
                  ┌─────────────
                  │  7,000  ← have
                  │     ?   ← 19,400
                  │ 26,400  ← balance needed
```

Next, a comprehensive example is presented that ties together nondrilling exploration costs and acquisition costs with an emphasis on option costs and delinquent taxes.

COMPREHENSIVE EXAMPLE # 2

a. Tyler Company obtained the following rights:

Undeveloped Area	Acres	Right Acquired	Cost
A	320	Shooting	$ 500
B	640	Shooting	1,500
C	2,560	Shooting, option to lease	4,000
D	2,400	Shooting, option to lease	3,600
E	1,000	Shooting, option to lease	5,000

Entries

AREA A

G&G expense	500	
Cash		500

AREA B

G&G expense	1,500	
Cash		1,500

AREA C—Tyler Company decided to treat the entire amount as option cost.

Property purchase suspense	4,000	
Cash		4,000

AREA D—Tyler Company decided to treat the entire amount as option cost.

Property purchase suspense	3,600	
Cash		3,600

AREA E—Tyler Company decided to treat the entire amount as option cost.

Property purchase suspense	5,000	
Cash		5,000

b. Tyler Company hired Seismo Company to perform G&G work on areas A–E and paid the company $500,000.

Entry

G&G expense	500,000	
Cash		500,000

c. As a result of the G&G work, Tyler Company decided to lease the following properties:

Lease (Area)	Acres Leased	Bonus per Acre	Total Bonus	Legal Costs, Recording Fees
A	320	$50	$ 16,000	$300
B	640	50	32,000	500
C	640	60	38,400	200
D	800	80	64,000	750
E	0	Option expires		
	2,400			

Entries—Tyler Company decided to apportion the option cost based on acreage leased.

LEASE A

Unproved property.	16,300	
Cash		16,300

LEASE B

Unproved property.	32,500	
Cash		32,500

LEASE C—leasing 1/4 of total acreage

Unproved property ($38,600 + 1/4 × $4,000).	39,600	
Surrendered lease expense (3/4 × $4,000)	3,000	
Property purchase suspense.		4,000
Cash		38,600

LEASE D—leasing 1/3 of total acreage

Unproved property ($64,750 + 1/3 × $3,600).	65,950	
Surrendered lease expense (2/3 × $3,600)	2,400	
Property purchase suspense.		3,600
Cash		64,750

LEASE E

Surrendered lease expense	5,000	
Property purchase suspense.		5,000

d. Landmen salaries and overhead relating to the above acquisitions were $16,800. The costs were allocated to each lease based on relative acres leased.

$$\frac{\$16,800}{2,400 \text{ (total acres leased)}} = \underline{\$7/\text{acre}}$$

Entries

LEASE A

Unproved property (320 × $7)...............	2,240	
Cash.............................		2,240

LEASE B

Unproved property (640 × $7)...............	4,480	
Cash.............................		4,480

LEASE C

Unproved property (640 × $7)...............	4,480	
Cash.............................		4,480

LEASE D

Unproved property (800 × $7)...............	5,600	
Cash.............................		5,600

At the end of the first year, no impairment was assessed on the properties.

e. During the second year, Tyler Company incurred and paid the following items:

1) Ad valorem taxes of $1,500 on Lease A

Entry

Ad valorem taxes expense (exploration)	1,500	
Cash.............................		1,500

2) Dry-hole contribution, $20,000

Entry

Test-well contribution expense (exploration)	20,000	
Cash.............................		20,000

3) Legal costs for title defense, $8,500 on Lease C

Entry

Legal expense (exploration)	8,500	
Cash		8,500

Note: In contrast, legal fees in connection with a title exam would be an acquisition cost and would be capitalized.

4) Bottom-hole contribution, $15,000

Entry

Test-well contribution expense (exploration)	15,000	
Cash		15,000

5) Nonrecoverable delinquent taxes on Lease A, $2,000 (specified in lease contract)

Entry

Unproved property (acquisition)	2,000	
Cash		2,000

6) Recoverable delinquent taxes on Lease B, $1,500

Entry

Receivable from lessor	1,500	
Cash		1,500

7) Delay rentals on Leases A, C, and D, $6,000

Entry

Delay rental expense (exploration)	6,000	
Cash		6,000

8) Delay rental on Lease B, $2,000

Entry

Delay rental expense (exploration)	2,000	
Receivable from lessor (delinquent taxes)		1,500
Cash		500

f. At the end of the second year, Tyler Company assessed the leases with the following results (assume all of the leases were individually significant):

Lease	Balance in Unproved Property	Percentage Impairment	Amount Impaired	Net Carrying Value after Impairment
A	$20,540 ($16,300 + $2,240 + $2,000)	25%	$ 5,135	$15,405
B	$36,980 ($32,500 + $4,480)	0	0	36,980
C	$44,080 ($39,600 + $4,480)	30	13,224	30,856
D	$71,550 ($65,950 + $5,600)	20	14,310	57,240

Entries

LEASE A

Lease impairment expense ($20,540 × 25%)	5,135	
Allowance for impairment		5,135

LEASE B
No entry

LEASE C

Lease impairment expense ($44,080 × 30%)	13,224	
Allowance for impairment		13,224

LEASE D

Lease impairment expense ($71,550 × 20%)	14,310	
Allowance for impairment		14,310

g. Early in the third year, Tyler Company abandoned Lease C and Lease D.

Entries

LEASE C

Surrendered lease expense.	30,856	
Allowance for impairment.	13,224	
Unproved property		44,080

LEASE D

Surrendered lease expense.	57,240	
Allowance for impairment.	14,310	
Unproved property		71,550

h. Also in the third year, Tyler Company discovered proved reserves on Lease A and Lease B.

Entries

LEASE A

Proved property	15,405	
Allowance for impairment.	5,135	
Unproved property		20,540

LEASE B

Proved property	36,980	
Unproved property		36,980

LAND DEPARTMENT

The land department of an oil company usually is responsible for property acquisition and property administration. The exploration and legal departments are also concerned with these functions. The exploration department is responsible for recommending property acquisition, retention, and development or abandonment. The legal department conducts title examinations and title litigation and approves or prepares any legal documents involved.

The land department acts on information obtained from the exploration department's activities and from land department scouts in acquiring properties. Subscription services and information exchanges provide additional information to the land department. The actual acquisition of properties is negotiated by landmen, who also promote trades, joint ventures, unitizations, and various types of sharing arrangements.

The land department is responsible for recording and maintaining basic records on properties, ensuring that all contractual obligations in the lease contract are fulfilled, and preparing various types of reports. An example of an important contractual obligation is the delay rental payment. It is very important that the record system give adequate notice of leases due for rental payment, because failure to pay delay rentals when due will result in the lease terminating. After the property becomes proved, the land department works closely with the division order department to ensure that revenues from the property are distributed to the various owners.

Problems

1. Define the following:

    top lease

    bonus

    option to lease

    purchase in fee

    delinquent taxes paid by the lessee

    impairment

    internal costs

2. On December 31, 2016, Dill Oil Company recognized impairment of $100,000 on an individually significant lease. Before the financial statements were issued early the next year, a well was drilled and proved reserves were found. Dill easily revised their financial statements so that no impairment was recognized on the property. Please comment.

3. Tharp Energy Company, which uses the successful efforts method of accounting, owns an individually significant lease, with a cost of $200,000. On December 31, 2017, the lease is not considered impaired. However, prior to completion of the audit, a well on adjacent property is abandoned as a dry hole, and the lease is now considered to be 40% impaired. Prepare any necessary adjusting entry.

4. Decade Oil Corporation paid $500/acre for a lease with a three-year primary term. During the next two years, Decade drilled three dry holes on the property. With one more year of the primary term left, Decade still intends to try one more time. Should any impairment be recognized on the property?

5. Rock Petroleum began operations in 2016 with the acquisition of four undeveloped leases, all individually significant. Give the entries assuming the following transactions. For simplicity, you may combine entries for the different leases for all items marked with an asterisk (*).

Year	Transaction	Lease A	Lease B	Lease C	Lease D
2016	Shooting rights*	$20,000	None	None	$15,000
	G&G costs, broad*	50,000	None	None	90,000
	Lease bonuses*	30,000	$40,000	$60,000	50,000
	G&G costs, detailed*	65,000	30,000	45,000	35,000
	Dry-hole contributions paid*	15,000	None	None	25,000
	Legal costs, title exams*	1,000	5,000	4,000	10,000
	Legal costs, title defense	None	None	50,000	None
	Delinquent taxes (in contract and nonrecoverable)	25,000	None	None	None
Dec 31	Impairment	10%	25%	—	40%
2017	Delay rentals*	2,000	10,000	3,000	8,000
	Property tax*	3,000	None	5,000	None
	Miscellaneous	Abandoned lease	Drilled & found oil	—	+ 10% Impaired

6. On December 31, 2015, Launch Oil Company's unproved property account for leases that are not individually significant had a balance of $800,000. The impairment allowance account had a balance of $75,000. Give the entries for each of the following transactions occurring in 2015, 2016, and 2017. (All transactions concern individually insignificant unproved leases.)

   a. Assuming Launch has a policy of maintaining a 55% allowance, i.e., 55% of gross unproved properties, give the entry to record impairment on December 31, 2015.
   b. During 2016, Launch surrendered leases that cost $300,000.
   c. During 2016, leases that cost $50,000 were proved.
   d. During 2016, leases costing $310,000 were acquired.
   e. Give the entry to record impairment on December 31, 2006.
   f. During 2017, leases costing $428,000 were surrendered.

7. Given the following data for Float Energy:

Unproved property—Lease A (significant)	$700,000
Allowance for impairment—Lease A	500,000
Unproved property—Lease B (insignificant)	30,000
Allowance for impairment, group basis	450,000

    a. Give the entries to record the abandonment of both Lease A and Lease B.

    b. Give the entries assuming instead that both Lease A and Lease B were proved, i.e., oil or gas was discovered on the leases.

8. Guarantee Oil Company's internal land department incurred costs of $150,000 in acquiring leases. Of the 1 million acres of prospects, only 450,000 acres were leased.

    a. How much, if any, of the $150,000 incurred by the land department should be capitalized?

    b. If capitalized, what account(s) should be debited?

9. Gusher Oil Corporation obtained shooting rights only for $10,000 on 5,000 acres owned by Mr. Q and shooting rights coupled with an option to lease for $12,000 on 4,000 acres owned by Mr. S. The 4,000 acres owned by Mr. S are located adjacent to the 5,000 acres owned by Mr. Q. Ignore any other acquisition costs.

    a. Give the entries to record the rights obtained, assuming there is no apportionment of the cost between the option and the shooting rights.

    b. Give the entry to record the rights obtained from Mr. S, assuming instead that the $12,000 was apportioned between the option and the shooting rights.

    c. Give the entry to record the leasing of all 4,000 of Mr. S's acres, assuming that the original cost of $12,000 was *not* apportioned between the option and the shooting rights.

    d. Give the entry to record the leasing of only 1,000 acres from Mr. S, again assuming that the original cost of $12,000 was not apportioned. Also assume Gusher did not apportion the amount in the suspense account based on the acreage leased.

    e. Give the entry to record the leasing of 1,000 acres from Mr. S, assuming that the original cost of $12,000 was not apportioned. Assume that Gusher Oil Corporation apportioned the amount in the suspense account based on relative acreage leased.

10. The following transactions relate to one lease:

    a. On March 10, 2017, Axis Petroleum paid delinquent property taxes of $2,000 on an undeveloped lease. Assume that these taxes are recoverable out of future delay rental or royalty payments. Give the entry to record payment.

    b. On February 15, 2018, a delay rental payment of $800 is due. Determine the amount of cash actually paid and give the entry to record payment.

    c. On July 21, 2018, Axis Petroleum decided to surrender the lease. Give the entry to record abandonment with respect to the delinquent property taxes. Ignore acquisition costs of the property.

d. Assume instead that the $2,000 payment of delinquent taxes was not recoverable and was made at the time Axis Petroleum was acquiring the lease. Give the entry to record the payment.

e. Assume instead that the $2,000 payment was not recoverable and was made by Axis Petroleum six months after acquiring the lease in order to protect Axis' investment. Give the entry to record the payment.

11. During 2015, Prosperity Oil Company acquired the following leases:

Lease	Acres	Bonus/Acre
A	2,000	$50
B	3,000	60
C	5,000	40

In acquiring and exploring these leases, Prosperity Oil Company incurred the following additional costs:

Shooting rights, $3,000

Salaries for G&G exploration activities, $40,000

Mapping costs for exploration activities, $10,000

Salary of in-house lawyer working on lease acquisition, $10,000

Minor repairs of G&G exploration equipment, $500

Salaries of landmen working on lease acquisition, $30,000

Prosperity Oil allocates internal costs relating to lease acquisition to specific leases. Assuming Lease A was abandoned at the end of the year, answer the following questions:

a. What was the total nondrilling exploration expense for all three leases for the year?

b. What was the surrendered lease expense?

c. How much was capitalized as unproved property for Lease B?

Hint: Some of these costs must be allocated to the individual leases on some reasonable basis.

12. Bryant Oil Corporation acquired a lease on October 15, 2015, for $200,000 cash. No drilling was done on the lease during the first year. Since Bryant wished to retain the lease, a delay rental of $10,000 was paid on October 15, 2016. During November and December of 2016, three dry holes were drilled on surrounding leases. Based on the dry holes, Bryant's management decided that the lease was 75% impaired. Bryant had still not started drilling operations by the end of the second year and so paid a second delay rental. During November 2017, with less than one year of the primary term left, Bryant drilled a dry hole on the lease and decided to abandon the lease. Because the end of Bryant's accounting period is December 31 and for income tax purposes, Bryant executed a quit claim deed and relinquished all rights to the lease the last day of November 2017. Give the entries.

13. Railway Oil and Gas Company owned the following unproved property as of the end of 2010:

Significant Leases		Individually Insignificant Leases	
Lease A	$300,000	Lease C	$ 50,000
Lease B	350,000	Lease D	25,000
Total	$650,000	Lease E	40,000
		Lease F	30,000
		Total	$145,000

Although no activity took place on Lease A during the year, Railway decided that Lease A was not impaired, because there were still three years left in that lease's primary term. Two dry holes were drilled on Lease B during the year; but because Railway intended to drill one more well on Lease B in the coming year, it decided that Lease B was only 40% impaired. With respect to the individually insignificant leases, past experience indicates that 70% of all unproved properties assessed on a group basis will eventually be abandoned. Railway's policy is to provide at year-end an allowance equal to 70% of the gross cost of these properties. The allowance account had a balance of $20,000 at year-end. Give the entries to record impairment.

14. Latitude Energy decided to explore some acreage in Texas before acquiring any leases. Latitude acquired shooting rights only on 15,000 acres owned by Mr. T for $0.10/acre. Latitude obtained shooting rights coupled with an option to lease on 10,000 acres owned by Mr. H for $0.25/acre. After completing G&G surveys at a cost of $85,000, Latitude decided to lease 5,000 acres from Mr. T, paying a bonus of $35/acre. Latitude also decided to lease 5,000 acres from Mr. H, paying the same bonus of $35/acre. Latitude's income statement has seen better days, and although it will not help much, Latitude decides to capitalize every possible cost. Give the entries.

15. Bear Oil Incorporated has two unproved leases with the following capitalized costs:

Lease C (600 acres)	$20,000
Lease D (1,200 acres)	$50,000

Bear Oil Incorporated negotiated a new lease on Lease C immediately following the end of the primary term. The lease bonus was $60/acre on the new lease. Before the end of the primary term, Bear Oil also obtained a new lease on Lease D at a lease bonus rate of $40/acre. This lease was to take effect prior to the end of the primary term. Prepare journal entries for the two leases.

16. Sauer Energy purchased land in fee for $100,000. The fair market values of the surface and mineral rights were determined by a qualified appraiser as follows:

Surface rights	$ 80,000
Mineral rights	40,000
	$120,000

Prepare a journal entry to record the purchase.

17. Critic Oil Company purchased three leases as follows:

July 1, 2014	Lease A	$100,000
August 15, 2014	Lease B	200,000
October 10, 2014	Lease C	300,000

All the leases are classified as individually significant.

a. On December 31, 2014, Lease A is determined to be 25% impaired. Lease B and Lease C are not impaired.

b. On December 31, 2015, Lease A is determined to be impaired a total of 75%, and Lease C, 60%. Lease B is not impaired.

c. On December 31, 2016, Lease A is considered to be 100% impaired and is abandoned. Lease B is 30% impaired, and a well on Lease C found proved reserves.

Prepare journal entries for all of the transactions except the initial purchase.

18. Seaside Oil Corporation has an offshore lease that cost $4,000,000. The total acreage is 40,000 acres. Proved reserves are found on a 10,000 acre tract that is included in the 40,000 acres. Management decides to reclassify only the proved area acreage.

Prepare the entry to reclassify the acreage.

19. Bartz Corporation leased 10,000 acres with a lease bonus of $80/acre. Delay rentals are to be at a rate of $5/acre. The lease specified that Bartz Corporation could abandon the lease in 1,000 acre portions. At the end of year 1, Bartz Corporation abandoned 4,000 acres and paid a delay rental on the remainder.

Prepare journal entries for the activities.

20. Ernest Oil Company's balance sheet, at 12/31/14, included account balances as follows:

	Cost	Allowance for Impairment
Lease A	$ 40,000	$ 10,000
Lease B	80,000	24,000
Lease C	100,000	0
Individually insignificant (Leases D–J)	200,000	120,000

During 2015, the following events related to the above unproved properties occurred:

a. Lease A is abandoned.
b. Lease B is surrendered.
c. Leases G & F (individually insignificant) in the amounts of $2,000 and $3,000, respectively, are abandoned.

Prepare the necessary entries.

21. Monarch Energy Corporation owned the following unproved property at 12/31/14:

	Individually Significant Leases			**Individually Insignificant Leases**	
	Cost	Allowance		Cost	Total Allowance
Lease A	$300,000	$200,000	Lease C	$20,000	
Lease B	400,000	100,000	Lease D	30,000	
			Lease E	15,000	
			Lease F	10,000	
				$75,000	$32,000

Prepare journal entries for 2015, assuming the following events:

a. Found proved reserves on Leases A & B.
b. Found proved reserves on Lease D.
c. Found proved reserves on Lease C.

22. Dwight Corporation has the following groups of individually insignificant leases at 12/31/15.

	Group A	Group B	Group C
Total costs	$200,000	$300,000	$400,000
Total allowance for impairment 12/31/15	40,000	0	80,000
Expected average percentage of impairment	60%	70%	65%

Prepare journal entries to record impairment for each of the groups at 12/31/15.

23. Give the entry to record abandonment in each of the following cases.

   a. Gaylene Energy Company abandoned an unproved property that cost $150,000. The property was considered significant and had been impaired $100,000 on an individual basis.

   b. Gaylene also abandoned an unproved property that was not considered individually significant. The property cost $10,000, and the group allowance account had a balance of $400,000 at the time of the abandonment.

24. Opaque Corporation purchased land in fee for $420,000. The land was located in a remote area in Oklahoma. An appraiser estimated the fair market value of the surface rights to be $200,000. The appraiser was not able to make an estimate of the value of the mineral rights.

   Prepare a journal entry to record the purchase.

25. On January 1, 2010, Local Petroleum entered into a concession agreement with the government of Egypt and paid a $3,000,000 signing bonus. The agreement covers 20,000 acres, has a term of five years, and requires that exploration begin immediately. At the end of the first three years, Local is required to begin relinquishing acreage at a rate of 25% of the contract area per year. However, Local is not required to relinquish proved acreage. On July 16, 2012, Local makes a commercial discovery and determines that a 1,000 acre block is proved. On December 31, 2012, 25% of the initial contract area is relinquished. Local estimates that only another 25% of the original contract acreage will be relinquished.

   Prepare all journal entries that would be required to account for the concession area from January 1, 2010 through December 31, 2012.

26. Lomax Company typically acquires a large number of individually insignificant properties each year. In computing impairment, Lomax groups these properties by year of acquisition. During 2015, Lomax acquired individually insignificant unproved property costing a total of $400,000. In the past, Lomax has abandoned 30% of all individually insignificant unproved properties without ever finding oil or gas. It takes Lomax an average of three years to determine whether a property will be proved or surrendered.

   Give the entry to record impairment.

REFERENCES

1. Institute of Petroleum Accounting. 2001. *2001 PricewaterhouseCoopers Survey of U.S. Petroleum Accounting Practices.* Denton, TX: Institute of Petroleum Accounting. p. 62.

5

DRILLING AND DEVELOPMENT COSTS—SUCCESSFUL EFFORTS

Chapter 3 discussed the *nondrilling* types of exploration costs. This chapter discusses the two types of exploration *drilling* costs and both drilling and nondrilling development costs. However, before discussing the financial accounting treatment of these costs, the tax treatment must be briefly discussed. The tax treatment of drilling costs differs significantly from the financial accounting treatment. The different treatment of drilling costs under tax versus financial accounting affects the manner in which the costs are entered into the accounting system. These costs must be entered with sufficient detail to provide the information needed for both financial accounting and income tax accounting.

INCOME TAX ACCOUNTING FOR DRILLING COSTS

For tax purposes, drilling and development costs are classified as either **intangible drilling costs (IDC)** or **equipment costs** (lease and well equipment). The distinction between IDC and equipment costs is very important, because for tax purposes, all or most of the IDC may be expensed as incurred, whereas equipment costs must be capitalized and depreciated. For operations in the United States, the taxpayer makes a one-time election to either expense or capitalize IDC. Regardless of whether the election is to expense or to capitalize IDC, IDC related to dry holes may be expensed when incurred. This election is per taxpayer and applies to all the taxpayer's properties. Rules for integrated producers are modified somewhat, in that a specified percentage of all productive IDC must be capitalized, regardless of the election made. Currently, integrated producers are permitted to expense 70% of their IDC incurred on productive wells, with the other 30% being written off over 60 months. IDC may be written off entirely if the well is dry. In contrast, independent producers may deduct all of their IDC, whether related to productive wells or dry holes, in the year incurred. Tax rules relating to E&P operations are discussed in more detail in chapter 11.

IDC is defined as expenditures for drilling that in themselves do not have a salvage value and are "incident to and necessary for the drilling of wells and the preparation of wells for the production of oil and gas" [U.S. Treas. *Reg. Section 1*, 612-4(a)]. In general, IDC includes the intangible or nonsalvageable costs of drilling up to and including the cost of installing the Christmas tree. The term **Christmas tree** refers to the valves, pipes, and fittings assembly that is used to control the flow of oil and gas from the wellhead. In many cases, the physical arrangement of the valves, pipes, and fittings resembles a Christmas tree. The Christmas tree is located at the top of the well on the surface of the land.

In general, equipment costs include all tangible or salvageable costs of drilling up to and including the Christmas tree, plus both intangible and tangible costs past the Christmas tree. In considering whether an item is before or after the Christmas tree, the physical flow of the oil and gas should be considered, not the time of incurrence. Note that neither the word *salvageable* nor the word *tangible* is entirely appropriate in defining which costs are IDC rather than equipment. For example, casing, which is tangible, is cemented into the wellbore and is not typically salvageable. Nevertheless, casing is equipment, not IDC. One distinction between IDC and equipment costs appears to be whether a tangible cost in itself has a salvage value. IDC and equipment costs may be diagramed as follows:

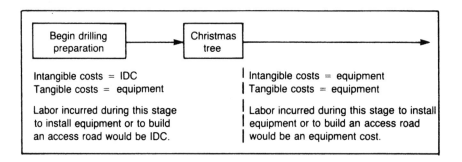

As can be seen from the preceding diagram, the purpose for constructing a lease road determines whether the costs of constructing the road are considered IDC.

EXAMPLE

IDC versus Equipment

a. Tyler Company incurred acquisition (purchase) costs of $40,000 for casing and installation costs of $5,000. Casing is subsurface well equipment and is, therefore, before the Christmas tree.

    IDC   = $5,000, installation costs
    Equipment = $40,000, purchase cost

b. Tyler Company purchased flow lines and storage tanks at a cost of $75,000. Installation costs were $10,000. Flow lines and tanks are nonwell equipment, are past the Christmas tree, and are not incident and necessary for drilling a well.

    IDC   = $0, all costs are considered to be equipment costs (since these costs are past the Christmas tree)
    Equipment = $85,000, purchase cost and installation costs

c. Tyler Company incurred $15,000 in labor costs in building an access road to a drillsite. This cost is related to drilling a well and is before the Christmas tree.

    IDC   = $15,000
    Equipment = $0

d. Tyler Company incurred $15,000 in labor costs in building an access road to a producing well. This cost is not related to drilling and therefore is not IDC.

    IDC   = $0, all costs are considered to be equipment costs (since this cost is past the Christmas tree)
    Equipment = $15,000

In financial accounting when using the successful efforts method, it is important to be able to distinguish an exploratory well from a development or service well. Note that for tax purposes, this distinction is not important.

Items 1 and 2 below list costs related to drilling an exploratory or development well that are usually considered IDC. These costs are given in the order in which they are normally incurred. Note that the costs listed in Item 2 are *past* the Christmas tree, but because they are intangible and directly related to the process of drilling a well, the costs are considered IDC. Item 3 lists other types of wells for which IDC is incurred.

1. Up to and including installation of the Christmas tree
   a. Prior to drilling
      - G&G to determine specific drillsite
      - Preparation of the site, such as leveling, clearing, and building access roads and disposal pits
      - Rigging up
   b. During drilling
      - Drilling contractor's charges (when the drilling contractor furnishes equipment such as casing, part of the charge is equipment)
      - Drilling mud, chemicals, cement, and supplies
      - Wages and fuel
      - Well testing, such as core analysis and analysis of cuttings
   c. At target depth and during completion
      - Well testing, such as well logs and drillstem tests
      - Perforating and cementing
      - Swabbing, acidizing, and fracturing
      - Labor related to installing subsurface equipment to the wellhead and installation of the Christmas tree
      - If the well is dry, plugging and abandoning costs
2. After the Christmas tree, following completion
   a. Removal of drilling rig
   b. Restoration of land and damages paid to the surface owner
3. Wells other than original exploration or development wells
   a. Intangible costs (those listed in items 1 and 2) incurred in deepening a well
   b. Intangible costs incurred in drilling a water or gas injection well
   c. Intangible costs of drilling a water supply or injection well where water is to be used for drilling an exploration or development well or for injection

In financial accounting, the cost of a machine, for example, includes not only the purchase cost, but also the intangible costs necessary to get the machine ready for use. Thus, for financial accounting purposes, the cost of a well includes both the tangible and intangible costs. Consequently, the distinction between IDC and equipment costs has no meaning for financial accounting. However, because of the importance of the classification of costs as either IDC or equipment costs for tax purposes, the distinction between IDC and equipment costs is usually made in financial accounting as well as in tax accounting.

Financial Accounting for Drilling and Development Costs

Under successful efforts accounting, a direct relationship is required between costs incurred and specific reserves discovered before costs are ultimately identified as assets. Consequently, with respect to exploration costs, only successful exploratory drilling costs are considered to be a part of the cost of the oil or gas reserves; unsuccessful exploratory drilling costs are written off. Nondrilling exploration costs are charged to expense as incurred. In contrast to exploration costs, the purpose of development activities is considered to be building a producing system of wells and related equipment and facilities rather than searching for oil and gas. Thus, both successful and unsuccessful development costs are capitalized as part of the cost of the oil or gas.

The board justified the different treatment of exploratory dry holes and development dry holes in paragraphs 205 and 206 of *SFAS No. 19* as follows:

In the Board's judgment, however, there is an important difference between exploratory dry holes and development dry holes. The purpose of an exploratory well is to search for oil and gas. The existence of future benefits is not known until the well is drilled. Future benefits depend on whether reserves are found. A development well, on the other hand, is drilled as part of the effort to build a producing system of wells and related equipment and facilities. Its purpose is to extract previously discovered proved oil and gas reserves. By definition (Appendix C, paragraph 274), a development well is a well drilled within the proved area of a reservoir to a depth known to be productive. The existence of future benefits is discernible from reserves already proved at the time the well is drilled. An exploratory well, because it is drilled outside a proved area, or within a proved area but to a previously untested horizon, is not directly associable with specific proved reserves until completion of drilling. An exploratory well must be assessed on its own, and the direct discovery of oil and gas reserves can be the sole determinant of whether future benefits exist and, therefore, whether an asset should be recognized. Unlike an exploratory well, a development well by definition is associable with known future benefits before drilling begins. The cost of a development well is a part of the cost of a bigger asset—a producing system of wells and related equipment and facilities intended to extract, treat, gather, and store known reserves.

Moreover, because they are drilled only in proved areas to proved depths, the great majority of development wells are successful; a much smaller percentage (22 percent in the United States in 1976), as compared to exploratory wells (73 percent in the United States in 1976) are dry holes. Development dry holes occur principally because of a structural fault or other unexpected stratigraphic condition or because of a problem that arose during drilling, such as tools or equipment accidentally dropped down the hole, or simply the inability to know precisely the limits and nature of a proven reservoir. Development dry holes are similar to normal, relatively minor "spoilage" or "waste" in manufacturing or construction. The Board believes that there is a significant difference between the exploration for and the development of proved reserves. Therefore, in the Board's judgment, it is appropriate to account for the costs of development dry holes different from exploratory dry holes.

Thus, the cost of drilling an exploratory well is expensed if the well is dry and capitalized if the well is successful. However, the cost of drilling a development well is always capitalized, regardless of the outcome of the well. Despite this disparity between the final accounting treatment for exploratory drilling and development drilling, preliminary accounting for the costs of drilling a well, i.e., accounting for drilling in progress, is the same regardless of whether the well is an exploratory well or a development well.

Well Classification

Exploratory and development wells are defined in *Reg. S-X 4-10* as follows:

Exploratory well. *A well drilled to find and produce oil or gas in an unproved area, to find a new reservoir in a field previously found to be productive of oil or gas in another reservoir, or to extend a known reservoir. Generally, an exploratory well is any well that is not a development well, a service well, or a stratigraphic test well as those items are defined below.*

Development well. *A well drilled within the proved area of an oil or gas reservoir to a depth of a stratigraphic horizon known to be productive.*

Service well. *A well drilled or completed for the purpose of supporting production in an existing field. Specific purposes of service wells include gas injection, water injection, steam injection, air injection, salt water disposal, water supply for injection, observation, or injection for in-situ combustion.*

Stratigraphic test well. *A drilling effort, geologically directed, to obtain information pertaining to a specific geologic condition. Such wells customarily are drilled without the intention of being completed for hydrocarbon production. This classification also includes tests identified as core tests and all types of expendable holes related to hydrocarbon exploration. Stratigraphic test wells are classified as (i) "exploratory type," if not drilled in a proved area, or (ii) "development type," if drilled in a proved area.*

Classification of a well as either a service well or a stratigraphic test well is straightforward. In contrast, classification of a well as either exploratory or development is not always as straightforward as it might first appear. The following discussion relates to exploratory versus development well classification for wells other than service wells or stratigraphic test wells.

The first step in accounting for the drilling of a well is to determine whether the well is exploratory versus development. Proper classification of wells is critical. Note that if a well does not meet the development well definition, the well is to be classified as an exploratory well. Thus, it is critical to understand the development well definition. In order for a well to be classified as a development well, the well must be drilled in a proved area. In *Reg. S-X 4-10*, a **proved area** is defined as the "part of a property to which proved reserves have been specifically attributed." The proved reserves must be either proved developed or proved undeveloped. If there are no proved reserves, then the area is not a proved area, and the well cannot be classified as a development well.

Reg. S-X 4-10 definitions of proved developed and proved undeveloped reserves are as follows:

> **Proved developed oil and gas reserves.** *Proved developed oil and gas reserves are reserves that can be expected to be recovered through existing wells with existing equipment and operating methods. Additional oil and gas expected to be obtained through the application of fluid injection or other improved recovery techniques for supplementing the natural forces and mechanisms of primary recovery should be included as "proved developed reserves" only after testing by a pilot project or after the operation of an installed program has confirmed through production response that increased recovery will be achieved.*

> **Proved undeveloped reserves.** *Proved undeveloped oil and gas reserves are reserves that are expected to be recovered from new wells on undrilled acreage, or from existing wells where a relatively major expenditure is required for recompletion. Reserves on undrilled acreage shall be limited to those drilling units offsetting productive units that are reasonably certain of production when drilled. Proved reserves for other undrilled units can be claimed only where it can be demonstrated with certainty that there is continuity of production from the existing productive formation. Under no circumstances should estimates for proved undeveloped reserves be attributable to any acreage for which an application of fluid injection or other improved recovery technique is contemplated, unless such techniques have been proved effective by actual tests in the area and in the same reservoir.*

If the reserves are proved developed, then the wells necessary to produce the reserves have, in all likelihood, already been drilled. Any additional wells drilled in such an area would normally be service wells or wells drilled to replace existing wells. Service wells include water supply wells drilled to support production of the proved developed reserves. Wells also are sometimes drilled to replace existing wells that for mechanical or other reasons are no longer able to produce. In either case, these wells would be classified as development and capitalized.

Classification of wells when a *new* drilling effort is directed at proved undeveloped reserves is more complicated. In order for a well that is being drilled on undrilled acreage within a productive area to be classified as development, one of the following two situations must apply:

1. The well is being drilled in an offset location, i.e., offsetting an existing productive well, where there is reasonable certainty of production when drilled.
2. If the well is not being drilled in an offset location, there must be *certainty* of continuity of production between the well and an existing productive formation.

If neither of these situations exists, the well should most likely be classified as an exploratory well.

Frequently, producing wells are "recompleted" to another depth. If the recompletion is aimed at accessing proved undeveloped reserves, then the cost of the recompletion is a development cost and is capitalized. If an existing well is reentered and additional drilling is undertaken in an effort to find new proved reserves, then that portion of the drilling would be classified as exploratory. If a well is reentered to fix a problem with no expected change in proved or proved developed reserves, the costs are charged to expense as a repair.

EXPLORATORY DRILLING COSTS

Figure 5–1 diagrams the accounting treatment of the five types of exploration costs. The first three types are nondrilling costs and are discussed in chapter 3. The last two types are (1) the costs of drilling and equipping exploratory wells and (2) the costs of drilling exploratory-type stratigraphic test wells.

Chapter 5 • Drilling and Development Costs—Successful Efforts   145

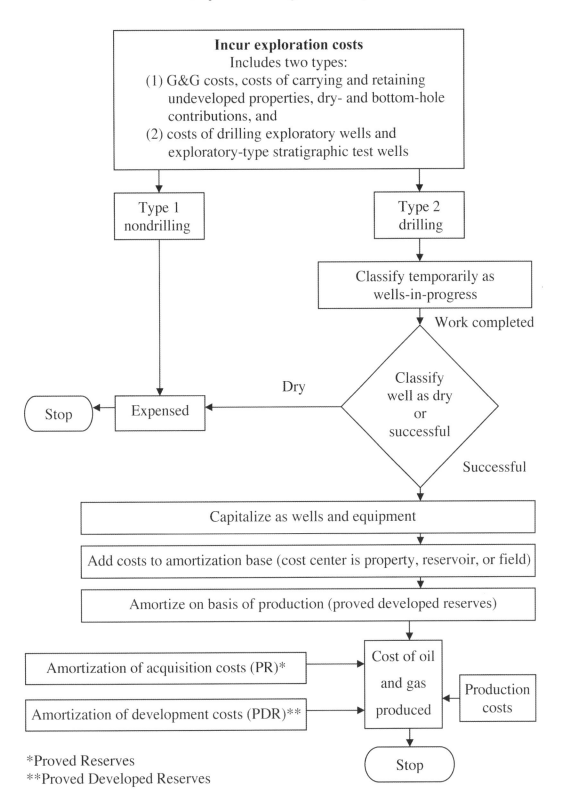

Fig. 5–1. Successful efforts, exploration costs

Exploratory wells and exploratory-type stratigraphic test wells are accounted for as follows, according to *SFAS No. 19*, par. 19:

> *The cost of drilling exploratory wells and the costs of drilling exploratory-type stratigraphic test wells shall be capitalized as part of the enterprise's uncompleted wells, equipment, and facilities, pending determination of whether the well has found proved reserves. If the well has found proved reserves (paragraphs 31–34), the capitalized costs of drilling the well shall become part of the enterprise's wells and related equipment and facilities (even though the well may not be completed as a producing well); if, however, the well has not found proved reserves, the capitalized costs of drilling the well, net of any salvage value, shall be charged to expense.*

The following example illustrates typical drilling and completion costs and their accounting treatment under successful efforts. As shown in the example, drilling costs are temporarily classified as (1) wells-in-progress—IDC or (2) wells-in-progress—lease and well equipment. (The term *wells-in-progress* is used in place of the FASB terminology *uncompleted wells, equipment, and facilities*. Note that wells-in-progress therefore includes not only unfinished wells, but also unfinished equipment and facilities.) The wells-in-progress accounts are not subject to DD&A.

When drilling reaches the targeted depth, a decision must be made as to whether the well has found proved reserves. If proved reserves have been found, the wells-in-progress account balances must be transferred to wells and equipment accounts. (The term *wells and equipment* is used in place of the FASB terminology *wells and related equipment and facilities*. Note that similar to wells-in-progress, wells and equipment includes not only completed well costs, but also completed equipment and facilities costs.) In addition, if the well is the first successful exploratory well drilled on the property, the unproved property account must be reclassified or transferred into a proved property account, because proved reserves have been found and are now attributed to that property. Both the proved property account and the wells and equipment accounts are subject to DD&A. (See chapter 6.)

If no proved reserves are found, the well must be plugged and abandoned. Equipment in the hole is salvaged when possible. However, installed casing cannot usually be removed because of either regulatory requirements or physical constraints. If the well is an exploratory well as in the following example, the costs of plugging and abandoning, in addition to the capitalized costs in the wells-in-progress accounts (net of any salvaged equipment), must be written off to dry-hole expense. If the lease is also abandoned, the net amount capitalized as unproved property will be written off as surrendered lease expense or charged to the allowance account, depending upon whether the property is individually significant or individually insignificant.

EXAMPLE

Exploratory Drilling Costs

a. On January 2, 2012, as a result of G&G work done in 2011, Tyler Company decided to lease 1,000 acres at $20/acre. The lease was undeveloped.

    Entry

Unproved property	20,000	
Cash		20,000

b. Early in 2012, Tyler Company decided to begin drilling operations and had additional seismic studies conducted at a cost of $100,000 to select a specific drillsite. (Even though G&G work has been done to locate a possible reservoir, additional seismic work may be necessary to select the drillsite. The cost of this seismic work is considered part of the cost of drilling the well and not G&G expense.)

    Entry

Wells-in-progress—IDC.	100,000	
Cash		100,000

c. In preparing the drilling site, Tyler Company incurred costs of $25,000 in clearing and leveling the site and in building an access road. (These activities normally would be performed by a drilling contractor.)

    Entry

Wells-in-progress—IDC.	25,000	
Cash		25,000

d. Additional preparation costs of $30,000 were incurred in digging a mud pit and installing a water line. Pipes for the water line cost $10,000. (These activities normally would be performed by a drilling contractor.)

    Entry

Wells-in-progress—IDC.	30,000	
Wells-in-progress—lease and well equipment (L&WE) .	10,000	
Cash		40,000

e. Tyler Company purchased casing for the well at a cost of $160,000.

    Entry

Wells-in-progress—L&WE	160,000	
Cash		160,000

f. Tyler Company had hired a drilling contractor on a footage-rate contract, and as is usual in such an agreement, payment was contingent upon the contractor drilling to a specified depth. The well was spudded (i.e., drilling was begun) early in June, and contract depth was reached in late July. Tyler Company paid the contractor $400,000. (In a footage-rate contract, a specified amount is paid per foot drilled. In contrast, in a day-rate contract, a specified amount is paid per day, typically a different amount for a drilling day versus a standby day.)

Entry

Wells-in-progress—IDC.	400,000	
Cash		400,000

g. In evaluating the well, Tyler Company incurred costs of $50,000. A well log was run and a drillstem test was made.

Entry

Wells-in-progress—IDC.	50,000	
Cash		50,000

h. Based on the well log and drillstem test, as well as other tests performed as the well was drilled, Tyler Company decided to complete the well. Casing was set (i.e., installed by cementing between the pipe and wellbore) at a cost of $135,000 for casing for the well and $60,000 for cementing services.

Entry

Wells-in-progress—IDC.	60,000	
Wells-in-progress—L&WE	135,000	
Cash		195,000

i. Tyler incurred acquisition costs of $8,000 and installation costs of $1,000 for a string of production tubing through which the oil and gas will be produced. (Although oil and gas can be produced through casing, tubing is usually used because it is much easier than casing to remove and repair.)

Entry

Wells-in-progress—IDC.	1,000	
Wells-in-progress—L&WE	8,000	
Cash		9,000

j. Tyler Company incurred acquisition (purchase) costs of $50,000 for a Christmas tree and installation costs of $13,000.

Entry

Wells-in-progress—IDC.	13,000	
Wells-in-progress—L&WE	50,000	
Cash		63,000

k. Tyler Company incurred $40,000 for perforating and acidizing services. (Perforating involves using a perforating gun to make perforations or holes through the casing and cement so that oil and gas can flow from the formation into the wellbore. Acidizing is a method used to increase the permeability of the formation by introducing acid into the formation.)

Entry

Wells-in-progress—IDC.	40,000	
Cash		40,000

l. The work on the well is finished and proved reserves have been found. Two entries are necessary: one to transfer the cost of the well from an unfinished goods account to a finished goods account, and one to reclassify the lease as proved.

Entry l-1

Wells and equipment—IDC.	719,000	
Wells and equipment—L&WE	363,000	
Wells-in-progress—IDC.		719,000
Wells-in-progress—L&WE		363,000

Wells-in-progress—IDC			Wells-in-progress—L&WE		
b.	100,000		d.	10,000	
c.	25,000		e.	160,000	
d.	30,000		h.	135,000	
f.	400,000		i.	8,000	
g.	50,000		j.	50,000	
h.	60,000			363,000	
i.	1,000				
j.	13,000				
k.	40,000				
	719,000				

Entry l-2

Proved property	20,000	
Unproved property		20,000

m. Tyler Company purchased pipes (flow line to lease tanks), storage tanks, and separators (to separate the gas from the oil) for a cost of $15,000. Installation costs were $1,000.

Entry

Wells and equipment—L&WE	16,000	
Cash		16,000

Note that the cost center under successful efforts is a property, reservoir, or field, not an individual well.

n. If after evaluating the well in part g, Tyler Company instead had decided the well was dry, only costs in parts a–g would have been incurred, and the entry to record the dry hole would have been:

Entry

Dry-hole expense—IDC	605,000	
Dry-hole expense—L&WE	170,000	
Wells-in-progress—IDC		605,000
Wells-in-progress—L&WE		170,000

o. Costs of $2,000 were incurred in plugging and abandoning the hole.

Entry

Dry-hole expense—IDC	2,000	
Cash		2,000

It is important to distinguish between abandonment of the well and abandonment of the lease. In this case, only the well has been abandoned, and therefore no entry would be made relating to the lease, i.e., the unproved or proved property account.

Development Drilling Costs

Development wells were defined earlier. Development drilling is accounted for as follows, according to *SFAS No. 19*, par. 22:

> *Development costs shall be capitalized as part of the cost of an enterprise's wells and related equipment and facilities. Thus, all costs incurred to drill and equip development wells, development-type stratigraphic test wells, and service wells are development costs and shall be capitalized, whether the well is successful or unsuccessful. Costs of drilling those wells and costs of constructing equipment and facilities shall be included in the enterprise's uncompleted wells, equipment, and facilities until drilling or construction is completed.*

Most of the entries in the previous example (entry a, recording property acquisition, through entry l-1, classifying the exploratory well as completed and successful) would have been the same if the well had been a development well. Only three entries—l-2, n, and o—would have been different. If the well had been a development well instead of an exploratory well, it would have, by definition, been drilled in a proved area. Therefore, entry l-2, reclassifying the acquisition cost of the property from the unproved to proved category, would not have been necessary. Instead, the reclassification to proved property would have been made at an earlier time when proved reserves were first discovered on the property.

For entries n and o, the well was assumed to be a dry hole. Entries n and o transferred to dry-hole expense the cost of drilling, evaluating, and plugging and abandoning the well. If the well had been a development well, these costs would have been capitalized, regardless of the well's outcome, as wells and equipment.

Figure 5–2 outlines the accounting for development costs. A brief example that illustrates the accounting for a development well follows.

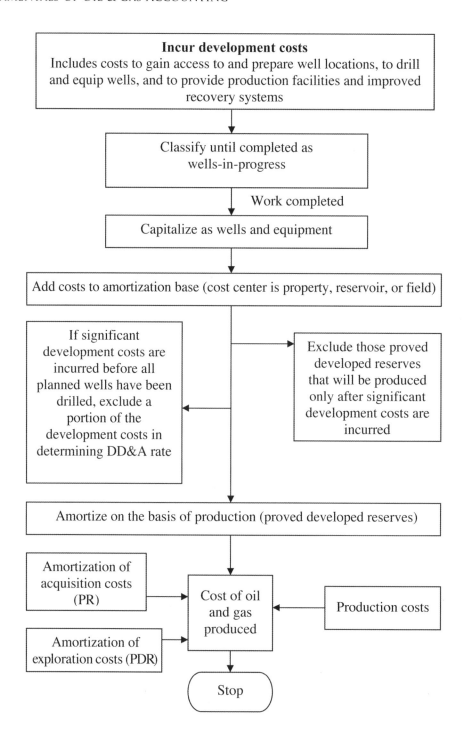

Fig. 5–2. Successful efforts, development costs

EXAMPLE

Development Drilling

During 2012 and 2013, Tyler Company drilled several successful exploratory wells on Lease A. As a result, Lease A was classified as a proved property, and the estimated boundaries of the reservoir were delineated. Tyler Company decided in 2014 to drill an additional well within the proved area—a development well—and hired a drilling contractor under a turnkey contract. The drilling contract specified that the contractor was to perform all services and furnish all materials up to completion.

a. The well is drilled and equipped to the point of completion, and Tyler Company pays the contractor under a turnkey contract the agreed-upon amount of $150,000. Of this $150,000, IDC was $120,000 and equipment costs were $30,000. (Unlike a footage-rate contract and a day-rate contract, under a turnkey contract, the drilling contractor assumes all the responsibility and furnishes all the necessary materials and equipment.)

Entry

Wells-in-progress—IDC.	120,000	
Wells-in-progress—L&WE	30,000	
Cash		150,000

b. Assume the well was determined to be dry and was plugged and abandoned for an additional $2,000.

Entry

Wells-in-progress—IDC.	2,000	
Cash		2,000

c. Work on the well is finished, and the wells-in-progress accounts are closed.

Entry

Wells and equipment—IDC ($120,000 + $2,000)	122,000	
Wells and equipment—L&WE	30,000	
Wells-in-progress—IDC.		122,000
Wells-in-progress—L&WE		30,000

d. Assume instead that the well was successful and that additional IDC of $15,000 and equipment costs of $70,000 were incurred to complete the well.

Entry to record completion costs

Wells-in-progress—IDC	15,000	
Wells-in-progress—L&WE	70,000	
Cash		85,000

Entry to record completion of work on well

Wells and equipment—IDC ($120,000 + $15,000). . . .	135,000	
Wells and equipment—L&WE ($30,000 + $70,000) . . .	100,000	
Wells-in-progress—IDC		135,000
Wells-in-progress—L&WE		100,000

STRATIGRAPHIC TEST WELLS

Today stratigraphic test wells are relatively rare in onshore sites in the lower-48 United States. For these locations, companies may find 3-D seismic to be a more cost-effective alternative. Outside the United States and in offshore locations, drilling exploratory-type stratigraphic test wells occurs with greater frequency. Exploratory-type stratigraphic test wells are often drilled offshore to determine the existence and quantity of proved reserves. Once proved reserves are discovered, development-type stratigraphic test wells may be drilled to determine the optimal location for installation of a permanent development drilling and production platform and to obtain information that can be used in designing the field development plan.

Classification of stratigraphic test wells as being exploratory-type or development-type requires the same process as described earlier. The rules for capitalizing or expensing costs of exploratory-type and development-type stratigraphic test wells are also the same as the rules for capitalizing or expensing the costs of exploratory and development wells. As a result, dry exploratory-type stratigraphic test wells are expensed, while successful exploratory-type stratigraphic test wells are capitalized. All development-type stratigraphic test wells, whether dry or successful, are capitalized. Thus, capitalization is not dependent upon whether the well will be completed as a producer, but upon whether the well has either found proved reserves or is a development well.

AFEs and Drilling Contracts

Costs to be incurred in drilling operations are generally budgeted and detailed in an Authorization for Expenditure (AFE). (See appendix A.) AFEs provide two useful functions:

- In joint operations, the operating agreement typically requires that the operator get approval from the nonoperators for expenditures relating to the drilling of wells. AFEs are used to fulfill this requirement by informing nonoperators as to drilling plans, providing cost estimates, and obtaining necessary approvals. (See chapter 12 for a discussion of situations where one or more nonoperators do not agree to approve the AFE.)
- In both joint and sole operations, AFEs are used by the operator for both internal budgeting and cost-control purposes.

An AFE should include enough detail for the nonoperators to determine the reasonableness of the estimated costs. Specifically, the AFE should include the following:

- Estimates for intangible drilling costs and equipment costs to be incurred in drilling the well
- Completion costs if the well is determined to be successful
- Plugging and abandonment costs if the well is determined to be dry

After drilling begins, the operator should monitor actual spending for each cost category. If actual costs exceed the budgeted and approved estimates by a certain amount—often 10% or the percentage specified in the joint operating agreement—then the nonoperators must be notified and approval for the cost overruns obtained. In addition to drilling wells, AFEs are used by operators to get approval from the nonoperators for facility construction or other projects where estimated costs exceed the single expenditure limits specified in the joint operating agreement.

After the nonoperators approve an AFE, the operator will typically contact a drilling contractor and negotiate a drilling contract. Drilling contract dollar amounts are usually based on a day rate or a footage rate. Under the day-rate contract, a stated amount is paid per day worked, regardless of the footage drilled. The drilling contractor generally provides a rig and crew and specified contractual services, while the operator provides all materials, supplies, equipment, etc. Under a footage-rate contract, a stated dollar amount is paid for each foot drilled. Normally, payment under a footage-rate contract is contingent upon a specified depth being reached. Under this contract, the drilling contractor generally provides the rig, crew, specified contractual services, and certain materials and supplies. Logging, core tests, drilling mud, and well equipment are typically supplied by the operator.

A drilling contract may also be on a turnkey basis, whereby the contractor agrees to drill to a specified depth for an agreed-upon dollar amount. Unlike the other types of contracts, under a turnkey contract, the contractor assumes all the responsibility and is in total charge of drilling and completion operations. The contractor provides all labor, equipment, and supplies. Turnkey contracts cost more and are not used as frequently as footage-rate and day-rate based contracts. Turnkey contracts may be used in domestic operations when several investors are buying an interest in the well by paying a designated amount ($100,000 or $500,000, etc.) and need to know the total well cost prior to investing.

In international operations, turnkey contracts may be used by an operator who is starting a drilling program in a new country or region and has yet to establish the infrastructure necessary to support drilling (i.e., employees, warehouses, suppliers, and contractors). In these cases, an operator may pay a drilling contractor to drill the first well or two on a turnkey basis, while making the necessary arrangements to establish the necessary infrastructure in the area and then afterward contract utilizing a day-rate or footage-rate basis.

Regardless of the type of drilling contract, the contractor bills the operator for services rendered as drilling progresses and when drilling operations are completed. The billing statements should, at a minimum, detail the drilling costs by categories of intangible drilling costs and equipment costs. This information allows the working interest owners to correctly classify these costs in their accounts.

Special Drilling Operations and Problems

Workovers

Workover operations generally involve using a special workover rig to restore or stimulate production from a particular well. A situation in which a workover may be necessary would be an open-hole completion where sand from the producing formation has clogged the tubing end, reducing or completely cutting off the fluid flow from the producing horizon. A workover may also be necessary when the casing has been perforated, and rock or sand particles have clogged the openings in the casing. Either of these cases may require a workover to restore production. These types of workover costs are expensed as production expense, specifically lease operating expense (chapter 8), because production has merely been restored. Workover operations may also involve recompletion in the same producing zone in an effort to restore production. This type of workover is also expensed as lease operating expense. The distinguishing feature of a workover is that there are no new proved reserves nor are reserves reclassified from proved undeveloped to proved developed.

In contrast, a drilling operation whose objective is to add new proved reserves or to access proved undeveloped reserves is to be accounted for as a new drilling, i.e., exploratory or development drilling. When these operations involve reentry into an existing well, they are typically referred to as **recompletions**. For instance, a recompletion operation may involve plugging back and completion at a shallower depth. For example, a well was producing at 8,000 feet. The reserves were depleted, so the well was plugged back to 5,000 feet (where there were reserves "behind the pipe") and completed in that zone. In another recompletion operation, a well was drilled to 8,000 feet, and casing was set to that depth. The well was then completed at 5,000 feet rather than 8,000 feet. Later, a workover rig was brought in to dually complete the well at 8,000 feet.

In both of these examples, the costs would be treated as drilling costs and would be subject to successful efforts *drilling* capitalization rules. These rules would apply because the purpose of the operation was to obtain production from a new formation, not merely to restore production from a formation already producing. In these cases, the costs would be capitalized because the operations were developmental in nature. Similarly, a well might be reentered and deepened below the casing point in the attempt to obtain production from a deeper horizon. Again, the costs would be treated as drilling costs, with the final accounting treatment dependent upon whether the attempt was successful and whether that portion of the well was classified as development or exploratory.

EXAMPLE

Workovers and Recompletions

Tyler Company had the following expenditures during July 2014:

July 10, 2014	Workover costs in connection with well #1036—cleaning and reacidizing producing formation. . .		$ 5,000
July 20, 2014	Recompletion costs on well #1097—testing, perforating, and completion at 8,000 feet. This depth is a new producing formation. Casing was previously set.		
		IDC	20,000
		Equipment	2,000
July 30, 2014	Recompletion completed in deepening well #1102 to a new unproved formation at 9,000 feet. Result was a dry hole at that depth. Tyler continued producing from the formation at 5,000 feet.		
		IDC	50,000
		Equipment	5,000

Entries

July 10, 2014:

Lease operating expense—#1036	5,000	
Cash		5,000

July 20, 2014:

Wells-in-progress—IDC—#1097	20,000	
Wells-in-progress—L&WE—#1097	2,000	
Cash		22,000
Wells and equipment—IDC—#1097	20,000	
Wells and equipment—L&WE—#1097	2,000	
Wells-in-progress—IDC		20,000
Wells-in-progress—L&WE		2,000

July 30, 2014:

Wells-in-progress—IDC—#1102	50,000	
Wells-in-progress—L&WE—#1102	5,000	
Cash		55,000
Dry-hole expense—IDC	50,000	
Dry-hole expense—L&WE	5,000	
Wells-in-progress—IDC—#1102		50,000
Wells-in-progress—L&WE—#1102		5,000

Damaged or lost equipment and materials

Equipment or materials may be damaged or lost during the drilling process. Some examples of damaged equipment would be twisted drillpipe or a broken bit. Examples of lost equipment include parts of the drill bit, hand tools (e.g., wrenches), and drillpipe twisted off downhole. Drilling mud may also be lost into a very porous formation with large cracks or fissures.

The costs of the damaged equipment, lost equipment, or lost drilling mud are costs incurred in the drilling process and are handled in the same manner as other drilling costs. Damaged or lost material or equipment, less salvage value for the damaged equipment, is capitalized if the well is a development well. If the well is an exploratory well, the costs are expensed or capitalized, depending upon whether the well is unsuccessful or successful.

Fishing and sidetracking

When equipment is lost in the hole, recovery of the lost equipment may be attempted through **fishing operations**. If fishing operations are not successful, drilling through or sidetracking around the lost equipment may be necessary. **Sidetracking** involves plugging the lower portion of a well and drilling around the obstruction. Sidetracking is possible because drillpipe is flexible and allows deviations from the vertical. Sidetracking costs are generally considered part of drilling costs and are capitalized or expensed, depending upon whether the well is a development well or an exploratory well and whether proved reserves are found. If the well is an exploratory well, expensing the costs of the abandoned portion of the well regardless of whether proved reserves are found appears consistent with the theory of successful efforts because that portion of the well was abandoned. However, difficulties may be encountered in attempting to allocate drilling costs between the abandoned portion of the well and the portion that is completed. Therefore in practice, the cost of the entire well is typically capitalized if the well is successful.

Abandonment of portions of wells

SFAS No. 19 specifies the treatment for "standard" exploratory wells or development wells. However, many drilling situations, such as those described in the previous paragraphs, do not fall under these guidelines. Several "nonstandard" situations are depicted in the drawings that follow (Fig. 5–3). In all the situations, the shaded portion of the well is the portion for which treatment is not explicitly specified in *SFAS No. 19*.

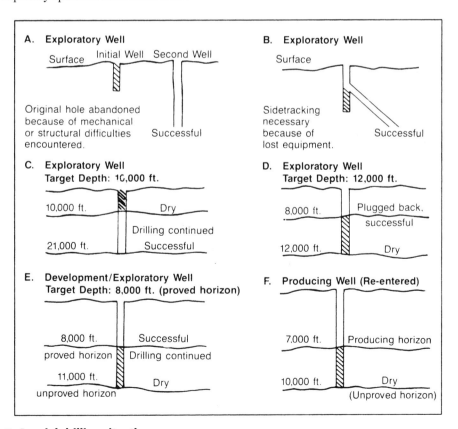

Fig. 5–3. Special drilling situations

In practice there is often disagreement as to whether to expense or capitalize the shaded portion of the wells. In Figure 5–3, situation A depicts an initial exploratory well that was abandoned because of mechanical or structural difficulties, with a second well drilled nearby that was successful. In this situation, the common treatment is to charge the cost of the initial drilling to expense and capitalize the cost of the second well.

Practice varies in relation to situation B. Some companies choose to expense the cost of the abandoned portion of an exploratory well that had to be plugged back and sidetracked around because of drilling difficulties. The argument for this treatment is that the abandoned portion of the well did not add to the value of the sidetracked hole that was completed. Some companies, on the other hand, choose to capitalize the entire cost of the drilling, since the entire drilling effort was necessary in order to find new proved reserves.

Situation D depicts a situation in which the exploratory well was dry at the target depth but was plugged back and completed at a more shallow depth. In practice many argue that the incremental costs of drilling beyond the depth at which the well was completed should be charged to dry-hole expense. The costs of drilling to the completed depth along with completion costs are capitalized. Similarly, in situation E, it can be argued that the incremental costs of drilling beyond the successful target depth to a dry, unproved horizon should be expensed.

If an operator enters a producing well and drills deeper to an unproved horizon but finds no other producing zones as shown in situation F, the costs of drilling to an additional depth should also be charged to expense. In all cases in which additional drilling finds another producible zone, the incremental drilling costs along with completion costs should be capitalized.

Situation C depicts a well that was dry at the target depth but successful at a greater depth. The entire cost should be capitalized because it is necessary to drill to the shallower target depth in order to reach the deeper producible zone.

If a portion of a well is expensed and a portion is capitalized, the question arises as to how the costs should be apportioned. Again, *SFAS No. 19* is silent. Practice is quite varied. Some companies allocate the drilling costs on a per-foot basis, some allocate on a drilling-day basis, some use actual costs incurred, and some use incremental costs, etc.

ADDITIONAL DEVELOPMENT COSTS

In the earlier development drilling example, only development costs directly relating to drilling and completing a well were discussed. *SFAS No. 19* specifies that all development costs should be capitalized as wells and equipment. *Reg. S-X 4-10* defines development costs as:

> **Development costs.** *Costs incurred to obtain access to proved reserves and to provide facilities for extracting, treating, gathering and storing the oil and gas. More specifically, development costs, including depreciation and applicable operating costs of support equipment and facilities and other costs of development activities, are costs incurred to:*

(i) Gain access to and prepare well locations for drilling, including surveying well locations for the purpose of determining specific development drilling sites, clearing ground, draining, road building, and relocating public roads, gas lines, and power lines, to the extent necessary in developing the proved reserves.

(ii) Drill and equip development wells, development-type stratigraphic test wells, and service wells, including the costs of platforms and of well equipment such as casing, tubing, pumping equipment, and the wellhead assembly.

(iii) Acquire, construct, and install production facilities such as lease flow lines, separators, treaters, heaters, manifolds, measuring devices, and production storage tanks, natural gas cycling and processing plants, and central utility and waste disposal systems.

(iv) Provide improved recovery systems.

Service wells, considered to be development costs per the earlier definition, are drilled to support production in an existing field. Examples of service wells include gas injection wells, water injection wells, saltwater disposal wells, and water supply wells. Gas or water injection wells may be used for saltwater disposal, pressure maintenance, or secondary or tertiary recovery purposes. Likewise, costs associated with secondary and tertiary recovery methods, i.e., improved recovery systems, are also considered to be development costs.

EXAMPLE

Additional Development Costs

Tyler Oil Company began installing flow lines. The flow lines cost $10,000, and installation charges were $2,000.

Entry

Wells-in-progress—L&WE	12,000	
Cash		12,000

Installation of the flows lines was completed late in July.

Entry

Wells and equipment—L&WE	12,000	
Wells-in-progress—L&WE		12,000

Support Equipment and Facilities

As discussed in chapter 3, operating costs and depreciation related to support equipment and facilities used in exploration, development, or production activities should be classified as an exploration, development, or production cost, as appropriate.

EXAMPLE

Support Equipment and Facilities

Depreciation on the automobile used by the drilling foreman totaled $5,000 for the year. Since the automobile is used to support the drilling of several wells, the depreciation must be allocated between all of the wells being drilled during the period. Using a mileage driven basis, $1,000 of the depreciation is allocated to Well No. 1. The entry to record the depreciation would be as follows:

Entry

Wells-in-progress—L&WE	1,000	
Accumulated depreciation		1,000

If Well No. 1 is a development well, the depreciation will ultimately be transferred to wells and equipment. If the well is an exploratory well, the costs will ultimately be recorded either as wells and equipment or dry-hole expense, depending upon whether the well finds proved reserves.

Drilling and Development Seismic

SFAS No. 19 indicates that seismic is G&G related and therefore should be charged to expense as incurred. *SFAS No. 19* does not specifically address situations where seismic costs are incurred in relation to drilling and development activities. For example, seismic may be used to determine a specific drillsite. In this situation, many companies treat the seismic as part of the cost of drilling the well. If the well is an exploratory well, the seismic costs would ultimately be recorded either as wells and equipment or dry-hole expense, depending upon whether the well finds proved reserves. If the well is a development well, the seismic costs would ultimately be recorded as wells and equipment. Seismic may also be used in the formulation of overall field development plans. When this occurs, the costs of the seismic are typically treated as a development cost and are capitalized. According to the *2001 PricewaterhouseCoopers Survey of U.S. Petroleum Accounting Practices*, 67% of responding successful efforts companies capitalize seismic costs associated with producing reservoirs as a development cost.[1]

POST–BALANCE SHEET EVENTS

As discussed in chapter 4, post–balance sheet events relate to information about conditions that (a) existed at balance sheet date and that (b) become known after the end of the period but before the financial statements are issued. The drilling of an exploratory well over a fiscal year-end is an example of a potential post–balance sheet event.

As discussed earlier, the costs associated with drilling an exploratory well are capitalized as wells-in-progress until it is determined whether the well has found proved reserves. If the well is dry, the capitalized costs are charged to expense. If the well is successful, the capitalized costs are reclassified as wells and equipment. Normally, all the costs of an exploratory dry hole must be written off in the year in which a well is determined to be dry. However, if a well is determined to be dry after year-end but before the financial statements are issued, the costs incurred prior to year-end should be written off in that fiscal year. Any costs incurred in the second year relating to the well should be charged to expense in the second year (FASB Interpretation 36 to SFAS No. 19).

EXAMPLE

Post–Balance Sheet Events

Tyler Company incurred the following drilling costs on Well #1:

Prior to December 31, 2015:
IDC	$235,000
Equipment	20,000

During January, 2016:
IDC (testing)	30,000

The well was determined to be dry on February 1, 2016, before the financial statements were published. No equipment was salvaged. The well is an exploratory well.

Disposition of the Drilling Costs:

IDC costs of $235,000 and equipment costs of $20,000 should be expensed as dry-hole costs for the year ended December 31, 2015. The $30,000 of IDC incurred in 2016 should be expensed in 2016.

Accounting for Suspended Well Costs

At the time that drilling is completed for an exploratory well or exploratory-type stratigraphic test well, substantial uncertainty may exist regarding whether or not the well has found proved reserves. Since the costs of drilling the well have been capitalized to a wells-in-progress account, it is necessary to make a decision as to whether the costs should stay capitalized or be moved to dry-hole expense. On occasion a well may find oil or gas, but the final decision regarding whether the reserves meet the definition of proved reserves (and thus whether the well is a success) may be delayed for quite some time while other drilling is undertaken or alternatives are evaluated.

In some situations, determination of whether the oil or gas that has been discovered is sufficient to justify the cost of completion and development of the reserves may involve a lengthy process. In some cases, a major capital expenditure such as a pipeline or platform may be required, and additional reserves must be established to justify the expenditure. Occasionally it is necessary to undertake extensive environmental or economic studies or to obtain governmental approvals or to secure project financing. In any event, the decision regarding whether the project can go forward may involve a lengthy period of time.

As originally written, *SFAS No. 19* required that determination of the outcome of drilling the well be made within one year unless fairly limiting criteria were met. If those criteria were met, costs could be capitalized beyond one year. In 2005, the FASB amended *SFAS No. 19* with the issuance of *FSP 19-1*, "Accounting for Suspended Well Costs."

According to *FSP 19-1*, the capitalized costs of drilling an exploratory well or an exploratory-type stratigraphic well may continue to be capitalized if two conditions are met. First, the well has found a sufficient quantity of reserves to justify its completion as a producing well. Second, sufficient progress is being made in assessing the reserves and the economic and operating viability of the project. Both of these criteria must be met. If either criterion is not met, the well is deemed to be impaired, and its costs, net of any salvage value, are charged to expense.

In paragraph 32 of *FSP 19-1*, the FASB provides additional guidance regarding the term *sufficient progress*:

> *All relevant facts and circumstances shall be evaluated when determining whether an enterprise is making sufficient progress on assessing the reserves and the economic and operating viability of the project. The following are some indicators, among others, that an enterprise is making sufficient progress. No single indicator is determinative. An entity should evaluate indicators in conjunction with all other relevant facts and circumstances.*
>
> *a. Commitment of project personnel who are at the appropriate levels and who have the appropriate skills*
>
> *b. Costs are being incurred to assess the reserves and their potential development*

c. An assessment process covering the economic, legal, political, and environmental aspects of the potential development is in progress

d. Existence (or active negotiations) of sales contracts with customers for the oil and gas

e. Existence (or active negotiations) of agreements with governments, lenders, and venture partners

f. Outstanding requests for proposals for development of any required facilities

g. Existence of firm plans, established timetables, or contractual commitments, which may include seismic testing and drilling of additional exploratory wells

h. Progress is being made on contractual arrangements that will permit future development

i. Identification of existing transportation and other infrastructure that is or will be available for the project (subject to negotiations for use).

The previous one-year limitation has been removed from the standard. However, it is not appropriate to delay the determination of the outcome of the well on the chance that the market price of oil or gas will go up or that technological advancements will make the project more operationally viable.

INTEREST CAPITALIZATION

SFAS No. 34, "Capitalization of Interest," requires capitalizing interest as part of the cost of assets that require a period of time to be prepared for their intended use. Essentially, *SFAS No. 34* requires interest capitalization for all qualifying assets, where qualifying assets are defined as assets constructed by an entity for its own use. It is not necessary for borrowings to relate specifically to the qualifying asset, only that interest is being incurred at some level within the enterprise. Either the interest rate on specific borrowings associated with the qualifying asset may be used, or if the borrowings are not related to the specific asset, a weighted average interest rate may be used. To obtain the amount of interest to capitalize each period, the interest rate is applied to the average amount of accumulated capital expenditures. Capitalized interest cannot exceed actual corporate-wide interest costs. The interest capitalization period begins when the three following conditions are met:

1. Expenditures for the asset have been made.
2. Activities necessary to get the asset ready for its intended use are in progress.
3. Interest cost is being incurred.

The term *activities*, used in condition 2, is to be construed broadly, encompassing technical and administrative activities such as obtaining permits. The interest capitalization period should end when the asset is substantially complete and ready for productive use.

Applying this statement to an industry as unique as the oil and gas industry creates interpretation problems. In fact, according to a survey of successful efforts companies, application of this statement has been quite varied.[2] The starting point used has ranged from the time a prospect is acquired to the spud-in date. The stopping point, which is less varied, has ranged from the time proved reserves are found to the time when production begins. Activity cost has included leasehold costs and tangible and intangible drilling costs or IDC and tangible equipment only.

Specifically, capitalized interest is computed as follows:

$$\text{Average accumulated expenditures during construction} \times \text{Interest rate} \times \text{Construction period}$$

The amount of average accumulated capitalized expenditures is computed by adding the beginning balance and ending balance of capitalized expenditures and dividing by two. A simple example of one interpretation of interest capitalization for an oil and gas company follows.

EXAMPLE

Interest Capitalization

Tyler Company acquired an unproved property, Lease A, costing $60,000 on January 1, 2014. During the year 2014, drilling costs are incurred on Lease A in the amount of $300,000. A 10%, $400,000 note is outstanding during the entire year.

Interest to Be Capitalized during 2014

Average accumulated expenditures: $\dfrac{\$60,000 + \$360,000}{2} \times 12/12^* = \underline{\$210,000}$

Interest costs to be capitalized: $\$210,000 \times 10\% = \underline{\$21,000}$

Entry

Wells-in-progress—IDC.	21,000	
Interest expense		21,000

*Because the property was acquired on January 1, 2014, the capitalization period is a full year or 12/12.

In the above entry, wells-in-progress—IDC is debited, although it is unlikely that the same amount would be considered IDC for tax. In practice, when taxes are prepared, companies generally make relevant adjustments to the accounts to accommodate tax and financial accounting differences such as this one.

OFFSHORE AND INTERNATIONAL OPERATIONS

The principles of accounting for offshore and international drilling and development costs are the same as for domestic onshore properties. However, the significant cost of operating offshore and/or in international locations creates special accounting concerns. Drilling is often substantially more expensive because of location and technical adaptations that may be necessary, especially for offshore locations. Drilling crews must be housed, fed, and transported to and from their base location to the drillsite. Costs related to support equipment and facilities such as warehouses, repair facilities, docks, boats, and helicopters for transportation are much higher than for domestic onshore operations. In addition, most, if not all, international operations are taxed according to local laws, and importation of equipment and supplies is typically subject to payment of customs fees and duties.

PROBLEMS

1. Define the following terms:

    dry hole

    wells and equipment—lease and well equipment

    wells and equipment—IDC

    day-rate contract

    footage-rate contract

    turnkey contract

    AFE

    Christmas tree

    sidetracking

2. Which of the following would be IDC?

    a. Labor costs to build a road to the drillsite

    b. Labor costs to build a road to a producing well

    c. Cost of a drillstem test

    d. Cost of surface casing

e. Installation costs of surface casing
   f. Cost of drilling mud
   g. Damages paid to landowner
   h. Cost of flow lines, tanks, and separators
   i. Installation costs of flow lines, tanks, separators
   j. Drilling contractor's charges under a footage-rate contract
   k. Drilling contractor's charges under a turnkey contract
   l. Cost of cement

3. List and briefly discuss the three basic types of drilling contracts.

4. An exploratory-type stratigraphic test well that was drilled offshore discovered proved reserves. However, it was decided that the permanent platform should be placed in a different location. How should the costs of the well be handled?

5. Lomax Oil Company drilled an exploratory well on a lease located in a remote area. The well found reserves, but not enough to justify building a necessary pipeline. The company does not plan to drill any additional exploratory wells at this time. How should the costs of the well be handled?

6. Intercontinental Oil and Gas Company drilled an exploratory well that found reserves, but the reserves could not be classified as proved at that time. No major capital expenditure was required. How should the costs of the well be handled, assuming the company attention is currently focused on other projects?

7. Near the end of 2017, Royalty Corporation drilled an exploratory well that found oil, but not in commercially producible quantities unless the price of oil went up from $100 per barrel to $150 per barrel. Royalty decided to defer classification of the well for up to one year because its financial advisors felt that there was a high probability that oil prices would go up during the next year. Please comment.

8. a. Wildcat Oil Corporation drills an exploratory well during 2016 that finds oil, but not in commercially producible quantities at current oil prices. Since proved reserves are not found, Wildcat expenses the cost of the well in 2016. Early in the next year, but after Wildcat's financial statements have been published, the price of oil goes up so that the reserves found by the exploratory well during 2016 become commercially producible. Should the costs of the well be reinstated?

   b. Assume the same situation except that the price of oil goes up early that next year before Wildcat's financial statements are published. Should the costs of the well be reinstated in this case?

9. Mountain Petroleum had an exploratory well in progress at the end of 2018. Total costs incurred by 12/31/18 were $300,000. During January 2019, drilling was continued, and costs of $200,000 were incurred. Total depth was reached, and the well was determined to be dry by the end of January. Assuming Mountain's financial statements are not published until early February, what costs, if any, should be expensed for 2018 and for 2019?

Chapter 5 • Drilling and Development Costs—Successful Efforts   169

10. Pessimistic Oil Corporation incurred the following costs during 2018:

   a. Began drilling an exploratory-type stratigraphic test well, incurred $50,000 of IDC
   b. Began drilling an exploratory-type stratigraphic test well, incurred $80,000 of IDC and $10,000 in equipment costs
   c. Began drilling a development-type stratigraphic test well, incurred $100,000 of IDC
   d. Began drilling a development-type stratigraphic test well, incurred $200,000 of IDC and $30,000 in equipment costs

   The following results were obtained late in 2018:

   a. The well was determined to be dry.
   b. The well found proved reserves.
   c. The well found proved reserves.
   d. The well was determined to be dry.

   Prepare the necessary entries.

11. Structure Petroleum began in 2015 with the acquisition of four individually significant unproved leases. Give the entries, assuming the following transactions. You may combine entries for items marked with an asterisk (*).

Year	Transaction	Lease A	Lease B	Lease C	Lease D
2015	Lease bonuses*	$ 50,000	$40,000	$ 70,000	$ 55,000
	G&G costs*	$ 60,000	$50,000	$ 75,000	$ 90,000
	Drilling costs IDC Equipment	 None None	Well 1 (exploratory) $300,000 125,000	Well 1 (exploratory) $250,000 50,000	Well 1 (exploratory) $150,000 40,000
	Drilling results	None	Well dry; property impaired 25%	Drilling completed; results undetermined⁺	Drilling not completed
2016	Delay rentals*	$ 4,000	None	$ 3,000	None
	Drilling costs IDC Equipment	Well 1 (exploratory) $275,000 50,000	Well 2 (exploratory) $275,000 50,000	None None	Well 1 $ 60,000 40,000
	Drilling results	Drilling completed; proved reserves	Dry; abandoned lease	Still undetermined; both criteria no longer met	Drilling completed; proved reserves

Year	Transaction	Lease A	Lease B	Lease C	Lease D
2017	Delay rentals*	None	X	3,000	None
	Drilling costs IDC Equipment	Well 2 (development) $300,000 80,000	X	None None	Well 2 (development) $250,000 100,000
	Drilling results	Dry	X	None	Drilling completed; proved reserves

⁺Assume both criteria for deferring classification of the well are met.

12. Record the following transactions:
    a. Landslide Energy incurred costs of $30,000 in preparing a drillsite.
    b. The contractor was paid $400,000 on a day-rate contract (all intangible).
    c. Equipment (casing) costs of $75,000 were incurred.
    d. Costs of $70,000 were incurred in evaluating the well.
    e. Landslide Energy decided to complete the well and incurred costs of $45,000 (perforating and fracturing), $60,000 (cementing), and $100,000 (equipment) in completing the well.

13. Optimistic Oil Company incurred the following costs during the years 2016 and 2017:

    2016
    a. Contracted and paid $50,000 for G&G surveys during the year.
    b. Leased acreage in four areas as follows:
        1) Williams lease—500 acres @ $50 per acre bonus; other acquisition costs, $2,000
        2) Van Dolah lease—800 acres @ $100 per acre bonus; other acquisition costs, $3,000
        3) Sauer lease—200 acres @ $60 per acre bonus; other acquisition costs, $500
        4) Raupe lease—600 acres @ $30 per acre bonus; other acquisition costs, $800

        Each lease had a delay rental clause requiring payment of $2 per acre if drilling was not commenced by the end of one year. Also, each of the above leases was considered individually significant.

    c. The company also leased 10 individual tracts for a total consideration of $60,000. The tracts are considered to be individually insignificant and are the first insignificant unproved properties acquired by Optimistic.

d. The company incurred $1,000 in costs to maintain lease and land records in 2016. Also, costs of $8,000 were incurred to successfully defend a title suit concerning the Williams lease.

e. During 2016, the company incurred the following costs in connection with the Williams lease when drilling an exploratory well:

Roads, location, damages, etc.	$ 18,000
G&G costs to locate the specific drillsite	2,000
Drillpipe	18,000
Conductor casing	10,500
Wellhead equipment	28,000
Contractor's charges and drilling fee (no equipment)	737,000
Equipment rentals	30,000
Water, fuel, power, lubricants	70,000
Drill bits	19,000
Electric logging	20,000
Cement	20,000
Cementing services	15,000
Casing crews	40,000
Surface casing	32,600

Completion costs in connection with the above well were as follows:

Production casing	$151,500
Christmas tree	22,000
Tubing	37,500
Labor for installing casing	20,000
Labor for installing Christmas tree	1,000
Perforating	40,000
Flow lines and tanks	45,000
Labor for installing flow lines	8,000

f. An exploratory well was drilled on the Van Dolah lease in 2016 on a turnkey basis to 9,000 feet. The contractor's charge was $300,000, which included $40,000 for casing. At the end of 2016, a decision had not been made to complete or abandon the well. Both criteria for delaying classification of the well were met.

g. At the end of 2016, the Raupe lease was impaired by 40%, and the Sauer lease by 20%. The company has a policy of maintaining an allowance for impairment equal to 60% of individually insignificant leases.

2017

a. Delay rentals were paid on the Sauer and Raupe leases.

b. Late in 2017, the company abandoned the Sauer lease and two of the individually insignificant leases, which cost a total of $8,000 when acquired. The Raupe lease is now considered to be a very valuable lease, because a large producer was found on adjacent property.

c. At year-end, the company still could not decide whether to complete or abandon the well on the Van Dolah lease, and both criteria for delaying classification of the well were no longer met.

Prepare journal entries for the Optimistic Oil Company's transactions.

14. During the calendar year 2015, Deep Corporation had the following transactions on an unproved property:

   Grant #1 was drilled, with IDC costs of $310,000 and equipment costs of $42,000. The well was determined to be dry and was plugged and abandoned at a cost of $10,000. Salvaged equipment placed in inventory was valued at $8,000.

   Prepare journal entries for the transactions.

15. During 2015, O'Neal Corporation incurred the following costs in connection with the Batch lease:

   a. Acquired the 800 acre Batch lease at a lease bonus of $70 per acre and other acquisition costs of $10,000.

   b. Incurred the following costs in connection with Batch #1:

G&G costs to locate wellsite	$ 3,000
Surface damages	15,000
Surface casing	7,000
Contractor's fee—day rate	175,000
Equipment rentals	200,000
Drilling fluids	35,000
Fuel	9,000
Drill bits	20,000
Cementing services	5,000
Casing crews	6,000
Roustabout labor	8,000
Hauling and transportation	7,500
Production casing	36,000
Tank battery	11,000
Lines and connections	5,500
Pumping unit motor and accessories	50,000

Casinghead and connections	6,500
Tubing	8,500
Separating and treating equipment	7,100
Measuring equipment	300
Downhole pump and rods	5,600

Record the above transactions. **Hint:** The type of equipment installed when the well reached target depth indicates whether the well was successful or dry.

16. The Cross Oil Corporation incurred the following costs and had the following other transactions for the years 2015 and 2016. The company uses the successful efforts method of accounting.

    2015
    a. Paid $100,000 for G&G costs during the year.
    b. Leased acreage in three individually significant areas as follows:
       1) Jones lease—1,000 acres @ $60 per acre bonus, and other acquisition costs of $3,000
       2) Batch lease—800 acres @ a lease bonus of $70 per acre, and other acquisition costs of $10,000
       3) Highland lease—600 acres @ $60 per acre bonus, and other acquisition costs of $8,000
    c. The company also leased 20 individual tracts for a total cost of $80,000. These leases are considered to be individually insignificant and are the first insignificant unproved properties acquired by Cross.
    d. Paid $5,000 in costs to maintain lease and land records in 2015. Also, paid $30,000 to successfully defend a title suit concerning the Batch lease.
    e. Paid the following costs in connection with Batch #1, a successful well:

G&G costs to locate a wellsite	$ 5,000
Location and road preparation prior to spudding-in the well	6,000
Surface damages	16,000
Surface casing	6,000
Contractor's fee (daywork rate)	200,000
Equipment rentals	100,000
Drilling fluids	40,000
Fuel	10,000
Drill bits	20,000
Cementing services	5,000
Roustabout labor	9,000
Hauling and transportation	8,000

Production casing	36,000
Tank battery	11,000
Flow lines and connections	5,000
Pumping unit motor and accessories	50,000
Casing head and connections	7,000
Tubing	13,000
Separating and treating equipment	12,000
Measuring equipment	1,000
Downhole pump and rods	4,000
Testing and acidizing	11,000

   f. An exploratory well was drilled on the Highland lease in 2015 on a turnkey basis to 8,000 feet. The contractor's charge of $400,000 was paid. The charge included $60,000 for casing. At the end of 2015, a decision had not been made to complete or abandon the well. Both criteria for delaying classification of the well were met. At the end of 2015, the Jones lease was impaired 60%, and the Highland lease by 30%. The company's policy is to maintain an allowance for impairment at 70% of the cost of insignificant leases.

2016

   a. Delay rentals of $2,000 were paid on the Jones lease, $1,200 on the Highland lease, and $3,000 on insignificant leases.

   b. During 2016, the Jones lease was abandoned, and three of the individually insignificant leases (cost $8,000) were also abandoned. The Highland lease is now considered to be a very valuable lease, because a large producer was discovered on adjoining land.

   c. At year-end (2016), the company had not made a decision to complete or abandon the Highland well. Both criteria for delaying classification of the well were no longer met.

   REQUIRED: Prepare journal entries for the Cross Oil Corporation's transactions.

17. Support equipment used to drill a development well cost $13,000 and has a 10-year life with a salvage value of $1,000. The equipment was used for three months in drilling Badger #1. Record depreciation. Ignore the wells-in-progress account and use the appropriate final account.

18. An exploratory well that was later determined to be dry had the following costs that are appropriately assigned to the well:

Depreciation on support equipment	$ 3,000
Operating costs of the support equipment (including fuel, maintenance, labor, etc.)	10,000

   Record the amounts.

19. Bartz Corporation paid a seismic crew $2,000 to complete a G&G survey to select a drillsite where Cellar #1 would be spudded-in. Record the above transaction.

20. Joust Oil Company had the following transactions in 2015. Record the transactions.

    a. Acquired an undeveloped lease, $40,000
    b. Paid a drilling contractor as follows:

Footage rate for drilling.	$250,000
Equipment costs (casing)	75,000
Equipment costs (tanks, flow lines, and labor to install equipment)	80,000

    c. Paid costs in evaluating the well, $20,000
    d. Completion costs for fracturing and perforating, $25,000

21. Decade Petroleum incurred and paid the following costs during 2016:

	Lease A Unproved	Lease B Unproved	Lease C Proved	Lease D Proved
Acquisition costs	$ 30,000	$ 35,000	Purchased in 2015	Purchased in 2015
Well:	Exploratory	Exploratory	Development	Development
Drilling contractor's charges—day-rate	180,000	200,000	$160,000	$190,000
Casing	30,000	35,000	28,000	22,000
Production equipment		75,000		80,000
Well logs	10,000	15,000	12,000	14,000
Drilling results	Well dry; abandon lease	Found proved reserves	Well dry	Found proved reserves

    Record Decade Petroleum's transactions.

22. The Kincaid Oil Company has unproved property costs of $40,000 at January 1, 2016. During 2016, Kincaid incurred $400,000 drilling costs on Lease A. An 8%, $500,000 note is outstanding during the entire year and was obtained to finance the drilling program. Compute the interest capitalization amount and record the interest.

23. Indicate whether the following types of expenditures are capitalized (C) or expensed (E) under the acceptable GAAP methods for each well drilled.

	Successful Efforts		Full Cost	
	IDC	L&WE	IDC	L&WE
Exploratory well: Successful				
Dry Hole				
Development well: Successful				
Dry Hole				

24. Duck Petroleum hires a drilling contractor to drill a well to the depth of 8,000 feet at a cost of $300,000. The $300,000 cost includes $5,000 for surface casing. Any drilling to be completed after reaching the 8,000 foot depth is to be paid at a day rate of $4,000 per day. The 8,000 foot depth is reached on September 14, 2017, and the additional drilling to 10,000 feet is completed on October 27, 2017.

Duck incurred additional costs as shown on the following schedule.

REQUIRED: Complete the schedule showing whether the cost is IDC or L&WE.

Description	Amount	IDC	L&WE
1. Rig drilling—turnkey contract	$ *		
2. Rig drilling—day rate	*		
3. Location and road preparation	8,000		
4. Drilling water	7,500		
5. G&G—select drillsite	22,400		
6. Electric logging	1,200		
7. Bits	13,000		
8. Surface damages	6,300		
9. Mud (after 8,000')	40,000		
10. Production casing	35,000		
11. Cementing services	11,000		
12. Casing crews	5,500		
13. Tank battery	11,500		
14. Flow lines and connections	5,000		
15. Separators	7,000		
16. Cement	48,000		
17. Installing separators	4,800		

18. Tubing	30,000
19. Christmas tree	20,000
20. Measuring equipment	500
21. Installing Christmas tree	3,000
22. Equipment rental, drilling	10,000

*To be determined from the information given in the problem statement.

25. Give the journal entries necessary to record the following expenditures made by Lomax Company during 2013:

    a. Recompletion costs on well #560 of $50,000 for IDC and $20,000 for equipment. The well was recompleted at 12,000 feet, which was a new producing formation.

    b. Recompletion costs on well #820 of $90,000 for IDC and $10,000 for equipment. The well was deepened to 15,000 feet to evaluate a new unproved horizon. Proved reserves were not found.

    c. Workover costs on well #310 of $15,000 necessary to restore production after sand had clogged the tubing.

References

1. Institute of Petroleum Accounting. 2001. *2001 PricewaterhouseCoopers Survey of U.S. Petroleum Accounting Practices.* Denton, TX: Institute of Petroleum Accounting. p. 71.

2. Gallun, Rebecca A., Della Pearson, and Robert Seiler. 1983. "Capitalization of Interest by Oil and Gas Companies." *Journal of Extractive Industries Accounting.* Spring. pp. 63–70.

6

PROVED PROPERTY COST DISPOSITION—SUCCESSFUL EFFORTS

The previous chapters outlined which costs should be capitalized and which costs should be expensed under successful efforts accounting. Chapter 4 discussed the acquisition and disposition of unproved property costs. In chapter 5, accounting for the incurrence of drilling and development costs was presented. This chapter deals with the disposition of capitalized costs of proved properties and wells and equipment.

The following T-accounts show the types of costs capitalized as proved properties and wells and equipment. Costs are moved to the proved property classification after proved reserves are discovered on the property. Costs are moved to the wells and equipment classification only after a well has been successfully completed or a dry development well has been plugged and abandoned. Therefore, the cost of wells-in-progress is not included in wells and equipment. Note that the wells and equipment account also includes the cost of installed equipment and facilities.

Proved Property		Lease & Well Equipment	
Acquisition cost of property classified as proved, net of any impairment if property significant		Completed development drilling costs	
		Completed successful exploratory drilling costs	
		Lease equipment and facilities	

Figure 6–1 is a flowchart summary of the four basic types of costs incurred in the exploration and production segment of oil and gas companies and their accounting treatment. Trace through this flowchart for further clarification or as a refresher of how costs become part of proved properties and wells and equipment.

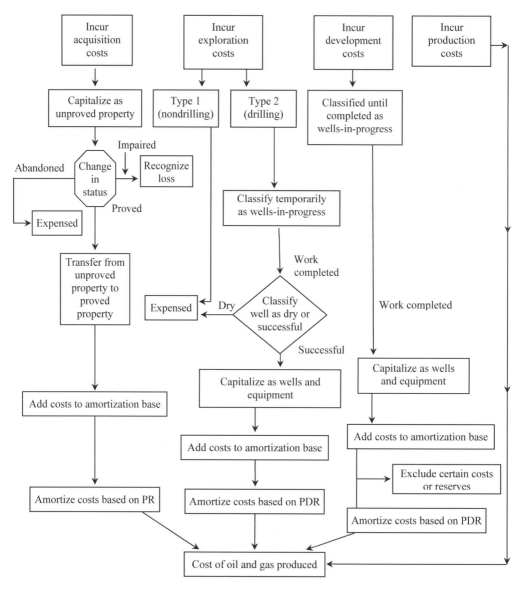

Fig. 6–1. Successful efforts summary

COST DISPOSITION THROUGH AMORTIZATION

In financial accounting, tangible property and equipment is depreciated, intangible property is amortized, and natural resources are depleted. For oil and gas financial accounting, industry professionals commonly refer to depreciation, depletion, and amortization of proved property and wells and equipment as DD&A. Both the SEC and the FASB use the term **amortization** in their written rules; however, the term **DD&A** is most commonly used in practice.

In particular, for successful efforts companies, acquisition costs of proved properties and the costs of wells and equipment are amortized to become part of the cost of oil and gas produced (*SFAS No. 19*, par. 27). Acquisition costs are amortized over **proved reserves**. (See definitions of reserve categories in chapter 2.) Wells and equipment are amortized over **proved developed reserves**.

Acquisition costs represent expenditures made on behalf of the entire cost center and thus apply to all proved reserves that will be produced from that cost center. Proved reserves are the reserves reasonably certain of being produced from a property and include both reserves that will be produced from wells already completed and from wells to be drilled in the future. Therefore, proved reserves should be used to amortize acquisition costs.

Proved developed reserves are reserves that are expected to be produced from existing wells and equipment. Wells and equipment should be amortized using proved developed reserves, because by definition, those are the reserves that will be produced as a result of the costs already incurred for completed wells and equipment. The remaining proved reserves, i.e., the proved undeveloped reserves, are excluded in amortizing wells and equipment, because those reserves will be produced only as a result of incurring additional future costs. When a property is fully developed, proved reserves and proved developed reserves are the same.

Reserves owned or entitled to

Both reserves and production data are used in computing DD&A. The reserves and production used in computing DD&A include only the portion of total proved or proved developed reserves and production that the working interest owner owns or is entitled to. In a basic leasing agreement, an ownership interest in the minerals is conveyed to the working interest owner. Any number of agreements may be then executed in relation to a mineral interest that may alter the amount of production and reserves that go to the working interest owner. A number of these arrangements are discussed in more detail in chapter 13. If the only contract in place is a lease agreement, the working interest would be entitled to an amount of production and reserves equal to one minus the royalty interest. For example, in a simple lease agreement with a royalty interest and a working interest, if the lease calls for a 1/8 royalty interest, the working interest would receive 7/8 of the production and report 7/8 of the reserves.

It is possible for more than one company to share in the working interest in the property. In those cases, each company's share of reserves and production is determined by multiplying the appropriate working interest percentage by the net-of-royalty reserves and production. Determining reserves and production in such a manner is referred to as the **working interest method**, which is illustrated in the following example.

EXAMPLE

Working Interest Method

Alec Brown owns the mineral rights in a property in Oklahoma. Early in 2010, he leased this mineral right, agreeing to a 1/8 royalty. Tyler Company owns 60% of the working interest (WI), and Sharp Oil Company owns 40% of the working interest in the property. In 2011, Tyler and Sharp drilled a successful exploratory well on the property, and by the end of 2012, the property was producing. Total proved reserves are estimated to be 900,000 barrels (bbl), and total proved developed reserves are 350,000 barrels. Total production during 2012 was 20,000 barrels.

Determination of reserves for DD&A:

WI	Gross Proved Reserves (bbl)	Royalty to Brown (bbl)	Net WI Share of Proved Reserves (bbl)	Net WI Share of Proved Developed Reserves (bbl)
Tyler (60%)	540,000	67,500	472,500	183,750
Sharp (40%)	360,000	45,000	315,000	122,500
	900,000	112,500	787,500	306,250

Determination of production for computing DD&A:

WI	Production (bbl)	Royalty to Brown (bbl)	Production in DD&A (bbl)
Tyler (60%)	12,000	1,500	10,500
Sharp (40%)	8,000	1,000	7,000
	20,000	2,500	17,500

In operations outside the United States, contractual arrangements vary widely. These are discussed in more detail in chapter 15. Oftentimes working interest owners earn an interest in reserves by incurring exploration, drilling, development, and production costs. Under these international agreements, the working interest owners do not own the reserves. Rather, they have an economic or entitlement interest in the reserves. In these cases, the working interest share of reserves and revenues are not calculated using the working interest method. Instead, the economic interest method is used. (For a discussion of the economic interest method, see *International Petroleum Accounting*.[1]) *SFAS No. 69* recognizes only the working interest method. However, in 2001, the SEC issued *Interpretations and Guidance*, of which par. (II)(F)(3)(l) indicates that use of the economic interest method is acceptable to the SEC staff.

The specific amounts of reserves and production that the working interest owner owns or is entitled to can be determined only by reference to the terms of the specific agreement. If the contract is a lease agreement, determination of the working interest owner's share of reserves is relatively straightforward.

In determining the reserves and production to use in calculating DD&A and in the discussion of amortization throughout this book, unless otherwise stated, proved reserves, proved developed reserves, and the current year's production refer to the working interest owners' proportionate share of reserves and production calculated using the working interest method.

DD&A calculation

Both acquisition costs and the costs of wells and equipment are amortized using the unit-of-production method, which is as follows:

Unit-of-production formula:

$$\frac{\text{Book value end of period}}{\text{Estimated reserves at beginnning of period}} \times \text{Production for period}$$

Equivalent formula:

$$\frac{\text{Production for period}}{\text{Estimated reserves at beginnning of period}} \times \text{Book value end of period}$$

In the unit-of-production formula, the book value at *period-end* is used. Book value at period-end is the total cost accumulated to period-end, minus accumulated DD&A at the beginning of the period. Consistency would suggest a book value at period-end would be used with a reserve estimate also made at period-end, so that all added reserves found by incurring the costs during the period would be included in the DD&A computation. Instead, the denominator calls for estimated reserves as of the *beginning of the period*. However, rather than using a reserve estimate determined as of the beginning of the period, the most current estimate should be used. Preferably this is an estimate determined as of the end of the period, which includes current period discoveries. The production during the period is then added to the period-end reserve estimate to convert the period-end reserve estimate of reserves in place at period-end into an estimate of reserves in place at the beginning of the period. The resulting estimate of beginning-of-the-period reserves would, therefore, include the reserves discovered by costs incurred during the period. In this way, the most up-to-date estimate is used, and one that reflects the additional reserves found by costs incurred during the period.

Reserve and production data are typically measured in either barrels (bbl) or thousand cubic feet (Mcf). Oil reserves and oil production are estimated and measured in terms of barrels. A barrel is 42 gallons of oil measured at a standard temperature of 60° Fahrenheit (F). Gas reserves and gas production are commonly estimated and measured in terms of Mcf. An Mcf is 1,000 cubic feet of gas measured at a temperature of 60° F under pressure of 14.73 psia (pounds per square inch absolute).

EXAMPLE 1

DD&A

Tyler Oil Company drilled the first successful well on Lease A early in 2012. The company plans to develop this lease fully over the next several years. Data for the lease as of December 31, 2012, are as follows:

Leasehold cost (acquisition costs—proved property)	$ 50,000
IDC (wells and equipment)	90,000
Lease and well equipment (wells and equipment)	30,000
Production during 2012	5,000 bbl
Estimated proved reserves, December 31, 2012	895,000 bbl
Estimated proved reserves recoverable from the well, December 31, 2012 (i.e., proved developed reserves)	95,000 bbl

DD&A Calculations

a. First, reserves as of the beginning of the year are determined.

Proved Reserves (bbl):

Estimated proved reserves, 12/31/12—end of year	895,000
Add: Current year's production	5,000
Estimated proved reserves, 1/1/12—beginning of year	900,000

Proved Developed Reserves (bbl):

Estimated proved developed reserves, 12/31/12—end of year	95,000
Add: Current year's production	5,000
Estimated proved developed reserves, 1/1/12—beginning of year	100,000

b. Next, DD&A is calculated using the first unit-of-production formula given previously.

For Leasehold (proved property):

$$\frac{\text{Book value at year-end}}{\text{Estimated proved reserves at beginning of year}} \times \text{Production for year}$$

$$= \frac{\$50,000}{900,000 \text{ bbl}} \times \$5,000 \text{ bbl} = \underline{\underline{\$278}}$$

Alternatively, using the equivalent formula the calculation would be:

$$\frac{\text{Production for year}}{\text{Estimated proved reserves at beginning of year}} \times \text{Book value at year-end}$$

$$= \frac{5,000 \text{ bbl}}{900,000 \text{ bbl}} \times \$50,000 = \underline{\underline{\$278}}$$

For IDC and Lease and Well Equipment (wells and equipment):

$$\frac{\text{Book value at year-end}}{\text{Estimated proved reserves at beginning of year}} \times \text{Production for year}$$

$$= \frac{\$120,000}{100,000 \text{ bbl}} \times 5,000 \text{ bbl} = \underline{\underline{\$6,000}}$$

Alternatively, using the equivalent formula the calculation would be:

$$\frac{\text{Production for year}}{\text{Estimated proved reserves at beginning of year}} \times \text{Book value at year-end}$$

$$= \frac{5,000 \text{ bbl}}{100,000 \text{ bbl}} \times \$120,000 = \underline{\underline{\$6,000}}$$

Entry to record DD&A

DD&A expense—proved property	278	
DD&A expense—wells	6,000	
Accumulated DD&A—proved property		278
Accumulated DD&A—wells		6,000

EXAMPLE 2

DD&A

Data for Tyler Oil Company's partially developed lease as of December 31, 2013, are as follows:

Cost Data:

Lease bonus.	$ 490,000
Other capitalized acquisition costs	50,000
Total leasehold costs at year-end.	$540,000
Accumulated DD&A on leasehold costs at beginning of year	$ 40,000
IDC at year-end	650,000
Accumulated DD&A on IDC at beginning of year.	120,000
Lease and well equipment at year-end.	275,000
Accumulated DD&A on equipment at beginning of year .	50,000

Reserve and Production Data:

Estimated proved developed reserves, 12/31/13	1,750,000 bbl
Estimated proved undeveloped reserves, 12/31/13.	2,200,000 bbl
Production during year.	50,000 bbl

DD&A Calculations

a. To calculate DD&A, first reserves as of the beginning of the year are determined. Proved reserves equal proved developed reserves plus proved undeveloped reserves.

Proved Reserves (bbl):

Estimated proved developed reserves, 12/31/13	1,750,000
Add: Estimated proved undeveloped reserves, 12/31/13	2,200,000
Estimated proved reserves, 12/31/13	3,950,000
Add: Current year's production	50,000
Estimated proved reserves, 1/1/13	4,000,000

Proved Developed Reserves (bbl):

Estimated proved developed reserves, 12/31/13	1,750,000
Add: Current year's production	50,000
Estimated proved developed reserves, 1/1/13	1,800,000

b. Second, year-end costs are determined.

Proved Property Costs:

Leasehold costs at year-end	$540,000
Less: Accumulated DD&A on leasehold costs	40,000
Net leasehold costs	$500,000

Wells and Equipment:

IDC at year-end	$650,000
Less: Accumulated DD&A on IDC	120,000
Net IDC	$530,000
Lease and well equipment at year-end	$275,000
Less: Accumulated DD&A on L&WE	50,000
Net lease and well equipment	$225,000

c. Third, DD&A is calculated.

For Proved Property Costs:

$$\frac{\text{Current year's production}}{\text{Estimated proved reserves 1/1/13}} \times \text{Book value at year-end}$$

$$= \frac{50,000 \text{ bbl}}{4,000,000 \text{ bbl}} \times \$500,000 = \underline{\underline{\$6,250}}$$

Wells and Equipment:*

$$\frac{\text{Current year's production}}{\text{Estimated proved reserves 1/1/13}} \times \text{Book value at year-end}$$

IDC: $\dfrac{50{,}000 \text{ bbl}}{1{,}800{,}000 \text{ bbl}} \times \$530{,}000 = \underline{\$14{,}722}$

L&WE: $\dfrac{50{,}000 \text{ bbl}}{1{,}800{,}000 \text{ bbl}} \times \$225{,}000 = \underline{\$6{,}250}$

Total DD&A Expense = $6,250 + $14,722 + $6,250 = $\underline{\$27{,}222}$

Entry to record DD&A

DD&A expense	27,222	
Accumulated DD&A—proved property		6,250
Accumulated DD&A—IDC		14,722
Accumulated DD&A—L&WE		6,250

---

* Although this example shows DD&A expense computed separately for IDC versus lease and well equipment, separate computation is not necessary.

DD&A on a fieldwide basis

SFAS No. 19 specifies that the cost center for accumulating costs to be amortized is a property or some reasonable aggregation. Any aggregation of properties must be based on a common geological structural feature or stratigraphic condition, such as a reservoir or field (*SFAS No. 19*, par. 30, 35). While DD&A may be computed on a property-by-property or reservoir basis, in practice, the most commonly used cost center is a field. According to *Reg. S-X 4-10*, par. 8, a field is defined as:

> *An area consisting of a single reservoir or multiple reservoirs all grouped on or related to the same individual geological structural feature and/or stratigraphic condition. There may be two or more reservoirs in a field which are separated vertically by intervening impervious strata, or laterally by local geologic barriers, or by both. Reservoirs that are associated by being in overlapping or adjacent fields may be treated as a single or common operational field. The geological terms "structural feature" and "stratigraphic*

condition" are intended to identify localized geological features as opposed to the broader terms of basins, trends, provinces, plays, areas-of-interest, etc.

Since this definition refers to a reservoir, it is important to also understand how a reservoir is to be defined. According to *Reg. S-X 4-10*, par. 9, a reservoir is defined as:

A porous and permeable underground formation containing a natural accumulation of producible oil or gas that is confined by impermeable rock or water barriers and is individual and separate from other reservoirs.

If all the properties in a fieldwide cost center are proved properties, then all of the leasehold costs will be aggregated and amortized or depleted over the total proved reserves of the field or reservoir. If the cost center contains some properties that are proved and some properties that are unproved, only the proved property leasehold costs would be amortized. The total capitalized costs for wells and equipment would also be aggregated and amortized over the total proved developed reserves of the field or reservoir.

EXAMPLE

Fieldwide DD&A

Tyler Company has interests in three leases, all obtained in January 2011, which cover the same field. Two of the leases are proved. The well on the third lease, Lease C, is incomplete and the lease remains unproved. (The costs of the well on Lease C would thus be in wells-in-progress accounts versus wells and equipment accounts.) Tyler decides to use the field as the cost center for DD&A. Information for each lease as of December 31, 2011, is as follows:

	Lease A	Lease B	Lease C
Leasehold costs	$ 50,000	$ 40,000	$43,000
IDC costs	102,000	81,000	80,000
Lease and well equipment costs	40,000	22,000	21,000
Production during 2011	8,000 bbl	6,000 bbl	0
Estimated proved reserves, 12/31/11	950,000 bbl	650,000 bbl	0
Estimated proved developed reserves, 12/31/11	100,000 bbl	50,000 bbl	0

Fieldwide Production and Reserves:

	Production	Proved Reserves, 12/31/11	Proved Developed Reserves, 12/31/11
Lease A	8,000 bbl	950,000 bbl	100,000 bbl
Lease B	6,000 bbl	650,000 bbl	50,000 bbl
Total	14,000 bbl	1,600,000 bbl	150,000 bbl

Proved Property DD&A: Leasehold costs to be amortized equal $90,000, the costs of Lease A and Lease B. Lease C is unproved as of 12/31/11 and so is not included in amortization.

$$\frac{\$90,000}{1,600,000 \text{ bbl} + 14,000 \text{ bbl*}} \times 14,000 \text{ bbl} = \underline{\$781}$$

Wells and Equipment DD&A: IDC and lease and well equipment costs to be amortized equal $245,000, the costs of IDC and equipment on Lease A and Lease B. The well on Lease C is incomplete and the property not proved, and as such, the well is not included in amortization.

$$\frac{\$245,000}{150,000 \text{ bbl} + 14,000 \text{ bbl*}} \times 14,000 \text{ bbl} = \underline{\$20,915}$$

---

*The 14,000 barrels was added in the denominator of both of these calculations in order to convert end of the year reserve estimates to beginning of the year reserve estimates.

DD&A when oil and gas reserves are produced jointly

Most reservoirs contain both oil and gas. In those cases, the following is necessary:

Convert   oil and gas reserves and oil and gas produced to a common unit of measure based on relative energy content. Energy content is measured by the **British thermal unit (Btu)** and varies somewhat from reservoir to reservoir. Most companies use a generally accepted industry average, converting the Btu content of 1 barrel of oil to the Btu content of 1 Mcf of gas at an approximate rate of 5.8 to 1. In other words, 1 barrel of oil is approximately equal to 5.8 Mcf of gas in terms of energy content. Some companies round up to 6 and use a conversion rate of 6 to 1. The conversion can be made by either dividing the quantity of gas in Mcf by 5.8 or 6 to get **barrels of oil equivalent (BOE)**, or multiplying the barrels of oil by 5.8 or 6 to get **equivalent Mcfs (Mcfe)**.

However, if   the relative proportion of oil to gas extracted in the current period from the reservoir is expected to remain the same, then amortization *may* be computed based on only one of the two minerals—either oil or gas.

Or, if   either oil or gas clearly dominates both the reserves and the current production based on relative energy content, then amortization *may* be computed based on the dominant mineral only (*SFAS No. 19*, par. 38).

Therefore, if oil and gas reserves are produced jointly, one of three different amortization methods may be used, assuming the conditions given previously are satisfied:

1. Common unit of measure—converting to common energy unit
2. Same relative proportion—using either oil or gas
3. Dominant mineral—using the dominant mineral

Additionally, measurement and pricing units other than Mcfs may be used for gas. For example, gas may be measured and priced in terms of million British thermal units (MMBtu) rather than Mcf. If a measurement unit other than Mcf is used for gas, the conversion ratio must be calculated based on the units of measurement being used. However, the underlying conversion process is the same process as described previously and as illustrated in the following example.

EXAMPLE

Joint Production DD&A

Tyler Oil Company has a *fully developed* producing lease that has both oil and gas reserves. Data for Tyler's cost and working interest share of reserves in the lease are as follows. (In a fully developed lease, the proved reserves and proved developed reserves are the same amount.)

Net capitalized costs, December 31	$2,200,000
Estimated proved developed reserves, December 31:	
Oil	400,000 bbl
Gas	1,800,000 Mcf
Production during the year:	
Oil	50,000 bbl
Gas	240,000 Mcf

1. Assume DD&A is determined based on a common unit of measure, either equivalent barrels or Mcfs. If BOEs are to be used, then gas would be converted to BOEs; if Mcfes are to be used, then oil would be converted to Mcfes. Both annual production and proved and proved developed reserves would be converted. Tyler converts using a ratio of 6 to 1. These alternative calculations are illustrated below:

BOE

Production during year:	
Oil	50,000 bbl
Gas (240,000/6)	<u>40,000 BOE</u>
Total	90,000 BOE
Proved developed reserves, 12/31:	
Oil	400,000 bbl
Gas (1,800,000/6)	<u>300,000 BOE</u>
Total	700,000 BOE

Mcfe

Production during year:	
Oil (50,000 × 6)	300,000 Mcfe
Gas	<u>240,000 Mcf</u>
Total	540,000 Mcfe

Proved developed reserves, 12/31:

Oil (400,000 × 6) . . . . 2,400,000 Mcfe

Gas . . . . 1,800,000 Mcf

Total . . . . 4,200,000 Mcfe

DD&A Calculation

Using BOE:

$$\frac{\$2,200,000}{700,000 \text{ BOE} + 90,000 \text{ BOE}} \times 90,000 \text{ BOE} = \underline{\$250,633}$$

Using Mcfe:

$$\frac{\$2,200,000}{4,200,000 \text{ Mcfe} + 540,000 \text{ Mcfe}} \times 540,000 \text{ Mcfe} = \underline{\$250,633}$$

Entry

DD&A expense 250,633

    Accumulated DD&A 250,633

2. Instead, assume the relative proportion of oil to gas extracted in the current period is expected to remain the same and that Tyler Company decides to calculate DD&A using only gas.

DD&A Calculation Computed Using Only Mcf

$$\frac{\$2,200,000}{1,800,000 \text{ Mcf} + 240,000 \text{ Mcf}} \times 240,000 \text{ Mcf} = \underline{\$258,824}$$

Entry

DD&A expense 258,824

    Accumulated DD&A 258,824

3. Instead, assume that oil is clearly the dominant mineral and that Tyler Company decides to calculate DD&A based on the dominant mineral. Therefore, only oil reserves should be used in computing DD&A.

DD&A Calculation Using Only Oil Barrels

$$\frac{\$2,200,000}{400,000 \text{ bbl} + 50,000 \text{ bbl}} \times 50,000 \text{ bbl} = \underline{\$244,444}$$

Entry

DD&A expense	244,444	
Accumulated DD&A		244,444

Estimated future dismantlement, site restoration, and abandonment costs

Dismantlement, restoration, and abandonment costs are costs a company incurs at the end of the life of a property when it is abandoned. At that time, equipment and facilities must be dismantled and the environment returned to its original condition. Originally, *SFAS No. 19* required companies to incorporate estimated future decommissioning costs, i.e., dismantlement, restoration, and abandonment costs, into their calculation of DD&A. This provision was amended in 2001 when the FASB issued *SFAS No. 143*, "Accounting for Asset Retirement Obligations (AROs)." *SFAS No. 143* applies to both successful efforts and full cost companies and requires up-front recognition of future AROs.

Under *SFAS No. 143*, a company is required to recognize all obligations associated with the future retirement of tangible long-lived assets. *SFAS No. 143* defines the term **retirement** as the other-than-temporary removal of a long-lived asset from service, including the sale, abandonment, or other disposal of the asset. Upon initial recognition of a liability for retirement obligations, a company must also capitalize the fair value of the estimated AROs as part of the cost basis of the related long-lived asset. As such, the ARO is allocated to expense through DD&A over the useful life of the asset. Changes in the estimated fair value that result from revised estimates of the amount or timing of the cash flows required to settle the future liability are recognized by increasing or decreasing the carrying value of the ARO liability and the related long-lived asset. When this occurs, the DD&A rate must be revised to reflect the change in the net book value of the underlying asset.

Changes in estimated AROs that result solely from the passage of time are referred to as **accretion of the discount**. Accretion of discount is recognized as an increase in the carrying value of the liability and as an expense, referred to as **accretion expense**, which is classified as an operating item in the income statement. Thus, changes in the liability related to the passage of time do not affect the carrying value of the asset or the related DD&A. *SFAS No. 143* is discussed in detail in chapter 9.

Exclusion of costs or reserves

Generally, the capitalized costs of all wells and equipment in a cost center are to be amortized over the proved developed reserves recoverable from that cost center. This requirement poses a problem in situations where significant costs of a development project have been incurred, but a portion of the proved reserves are still undeveloped.

For example, assume that an offshore platform capable of producing 5,000,000 barrels of oil is constructed at a cost of $5,000,000. However, at the time the platform goes into service, only 2 of the anticipated 10 wells (each capable of producing 1/10 of the proved reserves) have been drilled. The $5,000,000 cost would be capitalized; however, only 2/10 of the reserves are proved developed (8/10 of the proved reserves are proved undeveloped). Since the cost of wells and equipment are to be amortized over proved developed reserves, unless an adjustment is made, the $5,000,000 cost would be amortized over only a portion (2/10) of the related reserves. Consequently, unless a portion of the cost of the platform is excluded from the DD&A calculation, costs and reserves will be mismatched in the DD&A calculation.

Therefore, in situations where a significant development expenditure (i.e., the platform) has been made, but a portion of the related proved reserves are not yet developed, *SFAS No. 19* requires a portion of those development costs be excluded in determining the DD&A rate. The exclusion of a portion of the significant development costs is to continue until additional wells are drilled and all of the related reserves are proved developed (*SFAS No. 19*, par. 35).

Similarly, if proved developed reserves will be produced only after significant additional development costs are incurred, i.e., an improved recovery system, then those proved developed reserves *must* be excluded in determining the DD&A rate (*SFAS No. 19*, par. 35). In this case, the mismatching occurs because the reserves are known, but the costs have not yet been incurred. Note, however, that proved developed reserves are defined to be reserves that are expected to be recovered from existing wells with existing equipment and operating methods. Thus, if the proved reserves are developed, then the related development costs have normally already been incurred. However, when an improved recovery system is involved, reserves may be classified as proved developed reserves after testing by a pilot project or after the operation of an installed program has confirmed that increased recovery will result. Thus, by definition, the situation of proved developed reserves being produced only after significant future development costs are incurred should not arise unless an improved recovery system is involved.

EXAMPLE

Exclusion of Costs

Tyler Oil Company has constructed an offshore drilling platform costing $18,000,000. At the end of 2012, only 2 wells have been drilled at a cost of $3,000,000, with 16 more wells to be drilled in the future. Proved developed reserves at year-end were 3,000,000 barrels, and 300,000 barrels were produced during the year.

DD&A Calculation

A portion of the drilling platform must be excluded in computing DD&A. Of the 18 total wells to be drilled, only 2 have been completed; therefore, 16 will be completed in the future. In this case, 16/18 of the $18,000,000 platform cost would be excluded.

$$\text{DD\&A expense} = \frac{(\$18,000,000 \times 2/18) + \$3,000,000}{3,000,000 \text{ bbl} + 300,000 \text{ bbl}} \times 300,000 \text{ bbl} = \underline{\$454,545}$$

In the preceding example, the amount of development costs to be amortized was determined based on the ratio of wells already drilled over total wells, both drilled and yet to be drilled. Other reasonable methods that may be used include drilling costs incurred over total expected drilling costs or proved developed reserves over total proved reserves.

The following example illustrates the situation in which proved developed reserves will be produced only after significant additional development costs are incurred.

EXAMPLE

Exclusion of Reserves

Tyler Oil Company has an offshore lease that has proved developed reserves of 50,000,000 barrels at the beginning of the year. Of those 50,000,000 barrels, 10,000,000 are associated with significant development costs to be incurred in the future. Total net capitalized drilling and equipment costs (i.e., wells and equipment) at the end of the year are $3,000,000. Production during the year was 250,000 barrels.

DD&A Calculation

Proved developed reserves associated with the future development costs must be excluded in calculating DD&A.

$$\text{DD\&A expense} = \frac{\$3,000,000}{50,000,000 \text{ bbl} - 10,000,000 \text{ bbl}} \times 250,000 \text{ bbl} = \underline{\$18,750}$$

Note that because the reserve estimate is already as of the beginning of the year, production is not added in the denominator of the DD&A calculation.

Under no circumstances should *future* development costs be included in computing DD&A expense under successful efforts accounting.

Depreciation of support equipment and facilities

When support equipment and facilities are acquired, they must be analyzed to resolve two critical issues:

- Will the equipment or facility serve only one cost center?
- What activities (property acquisition, exploration, drilling, production, etc.) will the equipment or facility support?

Support equipment and facilities that service a particular field or other area constituting a cost center should be capitalized and depreciated using the unit-of-production method over the proved developed reserves of the cost center. This situation poses little difficulty either from a theoretical or practical standpoint. When support equipment and facilities cannot be identified with a single field or cost center, the unit-of-production method may not be appropriate. For example, a warehouse may service numerous fields. Since the warehouse is not related to a particular field or cost center, use of the unit-of-production method on a fieldwide basis is not feasible. Instead, the straight-line method, unit-of-output method (based on a relevant measure of usage), sum-of-the-years-digits method, or some other acceptable method should be used. In the past, the SEC has permitted companies to use total proved reserves aggregated over the fields being served (versus proved developed reserves) as an alternative means of depreciating such support equipment and facilities.

Similar issues arise when support equipment and facilities are used for more than one activity (i.e., property acquisition, exploration, development, and production) on multiple cost centers. For example, a warehouse might store equipment used in both drilling and production activities for all of the properties in a large geographical area. Similarly, a truck might be used by the foreman to travel between drilling locations and production locations covering a large geographical area. A field office building could be used to support all types of activities for numerous properties.

In each of these cases, it is not logical to depreciate the equipment or buildings using the unit-of-production method on a fieldwide basis. In such cases, use of a method other than unit-of-production is permitted. The depreciation would logically then be allocated to the activities served based on usage and either capitalized or expensed, depending on the activity. For example, if a truck is serving production activities, drilling activities, and development activities in multiple fields or cost centers, it would be depreciated using straight-line, unit-of-output, or some other method. The depreciation would then be allocated to the activities being served (i.e., production, drilling, and development) and to the different properties. The portion allocated to production activities would be written off as operating expense. The portion allocated to nondrilling development costs would be capitalized to wells and equipment for each of the cost centers the truck is serving. The cost would then be amortized along with the other capitalized costs for the cost center using the unit-of-production method. The portion allocated to drilling would be capitalized to wells-in-progress for the wells being drilled in the various cost centers being served. The wells-in-progress accounts would subsequently be cleared to dry-hole expense if the wells to which the cost was allocated were dry exploratory wells, or wells and equipment if the wells being drilled were development wells, service wells, or successful exploratory wells.

Support equipment and facilities serving multiple activities within a single cost center are typically capitalized directly to the cost center and amortized with along with the wells and equipment using unit-of-production over the life of the cost center. Depreciation of support equipment and facilities should theoretically be allocated to exploration, development, or production as appropriate. However, in practice, the depreciation on the support equipment and facilities is not always separated and allocated between multiple activities. Instead, it is often expensed as DD&A expense. This treatment may result in minor differences in the timing of expense recognition, but these differences are not likely to be material.

Buildings and equipment that cannot be related to specific activities (i.e., the home office building) should be depreciated using straight-line, unit-of-output, or some other method. Such depreciation may appropriately be treated as a component of general administrative overhead.

EXAMPLE

Depreciation of Support Equipment and Facilities

Assume Tyler owns the following assets that are used to support operations in various cost centers during 2015:

Warehouse: Purchased on 1/1/13 for $100,000 with an expected life of 20 years. Approximately 80% of the equipment stored at the warehouse is ultimately used in production operations and 20% in drilling operations. During the current year, 10 production operations were served, and five wells were drilled. Ignore salvage value.

$$\text{Depreciation for the year} = \frac{\$100,000}{20 \text{ years}} = \underline{\$5,000/\text{yr}}$$

The $5,000 depreciation would be allocated to production operations and drilling operations as follows:

Production:	$5,000	×	80%	=	$4,000
Drilling:	$5,000	×	20%	=	$1,000

The $4,000 related to production operations would then be allocated to the specific properties served and recorded as operating expense. The $1,000 related to drilling operations would be allocated to the specific wells being drilled. The $1,000 would be capitalized to wells-in-progress and cleared according to whether the wells were exploratory-type or development-type wells and whether they were successful or dry.

Automobile: Purchased on 5/3/13 for $30,000. It is estimated that the automobile will be driven a total of 100,000 miles during its useful life. Accumulated depreciation is $5,000. This year the automobile was driven 5,000 miles related to property acquisition (50% of the properties under consideration were leased), 10,000 miles related to production activities, and 5,000 miles related to the drilling of development wells.

$$\text{Depreciation per mile} = \frac{\$30,000}{100,000 \text{ miles}} = \underline{\$0.30/\text{mile}}$$

Property acquisition: 5,000 miles × $0.30/mile = $1,500. The $1,500 would be allocated to the properties that were considered and then would be capitalized if the properties were leased, or charged to expense if the properties were not leased.

Production: 10,000 miles × $0.30/mile = $3,000. The $3,000 would be allocated to the producing properties being served and then charged to operating expense.

Drilling: 5,000 miles × $0.30/mile = $1,500. The $1,500 would be allocated to the wells that were being drilled and then capitalized (since the wells were development wells).

Cost disposition—nonworking interests

Nonworking interests should theoretically be amortized over proved reserves using the same unit-of-production formula utilized for working interests in proved properties. However, individual nonworking interests may be insignificant in value. Further, the reserve quantity information necessary to compute unit-of-production amortization is often not obtainable from the related working interest owner. Paragraph 30 of *SFAS No. 19* states with regard to royalty interests:

> *When an enterprise has a relatively large number of royalty interests whose acquisition costs are not individually significant, they may be aggregated, for the purpose of computing amortization, without regard to commonality of geological structural features or stratigraphic conditions; if information is not available to estimate reserve quantities applicable to royalty interests owned (paragraph 50), a method other than the unit-of-production method may be used to amortize their acquisition costs.*

Thus, in contrast to working interests, nonworking interests may be amortized using a method other than unit-of-production, such as straight-line. Further, if the nonworking interests are not significant, the interests may be aggregated without regard to commonality. In practice, nonworking interests are commonly aggregated and amortized on a straight-line basis over 8 or 10 years [COPAS *Petroleum Industry Accounting Educational Training Guide* (TR-9)].[2]

The dollar amount of nonworking interest acquisition costs to be amortized depends upon the method of acquisition. Nonworking interests may be acquired in several different ways. When mineral rights are purchased separately from the surface rights, the purchase price is the amount to be amortized. If purchased in fee, the cost of the property acquired in fee should be allocated between the surface and the mineral rights on a fair market value basis. The cost allocated to the mineral rights is the amount to be amortized.

An overriding royalty interest (ORI) is created out of a working interest by being either retained or carved out. An ORI is carved out when the working interest owner sells an ORI to another party or when the working interest owner conveys an ORI to another party for some other reason, such as compensation for services rendered. If awarded for services rendered, the party receiving the ORI should assign the ORI the fair market value of the services rendered. An ORI is created by being retained when the working interest owner sells the working interest and retains an ORI, or when the working interest owner transfers the working interest to another party willing to develop the lease and retains an ORI. The amount assigned to a retained ORI depends upon how it was created. For more details, see chapter 13. An example of an ORI awarded as compensation follows.

EXAMPLE

DD&A on ORI

Paul Jones, a landman, received a 1/10 ORI for his services in obtaining a lease for Tyler Company. The fair market value of Jones's services as a landman is $4,000. Proved reserves related to the ORI at the end of the first year of production (2012) were 18,000 barrels. The ORI's share of production during the year was 2,000 barrels. Journal entries to record the acquisition of the ORI and DD&A for 2012 are as follows:

Entry to record ORI at acquisition

Investment in ORI	4,000	
Revenue		4,000

Entry to record DD&A

DD&A expense—ORI	400	
Accumulated DD&A—ORI		400

DD&A Computation: $\dfrac{2{,}000 \text{ bbl}}{18{,}000 \text{ bbl} + 2{,}000 \text{ bbl}} \times \$4{,}000 = \underline{\underline{\$400}}$

EXAMPLE

Multiple Nonworking Interests

Tyler Oil Company has multiple small nonworking interests located in Texas. The reserve information necessary to compute unit-of-production DD&A is not available. Further, the interests are individually insignificant, and the benefit received from computing DD&A for the interests on an individual basis does not justify the cost. The interests have a total cost of $60,000. Tyler amortizes such interests straight-line over 10 years.

DD&A Computation: $\dfrac{\$60,000}{10 \text{ years}} = \underline{\$6,000/\text{yr}}$

Another type of nonworking interest is a production payment interest. A production payment interest, which generally has a shorter life than the total life of the reservoir, is not always an economic interest for which amortization may be recognized. A production payment interest that is an economic interest is one that is acquired by purchasing a portion of proved reserves. The cost paid for the proved reserves should be amortized as production takes place, using total *purchased* reserves as the base. An illustration of a production payment interest follows. (Production payments are discussed in greater detail in chapter 13.)

EXAMPLE

DD&A for Production Payment Interest

Paul Jones purchased a 10,000 barrel production payment interest for $1,000,000 from Tyler Company. The production payment is to be paid at the rate of 1/4 of the working interest owner's share of production from Lease B. Lease B is burdened with a 1/6 royalty interest. Gross proved reserves at 12/31/12 were 40,000 barrels. Gross production from Lease B during 2012 was 6,000 barrels. DD&A and entries for Jones for 2012 are as follows:

DD&A Computations:

Production for 2012	6,000 bbl
Less: royalty (1/6 × 6,000).	<u>1,000 bbl</u>
Production net of royalty	5,000 bbl
Production to production payment interest (1/4 × 5,000)	<u>1,250 bbl</u>

$$\text{DD\&A for production payment interest:} \quad \frac{1{,}250 \text{ bbl}}{10{,}000 \text{ bbl}} \times \$1{,}000{,}000 = \underline{\$125{,}000}$$

In the DD&A computation, 10,000 barrels are used rather than a number based on the 40,000 barrels of proved reserves, because 10,000 is the number of barrels purchased by Paul Jones. Thus, DD&A is based on barrels received during the year relative to the total number to be received.

Entries

Investment in production payment interest	1,000,000	
Cash		1,000,000
DD&A expense—production payment.	125,000	
Accumulated DD&A—production payment.		125,000

Revision of DD&A rates

Reserves estimates that affect DD&A rates are required to be reviewed at least annually. Any resulting revisions to DD&A rates should be accounted for prospectively as a change in estimate (*SFAS No. 19*, par. 35). In other words, no changes should be made to adjust accounts to what they would have been had the new estimate been used throughout the period. If a company reports on a yearly basis, the revised reserve estimate is used in the year-end DD&A calculation. If a company reports on a quarterly basis, then the effect of a change in an accounting estimate should be accounted for in the period in which the change is made. The SEC has indicated that when reserve estimates are revised prior to the release of operating results for a quarter, they do not object to the reserve revisions being reflected in DD&A as of the beginning of that quarter, rather than delaying implementation until the following quarter. However, taking the reserve revisions back to earlier quarters is not appropriate [SEC *Accounting and Financial Reporting Interpretations and Guidance* (March 31, 2001) par. (II)(F)(6)].

The following example illustrates one method used in revising DD&A rates when a new reserve estimate is obtained. In the method shown, the fourth-quarter DD&A amount—the quarter in which the revised estimate was received—is determined by using the new reserve estimate converted to an estimate as of the beginning of the year. Note that, as would be the case with a change in estimate, DD&A amounts for the first three quarters remain unchanged. DD&A for the year equals the sum of the four quarterly amounts.

EXAMPLE 1

Revision of DD&A Rates

Tyler Oil Company reports on a quarterly basis. On December 2, 2012, the company received a new reserve report dated November 30, 2012, concerning a fully developed lease in Texas. The reserve report showed proved developed reserves of 450,000 barrels. The last reserve report, dated December 31, 2011, showed reserves of 400,000 barrels. Net capitalized costs as of December 31, 2011, were $1,000,000. Production and amortization through the third quarter of 2012 were as follows:

Quarter	Production	Amortization	Calculations
1	20,000 bbl	$50,000	$1,000,000/400,000 × 20,000
2	16,000 bbl	40,000	$1,000,000/400,000 × 16,000
3	22,000 bbl	55,000	$1,000,000/400,000 × 22,000

October and November production	10,000 bbl
December production	13,000 bbl
Total fourth-quarter production	23,000 bbl

DD&A Calculations

DD&A for fourth quarter:

Reserve estimate as of the beginning of the year using the new estimate:

Reserve estimate, November 30, 2012	450,000 bbl
Add: Production 1st quarter	20,000
Production 2nd quarter	16,000
Production 3rd quarter	22,000
Production during October and November	10,000
Reserve estimate, January 1, 2012	518,000 bbl*

* The new reserve estimate is an estimate of reserves in place as of November 30, 2012. Thus, to obtain estimated reserves as of the beginning of the year (518,000 barrels), production from the first of the year through November must be added to the new reserve estimate.

$$DD\&A \text{ for fourth quarter} = \frac{\$1,000,000}{518,000 \text{ bbl}} \times 23,000 \text{ bbl} = \$44,402$$

The fourth-quarter computation of DD&A expense treats the fourth quarter as a discrete time period.

DD&A for full year:

Amortization for the first three quarters:	$ 50,000
	40,000
	55,000
Amortization for the fourth quarter	44,402
Amortization for the full year	$189,402

Amortization for the year includes the first three quarters' computations (unchanged), plus the fourth quarter amount computed using the new reserve estimate.

Another widely accepted method for revising DD&A rates involves computing the entire DD&A for the year by using the new estimate. The previously recognized quarterly amounts are then subtracted to arrive at the current quarterly amount to be recognized. The following example illustrates this method using exactly the same data as in the previous example.

EXAMPLE 2

Revision of DD&A Rates

Tyler Oil Company reports on a quarterly basis. December 2, 2012, the company received a new reserve report dated November 30, 2012, concerning a fully developed lease in Texas. The reserve report showed proved developed reserves of 450,000 barrels. The last report, dated December 31, 2011, showed reserves of 400,000 barrels. Net capitalized costs as of December 31, 2011, were $1,000,000. Production and amortization through the third quarter of 2012 were as follows:

Quarter	Production	Amortization	Calculations
1	20,000 bbl	$ 50,000	$1,000,000/400,000 × 20,000
2	16,000 bbl	40,000	$1,000,000/400,000 × 16,000
3	22,000 bbl	55,000	$1,000,000/400,000 × 22,000
	58,000 bbl	$145,000	

October and November production	10,000 bbl
December production	13,000 bbl
Total fourth quarter production	23,000 bbl

Production for year:

Quarter	Production
1	20,000 bbl
2	16,000 bbl
3	22,000 bbl
4	23,000 bbl
Total production for year	81,000 bbl

DD&A Calculations

Reserve estimate as of the beginning of the year using the new estimate:

Reserve estimate, November 30, 2012	450,000 bbl
Add: Production 1st quarter	20,000
Production 2nd quarter	16,000
Production 3rd quarter	22,000
Production during October and November	10,000
Reserve estimate, January 1, 2012	518,000 bbl*

* The beginning of the year reserve estimate, 518,000 barrels, is computed exactly the same as in the previous example.

$$DD\&A \text{ for the full year} = \frac{\$1,000,000}{518,000 \text{ bbl}} \times 81,000 \text{ bbl} = \underline{\$156,371}$$

DD&A for fourth quarter:

Amortization for the first three quarters	$ 50,000
	40,000
	55,000
	$145,000
DD&A for year	$156,371
Less: Total DD&A for first three quarters	(145,000)
DD&A for fourth quarter	$ 11,371

DD&A for the first three quarters is unchanged. The fourth quarter DD&A amount is determined by subtracting total DD&A for the first three quarters from DD&A for the full year.

Cost Disposition through Abandonment, Retirement, or sale of Proved Property

When proved property or wells and equipment are abandoned, no gain or loss is normally recognized until the last well ceases to produce and the entire amortization base is abandoned. Until that time, any well, item of equipment, or lease that is abandoned and that is part of an amortization base should be treated as fully amortized and charged to accumulated DD&A. However, if the abandonment or retirement results from a catastrophic event, a loss should be recognized (*SFAS No. 19*, par. 41):

> *Normally, no gain or loss shall be recognized if only an individual well or individual item of equipment is abandoned or retired or if only a single lease or other part of a group of properties constituting the amortization base is abandoned or retired as long as the remainder of the property or group of properties continues to produce oil or gas. Instead, the asset being abandoned or retired shall be deemed to be fully amortized, and its cost shall be charged to accumulated depreciation, depletion, or amortization. When the last well on an individual property (if that is the amortization base) or group of properties (if amortization is determined on the basis of an aggregation of properties with a common geological structure) ceases to produce and the entire property or property group is abandoned, gain or loss shall be recognized. Occasionally, the partial abandonment or retirement of a proved property or group of proved properties or abandonment or retirement of wells or related equipment or facilities may result from a catastrophic event or other major abnormality. In those cases, a loss shall be recognized at the time of abandonment or retirement.*

EXAMPLE

Well Abandonment

Tyler Oil Company abandons a well with total capitalized costs of $500,000. The well is located on a lease with 10 other producing wells. Total accumulated DD&A for wells and equipment on this lease was $1,500,000. No equipment was salvaged.

Entry to record abandonment
Accumulated DD&A—wells   500,000
    Wells and equipment   500,000

Since the well is amortized as part of a larger group of equipment, it is assumed that the accumulated amortization related to that individual well is not determinable. Thus the entire cost of the well, net of any salvage value, is charged against the accumulated DD&A account. However, the entry would have been different if the well were the last one in the amortization base or cost center and the entire amortization base were being abandoned. See next example.

EXAMPLE

Lease Abandonment

Tyler Oil Company abandons a lease with capitalized acquisition costs of $70,000 and capitalized drilling and equipment costs of $200,000. Equipment worth $30,000 was salvaged.

a. The lease was part of a field that had total accumulated DD&A for proved property of $300,000 and total accumulated DD&A for wells and equipment of $500,000. This lease was *not* the last producing lease in the field.

Entry
Materials and supplies (salvage)   30,000
Accumulated DD&A—proved property   70,000
Accumulated DD&A—wells ($200,000 – $30,000) . . . 170,000
    Proved property   70,000
    Wells and equipment   200,000

Again, since this lease is amortized as part of a larger group of properties, it is assumed that the accumulated DD&A related to that lease is not determinable. Thus the entire cost of the lease, net of any salvage value, is charged against accumulated DD&A.

b. The lease was the last producing lease in the field. All other properties have already been abandoned. The balance in accumulated DD&A for proved property was $65,000 and accumulated DD&A for wells and equipment totaled $155,000.

Entry

Materials and supplies (salvage)	30,000	
Accumulated DD&A—proved property	65,000	
Accumulated DD&A—wells	155,000	
Surrendered lease expense	20,000	
Proved property		70,000
Wells and equipment		200,000

When the last well on the last lease in the field or cost center is plugged and the lease is abandoned, then all of the remaining asset accounts and the accumulated DD&A accounts are written off, and a gain or loss is recognized.

If a flood, fire, blowout, or some other catastrophic event occurs, loss recognition is required. In that case, the accumulated DD&A balances would be apportioned between the equipment and IDC that were lost and the remaining equipment and IDC. The net difference between the accumulated DD&A allocated to (a) the lost equipment and IDC and (b) the original cost of the lost equipment and IDC would be recognized as a loss.

When a portion of a proved property or group of properties being accounted for as a single cost center is sold by a successful efforts company, the sale may be treated as a normal retirement with no gain or loss recognized under one condition only. The nonrecognition of gain or loss must not have a material effect on the unit-of-production amortization rate for the cost center. If there would be a material change or distortion in the amortization rate, then a gain or loss should be recognized (SFAS No. 19, par. 47j).

EXAMPLE

Sale

Tyler Company sold a pump costing $2,000 for $200 salvage. The entry to record the sale of the pump would be:

Entry

Cash	200	
Accumulated DD&A—wells and equipment	1,800	
Wells and equipment		2,000

If a single piece of equipment is sold, gain or loss recognition is not required, assuming the amortization rate is not significantly affected. The piece of equipment would be treated as if it were fully depreciated and the cost charged against the accumulated DD&A account.

To illustrate the concepts relating to proved property cost disposition under successful efforts as well as to tie together the material learned in the previous chapters, the following comprehensive example is presented.

COMPREHENSIVE EXAMPLE

a. On February 2, 2011, Tyler Company acquired a lease burdened with a 1/6 royalty for $100,000. The property was undeveloped, and the acquisition costs were considered to be individually significant.

Entry

Unproved property	100,000	
Cash		100,000

b. On February 2, 2012, a delay rental of $3,000 was paid:

Entry

Delay rental expense	3,000	
Cash		3,000

c. Several dry holes were drilled on surrounding leases during 2012. As a result, on December 31, 2012, Tyler Company decided the lease was 40% impaired.

Entry

Impairment expense ($100,000 × 40%)............	40,000	
Allowance for impairment		40,000

d. During 2013, Tyler Company drilled a dry hole at a cost of $350,000 for IDC and $35,000 for equipment. No equipment was salvaged.

Entry to accumulate costs

Wells-in-progress—IDC..................	350,000	
Wells-in-progress—L&WE	35,000	
Cash		385,000

Entry to record dry hole

Dry-hole expense—IDC	350,000	
Dry-hole expense—L&WE	35,000	
Wells-in-progress—IDC.................		350,000
Wells-in-progress—L&WE		35,000

e. Undiscouraged, Tyler Company drilled another exploratory well in February at a cost of $500,000 for IDC and $175,000 for equipment. The well found proved reserves.

Entry to accumulate costs

Wells-in-progress—IDC..................	500,000	
Wells-in-progress—L&WE	175,000	
Cash		675,000

Entry to record completion of well

Wells and equipment—IDC.................	500,000	
Wells and equipment—L&WE	175,000	
Wells-in-progress—IDC.................		500,000
Wells-in-progress—L&WE		175,000

Entry to reclassify property as proved

Proved property	60,000	
Allowance for impairment	40,000	
Unproved property		100,000

f. A *total* of 12,000 barrels of oil was produced from the successful well during 2013. Related lifting costs were $4/bbl. (Lifting costs, which are production costs, should be expensed as lease operating expense. Note that production costs are based on total production from the lease, whereas revenue and the calculation of DD&A expense for Tyler would be based only on the working interest's share of production.)

   Entry

Lease operating expense ($4 × 12,000)	48,000	
Cash		48,000

g. During December Tyler Company began drilling a third exploratory well. Accumulated costs by December 31 were IDC of $100,000 and equipment costs of $15,000.

   Entry

Wells-in-progress—IDC.	100,000	
Wells-in-progress—L&WE	15,000	
Cash		115,000

h. The reserve report as of December 31, 2013 and production during 2013 for Tyler Company were as follows:

Proved reserves	900,000 bbl
Proved developed reserves.	300,000 bbl
Production	10,000 bbl

   DD&A Calculation (assume the lease constitutes a separate amortization base)

   $$\text{Acquisition costs:} \quad \frac{\$60,000}{900,000 \text{ bbl} + 10,000 \text{ bbl}} \times 10,000 \text{ bbl} = \underline{\$659}$$

   $$\text{Wells and equipment:} \quad \frac{\$675,000}{300,000 \text{ bbl} + 10,000 \text{ bbl}} \times 10,000 \text{ bbl} = \underline{\$21,774}$$

   Entry

DD&A expense—proved property	659	
DD&A expense—wells	21,774	
Accumulated DD&A—proved property		659
Accumulated DD&A—wells		21,774

   Note that the wells-in-progress accounts are not amortized.

i. During 2014, Tyler Company completed the third well at an additional cost of $300,000 for IDC and $175,000 for equipment. The well was successful.

Entry to record additional costs

Wells-in-progress—IDC.	300,000	
Wells-in-progress—L&WE	175,000	
Cash		475,000

Entry to record successful well

Wells and equipment—IDC.	400,000	
Wells and equipment—L&WE	190,000	
Wells-in-progress—IDC.		400,000
Wells-in-progress—L&WE		190,000

j. During 2014, a *total* of 36,000 barrels of oil was produced. Related lifting costs were $5/bbl (expense lifting costs).

Entry

Lease operating expense (36,000 × $5)	180,000	
Cash		180,000

k. The reserve report as of December 31, 2014 and production during the year for Tyler Company were as follows:

Proved reserves.	1,470,000 bbl
Proved developed reserves	970,000 bbl
Production	30,000 bbl

DD&A Calculation

Acquisition costs: $\dfrac{\$60{,}000 - \$659}{1{,}470{,}000 \text{ bbl} + 30{,}000 \text{ bbl}} \times 30{,}000 \text{ bbl} = \underline{\underline{\$1{,}187}}$

Wells and equipment: $\dfrac{\$675{,}000 + \$590{,}000 - \$21{,}774}{970{,}000 \text{ bbl} + 30{,}000 \text{ bbl}} \times 30{,}000 \text{ bbl} = \underline{\underline{\$37{,}297}}$

Entry

DD&A expense—proved property	1,187	
DD&A expense—wells	37,297	
Accumulated DD&A—proved property		1,187
Accumulated DD&A—wells		37,297

l. A disaster struck and Tyler Company abandoned the lease. No equipment was salvaged.

Entry

Accumulated DD&A—proved property	1,846	
Accumulated DD&A—wells	59,071	
Surrendered lease expense.	1,264,083	
Proved property		60,000
Wells and equipment—IDC.		900,000
Wells and equipment—L&WE		365,000

The following comprehensive example illustrates accounting for the disposition of costs on a fieldwide basis rather than a lease basis.

EXAMPLE

Comprehensive Field DD&A

Tyler Oil Company computes DD&A on a fieldwide basis. Balance sheet data as of 12/31/13 for Tyler's Texas field are as follows:

Unproved properties, net of impairment		$ 200,000
Proved properties	$ 500,000	
Less: Accumulated DD&A	200,000	
Net proved properties		300,000
Wells and equipment—IDC.	2,100,000	
Wells and equipment—L&WE	800,000	
Less: Accumulated DD&A—wells	850,000	
Net wells and equipment		2,050,000

Tyler's activities during 2014 were as follows:

Unproved properties acquired	$ 50,000
Delay rentals paid	6,000
Test-well contributions paid	30,000
Lease record maintenance, unproved properties	10,000
Title defenses paid	20,000
Unproved properties proved during 2014, net of impairment	60,000
Impairment of unproved properties	40,000
Exploratory dry hole drilled	300,000
Successful exploratory well drilled	500,000
Development dry hole drilled	350,000
Service well drilled	275,000
Tanks, separators, etc., installed	100,000
Development well, in progress 12/31/14	160,000

	Oil (bbl)	Gas (Mcf)
Production	50,000	300,000
Proved reserves, 12/31/14	900,000	3,000,000
Proved developed reserves, 12/31/14	500,000	1,800,000

Additional data: A truck serving this field in a production capacity was driven 4,000 miles during 2014. Total estimated miles for the truck are 50,000. The truck cost $12,000 and salvage value is estimated to be $0. The truck is support equipment and facilities.

Additional data: Tyler also owns a building that houses the corporate headquarters. The operations conducted in the building are general in nature and are not directly attributable to any specific exploration, development, or production activity. Since the building is not related to exploration, development, or production, the building is depreciated using straight-line depreciation. The depreciation is charged to general and administrative overhead and not to the field. The cost of the building was $400,000. The building has an estimated life of 40 years.

DD&A Calculations

	Production		Proved Reserves	
Oil	50,000 bbl	50,000 bbl	900,000 bbl	900,000 bbl
Gas	300,000 Mcf /6	50,000 BOE	3,000,000 Mcf /6	500,000 BOE
BOE		100,000 BOE		1,400,000 BOE

	Proved Developed Reserves	
Oil	500,000 bbl	500,000 bbl
Gas	1,800,000 Mcf /6	300,000 BOE
BOE		800,000 BOE

Leasehold

Costs to amortize:

Proved properties, net at 12/31/13	$300,000
Properties proved during 2014	60,000
	$360,000

$$\frac{\$360,000}{1,400,000 \text{ BOE} + 100,000 \text{ BOE}} \times 100,000 \text{ BOE} = \underline{\$ \ 24,000}$$

Wells

Costs to amortize:

Wells and equipment, net at 12/31/13	$2,050,000
New successful exploratory well	500,000
New development well	350,000
New service well	275,000
New tanks, etc.	100,000
Truck*	12,000
	$3,287,000

$$\frac{\$3,287,000}{800,000 \text{ BOE} + 100,000 \text{ BOE}} \times 100,000 \text{ BOE} = \underline{\$ \ 365,222}$$

*Truck

Since the truck serves only one cost center, in practice it would typically be capitalized directly to the cost center and depreciated using unit-of-production over the life of the cost center as shown above.

Building

$$\frac{\$400,000}{40 \text{ years}} = \underline{\$10,000/\text{yr}}$$

Entry

DD&A expense—proved property	24,000	
DD&A expense—wells	365,222	
G&A overhead expense—(building depreciation)	10,000	
Accumulated DD&A—proved property		24,000
Accumulated DD&A—wells		365,222
Accumulated depreciation—building		10,000

Successful Efforts Impairment

The potential exists in all industries for the recorded net value of assets to exceed their underlying value. The practice of recording impairment (i.e., writing assets down when their net book value exceeds their underlying value) has been required by the SEC for companies using the successful efforts method for a number of years. This requirement existed even though no specific standard was ever issued. In 1995, the FASB changed this situation by issuing *SFAS No. 121*, "Accounting for the Impairment of Long-Lived Assets and for Long-Lived Assets to Be Disposed Of." In 2001, the FASB replaced *SFAS No. 121* with *SFAS No. 144*, "Accounting for Impairment or Disposal of Long-Lived Assets." This standard requires companies in all industries to apply an impairment test. However, companies using the full cost method of accounting are not currently required to apply this standard. Accounting for impairment for full cost companies is discussed in detail in chapter 7. Application of *SFAS No. 144* to successful efforts companies is discussed in detail in chapter 9.

Problems

1. Define the following:

    future decommissioning costs (i.e., future dismantlement, restoration, and abandonment costs)

    common unit of measure based on energy

2. Beaver Oil Corporation drilled its first successful well on Lease A in 2016. Data for Lease A as of 12/31/16 are as follows:

Leasehold costs	$ 50,000
IDC of well	200,000
L&WE	75,000
Production during 2016	8,000 bbl
Total estimated proved reserves, 12/31/16	792,000 bbl
Total estimated reserves recoverable from well, 12/31/16	102,000 bbl

REQUIRED: Compute amortization for 2016.

3. Should amortization always be computed using a common unit of measure based on relative energy when oil and gas reserves are produced jointly? If not, under what circumstances would amortization not be based on units of energy? What basis would be used?

4. Both oil and gas are produced from Core Petroleum's lease in Oklahoma. Additional information as of 1/1/17:

Unrecovered IDC (unamortized IDC)	$900,000
Proved property costs, net	100,000
L&W equipment, gross	300,000
Beginning of year accumulated DD&A equipment	50,000
Estimated proved reserves, 12/31/17	
Oil	200,000 bbl
Gas	1,000,000 Mcf
Production during 2017	
Oil	10,000 bbl
Gas	300,000 Mcf

Assuming the lease is fully developed, compute amortization:

a. Assuming oil is the dominant mineral
b. Using a common unit of measure based on BOE

5. The following are costs incurred on Elizabeth lease:

Acquisition costs	$ 80,000
Well 1 costs	225,000
Well 2 costs	200,000
Well 3 costs	275,000
Tanks, separators, flow lines, etc	100,000
Total wells and equipment	800,000
Total lease, wells and equipment costs	880,000

Treat each of the following independently:

a. A fourth well, an exploratory well, was drilled at a cost of $175,000 and was determined to be dry. Give the entry to record the dry hole.

b. Give the entry to record abandonment of Well 2. Equipment costing $20,000 was salvaged. Accumulated DD&A on wells and equipment was $300,000. Wells 1 and 3 are still producing.

c. Give the entry to record abandonment of the entire Elizabeth lease. Assume the lease constituted a separate amortization base with accumulated DD&A on leasehold costs of $30,000 and accumulated DD&A on wells and equipment of $300,000.

d. Give the entry to record abandonment of the entire Elizabeth lease, assuming instead that amortization had been computed on a fieldwide basis with accumulated DD&A on leasehold costs of $400,000 and accumulated DD&A on wells and equipment of $2,500,000.

6. What purpose does a cost center serve?

7. During 2015, Post Oil and Gas Company completed the last well from its drilling and production platform off the coast of Louisiana. Unrecovered costs not including decommissioning costs on December 31, 2015, were $25 million, including $5 million in acquisition costs and $20 million in drilling and development costs. Total proved developed reserves were estimated to be 600,000 barrels as of January 1, 2015. Production during 2015 was 30,000 barrels. At the end of the life of the reservoir, decommissioning costs are estimated to be $14 million, and salvage value is estimated to be $1 million. Compute DD&A for 2015.

8. Describe how future decommissioning costs are accounted for.

9. Under what circumstances should development costs be excluded in determining DD&A? Under what circumstances should a portion of proved developed reserves be excluded in determining DD&A? Is the exclusion of the development costs or proved developed reserves dependent upon the choice of the company, or is it required by *SFAS No. 19*?

10. Upbeat Energy just completed (December 28, 2017) the successful testing of a tertiary recovery pilot project, and as a result has determined that 900,000 barrels of oil should be classified as proved developed reserves. However, 200,000 of the 900,000 barrels will be produced only after significant future development costs are incurred.

    Calculate DD&A for Upbeat's wells and equipment, assuming net capitalized drilling and equipment costs of $1,850,000 and production of 40,000 barrels.

11. During 2017, Aggie Oil Corporation constructed an offshore production platform at a cost of $25,000,000. In total, 16 wells are planned. As of 12/31/17, only 2 out of the 16 wells had been drilled. Calculate DD&A given the following information:

Leasehold costs	$ 300,000
Drilling costs	2,200,000
Platform costs	25,000,000
Proved reserves, 12/31/17	1,800,000 bbl
Proved developed reserves, 12/31/17	900,000 bbl
Production	100,000 bbl

12. The following information as of 12/31/16 relates to the first year of operations for Complex Oil Company. From the data, (1) prepare entries, and (2) prepare an income statement for Complex Oil Company for 2016, assuming revenue to the company from oil sales is $1,200,000. Expense lifting costs as lease operating expense.

Transactions, 2016	Lease A	Lease B	Lease C
*a. Acquisition costs of undeveloped leases (1/8 RI)	$ 60,000	$ 30,000	$ 20,000
*b. G&G costs	60,000	70,000	90,000
*c. Drilling costs	200,000	200,000	250,000
d. Drilling results:	Drilling completed	Drilling completed; dry	Drilling not completed
Proved reserves	700,000 bbl		
Proved developed reserves	300,000 bbl (as of 12/31)		
e. Production	10,000 bbl		
f. Lifting costs	$ 50,000		
g. December 31	Recorded DD&A	Impaired lease 40%	
Transactions, 2017			
h. Assume on January 2 of the second year (2017) that disaster struck both Lease A and Lease B. Give the entries to record abandonment of Lease A and Lease B. Assume equipment costing $15,000 was salvaged from Lease A. Assume this is not a post–balance sheet event that would give rise to changes in the balance sheets or income statements of previous years.			

*May combine entries for different leases.

13. Balance sheet data for Great Petroleum as of 12/31/15 is as follows for Lease A:

Leasehold costs	$ 100,000
Less: Accumulated DD&A	20,000
Net leasehold costs	$ 80,000
Wells and equipment—IDC	$1,100,000
Less: Accumulated DD&A—IDC	300,000
Net wells and equipment—IDC	$ 800,000
Wells and equipment—L&WE	$ 700,000
Less: Accumulated DD&A—L&WE	50,000
Net wells and equipment—L&WE	$ 650,000

Great Petroleum's activities during 2016 related to Lease A were as follows:

Exploratory dry hole drilled	$ 275,000
Development dry hole drilled	300,000
Tanks, separators, etc., installed	125,000
Production	100,000 bbl
Proved reserves, 12/31/16	1,020,000 bbl
Proved developed reserves, 12/31/16	900,000 bbl

Calculate DD&A for 2016, assuming Lease A constitutes a separate amortization base.

14. Dwight Oil Company computes DD&A on a fieldwide basis. Balance sheet data as of 12/31/15 for Dwight's Anadarko Basin field are as follows:

Unproved properties, net of impairment		$ 200,000
Proved properties	$ 500,000	
Less: Accumulated DD&A	(100,000)	
Net proved properties		400,000
Wells and equipment—IDC	3,000,000	
Wells and equipment—L&WE	1,400,000	
Less: Accumulated DD&A—Wells	(1,300,000)	
Net wells and equipment		$3,100,000

Dwight's activities during 2016 were as follows:

Unproved properties acquired	$ 35,000
Delay rentals paid on unproved properties	10,000
Test-well contributions paid on unproved properties	20,000
Title exams paid on unproved property	8,000
Title defenses paid	15,000
Unproved properties proved during 2016, net of impairment	50,000
Exploratory dry hole drilled	275,000
Successful exploratory well drilled	400,000
Development dry hole drilled	300,000
Service well	325,000
Tanks, separators, etc., installed	125,000
Development well, in progress 12/31/16	140,000

	Oil (bbl)	Gas (Mcf)
Production	100,000	500,000
Proved reserves, 12/31/16	1,020,000	5,000,000
Proved developed reserves, 12/31/16	900,000	4,700,000

REQUIRED: Using BOE:

a. Calculate DD&A for 2016.

b. Calculate DD&A for 2016, assuming that part of the field, a proved property with gross acquisition costs of $30,000, gross equipment cost of $70,000, and gross IDC of $215,000, was abandoned during 2016.

15. Tadpole Oil Company has the working interest in a fully developed lease located in Texas. As of 12/31/17, the lease had proved developed reserves of 1,200,000 barrels and unrecovered costs of $12,000,000. During the third quarter of 2018, a new reserve study was received that estimated proved developed reserves of 1,500,000 barrels as of August 1, 2018.

Calculate DD&A for each quarter, assuming the following production and using the first method illustrated in the chapter.

Quarter	Production
1	30,000 bbl
2	35,000 bbl
July	10,000 bbl
August	15,000 bbl
September	12,000 bbl
4	40,000 bbl

16. Van Dolah Petroleum had the following account balances for the years shown relating to a proved property:

	12/31/15	12/31/16
Proved property cost	$ 30,000	$ 30,000
Accumulated DD&A—proved property	6,000	
Wells and equipment—IDC	350,000	450,000
Accumulated DD&A—Wells and equipment—IDC	60,000	
Wells and equipment—L&WE	250,000	325,000
Accumulated DD&A—Wells and equipment—L&WE	55,000	

	2015	2016
Proved reserves, 12/31	800,000 Mcf	900,000 Mcf
Proved developed reserves, 12/31	500,000 Mcf	700,000 Mcf
Production during 2015 and 2016	40,000 Mcf	60,000 Mcf

Compute DD&A for the year ended 12/31/16.

17. Bayou Oil Corporation had the following information and account balances for the years shown relating to Lease No. 1.

	12/31/15	12/31/16
Proved property—cost	$ 40,000	$ 40,000
Accumulated DD&A—proved property	4,000	
Wells and equipment—IDC	400,000	600,000
Accumulated DD&A—wells and equipment—IDC	60,000	
Wells and equipment—L&WE	300,000	420,000
Accumulated DD&A—wells and equipment—L&WE	45,000	

	2015	2016
Proved reserves, 12/31—Oil	30,000 bbl	50,000 bbl
Gas	450,000 Mcf	600,000 Mcf
Proved undeveloped reserves, 12/31—Oil	10,000 bbl	12,000 bbl
Gas	200,000 bbl	120,000 bbl
Production during 2015 and 2016—Oil	5,000 bbl	7,000 bbl
Gas	50,000 Mcf	70,000 Mcf

REQUIRED: Compute DD&A for the year ended 12/31/16 using:

a. A common unit of measure based on equivalent Mcf

b. Gas as the dominant mineral

c. Same relative proportion

18. Charles Bartz purchases a 1/8 ORI for $10,000. Proved reserves at year-end 2016 are 20,000 barrels, and production for 2016 was 3,000 barrels.

   Prepare journal entries for Charles Bartz and compute DD&A for 2016.

19. Unlikely Corporation has a working interest in Lease A. As of 12/31/15, the lease had reserves as follows:

Proved developed reserves, oil:	120,000 bbl
Proved reserves, oil:	180,000 bbl

   Account balances at 12/31/15 were as follows:

Proved property	$ 120,000
Less: Accumulated DD&A—proved property	40,000
Net proved property	$ 80,000
Wells and equipment	$1,600,000
Less: Accumulated DD&A—wells and equipment	300,000
Net wells and equipment	$1,300,000

   A new reserve estimate on October 31, 2016 was as follows:

Proved developed reserves, oil:	100,000 bbl
Proved reserves, oil:	150,000 bbl

   Calculate DD&A for each quarter of 2016, assuming the following production, using both methods described in the book.

Quarter	Production
1	10,000 bbl
2	12,000 bbl
3	14,000 bbl
October	2,000 bbl
November	4,000 bbl
December	6,000 bbl

20. The Caleb Corporation had the following costs at 12/31/15 relating to Lease A:

Proved property	$ 50,000
Accumulated DD&A—proved property	30,000
Well No. 1: Wells and equipment	400,000
Well No. 2: Wells and equipment	350,000
Accumulated DD&A—wells	500,000

In January, 2016, Well No. 1 quit producing and was abandoned. In March 2016, Well No. 2 quit producing, and the well and the lease were abandoned.

REQUIRED: Prepare journal entries for the abandonments.

21. On January 1, 2018, Gaylene Raupe purchased a 40,000 barrel production payment interest from Watershed Oil Company for $700,000. The production payment will be paid out of 1/5 of the working interest owner's share of production from Lease #1003. The lease is burdened with a 1/7 RI. Gross production from Lease #1003 during 2018 was 28,000 barrels. Gross proved reserves at 12/31/18 were 130,000 barrels.

Give the entries by Gaylene Raupe to record the purchase of the production payment interest and DD&A expense for 2018.

22. Port Oil Corporation uses the successful efforts method. Recently, the company acquired a truck costing $60,000 with an estimated life of five years (ignore salvage value). The foreman drives the truck to oversee operations on seven leases, all in the same general geographical area. The foreman keeps a log of his mileage in order to determine how the truck is utilized. Analysis of the log indicated that 1/3 of the mileage driven was related to travel to drilling operations, 1/3 was related to travel to G&G exploration areas, and 1/3 was related to travel to production locations.

REQUIRED:

a. Give any entries necessary to record depreciation on the truck for the first year that it was in service, assuming the seven leases were located on different reservoirs.

b. Give any entries necessary to record depreciation on the truck for the first year that it was in service, assuming the seven leases were on the same reservoir. The total of net wells and equipment, including the truck above, was $900,000. Proved reserves at the end of the year were 300,000 barrels, proved developed reserves at the end of the year were 120,000 barrels, and production during the year was 30,000 barrels.

23. In 2009, Fitzgerald Smith leased the mineral rights in a property in Wyoming, agreeing to a 1/6 royalty. The joint working interest in the property is owned by the following companies:

Company:	WI
Black Company	70%
Bailey Oil	20%
HC Inc.	10%

In 2010, a successful exploratory well was drilled on the property. On December 31, 2011, the property was producing. On that date, total proved reserves are estimated to be 1,200,000 barrels, and total proved developed reserves are estimated to be 180,000 barrels. Total production during 2011 was 60,000 barrels.

REQUIRED:

a. Compute the proved reserves and proved developed reserves that belongs to the working interest and royalty interest owners.
b. Determine the production that each of the working interest owners would use in computing DD&A.

References

1. Wright, Charlotte, and Rebecca Gallun. 2005. *International Petroleum Accounting.* Tulsa, OK: PennWell.

2. Council of Petroleum Accountants Societies. 1990. *Petroleum Industry Accounting Educational Training Guide* (TR-9). Denver, CO: Council of Petroleum Accountants Societies. April. p. 90.

7
FULL COST ACCOUNTING

Under the full cost accounting method, both successful and unsuccessful costs incurred in the search for oil and gas are considered necessary to finding oil or gas. In other words, it is necessary to explore both productive and unproductive acreage and drill dry holes as well as successful wells in order to find oil and gas reserves. According to this reasoning, both successful and unsuccessful expenditures should be capitalized and amortized over production as part of the cost of the oil and gas. Thus, directly relating costs incurred to specific reserves is not relevant under full cost accounting. In contrast, under successful efforts accounting, only those exploration costs that can be directly related to specific reserves are capitalized.

Some would argue that full cost is a departure from traditional historical cost accounting since nonproductive exploration costs, having no future economic benefit, are capitalized. Others contend that full cost actually produces more meaningful financial statements. Unlike other industries, the primary assets of oil and gas companies are not the property, plant, and equipment, but the oil and gas in the ground. Full cost financial statements are based on the fact that unsuccessful expenditures are a necessary and unavoidable part of discovering those assets.

Recall from chapter 2 that the SEC's position is that neither the full cost method nor the successful efforts method provides sufficient information critical to investors in estimating E&P companies' future cash flows. The SEC indicated that their major objection relates to the fact that both methods focus on the historical cost of finding oil and gas reserves. As such, the financial statements produced by either method do not reflect the future economic benefits to be obtained by producing and selling oil and gas reserves. Consequently, while the SEC permits registrants to use either full cost or successful efforts, their financial statements must be supplemented by the substantial disclosures promulgated by *SFAS No. 69*. (See chapter 14.) The SEC's full cost accounting rules are provided in SEC *Reg. S-X 4-10*.

Under full cost, all costs associated with property acquisition, exploration, and development activities are capitalized. Consequently, G&G studies, delay rentals, and exploratory dry holes are capitalized. Even when property is impaired or abandoned, the costs of impairment or abandonment continue as part of the capitalized costs of the cost center. Any portion of general and administrative costs directly related to acquisition, exploration, and development activities, such as the internal costs of the land department, may also be capitalized. Costs related to production, general corporate overhead, and similar activities are expensed [*Reg S-X 4-10*, 3.18(i)]. General corporate overhead cost normally includes all costs incurred in maintaining the various administrative offices throughout the company, unless those office costs can be identified with acquisition, exploration, and development activities. Examples of general corporate overhead include executive compensation and offices, salaries of accounting personnel and the costs of their offices, and other administrative costs not related to acquisition, exploration, and development activities.

Under full cost accounting, the cost center for accumulating capitalized costs to be amortized is a country, e.g., the United States, Canada, etc. A country also includes any offshore area under the country's legal jurisdiction. Theoretically, only one capital account per cost center is necessary to account for the activities of a full cost company. However, because of management, tax, and regulatory requirements, detailed records similar to those of a successful efforts company are normally kept. As discussed later in this chapter, more detailed record-keeping is also dictated by certain full cost rules that allow exclusion of specific costs from amortization and require a ceiling test to be performed. Exclusion of certain costs from amortization also necessitates identification of unsuccessful costs, such as dry holes and impairment or abandonment of unproved properties.

Figure 7–1 shows the accounting treatment of the four basic types of costs incurred in exploration and production activities under full cost. These four costs—acquisition, exploration, development, and production—are defined for full cost as they are for successful efforts.

Chapter 7 • Full Cost Accounting 229

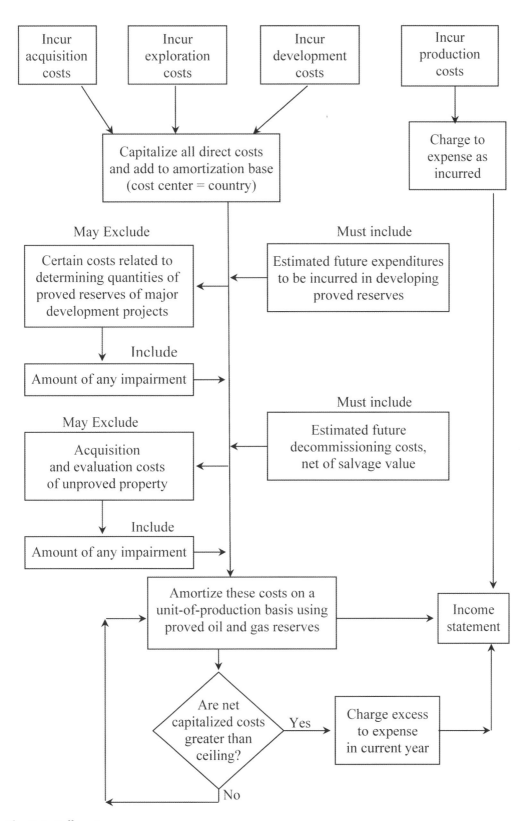

Fig. 7–1. Full cost summary

The following is an example of accounting for costs using the full cost method. In studying the example, compare the accounting treatment under full cost to that under successful efforts—which costs are capitalized versus expensed, as well as account titles used. Because the full cost rules allow exclusion of certain costs from the amortization base, all nondrilling costs that can be associated, when incurred, with specific unproved properties are debited to an unproved property control account. Drilling costs are initially charged to wells-in-progress (exploratory or development) and, if proved reserves are found, transferred to a wells and equipment account. If proved reserves are not found, exploratory wells-in-progress are transferred to exploratory dry holes. Although different accounts are used for successful exploratory wells versus exploratory dry holes, both accounts are asset accounts under full cost accounting and are written off to expense through DD&A. Development wells-in-progress that do not find proved reserves are transferred to a wells and equipment account in the same manner as under successful efforts.

EXAMPLE

Full Cost Entries

a. Tyler Company acquired shooting rights to 10,000 acres, paying $5,000.

Entry

G&G costs—nondirect	5,000	
Cash		5,000

Note that although these G&G costs are recorded as capitalized costs, they are not debited to unproved property control. They are instead recorded as nondirect G&G costs, because they could not be associated with a specific unproved property when incurred.

b. Tyler Company then hired Davis Company to conduct the G&G work and paid the company $30,000.

Entry

G&G costs—nondirect	30,000	
Cash		30,000

c. As a result of the G&G work, Tyler Company decided to lease the following properties:

Lease	Acres	Bonus $/acre	Legal Costs & Recording Fees
A	1,000	$25	$ 500
B	1,500	20	1,000

Entries

Unproved property—Lease A (1,000 × $25 + $500). . .	25,500	
Cash		25,500
Unproved property—Lease B (1,500 × $20 + $1,000). . .	31,000	
Cash		31,000

Note that although the cost center under full cost is a country and detailed records are not required, Tyler's accounting records are kept on a lease basis. When DD&A is computed, all lease costs are aggregated for the specific cost center.

d. During the following year, Tyler Company incurred and paid the following items:

1) Ad valorem taxes of $1,500 on Lease A

Entry

Unproved property—ad valorem taxes—Lease A	1,500	
Cash		1,500

Note that these ad valorem taxes and many of the costs below are nondrilling costs that can be associated with an unproved property when incurred. Therefore, they are debited to an unproved property control account.

2) Dry-hole contribution of $10,000 on Lease B

Entry

Unproved property—test well contribution— Lease B .	10,000	
Cash		10,000

3) Delay rentals on Leases A and B of $3,000 and $4,000, respectively

Entries

Unproved property—delay rental—Lease A.	3,000	
Cash		3,000
Unproved property—delay rental—Lease B.	4,000	
Cash		4,000

4) Nonrecoverable delinquent taxes on Lease B, $2,000 (not specified in lease contract)

Entry

Unproved property—exploration costs—Lease B . . .	2,000	
Cash		2,000

5) Costs of maintaining lease records for Leases A and B were $800 for the year [allocated based on relative acreage: $800/(1,000 + 1,500) = $0.32 per acre]

Entry

Unproved property—maintenance cost—Lease A ($0.32 × 1,000)	320	
Unproved property—maintenance cost—Lease B ($0.32 × 1,500)	480	
Cash		800

Note that the cost of the lease records department should be allocated between lease acquisition and maintenance activities versus those activities that are of a general administrative nature. The costs allocated to lease acquisition and maintenance should be capitalized, and the cost allocated to general administration should be expensed as part of general corporate overhead.

e. At the beginning of the next year, Tyler Company began drilling operations on both Lease A and Lease B. The company incurred the following costs:

Lease	Lease & Well Equipment	IDC
Lease A	$50,000	$200,000
Lease B	20,000	130,000

Entries

Lease A

Wells-in-progress—L&WE	50,000	
Wells-in-progress—IDC	200,000	
Cash		250,000

Lease B

Wells-in-progress—L&WE	20,000	
Wells-in-progress—IDC	130,000	
Cash		150,000

As with the successful efforts method, drilling costs are accumulated in the wells-in-progress accounts until drilling is complete.

f. Assume the well on Lease A is successful, and additional IDC of $10,000 and additional equipment costs of $70,000 are incurred to complete the well.

Entry

Wells-in-progress—L&WE	70,000	
Wells-in-progress—IDC	10,000	
Cash		80,000

Entry to record completion of work on well

Wells and equipment—L&WE	120,000	
Wells and equipment—IDC	210,000	
Wells-in-progress—L&WE		120,000
Wells-in-progress—IDC		210,000

Entry to reclassify lease as proved

Proved property	30,320	
Unproved property		30,320

Unproved Property

25,500	
1,500	
3,000	
320	
30,320	

The separation of property into proved and unproved and the separation of drilling costs into completed and uncompleted is not required under full cost accounting theory. However, this distinction is useful for management needs, income tax accounting, etc.

g. Assume the well on Lease B is determined to be dry and is plugged and abandoned for an additional $3,000.

Entry to record additional costs

Wells-in-progress—IDC	3,000	
Cash		3,000

Entry to record completion of work on well

Exploratory dry holes—L&WE	20,000	
Exploratory dry holes—IDC	133,000	
Wells-in-progress—L&WE		20,000
Wells-in-progress—IDC		133,000

Remember that under full cost, all exploration costs are capitalized whether successful or unsuccessful. However, the costs of the exploratory dry hole are transferred into an account identified as unsuccessful rather than into wells and equipment. This transfer facilitates the use of certain procedures related to exclusion of costs from amortization. The separation of exploratory dry-hole costs is also beneficial for management decision-making purposes and for regulatory purposes.

h. Tyler Company decides to abandon Lease B.

   Entry

   Abandoned leasehold costs   47,480
       Unproved properties 47,480

Unproved Property	
31,000	
10,000	
4,000	
2,000	
480	
47,480	

Remember that abandoned leasehold costs or impairment costs are not expensed but remain capitalized.

Disposition of Capitalized Costs

Under the theory of full cost accounting, all successful and unsuccessful costs associated with acquisition, exploration, and development activities are considered to be part of the cost of any oil or gas found and produced. As such, these costs should be amortized over production. All capitalized costs within a cost center are, with certain exceptions, amortized over *proved reserves* using the unit-of-production method.

If oil and gas are produced jointly, the oil and gas are converted to a common unit of measure based on relative energy content. However, oil and gas prices may be so disproportionate relative to their energy content that the unit-of-production method would result in an improper matching of the cost of oil and gas production against related revenue received. When that is the case, the unit-of-revenue is a more appropriate method of computing amortization. (The revenue method is not allowed under successful efforts accounting.) If the unit-of-revenue method is used, the actual selling price of the oil and gas should be used to value the production during the period. Current prices (period-end) rather than future prices should generally be used in valuing the proved reserves. Future prices should be used only when provided by contractual arrangements.

The basic formula for computing amortization using unit-of-production is essentially the same as the one used for computing amortization under successful efforts:

$$\frac{\text{Capitalized costs at end of period}}{\text{Estimated proved reserves at beginning of period}} \times \text{Production during period}$$

The formula for computing amortization using unit-of-revenue is the following:

$$\frac{\text{Capitalized costs at end of period}}{\text{Estimated proved reserves at beginning of period valued at end of period prices}} \times \text{Production during period valued at actual selling price}$$

The following example illustrates both DD&A computed using unit-of-production and DD&A computed using unit-of-revenue.

EXAMPLE

Full Cost DD&A

Tyler Company has oil and gas properties located only in the United States. Data for Tyler Company as of December 31, 2011, are as follows:

Costs to be amortized	$2,700,000
Production during 2011	
Oil	50,000 bbl
Gas	120,000 Mcf
Proved reserves, 12/31/11	
Oil	800,000 bbl
Gas	3,000,000 Mcf
Selling price at 12/31/11	
Oil	$100/bbl
Gas	$10/Mcf
Average selling price during 2011	
Oil	$90/bbl
Gas	$8/Mcf

a. Computation of DD&A based on equivalent energy units for 2011 is as follows:

Equivalent Units
Production:

Oil	50,000 bbl
Gas (120,000/6)	20,000 BOE
Total BOE	70,000 BOE

Proved reserves:

Oil	800,000 bbl
Gas (3,000,000/6)	500,000 BOE
Total BOE	1,300,000 BOE

DD&A

$$\frac{\$2,700,00}{1,300,000 \text{ BOE} + 70,000 \text{ BOE}} \times 70,000 \text{ BOE} = \underline{\$137,956}$$

b. Computation of DD&A based on unit-of-revenue is as follows:

Unit-of-Revenue
Production:

Oil (50,000 × $90)	$ 4,500,000
Gas (120,000 × $8)	960,000
Total revenue	$ 5,460,000

Proved reserves:

Oil (800,000 × $100)	$ 80,000,000
Gas (3,000,000 × $10)	30,000,000
Total	$110,000,000

DD&A

$$\frac{\$2,700,00}{\$110,000,000 + \$5,460,000} \times \$5,460,000 = \underline{\$127,681}$$

c. Which would be the appropriate basis in this case? One barrel of oil is approximately equivalent to 6 Mcf of gas, but the current price of 1 barrel of oil is $100 compared to $60 for 6 Mcf of gas. The average selling price of 1 barrel of oil during the year was $90 compared to $48 for 6 Mcf of gas. This analysis indicates that the DD&A amount based on relative energy units will be different from the DD&A amount based on unit-of-revenue. Therefore, if the difference is considered to be material, unit-of-revenue would be the appropriate basis.

Entry

DD&A expense	127,681	
Accumulated DD&A		127,681

Reg. S-X 4-10 states that reserves should be valued at end of period prices in computing amortization based on unit-of-revenue. However, it also states that if a significant (material) price change occurs, the change should be reflected in the quarter following the price change. For example, if a significant price increase occurs in the third quarter, the third-quarter amortization is computed using the old price, and the fourth-quarter computation uses the new price.

Inclusion of estimated future development expenditures

Net capitalized costs, including both leasehold costs and drilling and development costs, are amortized over proved reserves under full cost accounting. *Reg. S-X 4-10* indicates that proved reserves consist of proved developed and proved undeveloped reserves. The definitions of proved developed and proved undeveloped reserves were discussed in detail in chapter 2. Generally, proved developed reserves include the proved reserves that will be produced through existing wells and equipment. In contrast, proved undeveloped reserves include the portion of proved reserves that will be produced through future wells or require major future development activities in order to be producible, or both.

When the proved reserves are not fully developed, a portion of the proved reserves—which make up the denominator in the DD&A calculation—will be producible only as a result of relatively major future development costs. Consequently, those future development costs *must* be included in the costs being amortized to avoid a distortion of the DD&A rate, i.e., a mismatching of costs and reserves. Estimates of future development costs should be based on current costs.

Inclusion of estimated future decommissioning costs

Reg. S-X 4-10 par. (c)(3)(i) indicates that estimated future decommissioning costs (i.e., dismantlement, reclamation, abandonment, and reclamation costs) net of estimated salvage values are to be included in the calculation of DD&A. However, it does not require recognition of the liability. SFAS No. 143, issued in 2001, mandates the recognition of future asset retirement obligations (AROs) when wells, equipment, and facilities are installed. (This treatment is required in situations in which the company has an obligation to remove the wells, equipment, and facilities and restore the environment at the end of production.) Both an asset and a liability in the amount of the estimated ARO must be recorded. Therefore, to the extent that an ARO has been incurred as a result of acquisition, exploration, and development activities, the estimated future decommissioning costs described in Reg. S-X 4-10 par. (c)(3)(i) have already been included in the capitalized costs.

Recall that future development costs related to proved reserves are to be estimated and included in the DD&A calculation. Such future development activities may result in additional

AROs being capitalized when the wells, equipment, and facilities are actually installed. As a result of the issuance of SFAS No. 143, the SEC issued Staff Accounting Bulletin (SAB) 106 to clarify the treatment of estimated future decommissioning costs in the calculation of full cost DD&A. According to SAB 106, companies are to estimate the amount of future decommissioning costs that will be incurred as a result of future development activities. Those amounts are to be included in the costs to be amortized. SFAS No. 143 is discussed in detail in chapter 9.

EXAMPLE

DD&A: Future Development Costs

Data for Tyler Company as of December 31, 2012, is as follows:

Unrecovered costs (net capitalized costs).	$2,000,000
Production during 2012.	100,000 bbl
Proved reserves, December 31, 2012	6,000,000 bbl

Tyler Company estimates that future expenditures to develop partially developed properties (proved properties) will be $5,000,000. Tyler Company also estimates that the related future decommissioning costs will be $1,000,000, with salvage of $100,000. (Decommissioning costs related to equipment and facilities already installed have been previously recorded and would be included in the $2,000,000 of unrecovered costs.)

DD&A Calculations

Costs to be amortized:

Unrecovered costs	$2,000,000
Add: Future development costs	5,000,000
Add: Future decommissioning costs related to future development costs	1,000,000
Less: Future salvage	(100,000)
Total costs to amortize	$7,900,000

DD&A

$$\frac{\$7,900,000}{6,000,000 \text{ bbl} + 100,000 \text{ bbl}} \times 100,000 \text{ bbl} = \underline{\$129,508}$$

Exclusion of costs

Amortization for full cost is based on proved reserves. Consequently, amortization of unproved properties, which by definition have no proved reserves attributable to them, can result in a material distortion of the amortization rate. Similarly, amortization of major development projects expected to involve significant costs in order to determine the quantities of proved reserves can also distort the amortization rate. In recognition of this problem, the SEC allows unproved properties or development projects entailing significant future development costs to be excluded from the amortization base. *Reg. S-X 4-10* par. (3)(ii) indicates the following:

1. All costs directly associated with the acquisition and evaluation of unproved properties *may* be excluded until it is determined whether or not proved reserves can be assigned to the properties. However, exploratory dry holes must be included in the amortization base as soon as a well is determined to be dry. Further, any G&G costs that cannot be directly associated with specific unproved properties should be included in the amortization base as incurred. In practice, most companies are excluding at least a portion of these G&G costs.

2. Some major development projects are expected to entail significant costs in order to ascertain the quantities of proved reserves attributable to the properties under development. In this case, certain costs associated with major development projects **may** be excluded if they have not previously been included in the amortization base. Excludable costs include both costs already incurred and future costs. The portion of common costs (i.e., costs common to both the known reserves and the reserves yet to be determined) excluded should be based on a comparison of existing proved reserves to total expected proved reserves. Alternately, the exclusion can be based on a comparison of the number of wells to which proved reserves have been assigned to the total number of wells expected to be drilled. Both the excluded costs and the related proved reserves should be transferred into the amortization base on a well-by-well or property-by-property basis as proved reserves are established or as impairment is determined.

The above costs may be excluded from the amortization base until proved reserves are established or until impairment (or abandonment) occurs. Any impairment is not expensed but is transferred into the amortization base and recovered, i.e., written off to expense, through DD&A. Note that if proved reserves are established, the costs must be included in amortization. Thus, even though production may not yet have begun (e.g., construction of a pipeline is necessary before any oil or gas is produced), the costs are amortized.

The following example illustrates the basic computations involved when costs associated with unproved properties are included and excluded from amortization.

EXAMPLE

Unproved Property Inclusion and Exclusion

Costs incurred by Tyler Company on oil and gas properties located in the United States as of December 31, 2011, were as follows:

Proved property costs	$ 40,000
Unproved property costs	70,000
Nondrilling exploration costs, proved properties	80,000
Nondrilling exploration costs, unproved properties	100,000
Drilling costs, proved properties	400,000
Wells-in-progress, unproved properties	600,000
Dry holes, unproved properties	900,000
Total accumulated DD&A	(490,000)
Total capitalized costs, net of DD&A	$1,700,000

Other Data:

Future development costs	$ 300,000
Future decommissioning costs (on future development projects)	400,000
Proved reserves, 12/31/11, bbl	500,000
Production during year, bbl	100,000

a. DD&A assuming all possible costs are included in the amortization base:

Costs to Be Amortized

Total net capitalized costs	$1,700,000
Plus: Future development costs	300,000
Plus: Future decommissioning costs related to future development costs	400,000
Total costs to amortize	$2,400,000

DD&A

$$\frac{\$2,400,000}{500,000 \text{ bbl} + 100,000 \text{ bbl}} \times 100,000 \text{ bbl} = \underline{\$400,000}$$

b. DD&A assuming all possible costs are excluded from the amortization base:

Costs to Be Amortized

Proved property costs	$ 40,000
Nondrilling exploration costs, proved properties	80,000
Drilling costs, proved properties	400,000
Dry holes, unproved properties	900,000
Total accumulated DD&A	(490,000)
Total capitalized costs to be amortized, net of DD&A	$ 930,000
Plus: Future development costs	300,000
Plus: Future decommissioning costs (on future development projects)	400,000
Net costs to be amortized	$1,630,000

Costs to Exclude

Unproved property costs	$ 70,000
Nondrilling exploration costs, unproved properties	100,000
Wells-in-progress, unproved properties	600,000
Total costs excluded	$ 770,000
Total costs included and excluded	$2,400,000

DD&A

$$\frac{\$1,630,000}{500,000 \text{ bbl} + 100,000 \text{ bbl}} \times 100,000 \text{ bbl} = \underline{\$271,667}$$

An alternative approach to computing the costs to be amortized is to start with the net capitalized costs and subtract the costs that may be excluded from amortization:

Net capitalized costs	$1,700,000
Plus: Future development costs	300,000
Future decommissioning costs	400,000
Less: Unproved properties	(70,000)
Nondrilling exploration costs (unproved)	(100,000)
Wells-in-progress, unproved properties	(600,000)
Net costs to be amortized	$1,630,000

The following example presents the exclusion of incurred costs and future costs associated with a major development project, for which significant future costs are necessary to ascertain quantities of proved reserves attributable to the properties under development.

EXAMPLE

Major Development Project Exclusion

Tyler Company has an offshore lease on which proved reserves have been found. Preliminary estimates indicate 20,000,000 barrels of proved reserves. Costs incurred to date relating to this lease total $30 million. To develop this lease fully and to determine the quantity of reserves, Tyler Company plans to install an offshore drilling platform from which development wells will be drilled. It is estimated that the platform and the wells will cost $800 million.

Total unrecovered costs at year-end for Tyler Company's other properties in the cost center are $20 million, with year-end proved reserves of 400,000 barrels. Production for the year was 100,000 barrels of oil.

All costs and reserves relating to this major development project may be excluded from the DD&A calculation.

$$\frac{\$20,000,000}{400,000 \text{ bbl} + 100,000 \text{ bbl}} \times 100,000 \text{ bbl} = \underline{\$4,000,000}$$

Total costs excluded: $30,000,000 + $800,000,000 = $\underline{\$830,000,000}$

Impairment of unproved properties costs

Only those unproved properties excluded from the amortization base should be impaired. The unproved properties excluded from the amortization base should be assessed at least annually. As with successful efforts impairment, properties whose costs are individually significant must be assessed individually, while properties whose costs are not individually significant may be assessed on a group basis. Although not defined in the full cost rules, the SEC suggests that, in general, individual properties or projects are individually significant if their costs exceed 10% of the net capitalized costs of the cost center.

The procedure for determining impairment for significant unproved properties, as well as the factors considered, are the same as under successful efforts. The same methods used to impair individually insignificant (group) properties under successful efforts accounting may also be used to impair individually insignificant properties under full cost accounting. Individually insignificant properties may be grouped and assessed in one or more groups per cost center.

In contrast to successful efforts impairment, any impairment under full cost is *not* expensed but is transferred into the amortization base and continues to be capitalized until recovered (i.e., expensed) through DD&A. The impairment amount is recorded in an impairment account. An allowance for impairment account is also recorded. As with successful efforts accounting, this account is netted against the unproved property account to obtain the net carrying value of the unproved property, assuming the property is significant. The impairment account under full cost accounting is an asset account that is included in the costs to be amortized. The allowance for impairment account is a contra-asset that reduces the amount of unproved property costs excluded from amortization. An example follows that illustrates impairment under full cost for significant and individually insignificant properties.

EXAMPLE

Impairment of Individually Significant and Insignificant Properties

Tyler excludes all possible costs relating to unproved properties from the amortization base. Tyler's acquisition and evaluation costs related to unproved properties as of December 31, 2012, are as follows:

Individually Significant Unproved Properties:

	Lease A	Lease B
Unproved property, control	$500,000	$800,000
Allowance for impairment	100,000	0

Individually Insignificant Unproved Properties:

	Lease C	Lease D	Lease E	Total
Unproved property, control	$40,000	$60,000	$80,000	$180,000
Allowance for impairment, group				30,000

During 2013, wells on properties surrounding Lease A and Lease B are drilled and determined to be dry. Further, the primary term on Lease A is nearly over, and Tyler has no firm plans to drill on Lease A. As a result, Tyler decides Lease A is impaired a total of 80%, and Lease B is impaired 20%.

For individually insignificant unproved properties, Tyler has a policy based on historical experience of providing at year-end an allowance equal to 70% of the gross cost of the individually insignificant unproved properties.

Impairment for Individually Significant Unproved Properties

Lease	Total Impairment	Impairment from Prior Periods	Impairment to be Recognized in Current Period
A	80% × $500,000 = $400,000	$100,000	$300,000
B	20% × 800,000 = 160,000	0	160,000

Entry

Impairment for unproved properties	460,000	
Allowance for impairment—Lease A.		300,000
Allowance for impairment—Lease B.		160,000

Impairment for Individually Insignificant Properties

70% × $180,000 = $126,000

```
              Allowance
          ┌──────────────┐
          │    30,000    │  ← have
          │        ?     │  ← 96,000
          │   126,000    │  ← want
          └──────────────┘
```

Entry

Impairment for unproved properties—group.	96,000	
Allowance for impairment, group		96,000

Any costs associated with major development projects excluded from the amortization base must also be evaluated for impairment. Impairment is computed as it is for unproved properties, and as on unproved properties, any impairment is transferred into the amortization base and recovered through DD&A.

Abandonment of properties

In full cost accounting, all costs, both successful and unsuccessful, are considered necessary to finding oil or gas and consequently are considered part of the cost of oil or gas. Thus when a property, either proved or unproved, is abandoned, the costs generally are not expensed but remain capitalized and are recovered through DD&A. The abandonment of a property can be handled in several ways, depending upon the accounts used by a company and whether costs are excluded from the amortization base:

- No entry is made.
- Accumulated DD&A or an abandoned cost account is charged.
- If the abandonment significantly affects the amortization per unit, however, the costs are removed, and a gain or loss is recognized.

In this book, abandoned properties are generally charged to abandoned costs. When a proved property is abandoned, *all* the related costs are generally charged to abandoned costs. An individual property will not have accumulated DD&A associated with it specifically because the cost center under full cost is a country and any accumulated DD&A relates to the entire cost center and not an individual property. Thus, when a proved property is abandoned, only the accounts associated with the specific proved property are closed out and charged to abandoned costs.

EXAMPLE

Abandonment of Proved Property

Tyler Company abandoned a proved property late in 2011. Costs relating to the property are as follows:

Proved property	$ 50,000
Wells and equipment	600,000
Exploratory dry holes	400,000

Entry

Abandoned costs	1,050,000	
Proved property		50,000
Wells and equipment		600,000
Exploratory dry holes		400,000

The specific accounting entry to record abandonment of an unproved property depends on whether the property is considered to be significant or insignificant. If significant, both the unproved property account and the allowance for impairment account should be closed out, with the net balance charged to abandoned costs. If the property is not individually significant, the unproved property account should be charged to the allowance account. (See the next example.)

Reclassification of properties

When proved reserves are discovered on an unproved property, the property should be reclassified from unproved to proved. As with the abandonment of unproved properties, the specific accounting entry depends on whether or not the property is considered to be individually significant. For an individually significant property, both the unproved property account and the allowance for impairment account should be closed out, with the net carrying value transferred to a proved property account. For a property that is not individually significant and therefore assessed on a group basis, the gross unproved property account balance should be transferred because, as with successful efforts accounting, a net carrying value cannot be determined on a separate property basis for properties assessed on a group basis.

The following example illustrates impairment, abandonment, and reclassification for significant unproved properties.

EXAMPLE

Impairment, Abandonment, and Reclassification of Significant Unproved Properties

Tyler Company excludes all possible costs relating to unproved properties from the amortization base. Tyler's direct acquisition and evaluation costs at December 31, 2011, related to two significant, unproved properties are as follows:

	Lease A	Lease B	Total
Unproved property, control..........	$100,000	$ 75,000	$175,000
Wells-in-progress	300,000	200,000	500,000
Total capitalized costs	$400,000	$275,000	$675,000

During 2012, the wells on each lease are determined to be dry. As a result, Tyler impairs Lease A $40,000 and Lease B $25,000.

Entries

Exploratory dry holes ($300,000 + $200,000)	500,000	
Wells-in-progress—Lease A.............		300,000
Wells-in-progress—Lease B.............		200,000
Impairment for unproved properties ($40,000 + $25,000)	65,000	
Allowance for impairment—Lease A..........		40,000
Allowance for impairment—Lease B..........		25,000

Costs to Be Amortized Relating to above Leases at 12/31/12

	Lease A	Lease B	Total
Impairment	$ 40,000	$ 25,000	$65,000
Dry holes	300,000	200,000	500,000
	$340,000	$225,000	$565,000

Costs That May Be Excluded

	Lease A	Lease B	Total
Unproved property, control	$100,000	$75,000	$175,000
Less: Allowance	(40,000)	(25,000)	(65,000)
	$ 60,000	$50,000	$110,000
Total Capitalized Costs			$675,000

Note the $40,000 impairment amount for Lease A is included in the costs to be amortized, and the $40,000 allowance for impairment is shown as a reduction to the unproved property amount, reducing the amount excluded from amortization.

During 2013, Tyler abandons Lease A and drills a successful well on Lease B at a cost of $450,000.

Entry—Lease A

Abandoned costs.	60,000	
Allowance for impairment—Lease A	40,000	
Unproved property—Lease A		100,000

Note that the entry to record abandonment of an impaired property under full cost is similar to the entry to record abandonment of an impaired property under successful efforts. The difference is that under full cost, the abandoned amount continues to be capitalized, while under successful efforts, the amount is expensed.

Entries—Lease B

Wells and equipment.	450,000	
Cash		450,000
Proved property	50,000	
Allowance for impairment—Lease B	25,000	
Unproved property—Lease B.		75,000

Note that the entry to reclassify a property from unproved to proved is the same under full cost and successful efforts accounting.

Costs to Be Amortized Relating to above Leases at 12/31/13

	Lease A	Lease B	Total
Impairment	$ 40,000	$ 25,000	$ 65,000
Dry holes	300,000	200,000	500,000
Abandoned costs	60,000		60,000
Proved property		50,000	50,000
Wells and equipment		450,000	450,000
Total capitalized costs	$400,000	$725,000	$1,125,000

The above example illustrates that as properties are impaired, the impairment amounts are put into an asset account that is included in the amortization base. The impairment account remains untouched when the unproved property is either abandoned or is reclassified as proved. If a significant unproved property is abandoned, the unproved property and related allowance account are eliminated, with the net carrying value of the property being charged to abandoned costs, an asset account. If proved, the unproved property and allowance accounts are also eliminated, with the net carrying value transferred to proved property. It is therefore probable that the impairment asset account will have a greater balance than the contra-asset allowance for impairment account. The following example illustrates such a situation.

EXAMPLE

Impairment Account versus Allowance Account

Tyler Company's direct acquisition and evaluation costs at December 31, 2009, related to two significant unproved properties are as follows:

	Lease A	Lease B	Total
Unproved property, control...............	$100,000	$75,000	$175,000

During 2010, wells on properties surrounding each lease are determined to be dry. As a result, Tyler impairs Lease A $60,000 and Lease B $25,000.

Entry

Impairment for unproved properties	85,000	
Allowance for impairment—Lease A.		60,000
Allowance for impairment—Lease B.		25,000

During 2011, Tyler abandons Lease A. Lease B remains classified as unproved, and no further impairment is made.

Entry

Abandoned costs.	40,000	
Allowance for impairment—Lease A	60,000	
Unproved property—Lease A		100,000

The balance in the unproved property control account is now $75,000, and the allowance for impairment control account has a balance of $25,000. However, the impairment asset account has a balance of $85,000, $60,000 greater than the allowance for impairment account.

Costs to Be Amortized

Impairment for unproved properties	$ 85,000
Abandoned costs	40,000
Total costs to be amortized	$125,000

Costs to Be Excluded

Unproved property—Lease B	$ 75,000
Less: Allowance for impairment	25,000
Total costs to be excluded	$ 50,000
Total capitalized costs	$175,000

The following example illustrates impairment, abandonment, and reclassification for individually insignificant properties.

EXAMPLE

Impairment, Abandonment, and Reclassification of Individually Insignificant Unproved Properties

On December 31, 2012, Tyler's unproved property account containing leases not considered individually significant had a $500,000 balance, and the allowance for impairment account had a $60,000 balance. Based on historical experience, Tyler has a policy of providing at year-end an allowance equal to 60% of gross unproved properties.

Entry

Impairment for unproved properties—group.	240,000	
Allowance for impairment, group		240,000

Calculations

$500,000 \times 0.60 = \$300,000$

Allowance		
	60,000	← have
	?	← 240,000
	300,000	← balance needed

On March 15, leases with a cost of $30,000 were abandoned.

Entry

Allowance for impairment, group.	30,000	
Unproved property		30,000

On April 15, proved reserves were found on a lease with capitalized costs of $8,000.

Entry

Proved property	8,000	
Unproved property		8,000

The following example illustrates a complex DD&A problem in which abandoned costs and impairment are included. Part a illustrates the computation of DD&A when all possible costs are amortized, and part b illustrates the computation of DD&A when all possible costs are excluded from amortization. In the example, the column headed *I* relates to part a of the problem, where all possible costs are included in the amortization base. The column headed *E* relates to part b of the problem, where all possible costs are excluded from the amortization base. An *I* in the columns indicates that the cost must be included in amortization. An *E* in the columns indicates that the cost may be excluded from amortization.

EXAMPLE

Complex DD&A

Tyler Company has oil and gas properties located only in the United States. Costs for Tyler Company as of December 31, 2012, were as follows:

I	E		
I	I	Proved properties	$ 100,000
I	E	Unproved properties, acquisition costs...........	105,000
I	I	Wells-in-progress, proved properties	250,000
I	E	Wells-in-progress, unproved properties	550,000
I	I	Nondirect G&G costs..................	70,000
I	I	Direct G&G costs, proved properties	120,000
I	E	Direct G&G costs, unproved properties	210,000
I	I	Dry holes, proved properties	300,000
I	I	Dry holes, unproved properties	700,000
I	I	Delay rentals, proved properties	25,000
I	E	Delay rentals, unproved properties	75,000
I	I	Successful drilling costs................	600,000
I	I	Abandoned costs of unproved properties	95,000
I	I	Total accumulated DD&A	(500,000)
I	I	Future development costs	400,000
		Net costs, plus future development costs	$3,100,000
		Total capitalized costs, plus future development costs, net of DD&A	$3,100,000
		Proved reserves, 12/31/12	4,240,000 bbl
		Production during year	245,000 bbl

Note in the preceding data, several costs associated with proved properties, such as delay rentals, are incurred when the associated property is unproved. Under full cost, these costs are capitalized instead of being expensed as in successful efforts, and so are still on the books when the property becomes proved. Also recall that only unproved properties excluded from amortization are assessed for impairment. In part a, all properties are included in amortization. Consequently, the unproved properties have not been impaired, and there are no impairment or allowance for impairment accounts.

a. DD&A is computed as follows, assuming all possible costs are included in the amortization base:

    Costs to Amortize

    All of the above costs, less accumulated DD&A, would be included in computing DD&A. (The costs to be amortized are indicated above by a letter I in the column headed I.)

Total capitalized costs	$3,200,000
Plus: Future development costs	400,000
Less: Accumulated DD&A	(500,000)
Costs to be amortized	$3,100,000

    DD&A

    $$\frac{\$3,100,000}{4,240,000 \text{ bbl} + 245,000 \text{ bbl}} \times 245,000 \text{ bbl} = \underline{\$169,342}$$

b. DD&A is next computed assuming that all possible costs are excluded from the amortization base. It is also assumed that the allowance for impairment for the above unproved properties is $20,000 and that unproved properties have been impaired a total of $80,000. (As discussed, when a significant unproved property that has been impaired is either abandoned or proved, the costs of the property, net of impairment, are transferred to abandoned costs or proved property. In either case, the allowance for impairment account is closed out for that property. For abandoned individually insignificant unproved properties, the allowance for impairment, group account would be debited. In contrast, once the impairment amount is put on the books, it becomes a part of the total capitalized costs and is not removed. Thus, it is possible for the allowance account for unproved properties to be smaller than the impairment account.)

    Total Costs under New Assumptions

Net costs plus future development costs under old assumptions	$3,100,000
Less: Allowance for impairment*	(20,000)
Plus: Impairment	80,000
Total costs, new assumptions	$3,160,000

    *Contra-account to unproved property, which is included at gross in the $3,100,000.

Costs to Amortize

All of the above costs associated with unproved properties except dry-hole costs, abandoned costs, and impairment would be excluded from amortization. The costs to be amortized are indicated above by a letter *I* in the column headed *E*. The costs indicated by a letter *E* should be excluded.

Proved properties	$ 100,000
Wells-in-progress, proved properties	250,000
Nondirect G&G costs	70,000
Direct G&G costs, proved properties	120,000
Dry holes, proved properties	300,000
Dry holes, unproved properties	700,000
Delay rentals, proved properties	25,000
Successful drilling costs	600,000
Abandoned costs of unproved properties	95,000
Impairment for unproved properties	80,000
Total accumulated DD&A	(500,000)
Future development costs	400,000
Net costs to be amortized	$2,240,000

Costs to Exclude

Unproved properties	$105,000	
Less: Allowance for impairment	(20,000)	
Net unproved properties		$ 85,000
Wells-in-progress, unproved properties		550,000
Direct G&G costs, unproved properties		210,000
Delay rentals, unproved properties		75,000
Costs to exclude		$ 920,000
Total costs included and excluded		$3,160,000

DD&A

$$\frac{\$2,240,000}{4,240,000 \text{ bbl} + 245,000 \text{ bbl}} \times 245,000 \text{ bbl} = \underline{\$122,363}$$

Support equipment and facilities

Support equipment and facilities are defined and handled similarly under full cost accounting and successful efforts accounting. The depreciation and operating costs of support equipment and facilities should be allocated to exploration, development, or production as appropriate. The portion of the costs allocated to exploration and development would be capitalized; the portion allocated to production would be expensed. As with successful efforts accounting, when support equipment and facilities serve only one cost center, the support equipment and facilities should be depreciated using the unit-of-production method. Since the cost center for full cost accounting is a country, it is highly unlikely that support equipment and facilities would support more than one cost center. As a result, support equipment and facilities under full cost accounting would almost always be depreciated using unit-of-production.

Depreciation of support equipment and facilities should theoretically be allocated to exploration, development, or production as appropriate. In practice, however, support equipment and facilities are typically capitalized directly to the cost center and amortized with other capitalized costs using unit-of-production over the life of the cost center. The depreciation on the support equipment and facilities is not separated and allocated between multiple activities; instead, it is expensed as DD&A expense. This treatment may result in minor differences in the timing of expense recognition. However, these differences are not likely to be material.

DD&A under successful efforts versus full cost

A comparison of the DD&A computation for successful efforts and full cost is presented in the following example.

EXAMPLE

Successful Efforts Compared to Full Cost

Tyler Company, a new company, has only two oil and gas properties, both located in the United States, and both considered significant. Lease A was acquired in 2010, and production began on Lease A on January 1, 2011. Lease B was acquired early in 2012 and is still unproved. No equipment was salvaged from any of the dry holes drilled. Data for Tyler as of 12/31/14 are given on the next page, along with columns indicating whether each cost would be:

a. Amortized (I) under successful efforts (SE)

b. Amortized (I) under full cost (FC) when all possible costs are included in amortization (FC-I)

c. Amortized (I) under full cost when all possible costs are excluded from amortization (FC-E)

A letter *E* indicates the cost may be excluded from amortization under full cost.

254 Fundamentals of Oil & Gas Accounting

Item	Lease A, Proved				Lease B, Unproved			
	Cost	Amortize Under (I)			Cost	Amortize Under (I)		
		SE	FC-I	FC-E		SE	FC-I	FC-E
Bonus, before impairment					$ 60,000		I	E
Allowance for impairment					(10,000)		NA	*
Impairment for unproved property, 2013					10,000		NA	I
Bonus, net of impairment	$ 50,000	I	I	I				
Delay rentals, cumulative	25,000		I	I	20,000		I	E
Direct G&G, cumulative	90,000		I	I	70,000		I	E
Wells-in-progress—IDC	450,000		I	I	300,000		I	E
Wells-in-progress—L&WE	100,000		I	I	100,000		I	E
Exploratory dry holes—IDC	500,000		I	I	425,000		I	I
Exploratory dry holes—L&WE	200,000		I	I	175,000		I	I
Development dry holes—IDC	310,000	I	I	I	NA			
Development dry holes—L&WE	120,000	I	I	I	NA			
Successful wells—IDC	475,000	I	I	I	NA			
Successful wells—L&WE	225,000	I	I	I	NA			
Storage tanks, 18-year	150,000	I	I	I	NA			
Future development costs	400,000		I	I	NA			
Estimated salvage	(90,000)	(I)	(I)	(I)	0			
Total, net of salvage value	$3,005,000				$1,150,000			

*Acquisition costs net of allowance for impairment *may* be excluded from amortization.

Other Data, Lease A	Oil, bbl	Gas, Mcf
Proved reserves, 12/31/14	200,000	600,000
Proved developed reserves, 12/31/14	80,000	240,000
Production	10,000	24,000

a. DD&A under successful efforts is calculated in the simplest manner possible assuming gas is the dominant mineral. Assume accumulated DD&A for leaseholds is $5,000 and accumulated DD&A for wells and equipment is $430,000.

Leasehold: $$\frac{\$50{,}000 - \$5{,}000}{600{,}000 \text{ Mcf} + 24{,}000 \text{ Mcf}} \times 24{,}000 \text{ Mcf} = \underline{\$1{,}731}$$

Wells and Equipment: $$\frac{\$1{,}190{,}000^* - \$430{,}000}{240{,}000 \text{ Mcf} + 24{,}000 \text{ Mcf}} \times 24{,}000 \text{ Mcf} = \underline{\$69{,}091}$$

Total DD&A: $1,731 + $69,091 = $\underline{\$70{,}822}$

*Costs to be amortized:

$310,000 + $120,000 + $475,000 + $225,000 + $150,000 − $90,000 = $\underline{\$1{,}190{,}000}$

b. DD&A under full cost is calculated using the energy basis. All costs that may be included in the amortization base are included. Assume accumulated DD&A is $930,000. Remember if Lease B, an unproved property, is included in DD&A, the lease would not have been impaired. Note that the costs to amortize from Lease A include future development costs.

	Production		Proved Reserves	
Oil:		10,000 bbl		200,000 bbl
Gas:	24,000/6 =	4,000 BOE	600,000/6 =	100,000 BOE
		14,000 BOE		300,000 BOE

Costs to Amortize

$3,005,000 (Lease A) + $1,150,000 (Lease B) − $930,000 = $\underline{\$3{,}225{,}000}$

$$\frac{\$3{,}225{,}000}{300{,}000 \text{ BOE} + 14{,}000 \text{ BOE}} \times 14{,}000 \text{ BOE} = \underline{\$143{,}790}$$

c. DD&A under full cost is calculated using the revenue method, assuming prices at the end of the year were $100/bbl and $7/Mcf, and that oil sold during the year for $90/bbl and gas sold for $12/Mcf. All costs that may be excluded from the amortization base are excluded. Assume accumulated DD&A is $630,000.

	Production				Proved Reserves		
Oil:	10,000 ×	$90 =	$ 900,000	200,000 ×	$100 =	$20,000,000
Gas:	24,000 ×	$12 =	288,000	600,000 ×	$ 7 =	4,200,000
			$1,188,000			$24,200,000

Costs to Amortize from Lease B

Net capitalized costs	$ 1,150,000
Less: Bonus, net of impairment	50,000
Delay rentals	20,000
Direct G&G	70,000
Wells-in-progress—IDC	300,000
Wells-in-progress—L&WE	100,000
Net costs to amortize	$ 610,000

Costs to Amortize from Lease B (alternate approach)

Impairment	$ 10,000
Exploratory dry holes—IDC	425,000
Exploratory dry holes—L&WE	175,000
Costs to amortize	$ 610,000

Costs to Amortize from Lease A and Lease B

$3,005,000 (Lease A) + $610,000 (Lease B) = $3,615,000

$$\frac{\$3,615,000 - \$630,000}{\$24,200,000 + \$1,188,000} \times \$1,188,000 = \underline{\$139,679}$$

Reserves in Place—Purchase

Often, rather than exploring and drilling for oil and gas, companies opt to purchase oil and gas reserves in place. The cost of purchased reserves should be added to the full cost cost pool, and the reserve barrels and Mcf added to total proved reserves when computing DD&A, unless substantially different lives are involved. In other words, the costs and reserve amounts are, in most cases, treated the same for purchased reserves as for developed reserves.

INTEREST CAPITALIZATION

SFAS No. 34 requires all companies to capitalize interest for assets under construction. The interest capitalization rules relating to basic computations and definitions are essentially the same for successful efforts and full cost. As described in chapter 5, interest capitalization for successful efforts companies is, in practice, varied and subject to wide interpretation. However, for full cost companies, the capitalization procedure is clearer. Under *FASB Interpretation No. 33* (to *SFAS No. 34*), full cost companies capitalize interest only on assets that have been excluded from the amortization base. Once costs are included in the amortization base, they lose their separate identity and cannot be identified as being *under construction*. Costs in the amortization base and being amortized are considered to be completed, and therefore, interest cannot be capitalized on those assets.

Capitalized interest is computed as follows:

$$\text{Average accumulated expenditures during construction} \times \text{Interest rate} \times \text{Construction period}$$

Interest capitalization begins when the initial expenditure related to the asset is made during the construction period. In oil and gas activities, capitalization of interest may begin with the first drilling expenditures or the first expenditures made in preparation of drilling.

A simple example follows.

EXAMPLE

Interest Capitalization

Tyler Company has unproved property costs of $60,000 for Lease A at January 1, 2010. During 2010, an exploratory well is drilled in the amount of $240,000. A 10%, $400,000 note is outstanding during the entire year and was obtained specifically for the acquisition and drilling program related to Lease A. Work on the well began on January 1, 2010, and ended on June 30, 2010. The well was successful, and the property was reclassified as proved property on June 30, 2010. Tyler Company excludes all unproved property costs from the amortization base. Tyler Company should capitalize interest on the unproved property prior to reclassification, computed as follows:

$$\text{Average accumulated expenditures:} \quad \frac{\$60{,}000 + \$300{,}000}{2} = \$180{,}000$$

Capitalize: $\$180{,}000 \times 0.10 \times 6/12^* = \$9{,}000$

*Note that because interest capitalization ends when the asset is included in the amortization base, the construction period for interest capitalization for 2010 is only six months.

Entry

Wells-in-progress—IDC.	9,000	
Interest expense		9,000

In the above entry, wells-in-progress—IDC is debited, although it is unlikely that the same amount would be considered IDC for tax. In practice, when taxes are prepared, companies generally make relevant adjustments to the accounts to accommodate tax and financial accounting differences such as this one.

Limitation on Capitalized Costs—A Ceiling

A ceiling test is required under the full cost rules to ensure that net capitalized costs of full cost companies do not exceed the underlying value of the company. In applying this test, which must be performed quarterly, net capitalized costs are compared to a cost ceiling value. If the capitalized costs exceed the ceiling, capitalized costs must be permanently written down. Specifically, capitalized costs for each cost center, less accumulated DD&A and related deferred income taxes, must not exceed a cost ceiling, which is defined as the sum of the following:

	a. The present value of future net revenues from proved reserves
Plus:	b. The cost of properties not being amortized (i.e., unproved properties and costs of major development projects for which the quantity of proved reserves has not yet been determined)
Plus:	c. The lower of cost or market of unproved properties being amortized
Less:	d. The tax on the sum of a, b, and c minus the tax basis of the property

If total net capitalized costs exceed the cost ceiling, the excess should be expensed and may not be reinstated in the future. This test should be performed each period after recording the current period's DD&A.

In calculating net capitalized costs, only deferred taxes related to oil and gas activities should be included. Deferred taxes are subtracted, assuming a net deferred tax liability, as is the usual case. These taxes are subtracted because deferred taxes are tax amounts that have already been recognized (expensed) on the financial statements but will be paid to the IRS in the future.

In calculating the ceiling, the present value of future net revenues is calculated as follows:

	Future gross revenue (estimated future production at period-end prices)
Less:	Future development costs (estimated using costs at period-end)
Less:	Future production costs (estimated using costs at period-end)
Equals:	Future net revenue, undiscounted
	Discounted at 10%
Equals:	Future net revenue, discounted

The cost of properties not being amortized would include all the related costs that are excluded from amortization net of impairment. Thus, unproved property costs excluded would include not only the acquisition costs, but also related direct G&G, wells-in-progress, etc.

EXAMPLE

Ceiling Test

Data for Tyler Oil Company are as follows:

Present value of future gross revenues		$85,000,000
Present value of future related costs		45,000,000
Capitalized costs of proved properties		
Acquisition costs	$ 2,000,000	
Wells-in-progress	13,000,000	
Wells and equipment	15,000,000	
Exploratory dry holes	20,000,000	
Total		50,000,000
Unproved properties not being amortized		
Acquisition costs	800,000	
Test-well contributions	200,000	
G&G costs	1,000,000	
Wells-in-progress	6,000,000	
Total		8,000,000

Unproved properties being amortized
    Acquisition costs 500,000
    Wells-in-progress 1,500,000
    Total 2,000,000

Accumulated DD&A 5,000,000
Deferred income taxes 1,000,000
Income tax effects—books versus tax 1,000,000
Market value of unproved property being amortized . . 2,300,000

Calculations

Capitalized Costs Less Accumulated DD&A and Deferred Taxes:

Capitalized costs of proved properties	$50,000,000
Add: Unproved properties not being amortized	8,000,000
Add: Unproved properties being amortized	2,000,000
Less: Accumulated DD&A	(5,000,000)
Less: Deferred income taxes	(1,000,000)
Total	$54,000,000

Ceiling:

Present value of future net revenues ($85,000,000 – 45,000,000)	$40,000,000
Add: Unproved properties not being amortized	8,000,000
Add: Lower of cost or market of unproved properties being amortized	2,000,000
Less: Income tax effects—books versus tax	(1,000,000)
Total	$49,000,000

Entry

Loss on producing properties 5,000,000
    Accumulated capitalized cost reduction 5,000,000

Asset retirement obligations

Prior to the issuance of *SFAS No. 143*, "Accounting for Asset Retirement Obligations," future decommissioning costs were estimated, but no asset or liability was actually recorded. *SFAS No. 143* requires the recognition of a liability for asset retirement obligations (AROs) at fair value in the period in which the obligation is incurred, if a reasonable estimate of fair value can be made. When the ARO liability is recognized, a like amount is capitalized and added to the cost of the related long-lived asset. Once the ARO is added to the cost center, it is subject to the full cost ceiling test under *Reg. S-X 4-10*.

The full cost ceiling test requires that future decommissioning costs be estimated and treated as a reduction of future net cash flows. In *SAB 106*, the SEC modified the treatment of future decommissioning costs in order to avoid them being "double counted" in the ceiling test calculation. The SEC noted that application of *SFAS No. 143* would result in an increase in the capitalized cost of the underlying asset, while simultaneously resulting in a reduction of the expected future net cash flows. In order to avoid this doubling effect, the SEC addressed future cash outflows associated with AROs that have been accrued on the balance sheet. The SEC indicated that these cash outflows are to be excluded from the computation of the present value of estimated future net cash flows in the calculation of the full cost ceiling calculation.

SAB 106 also indicates that in order to fully inform financial statement users of the interaction of *SFAS No. 143* and the full cost rules, a company must provide full disclosure of the effect of *SFAS No. 143* in the financial statement footnotes and management's discussion and analysis. At a minimum, this disclosure should include an explanation as to how the ceiling limit calculation is affected by the adoption of *SFAS No. 143*.

Deferred taxes

The SEC's full cost rules were written at a time when deferred tax assets were not allowed. Since these rules were written, GAAP has been changed to require the recognition of deferred tax assets subject to a valuation allowance. Recognizing a loss due to a ceiling test write-down could generate recognition of a deferred tax asset. In applying the ceiling test, net costs, which are compared to the full cost ceiling, must include related deferred taxes. However, including a deferred tax asset increases net costs and would result in net costs again exceeding the full cost ceiling. In such a situation, there are two unknowns: the amount of the write-down and the amount of the deferred tax asset.

Income tax effects

The income tax effects that should be considered in computing the amount to be deducted from estimated future net revenues are the following:

- Tax basis of oil and gas properties
- Net operating loss carryforwards
- Foreign tax credit carryforward
- Investment tax credits (ITC)
- Minimum taxes on tax preference items
- Impact of percentage depletion

Although investment tax credits are not currently allowed by the tax code, they frequently have been in the past and may well be allowed in the future. Therefore, investment tax credits are included in the discussion of the ceiling test and in the example showing the income tax effects.

Only net operating loss (NOL) carryforwards that are directly applicable to oil and gas operations should be used in the above computation. Also, only ITC carryforward and foreign tax credit carryforwards clearly attributable to oil and gas operations should be used in computing the income tax effects. The computation of the income tax effects by using the short-cut method presented in *Reg. S-X 4-10* is illustrated below.

EXAMPLE

Ceiling Test Income Tax Effect

Assumptions

Capitalized costs of oil and gas assets		$2,000,000
Accumulated DD&A		(400,000)
Book basis of oil and gas assets		1,600,000
Related deferred income taxes		(90,000)
Net book basis to be recovered		$1,510,000
NOL carryforward*		$ 50,000
Foreign tax credit carryforward*		3,000
ITC carryforward*	$5,000	
Present value of ITC relating to future development costs	2,000	7,000
Estimated preference (minimum) tax on percentage depletion in excess of cost depletion		2,000
Tax basis of oil and gas assets		1,200,000
Present value of statutory depletion attributable to future deductions		30,000
Statutory tax rate		46%
Present value of future net revenues from proved oil and gas reserves		1,300,000
Cost of properties not being amortized		75,000
Lower of cost or estimated fair value of unproved properties included in costs being amortized		65,000

*All carryforward amounts in this example represent amounts that are available for tax purposes and that relate to oil and gas operations.

Calculations

Present value of future net revenue		$1,300,000
Cost of properties not being amortized		75,000
Lower of cost or estimated fair value of unproved properties included in costs being amortized		65,000
Tax effects:		
Total of above items	$1,440,000	
Less: Tax basis of properties	$(1,200,000)	
Statutory depletion	(30,000)	
NOL carryforward	(50,000) (1,280,000)	
Future taxable income	$ 160,000	
Tax rate	× 46%	
Tax payable at statutory rate	$ (73,600)	
ITC	7,000	
Foreign tax credit carryforward	3,000	
Estimated preference tax	(2,000)	
Total tax effects		(65,600)
Cost center ceiling		$ 1,374,400
Less: Net book basis		(1,510,000)
Required Write-off, net of tax**		$ (135,600)

**For accounting purposes, the gross write-off should be recorded to adjust both the oil and gas properties account and the related deferred income taxes.

An exemption to the cost ceiling limitation may be obtained when the cost of purchased reserves in place exceeds the present value of estimated future net revenues from the sale of such reserves. The primary reason for paying more for reserves in place than the expected future benefit as calculated per SEC rules is that the purchaser expects oil and gas prices to escalate. Further, a value is usually placed on probable and possible reserves. In such cases, the excess cost would not be written off if an exemption is obtained from the SEC.

Assessment of the ceiling test

Costs are capitalized under the full cost method without regard to future benefits. The SEC consequently felt that the capitalized costs of a full cost company could well exceed its future net revenues and therefore mandated the full cost ceiling test. The full cost ceiling test is an unusual and rigid test in many respects. First, the test must be performed quarterly using end of period prices, but restoration of quarterly write-downs is not allowed even if prices improve before year-end. This rule is especially harsh for companies producing predominantly natural gas. Gas pricing is highly seasonal, with the lowest prices being in the summer months. Regardless, permanent write-downs must be made at quarter-end even if the average gas price for the year or the year-end price is expected to be much higher than the seasonal quarter-end price.

Oil prices, while not seasonal, are highly volatile. The price of oil can be significantly different just weeks before or after period-end as compared to the period-end price. Potentially huge write-downs thus depend in part on how fortunate a company is in relation to the timing of its period-end as compared to the current monetary oil price. Figures 7–2 and 7–3 below illustrate the wide variation between period-end prices and the annualized average prices for both oil and gas, highlighting the inherent problem of a ceiling test that requires the use of period-end prices.

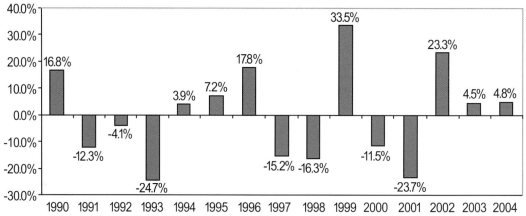

Fig. 7–2. End of year pricing vs. annual average pricing—crude oil
Source: Compiled by Ryder Scott Company, L.P. from published data.

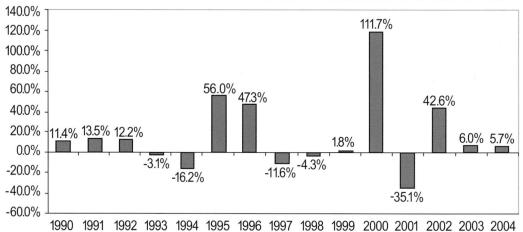

Fig. 7–3. End of year pricing vs. annual average pricing—natural gas
Source: Compiled by Ryder Scott Company, L.P. from published data.

Second, in order to compute the cost ceiling, the amount and timing of production of proved reserves must be obtained from petroleum engineers. These reserve estimates are imprecise. The lack of precision of the reserve estimates causes the ceiling test results to be questionable. Furthermore, only proved reserves are considered when computing the cost ceiling. Some portion of probable and possible reserves will normally be transferred to the proved category at some future date, but the ceiling test places no value on these reserves.

SFAS No. 144

As discussed in chapter 6, successful efforts companies must apply *SFAS No. 144*, "Accounting for the Impairment or Disposal of Long-Lived Assets." Because full cost companies are already required to apply the full cost ceiling test, full cost companies are not required to apply *SFAS No. 144* to their productive oil and gas properties. However, full cost companies must apply *SFAS No. 144* to their long-lived non–oil and gas assets. *SFAS No. 144* is discussed in detail in chapter 9.

Some of the more prominent differences between the application of the full cost ceiling test and the impairment test mandated by *SFAS No. 144* are as follows:

Full cost ceiling test	*SFAS No. 144*
Applied based on discounted net revenue	Applied only if fails recoverability test based on undiscounted net revenues
Applied using end of period prices	Applied using expected future prices
Applied on an interim and annual basis	Applied only when circumstances indicate possible impairment exists
Applied using only proved reserves	Applied using proved, probable, and even possible reserves
Applied on a countrywide basis	Applied on a cash-generating unit basis (typically a field)
Cash flows discounted at 10% rate	No single discount rate; future cash flows measured at fair value

Post–balance sheet events and the ceiling test

Two types of events occurring after year-end but prior to publication of the financial statements can have a major impact on the full cost ceiling test. First, if proved reserves are discovered after year-end but before the financial statements are published, the ceiling test should be revised to include the new reserves. This is the type of subsequent event for which GAAP requires an adjustment to the financial statements, since the reserves existed at the balance sheet date but were unknown at that time. Second, the SEC apparently allows some changes in the price of oil or gas after year-end but before the financial statements are issued to be used in calculating the full cost ceiling test. However, because a change in price is not an underlying condition as of the balance sheet date, the allowed treatment by the SEC is inconsistent with GAAP. GAAP allows an adjustment to the financial statements only when the subsequent event provides evidence of a condition that existed at the balance sheet date.

Problems

1. Indicate whether each of the following costs would be expensed (E) or capitalized (C) under full cost (FC) and successful efforts (SE) accounting.

Cost	SE	FC
a. Aerial magnetic study—an area of interest is identified		
b. Seismic studies on 20,000 acres—no land is leased		
c. Brokers' fees		
d. Bottom-hole contribution Well is productive		
Well is dry		
e. Dry-hole contribution		
f. Delay rental payment		
g. Cost of landmen in acquiring properties		
h. G&G to select specific drillsite		
i. Cash bonus		
j. Productive exploratory drilling Intangible costs		
Tangible costs		
k. Lease record maintenance on unproved properties		
l. Exploratory dry-hole drilling Intangible costs		
Tangible costs		
m. Shooting rights		
n. Development dry-hole drilling Intangible costs		
Tangible costs		
o. Production costs		

2. Given the following, compute DD&A assuming no exclusions from the amortization base:

Unrecovered costs	$700,000
Net salvage value of properties	60,000
Estimated future development costs for proved reserves. . . .	100,000
Proved reserves, 12/31	100,000 bbl
Proved developed reserves, 12/31	75,000 bbl
Production	20,000 bbl

3. Data for Dignity Petroleum for 12/31/16 are as follows:

Unrecovered costs	$700,000
Estimated future development costs for proved reserves	100,000
Estimated future decommissioning costs.	200,000
Net salvage value of properties	60,000

	Oil bbl	Gas Mcf
Proved reserves, 12/31/16	100,000	300,000
Proved developed reserves, 12/31/16	75,000	120,000
Production during year	20,000	50,000
Selling price during year.	$60/bbl	$5.00/Mcf
Selling price, 12/31/16	70/bbl	7.50/Mcf
Expected selling price, 2017.	74/bbl	8.00/Mcf

   a. Compute DD&A using a common unit of measure based on BOE.

   b. Compute DD&A using the unit-of-revenue basis.

   c. Which would be the appropriate basis?

4. When oil and gas are produced jointly, what basis or units of measure are allowed for full cost amortization? When oil and gas are produced jointly, what basis or units of measure are allowed for successful efforts amortization?

5. Which special costs or reserves must or may be excluded from amortization for full cost? Which costs or reserves must or may be excluded for successful efforts?

6. Which special costs or reserves must or may be included in amortization for full cost? Which costs or reserves must or may be included for successful efforts?

7. In full cost and successful efforts accounting, costs are amortized over either proved reserves (PR) or proved developed reserves (PDR). Fill in the following table to indicate which reserves (PR or PDR) should be used to amortize the costs under each accounting method. If the costs should not be amortized, put an X to indicate no amortization.

Cost	SE	FC
a. Proved property acquisition		
b. Unproved property acquisition		
c. Successful exploratory drilling		
d. Successful development drilling		
e. Unsuccessful exploratory drilling		
f. Unsuccessful development drilling		
g. G&G costs		
h. Wells-in-progress—proved property		

8. Data for Dora's U.S. properties:

I	E		
		Acquisition costs of unproved properties, net of impairment	$100,000
		Acquisition costs, proved properties	150,000
		Delay rentals, unproved properties—cumulative	30,000
		Delay rentals, proved properties—cumulative	50,000
		Dry holes, proved properties	300,000
		Dry holes, unproved properties	160,000
		Abandoned costs	40,000
		Successful well costs	400,000
		Wells-in-progress—unproved properties	175,000
		Wells-in-progress—proved properties	125,000
		Major development project necessary to determine quantity of proved reserves	500,000
		Future development costs (proved properties)	300,000
		Accumulated DD&A	(800,000)
		Production, bbl	50,000
		Proved reserves, beginning of year, bbl	600,000

For the column headed *I* above, assume all possible costs are included in the amortization base. For the column headed *E*, assume all possible costs are excluded from the amortization base. Place an *I* in the columns above if a cost must be included in amortization. Place an *E* in the columns above if the cost may be excluded from amortization.

Calculate DD&A assuming the following:

a. Inclusion of both unproved property costs and the major development project.
b. Exclusion of both unproved property costs and the major development project.
c. Inclusion of unproved property costs and exclusion of the major development project.
d. Exclusion of unproved property costs and inclusion of the major development project.
e. How would impairment for unproved properties have been handled?

9. Ernest Petroleum began operations January 1, 2015. Transactions for the first three years include the data below. Using that data:

a. Prepare journal entries assuming full cost (ignore revenue entries and assume no exclusions from the amortization base).
b. Prepare income statements under full cost and successful efforts for all three years, again assuming no exclusions from the full cost amortization base. Ignore severance tax. Assume a 1/8 royalty interest.
c. Recalculate DD&A assuming Ernest is a full cost company and that Ernest excludes all possible costs from the amortization base.
d. Which of the journal entries given in part a above would have been different if Ernest had been excluding all possible costs from the amortization base rather than including all costs?

Data:

2015—Acquired three leases; assume a 1/8 royalty interest.

Lease	Bonus	G&G Costs, Direct
A	$20,000	$30,000
B	15,000	70,000
C	30,000	50,000

2016

1) A delay rental of $2,000 is paid for Lease B.
2) A delay rental of $4,000 is paid for Lease A.
3) Drilling costs of $120,000 are paid on Lease C. Proved reserves are found and estimated to be 200,000 total gross barrels, and proved developed reserves are estimated to be 70,000 total gross barrels as of December 31, 2016. During 2016, 20,000 total gross barrels of oil are produced and sold. Lifting costs are $15/bbl, and the selling price is $65/bbl. (Expense lifting costs as lease operating expense.) Future development costs are estimated to be $100,000.

2017

1) Lease B is surrendered.

2) A dry hole is drilled on Lease A at a cost of $250,000. As a result, Ernest feels that Lease A is worth only 1/4 of the amount capitalized as unproved property. (Note: successful efforts and full cost impairment amounts will be different.)

3) An additional well (development) is drilled on Lease C at a cost of $300,000. Proved reserves at 12/31/17 are estimated to be 230,000 total gross barrels, and proved developed reserves are estimated to be 90,000 total gross barrels.

During 2017, 25,000 total gross barrels of oil are produced and sold for $70/bbl. Lifting costs are $14/bbl. Future development costs are estimated to be $150,000.

10. Cowboy Oil Corporation began operations in 2016. Give the entries, assuming the following transactions in the first three years of operations. Calculate DD&A twice, once assuming no exclusions and once assuming all possible exclusions from the amortization base. Ignore the ceiling test for 2016, but apply it for 2017 and 2018. Calculate the ceiling test twice for each of those years, once assuming no exclusions and once assuming all possible exclusions from the amortization base. Assume all leases are located in the United States. You may combine entries. Note that problem 11 in chapter 5 is similar to this problem. Compare these two problems for a comparison of successful efforts and full cost accounting.

Year	Transaction	Lease A	Lease B	Lease C
2016	Lease bonuses	$ 50,000	$ 40,000	$ 55,000
	G&G costs—direct	60,000	50,000	90,000
	Lease record maintenance	2,000	5,000	1,000
	Legal costs for title defense	15,000	None	10,000

	Drilling costs
	  IDC
	  Equipment
		
		None
		None
			Well 1, exploratory
			300,000
			125,000
				Well 1 exploratory
				150,000
				40,000

	Drilling results	None	Well dry	Drilling uncompleted
2017	Delay rental	4,000	None	None

	Drilling costs
	  IDC
	  Equipment
		Well 1, exploratory
		275,000
		50,000
			Well 2, exploratory
			225,000
			50,000
				Well 1
				
				60,000
				40,000

	Drilling results	Drilling completed, found PR of 100,000 bbl	Drilling uncompleted	Drilling completed, found PR of 300,000 bbl

	Installed flow lines, tanks, etc.
	  Installation costs
	  Purchase costs
		
		5,000
		30,000
			
			None
			None
				
				3,000
				45,000

	Future development costs	600,000	None	400,000
	Production during year	4,000 bbl	None	6,000 bbl
	PV of future net revenue as of 12/31/17	500,000	FMV of Lease B = $200,000	800,000

2018	Drilling costs
	  IDC
	  Equipment
		Well 2, development
		300,000
		80,000
			Well 2
			
			5,000
			
				Well 2 development
				250,000
				100,000

	Drilling results	Well dry	Well dry, abandon lease	Drilling completed, found PR of 200,000 bbl
	Future development costs	200,000	NA	0
	Production during year	5,000 bbl	None	20,000 bbl
	PV of future net revenue as of 12/31/18	1,000,000	FMV of Lease B = $0	3,000,000

PR = proved reserves

11. Data as of 12/31/15 for Dry Hole Oil Company's U.S. properties are as follows: (This problem is similar to problem 14 in chapter 6.)

Unproved properties, control	$ 200,000
Wells-in-progress—unproved properties	350,000
Dry exploratory wells on unproved properties	425,000
Abandoned costs of unproved properties	275,000
Proved properties	190,000
Wells and equipment	600,000
Accumulated DD&A	375,000

Dry Hole's activities during 2016 were as follows:

Unproved properties acquired	$ 35,000
Delay rentals paid	10,000
Test well contributions paid on unproved properties	20,000
Title exams paid on unproved properties	8,000
Title defenses paid on unproved properties	15,000
Unproved properties proved during 2016	30,000
Unproved properties abandoned	20,000
Exploratory dry hole drilled on unproved property (Total cost of well was $375,000; $100,000 was in progress at 12/31/15, and therefore not incurred during 2016; $275,000 was incurred in 2016.)	275,000
Successful exploratory well drilled	400,000
Development dry hole drilled	300,000
Service well drilled	325,000
Tanks, separators, etc., installed	125,000
Development well partially drilled, in progress 12/31/16	140,000
Estimated future development costs	225,000

	Oil (bbl)	Gas (Mcf)
Production	100,000	500,000
Proved reserves, 12/31/16	1,020,000	5,000,000
Proved developed reserves, 12/31/16	900,000	4,700,000

a. Use T-accounts to accumulate costs.

b. Calculate DD&A for 2016, assuming no cost exclusions and using a common unit of measure based on BOE.

c. Calculate DD&A for 2016, assuming all possible cost exclusions and using a common unit of measure based on BOE. In addition, assume impairment for unproved properties was $75,000 and the allowance for impairment was $25,000 at 12/31/16.

12. Core Petroleum started its oil and gas exploration and production business in 2015. During the years 2015 and 2016, the company provided the following information relating to leases located both in the United States and in Canada:

Year		Lease A (U.S.)	Lease B (U.S.)	Lease C (Canada)
2015	Leasehold costs	$ 40,000	$ 30,000	
	Well number	Well 1	Well 1	
	IDC	300,000	500,000	
	L&WE	0	350,000	
	Results	Well dry, abandon well	Proved	
	Future development costs		200,000	
	Production: Oil		4,000 bbl	
	Gas		20,000 Mcf	
	Estimated reserves:			
	Oil, PR—12/31/15		50,000 bbl	
	Gas, PR—12/31/15		300,000 Mcf	
	Oil, PDR—12/31/15		30,000 bbl	
	Gas, PDR—2/31/15		200,000 Mcf	
	Selling price: Oil		$60/bbl	
	Gas		$4.50/Mcf	
	Price (12/31/15): Oil		$66/bbl	
	Gas		$6.00/Mcf	
2016	Leasehold costs			$60,000
	Well number	Well 2		Well 1
	IDC	400,000		600,000
	L&WE	300,000		400,000
	Results	Proved		Proved
	Future development costs	100,000	150,000	175,000
	Production: Oil	5,000 bbl	8,000 bbl	8,000 bbl
	Gas	20,000 Mcf	30,000 Mcf	
	Estimated reserves:			
	Oil, PR—12/31/16	40,000 bbl	40,000 bbl	80,000 bbl
	Gas, PR—12/31/16	250,000 Mcf	280,000 Mcf	
	Oil, PDR—12/31/16	30,000 bbl	28,000 bbl	30,000 bbl
	Gas, PDR—12/31/16	200,000 Mcf	180,000 Mcf	

a. Record the above information for both years. Ignore revenue entries.

b. Compute and record DD&A for both years. Assume the revenue method may be ignored in the second year. If there is not a significant difference between the revenue basis and energy basis in the first year (less than $150,000), use the energy basis (equivalent Mcf). Assume that all possible costs are included in DD&A.

c. Compute DD&A using a common unit of measure based on Mcf, assuming Core Oil Company used successful efforts accounting instead of full cost accounting.

13. Indicate by a *C* whether the costs given below should be capitalized. Indicate by an *I* if the costs must be included in computing amortization or by an *E* if the cost may be excluded from amortization under full cost. Indicate by an *I* if the costs should be amortized under successful efforts. (U/P is unproved property, and P/P is proved property.)

	Costs to Capitalize		Costs to Amortize		
				FC	
Cost	SE	FC	SE	P/P	U/P
a. Bonus on unproved property					
b. Bottom-hole contribution					
c. Ad valorem taxes, paid on U/P					
d. Legal costs for title defense, U/P					
e. Nondirect G&G, cumulative					
f. Delay rentals, cumulative					
g. Wells-in-progress					
h. Dry exploratory wells					
i. Successful exploratory wells					
j. Successful development wells					
k. Dry development wells					
l. Abandoned costs					
m. Impairment of unproved properties					
n. Future development costs					
o. Estimated decommissioning costs					

14. The Clarence Oil Company provides the following information for the year ended December 31, 2017:

PV of future gross revenues	$60,000,000
PV of future related costs	15,000,000
Capitalized costs of proved properties	50,000,000
Unproved properties not being amortized	11,000,000
Accumulated DD&A	(4,000,000)

REQUIRED:

a. Prepare a ceiling test and an entry, if necessary, for the write-off of capitalized costs. Ignore income taxes.

b. Assuming the PV of future gross revenues is $70,000,000, repeat the requirements for part a.

15. Virginia Oil Corporation, a full cost company, incurs the following costs during 2016:

G&G costs—nondirect	$25,000
Acquisition costs:	
Lease A	26,000
Lease B	32,000
G&G costs—direct to Lease A	15,000
Ad valorem taxes on Lease B	1,200
Dry-hole contribution on land adjacent to Lease B	40,000
Nonrecoverable delinquent taxes on Lease A paid at the time of acquiring the lease	2,100
Cost of maintaining lease records:	
Lease A	500
Lease B	600

During 2017, the following costs were incurred:

Delay rentals were paid on Lease A, $2,000, and Lease B, $3,000.

In the latter half of 2017, drilling operations were commenced on both leases, and costs were incurred as follows:

	Lease A	Lease B
G&G costs to locate wellsite	$5,000	$ 6,000
Contractor's fee:		
IDC	325,000	350,000
L&WE	110,000	125,000
Flow lines:		
Equipment	3,000	5,000
Labor to install	2,000	3,000
Tank battery	15,000	17,000

REQUIRED: Record the above transactions.

16. The Jumper Oil Corporation incurs unproved property (Lease A) costs of $60,000 on April 1, 2015. An 8% loan is obtained on April 1, 2015 for $500,000 to finance a drilling program. Jumper started a well on Lease A on June 1, 2015, and the well is still in progress at 12/31/15. Drilling costs to 12/31/15 are $300,000. The company excludes all possible costs from the cost pool.

REQUIRED: Compute the interest capitalization amount and record the interest.

17. Determine the amount of the total amortization cost base, assuming (1) no exclusions from the amortization base, and (2) all possible costs are excluded from the amortization base.

	(1) Amount—no exclusions	(2) Amount—all exclusions
Unproved property—cost $50,000, amount impaired $20,000 (if excluded)		
Unproved property—abandoned—cost $22,000, no impairment		
Unproved property—found proved reserves—cost $18,000, no impairment		
Unproved property purchased—cost $40,000		
Wells-in-progress on unproved property—$100,000		
Well completed on unproved property—dry-hole cost—$220,000		
Total	?	?

18. Paddle Petroleum has the following account balances at 12/31/15:

    Lease A: unproved property . . . . $ 20,000
    Lease B: proved property 30,000
    Wells and equipment—IDC 250,000
    Wells and equipment—L&WE 150,000

    The above properties are abandoned in 2016. Record the abandonment entry.

19. Aaron Energy Corporation has the following account balances at 12/31/15:

    Unproved property—Lease A . . . . $ 60,000
    Unproved property—group 220,000
    Allowance for impairment, group 12,000

    At 12/31/15, Lease A is considered to be 40% impaired. Aaron Energy's estimated abandonment rate of insignificant unproved properties is 60% (i.e., the impairment rate is 60%).

    Prepare entries to record impairment.

20. Use the same facts as problem 19 and prepare entries using the following independent assumptions:

    a. Lease A is abandoned in 2016.
    b. Lease A is proved in 2016.

c. Insignificant Lease Y, with a cost of $3,000, is abandoned.

d. Insignificant Lease X, with a cost of $5,000, is proved.

21. Dwight Oil and Gas Company has the following information at 12/31/15.

Capitalized costs, plus future development costs.	$2,000,000
Accumulated DD&A	800,000
Proved reserves—Oil	200,000 bbl
—Gas	800,000 Mcf

    Production in 2015:

Oil	15,000 bbl	Sold at average price of $70/bbl
Gas	70,000 Mcf	Sold at average price of $5.00/Mcf

    The current prices at 12/31/15 are:

Oil	$75.00/bbl
Gas	$ 5.50/Mcf

    Compute DD&A using the unit-of-revenue method.

22. Complex Corporation started operations on 1/1/15. At 12/31/15, the company owned the following leases in Canada:

Lease A—Proved property	$ 75,000
—Wells and equipment.	500,000
Lease B—Proved property	90,000
—Wells and equipment.	800,000
Lease C—Unproved property.	60,000
—Wells-in-progress—Lease C	125,000
Proved Reserves—Oil	200,000 bbl
—Gas	900,000 Mcf
Production—Oil.	10,000 bbl
—Gas	50,000 Mcf

    The production was sold at $70/bbl and $5.50/Mcf. Current prices at 12/31/15 are $75/bbl and $6.00/Mcf.

    Compute DD&A for Canada, assuming the following:

    a. No exclusions from the amortization base, and using unit-of-production converted to a common unit of measure based on energy (equivalent Mcf).

b. No exclusions from the amortization base, and using unit-of-revenue method.

c. All possible costs are excluded from amortization, and using a common unit of measure based on energy (equivalent Mcf).

d. All possible costs are excluded from amortization, and using unit-of-revenue method.

23. Data for Porch Oil Company is as follows for all U.S. properties:

Acquisition costs of unproved property, net of impairment	$ 300,000
Nondrilling evaluation costs of unproved property—direct	400,000
Abandoned costs	100,000
Dry holes on unproved property	600,000
Capitalized costs of proved property	2,400,000
Costs of major development project (costs not included above)	800,000
Accumulated DD&A (assume same for part a and part b)	700,000
Deferred income taxes	500,000
PV of future gross revenue	4,100,000
PV of future related costs	1,400,000
Income tax effects—books versus tax	500,000
LCM of unproved properties	550,000

a. Apply the full cost ceiling test and record any entries necessary, assuming that all possible costs are excluded from amortization.

b. Apply the full cost ceiling test and record any entries necessary, assuming that all possible costs are being amortized.

24. Alan Oil Corporation, located in Denver, Colorado, has been operating for three years. Alan uses full cost accounting and excludes all possible costs from the amortization base. The following account balances are as of the end of 2020.

	Insignificant Leases			Significant Leases	
	Cost	Total Impairment		Cost	Impairment
Lease A	$ 50,000		Lease D	$400,000	$100,000
Lease B	60,000		Lease E	300,000	200,000
Lease C	30,000		Lease F	600,000	400,000
	$140,000	$84,000			

REQUIRED: Give any entries necessary for the following events and transactions:

a. Lease A was abandoned on March 20, 2021.

b. Lease B was proved on May 13.

c. Insignificant leases costing $80,000 were acquired during 2021.

d. Lease E was abandoned early in July.

e. Lease F was proved October 21.

f. At December 31, Alan decides that Lease D should be impaired an additional $150,000.

g. At December 31, Alan decides to continue its policy of maintaining an allowance for impairment equal to 60% of unproved insignificant leases.

25. Gusher Oil Company began operations on 1/5/2011 and has acquired only two properties. The two properties, which are both considered significant, are located in different states. Lease B was proved on 1/1/2013. Costs incurred from 1/5/2011 through 12/31/2013 are as follows:

	Lease A	Lease B	Unallocated
Seismic studies, nondirect			$70,000
Bonus	$ 50,000	$ 60,000	
Title exams	10,000	5,000	
G&G costs, direct	90,000	80,000	
Test-well contributions	15,000	18,000	
Insurance	2,000	3,000	
Exploratory dry holes—IDC	220,000	250,000	
Exploratory dry holes—L&WE	30,000	40,000	
Exploratory wells-in-progress—IDC	100,000	180,000	
Exploratory wells-in-progress—L&WE	22,000	19,000	
Wells and equipment—IDC	—	700,000	
Wells and equipment—L&WE	—	260,000	
Tanks and separators	—	110,000	
Lease operating costs	—	134,000	
Total	$539,000	$1,859,000	$70,000

Additional data

Future development costs	—	$500,000
Proved reserves, 1/1/2013	—	600,000 bbl
Proved developed reserves, 1/1/2013	—	200,000 bbl
Production during 2013	—	40,000 bbl

Other information:

The company also owns a building that it purchased 1/1/2011 at a cost of $500,000. The building houses the corporate headquarters and has an estimated life of 20 years (ignore salvage). The operations conducted in the building are general in nature and are not directly attributable to any specific exploration, development, or production activities. Since the building is not related to exploration, development, or production, it is depreciated using straight-line depreciation for financial accounting.

REQUIRED:

a. Give the entry to record DD&A for 2013 under full cost accounting, assuming all possible costs are excluded from amortization.

b. Give the entry to record DD&A for 2013 under successful efforts accounting.

26. Stadium Oil Corporation uses the full cost method. Recently, the company acquired a truck costing $60,000 with an estimated life of five years (ignore salvage value). The foreman drives the truck to oversee operations on seven leases, all in the same general geographical area but on multiple reservoirs. The foreman keeps a log of his mileage in order to determine how the truck is utilized. Analysis of the log indicated that 1/3 of the mileage driven was related to travel to drilling operations, 1/3 was related to travel to G&G exploration areas, and 1/3 was related to travel to production locations. Total net capitalized costs, including the truck above, are $2,300,000. Proved reserves at the beginning of the year were 300,000 barrels, and production during the year was 30,000 barrels. Stadium amortizes all possible costs.

REQUIRED:

Give any entries necessary to record depreciation on the truck for the first year that it was in service.

8

ACCOUNTING FOR PRODUCTION ACTIVITIES

This chapter discusses the accounting treatment of costs incurred in producing oil and gas. Accounting for these costs is essentially the same whether a company is using the successful efforts method or the full cost method.

After a well has been completed, and flow lines, heater-treaters, separators, storage tanks, etc., have been installed, production activities begin. Production activities involve lifting the oil and gas to the surface and then gathering, treating, processing, and storing the oil and gas (*SFAS No. 19*, par. 23). Costs incurred in these activities are called **production costs, lease operating costs, or lifting costs**. They are defined under both successful efforts and full cost accounting as follows per SEC *Reg. S-X 4-10*, par. 17:

> *(i) Costs incurred to operate and maintain wells and related equipment and facilities, including depreciation and applicable operating costs of support equipment and facilities and other costs of operating and maintaining those wells and related equipment and facilities. They become part of the cost of oil and gas produced. Examples of production costs (sometimes called lifting costs) are:*
>
> *(A) Costs of labor to operate the wells and related equipment and facilities.*
>
> *(B) Repairs and maintenance.*
>
> *(C) Materials, supplies, and fuel consumed and services utilized in operating the wells and related equipment and facilities.*

(D) Property taxes and insurance applicable to proved properties and wells and related equipment and facilities.

(E) Severance taxes.

(ii) Some support equipment or facilities may serve two or more oil and gas producing activities and may also serve transportation, refining, and marketing activities. To the extent that the support equipment and facilities are used in oil and gas producing activities, their depreciation and applicable operating costs become exploration, development or production costs, as appropriate. Depreciation, depletion, and amortization of capitalized acquisition, exploration, and development costs are not production costs but also become part of the cost of oil and gas produced along with production (lifting) costs identified above.

ACCOUNTING TREATMENT

The definition of production costs for both successful efforts and full cost states that production costs become part of the cost of the oil and gas produced. However, rarely, if ever, do companies report production costs as "cost of goods sold." Instead, production costs are listed on the income statement as production expense.

According to SEC *Reg. S-X 4-10*, production operations cease very near the point of sale or delivery. *Reg. S-X 4-10,* par. 1 indicates the following:

For purposes of this section, the oil and gas production function shall normally be regarded as terminating at the outlet valve on the lease or field storage tank; if unusual physical or operational circumstances exist, it may be appropriate to regard the production functions as terminating at the first point at which oil, gas, or gas liquids are delivered to a main pipeline, a common carrier, a refinery, or a marine terminal.

Since in oil and gas operations the point of production is frequently very close to the point of sale, the amount of oil or gas in inventory may be quite small. Historically, many companies chose to simply treat all production costs as relating to the accounting period, and as a consequence, did not record inventory. In recent years this practice has evolved, and recognition of inventories has become more common, thus necessitating the allocation of production costs between the oil or gas in inventory versus the oil or gas sold.

If all production costs are expensed as incurred, no inventory would be recorded. The full amount of production costs would be treated as a cost of operations for the period and charged to expense as shown in the following example and in Figure 8–1.

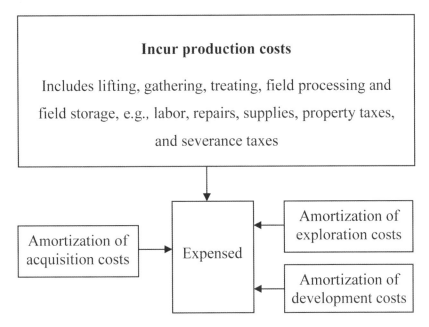

Fig. 8–1. Production costs

EXAMPLE

Production Costs

Tyler Company paid wages of $1,500 to employees engaged in operating the wells and equipment solely on Lease A.

Entry
Lease operating expense	1,500	
Cash		1,500

Cost of production versus inventory

When natural gas is produced, it often goes directly from the well into a gathering line for delivery to a purchaser or to the producer's downstream operations. Sometimes natural gas may require treatment, which occurs in the field. For example, as natural gas exits the wellhead, it may be run through a separator in order to separate the gas from the liquids or sediment contained in the raw gas. The natural gas may then be moved a short distance to a treatment plant, where more extensive processing is performed prior to placing the natural gas into a sales line. Regardless of the extent of processing required, the molecules of gas typically flow from the well, through the processing equipment, and into a gathering line for delivery on a continuous basis. It is an operational necessity to have gas in the lines in order to move gas production through the system. The working interest owners may own the natural gas pipeline that moves the gas from near the point of production to a processing plant or to market. The estimated amount of gas in the system at any given time is referred to as *fill gas* or *line fill*.

Technically, fill gas or line fill is inventory. Depending on the operation and such factors as size and length of the pipeline, the amount of gas in the system at any given time may range from a nominal quantity up to substantial quantities of natural gas. Typically, the volume is not material. However, in some cases it may represent several days, weeks, or even months worth of production. Some companies choose to treat this gas as a form of "permanent" inventory. That is, once the gas is in the system, its cost is reflected in inventory on a last-in, first-out (LIFO) basis until such time as the operation ceases. Other companies choose to ignore the inventory, essentially treating the gas as being sold at the point of production.

A somewhat different situation occurs with oil production. Sometimes oil is produced and sold at or near the wellhead. More often, the oil flows from wells to storage tanks, where it stays until it is lifted by the buyer or for delivery into the producer's downstream operations. Some fairly limited processing for the removal of gas, water, sediment, etc. may also be performed at or near the point of production. At any given point in time, the amount of oil in the storage tanks is oil inventory. The accumulation of oil into tanks is also a fairly continuous process, with tank capacity usually being limited to only a few days' production. Since there is normally no business reason for the producer to accumulate large stocks of oil inventories, producers typically seek to minimize the amount of oil in tanks and storage facilities.

For many years E&P companies have opted to recognize oil sales revenue at the point of production rather than attempt to recognize as inventory the oil that has been produced but not yet sold. A number of factors have led to this practice. First, it is generally not a matter of whether the oil will be sold, but rather a matter of when the buyer will take delivery. Second, there is an active market for oil, and therefore it is possible to determine the sales value without necessarily having physically delivered the oil to a buyer. Finally, in order to record the oil produced but not yet sold as inventory, it is necessary to know the cost of the oil produced. In E&P accounting, companies do not allocate material, labor, and overhead to production in the same manner that these costs are tracked and allocated in manufacturing operations.

Recognition of inventories

In order to report oil or gas inventories on the balance sheet, it is necessary for the producer to implement a process for measuring the physical quantities of oil and gas that have been produced but not yet transferred to the buyer at the end of each reporting period (e.g., oil and gas in tanks or storage facilities, pipelines, and processing vessels). In the *2001 PricewaterhouseCoopers Survey of U.S. Petroleum Accounting Practices*, of all responding companies, 30% recognize crude oil in lease tanks and 20.7% recognize crude oil in downstream storage facilities as inventory in their financial statements. Additionally, 26.7% recognize the natural gas held in storage as inventory.[1]

Lower-of-cost-or-market valuation

In situations where inventories are recorded, financial accounting principles typically require that the inventories be recorded at the cost of producing the products and getting them to a condition and location ready for sale. Due to the nature of oil and gas production, such costs are not typically tracked on a per unit basis or even divided between the cost of producing oil versus natural gas during a period. Instead, the cost of producing oil and gas are joint costs, typically reported on the income statement as production costs rather than inventory or cost of goods sold. In addition, oil and gas companies' accounting systems may not be designed to track or assign costs to individual products or to track the costs per unit of joint product produced as the products flow through the production process.

One popular approach to assigning costs if inventory is recorded is to compute the cost per BOE or Mcfe produced by taking the total production cost during the past period and dividing it by the equivalent number of units produced. The cost of production typically includes all direct costs such as labor, fuel, workovers, repairs, field-level overhead, etc. However, some differences of opinion exist between the companies that do report oil and gas inventories regarding whether DD&A or general corporate overhead should be included in the total cost figure. The *2001 PricewaterhouseCoopers Survey of U.S. Petroleum Accounting Practices* reveals that while most companies that report inventories include all direct costs, only 57.1% of those companies report that they include DD&A. Only 42.9% report that they include allocated home office or corporate overhead.[2]

If the actual costs incurred are used in valuing the inventory, the final valuation is based on the principle of lower-of-cost-or-market. This valuation requires a comparison of the estimated historical cost of the units in inventory with their net realizable value. If the net realizable value is higher than historical cost, then the units are valued at their historical cost. If the net realizable value is below cost, the units are valued at net realizable value.

For U.S. GAAP, *Accounting Research Bulletin (ARB) No. 43*, "Restatement and Revision of Accounting Research Bulletins," indicates that in certain circumstances it may be acceptable to carry inventory at an amount above cost. Specifically, *ARB No. 43* indicates that in order for inventory to be carried above its cost, all of the following criteria must be met:

1. Inability to determine appropriate approximate costs
2. Immediate marketability at quoted market price
3. Units completely interchangeable

As a consequence of *ARB No. 43*, in the past, some companies elected not to attempt to determine historical cost and instead used net realizable value to value their inventory. Obviously, use of net realizable value involves adjusting market values to reflect the approximate costs of processing, transportation, marketing, etc. that are incurred between the point of production and the point of sale.

However, the SEC's March 2001 *Interpretations and Guidance*, par (II)(F)(2) indicates that only in exceptional cases may inventory properly be stated at an amount above cost. As a consequence, many companies use lower-of-cost-or-market, while other companies continue to use net realizable value, claiming that their operations clearly meet the *ARB No. 43* criteria.

Accumulation and Allocation of Costs

As with acquisition, exploration, and development costs, production costs must be accumulated by cost center. Therefore, a company using successful efforts accumulates costs by an individual lease, reservoir, or field. A company using full cost accounting accumulates costs by country. However, a full cost company (as well as a successful efforts company) also normally accumulates costs on a smaller cost center, such as a lease or even on an individual well basis, because of tax, regulatory, and contractual obligations, as well as management requirements. Thus, even though the accounting method itself does not require allocation of costs to a well or lease, a company typically allocates costs to individual wells or leases.

Production costs can be divided into those that are directly attributable to a specific well or lease and those that must be assigned to the well or lease through some method of allocation. For costs such as repairs to a specific well or wages and benefits of an employee working solely on one lease, accumulation of production costs by lease is straightforward. The costs are simply charged directly to the well or lease involved. Allocable costs such as the cost of a saltwater disposal facility serving multiple leases must be allocated to each well or lease on some reasonable basis. Common allocation bases include the number of wells or number of barrels produced. Other reasonable allocation bases for different types of allocable costs include the following:

- Number of direct labor hours
- Amount of direct labor costs
- Number of miles driven for transportation and hauling
- Gallons of water used for waterflooding
- Miles traveled by boat for transportation

Table 8–1 provides several examples of production-related costs that are directly attributable to a particular well or lease versus those costs that must be assigned to a well or lease through cost allocation.

Table 8–1. Directly attributable versus allocable production costs

Directly Attributable Costs	**Allocable Costs**
a. Direct materials, supplies, and fuel—wells and leases involved identified in the invoices	a. Field offices and facilities serving several leases
b. Direct labor (pumpers), gaugers, etc.—employees who work on one lease only or who designate hours worked on certain wells or leases	b. Salaries and fringe benefits of field or operations center supervisors of several leases or fields
c. Contract labor or services for oxidizing, refracturing, scrubbing, etc.—invoices indicate wells	c. Depreciation—support facilities, gathering systems, treatment systems—several leases involved
d. Repairs and maintenance that can be traced to individual wells and leases	d. Transportation and hauling—several leases involved
e. Property taxes and insurance traceable from tax receipts or property descriptions on insurance policies	e. Operating costs of saltwater disposal system—several leases involved
f. Production or severance taxes—reports to the state identify these taxes to specific leases	f. Boat and fuel expenses, offshore operations—several leases involved
g. Operating costs of waterflooding systems—only one lease involved	g. Operating costs of waterflooding system—several leases involved

EXAMPLE

Cost Allocation

Expenses for Field Office A of Tyler Oil Company amounted to $10,000 for the month of May. The field office has supervision over the following leases and wells:

Lease	Number of Wells	Barrels of Oil Produced
A	1	1,000
B	3	500
C	4	2,000
D	2	1,500
Total	10	5,000

a. If the field office expense is allocated on the basis of the number of barrels (bbl) produced, each lease would be charged the following amount:

Lease	Computations	Amount
A	($10,000 × 1,000 bbl/5,000 bbl)	$ 2,000
B	($10,000 × 500/5,000)	1,000
C	($10,000 × 2,000/5,000)	4,000
D	($10,000 × 1,500/5,000)	3,000
Total		$10,000

b. If the field office expense is allocated on the basis of the number of wells supervised, each lease would be charged the following amount:

Lease	Computations	Amount
A	($10,000 × 1 well/10 wells)	$ 1,000
B	($10,000 × 3/10)	3,000
C	($10,000 × 4/10)	4,000
D	($10,000 × 2/10)	2,000
Total		$10,000

Individual Production Costs

Labor costs to operate wells and equipment include salaries, wages, and employee benefits of first-level supervisors and field employees who work on the leases. First-level supervisors generally supervise employees and contract laborers who are directly employed on a property in maintaining and operating the wells and equipment. First-level supervisor costs are directly related to the operation of the leases and are typically charged to the leases on the basis of detailed time records that show the amount of time worked on each property. If detailed time records are not kept, the costs are apportioned on some equitable basis. Field employees perform routine maintenance and provide supervision of the measurement process. Field employees include pumpers, gaugers, field technicians, etc. If field work covers more than one lease or well, detailed time sheets should be maintained for field employees so their salaries and benefits can be charged to the individual leases or wells served.

Labor costs include **employee benefits**, which are charged to leases along with other labor costs. Generally, companies estimate their total employee benefits and direct labor costs, and then compute the ratio of employee benefits to direct labor costs. For example, a company may estimate that employee benefits will equal 45% of direct labor costs. Then, as direct labor costs are charged to leases, an additional 45% is charged representing employee benefits.

Repairs and maintenance include normal repairs and maintenance, certain workover operations, and certain recompletions. Normal repair and maintenance may be necessary for all lease equipment such as tanks and flow lines, lease roads, and buildings on the lease. Examples of repair and maintenance include replacing a defective meter, repairing a broken flow line, and lubricating production equipment. Repair and maintenance cost is expensed unless the useful life of the asset is materially lengthened or the productivity of the asset is materially increased. Invoices should indicate which wells or leases to charge. Consequently, these costs are directly attributable to particular wells or leases.

Often, repair and maintenance costs, plus costs of operation of the wells and equipment, may be incurred on a contract basis, with an independent contractor providing this pumping service. **Pumping service** includes such items as routine maintenance, meter reading, and gauging operations (i.e., measuring the level of oil in a tank). The invoice from the independent contractor for pumping service should identify the well or lease to be charged and provide the data for the charges to particular properties.

Production operations also include many types of workovers and recompletions. **Workover operations** may include repairing sucker rods, tubing, casing, and leaks, cleaning out sand-filled perforations, acidizing, etc. If the workover cost is incurred for the purpose of stimulating or restoring production in the same producing horizon, it should be expensed as a production cost. If the work involves deepening a well or plugging back to a shallower horizon to access proved undeveloped reserves or to increase proved reserves, the operation is typically referred to as a recompletion. It then is considered to be exploratory or development drilling costs rather than production. Consequently, the costs should be capitalized or expensed, depending upon the financial method of accounting being used, the classification of the well, and the outcome of the drilling. A recompletion may also involve recompleting in a producing zone or in a zone that produced previously. If the purpose is to restore or increase production without an increase in proved reserves, the costs should be charged to expense.

Costs of materials, supplies, and fuel used in well operations include materials and supplies used in routine repair and maintenance as well as fuel used to operate the production equipment, such as separators, heater-treaters, etc. These costs are considered production costs and can usually be identified with a particular lease.

Property taxes and insurance on proved properties are also lease operating costs. **Property taxes** are ad valorem taxes levied by governmental agencies. Property taxes on proved properties are typically significant in amount. [Property taxes assessed on unproved properties are generally insignificant in amount and are considered exploration costs rather than production costs (chapter 3)]. Types of **insurance coverage** on proved property would include general liability, worker's compensation, and fire and casualty. Both property taxes and insurance are easily identified with a particular property.

Shut-in payments normally must be paid to the royalty owner when a well that is capable of producing is shut in. Companies using either successful efforts accounting or full cost accounting would normally expense shut-in payments as production costs if they were not recoverable from future production. When shut-in payments are recoverable from future production, a receivable from the royalty owner is recorded for the amount of the payments by both full cost and successful efforts companies.

Depreciation and operating costs of support equipment and facilities should theoretically be allocated to exploration, development, or production as appropriate. The portion allocated to production is expensed as a cost of production. Depreciation of support equipment and facilities is discussed in chapters 6 and 7.

General administrative overhead costs are not directly related to oil and gas acquisition, exploration, development, and production activities and so are not considered production costs. However, they are also expensed as incurred. This type of overhead typically includes home office expenses such as officers' salaries, legal fees, accounting, etc. Since general administrative overhead costs are not directly related to oil and gas producing activities and are not production costs, they are not allocated to individual leases or wells for financial accounting and reporting purposes. Instead, such costs are reported as general and administrative expense.

Secondary and tertiary recovery

Installation of secondary and tertiary recovery systems frequently requires large expenditures, e.g., the cost of drilling multiple injection wells, converting producing wells to injection wells, and the cost of purchasing injection equipment. The expenditures for the wells and equipment are considered part of the development phase and are therefore capitalized and amortized using the unit-of-production method. The costs of any chemical injectants that are utilized are commonly accounted for using one of two methods. First, the injectants may be considered depreciable property and recorded as wells and equipment. The costs are then amortized along with other wells and equipment costs for the cost center. Second, the costs of injectants may be charged directly to production expense.

Routine maintenance and operating costs of secondary and tertiary recovery systems are considered production costs and are expensed as incurred.

The following example illustrates drilling and operating a secondary recovery system.

EXAMPLE

Secondary Recovery System

Tyler Company incurred $3,000,000 in drilling and equipping a waterflood secondary recovery system. In the month following completion of the installation, $4,000 was incurred for supplies and fuel to operate the system. In addition, the company incurred a $10,000 cost for water that was purchased and injected into the formation.

Entries

Secondary recovery system	3,000,000	
Notes payable		3,000,000
Lease operating expense ($4,000 + $10,000)	14,000	
Cash		14,000

Gathering systems

A gathering system begins with pipelines that move the oil and gas produced from individual wells to a central point where the oil and gas are separated and the basic sediment and water (BS&W) is removed. After treatment, the oil and gas is then transferred either to tanks or into a pipeline to be transported to a refinery or gas processing plant, etc. An oil-gathering system includes equipment such as oil and gas separators, heater-treaters, and gathering tanks. A gas-gathering system includes equipment such as compressors and dehydrators. The term gathering system is also used to refer to just the network of flow lines.

Attendant costs incurred for gathering-system installation are development costs subject to DD&A. The operating costs, however, are considered production costs and are expensed as incurred.

Saltwater disposal systems

Saltwater, an unwanted by-product, is normally produced along with oil and gas. The saltwater, which must be disposed of, may be reinjected back into the formation as a means of disposal that does not harm the environment and that helps to maintain natural reservoir pressures. A saltwater disposal system gathers the saltwater, treats it to remove chemicals and corrosive materials, and then reinjects it into the earth. The cost of the system is capitalized and depreciated using either the unit-of-production method—if the system serves a single field—or any other acceptable method if the system serves multiple fields. If, as would normally be the case, the disposal system covers several leases, the operating costs must be allocated to the individual properties served.

In situations where the wells are producing approximately equal volumes of saltwater, the relative number of wells connected to the disposal system is a commonly used allocation basis. When the individual wells or properties are producing significantly different volumes of saltwater, the saltwater coming in from each lease will be metered. Relative saltwater throughput will then be used as the allocation basis. If only one lease is served by a saltwater disposal system, the operating costs are charged to lease operating expense for that lease.

Tubular goods

Another capital expenditure that gives rise to production costs involves tubular goods. In the oil and gas industry, the term **tubular goods** refers to casing and tubing. When a well or facility is originally drilled or constructed, the purchase and installation costs of tubular goods are capitalized. Subsequent repair and replacement of tubular goods are considered production costs and are expensed as incurred. In general, the price of tubular goods includes the invoice amount plus transportation costs.

The following example presents the purchase and installation of tubing for replacement of damaged tubing.

EXAMPLE

Tubular Goods

Tyler Company obtained a new string of tubing to replace damaged tubing in a producing well on Lease A. The net cost of the new tubing was $100,000, and transportation costs were $10,000. Installation costs of the new tubing were $20,000.

Entry

Lease operating expense.	130,000	
Cash		130,000

Severance taxes

Severance tax is a tax commonly levied by state governments on oil and gas production. Severance taxes, also called **production taxes**, are typically a specified percentage of the selling price of oil or gas. If not a specified percentage of the selling price, severance taxes are based on the quantity of oil or gas sold, e.g., $2/bbl. Outside of the United States, production taxes are also common. For example, value added taxes are often assessed on production.

Production costs are generally recorded as expenses when incurred. In contrast, severance taxes, which are usually based on the selling price of the product, are incurred in conjunction with the sale of oil or gas. As a result, this type of production cost is recorded when the related revenues are recorded. Recording severance taxes is illustrated in chapter 10.

PRODUCTION COSTS STATEMENTS

Each month, primarily for internal management purposes, operating statements similar to an income statement are prepared for each property. These statements are referred to as production costs statements. The statements normally include income and expense but may show expense only. Current-month along with year-to-date figures for revenues and expenses are usually reported.

Joint Interest Operations

When more than one company owns an undivided working interest in a property, the companies must enter into a joint operating agreement (JOA). One of the companies is designated the **operator**, typically the party with the largest interest. In production activities, the operator typically pays all of the direct costs related to operating the property and then bills the nonoperators for their proportionate share of the cost. Indirect costs, including corporate and administrative overhead incurred by the operator, are recovered through overhead charges to the nonoperators, as provided for in the JOA. (This topic is discussed in detail in chapter 12.)

Now that the costs involved in drilling, completing, and operating a well are known and understood, attention can be turned to the decision of whether to complete a well and whether a well or property is ultimately profitable.

Decision to Complete a Well

In making a decision whether to complete a well, the incremental costs to complete the well should be compared with the future net cash flows expected to be received from the sale of oil or gas produced from the well. This comparison entails estimating the following items:

- Quantity of oil or gas recoverable from the reservoir
- Timing of future production of the oil or gas
- Future selling price of the oil or gas
- Future production costs of the oil or gas, including severance taxes
- Completion costs
- Cost of capital

If the expected future net revenue is greater than the expected completion costs, the well is usually completed.

Estimation of total recoverable reserves is an inexact science, especially for the first well discovering oil or gas in a field. Reserve estimates are usually made by a reservoir engineer, who may be an employee of the company or an independent contractor. Some factors or characteristics taken into account by the reservoir engineer in preparing a reserve estimate are as follows:

- Size of the reservoir
- Porosity and permeability of the reservoir
- Pressure and temperature in the reservoir
- Oil, gas, and water contained in the reservoir pores

In addition to estimating the total recoverable reserves, the reserve engineer also must estimate the *timing* of production of the oil and gas reserves. Reserve recovery timing depends on such factors as the characteristics of the reservoir, product demand, and government regulations. Product demand is especially important for natural gas, because gas is difficult and expensive to store.

Another important consideration affecting the decision to complete a well is the future product price and the cost of producing the product. Both items must be estimated before net revenue can be determined. Estimating the future price of the product is a difficult and imprecise process. The price of oil or gas is affected by product supply and demand probably more than any other factor. Governmental intervention, whether domestic or foreign, also affects the price and is usually impossible to predict. On the other hand, lifting costs can generally be estimated with a relatively high degree of accuracy. Severance taxes are part of production costs and must also be estimated.

Finally, completion costs must be estimated and compared to future net cash inflows. Completion costs can typically be estimated relatively easily and accurately. An example follows showing the comparison of net cash inflows to completion costs necessary in the decision of whether to complete a well. Costs already incurred, i.e., sunk costs, are not relevant to that decision.

EXAMPLE

Completion Decision

Tyler Company's data follow:

Proved property cost (acquisition cost)	$ 40,000
Drilling cost incurred to date	500,000
Estimated completion cost	260,000
Estimated selling price per bbl	$50
Estimated lifting costs per bbl	$25
State severance tax	5%
Working interest share of revenue	90%
Royalty interest percentage	10%

REQUIRED: Should the well be completed assuming the following total gross production?

Production

Case A:	7,500 bbl
Case B:	15,000 bbl
Case C:	30,000 bbl

Computations

	Case A	Case B	Case C
Total revenue (bbl × $50)	$375,000	$750,000	$1,500,000
Less: Royalty (10%)	(37,500)	(75,000)	(150,000)
Net WI revenue	337,500	675,000	1,350,000
Less: Severance tax (5% × net WI revenue)	(16,875)	(33,750)	(67,500)
Net revenue before lifting costs	320,625	641,250	1,282,500
Less: Lifting costs ($25 × bbl)	(187,500)	(375,000)	(750,000)
Net cash flow to WI owner	$133,125	$266,250	$ 532,500

Case A: The well should not be completed. The net cash flow to the working interest owner is only $133,125, and estimated completion costs are $260,000. Reserve estimates for new production are often higher than actual. Therefore, the actual net cash inflow may be even less.

Case B: Further analysis of reserve estimates may be needed. The net cash inflow is projected to be $266,250 as compared with completion costs of $260,000. If the reserve estimate is off on the downside, completion costs would not be recovered.

Case C: The well should be completed. The estimated cash flow is $532,500, and completion costs are $260,000.

Even though a decision is made to complete a well because future net revenue is expected to be greater than completion costs, the well may still not be profitable. In order to be profitable, the net revenue from the well must exceed not only the completion costs but all other costs as well. The costs incurred before deciding to complete a well, such as drilling costs, etc., are sunk costs and do not enter into the completion decision. These costs are relevant, however, in determining whether the well has ultimately been profitable.

In Case C of the previous example, the well would be completed. However, the total well costs of $760,000 ($500,000 + $260,000) exceed the estimated future cash flows of $532,500. Thus, even though completed, the well would not be profitable. If all estimates were correct, a loss of $227,500 would be incurred. However, note that if the well were not completed, the loss would be even greater: $500,000 versus $227,500. Another important point to note is that if a well is completed, it is classified as successful. But, even though classified as successful, as seen in Case C, the well may still be unprofitable.

In the previous example, the timing of reserve recovery was not addressed, and the time value of money was not discussed. The estimated future net cash inflows should be discounted to present value, using an appropriate interest rate. Discounting is especially important in this type of analysis because many of the costs are up front, while the revenues are received over time. For example, in Case B above, the estimated future net cash inflows are only slightly more than the completion costs. If the net cash inflows are received over a period of time, the completion costs will in all likelihood exceed the discounted future net cash inflows.

The cost of production facilities was also ignored in the preceding example. In fields where these facilities are not already in place, the cost of the production facilities would have to be considered in the completion decision. In some cases, it might be necessary to drill additional exploratory wells to determine the extent of the reserves and whether those reserves justify the facilities.

Project Analysis and Investment Decision Making

Companies are continually faced with various decisions regarding which projects to invest in and which to bypass. Any number of methods may be employed in practice to make these decisions. Some of the most common methods are payback method, accounting rate of return, net present value, internal rate of return, and profitability index. These methods are often used as preliminary screening measures, so that those projects that do not meet certain minimal requirements may be screened out early in the evaluation process. Those projects that pass these preliminary tests are then taken to the next level of evaluation.

Payback method

This method simply compares the estimated net cash outflow at the time of the investment with the net cash inflows that are estimated to result. The method measures the estimated amount of time that will be required in order for the company to recoup its initial cash outlay on a project.

EXAMPLE

Payback Method

Tyler Company has two proved properties in which it owns a 100% WI. It has identified two wells that could be drilled. However, Tyler does not have sufficient cash to drill both wells and must decide which of the two wells to drill. Information related to each of the wells follows:

	Well A	Well B
Drilling cost	$200,000	$500,000
Completion cost	560,000	450,000
Selling price per bbl	$52	$52
Lifting costs per bbl	$28	$28
State severance tax	5%	5%
Royalty interest percentage	10%	10%
Estimated monthly production	1,000 bbl	2,000 bbl

REQUIRED: Using the payback method, determine which well Tyler should drill.

Computations

	Well A	Well B
Total monthly revenue (bbl × $52)	$52,000	$104,000
Less: royalty (10%)	(5,200)	(10,400)
Net WI revenue	46,800	93,600
Less: Severance tax (5% × net WI revenue)	(2,340)	(4,680)
Net revenue before lifting costs	44,460	88,920
Less: Lifting costs ($28 × bbl)	(28,000)	(56,000)
Net cash flow to Tyler	$16,460	$32,920

Payback period

Well A: $\dfrac{\$760,000}{\$16,460} = 46.17$ months

Well B: $\dfrac{\$950,000}{\$32,920} = 28.86$ months

Either well might be a viable investment. However, given that Well B should return its cost to the owner much more quickly than Well A, Well B would be the most likely well to be drilled.

One attractive aspect of the payback method is that it permits the ranking of several competing investment options. Given that money is at risk, the shorter the payback period, the better. On the other hand, a serious shortcoming of the method is that it ignores the cash flows that occur after payback. For example, one project could have high cash flows initially but a relatively short overall life. A second project could have a lower cash flow stream that extends over a longer period of time. The first project might have the shorter payback period, while the second project generates a greater amount of cash over its lifetime.

Another criticism of the payback method is that it does not incorporate the time value of money. In other words, it does not incorporate present value principles, i.e., the fact that a dollar received today is worth more than a dollar received at some time in the future. This is a serious consideration since, in most E&P projects, such as developing an oil field, positive net cash flows may not commence until a number of years into the project life. Furthermore, net cash flows may continue many years beyond the payback period.

Payback method is a rudimentary tool for screening alternative projects. Nevertheless, the payback method is popular in evaluating E&P projects.

Accounting rate of return

Another commonly used screening method is the accounting rate of return. This measure is based on accounting net income rather than on cash flows. The calculation involves dividing the estimated average annual net income by the investment amount:

$$\text{Accounting rate of return} = \frac{\text{Average annual income}}{\text{Project investment}}$$

This method is simple and relatively easy to understand but has several shortcomings. Like the payback method, this approach does not incorporate the time value of money or present value principles. Further, since it does not utilize net cash inflows, there is some question about whether it is an effective indicator of the true economic consequences of the investment.

EXAMPLE

Accounting Rate of Return

Tyler is considering the installation of production equipment on a well and has identified two completion programs. Alternative A costs $100,000, and Alternative B costs $100,000. The estimated net income generated from each is below:

Year	Net Income Alternative A	Net Income Alternative B
1	$ 10,000	$ 20,000
2	10,000	20,000
3	10,000	15,000
4	10,000	15,000
5	10,000	10,000
6	10,000	5,000
7	10,000	5,000
8	10,000	5,000
9	10,000	2,500
10	10,000	2,500
Total:	$100,000	$100,000

Alternative A: Average annual income: $\dfrac{\$100,000}{10 \text{ years}} = \$10,000$ per year

Accounting rate of return: $\dfrac{\$10,000}{\$100,000} = 10\%$

Alternative B: Average annual income: $\dfrac{\$100{,}000}{10 \text{ years}} = \$10{,}000$ per year

Accounting rate of return: $\dfrac{\$10{,}000}{\$100{,}000} = 10\%$

Based on this example, using only the accounting rate of return, Tyler would find both alternatives to be equally acceptable. However, closer inspection of the net income over the years indicates that Alternative B would result in a higher income in the earlier years and a lower income in the later years. From the standpoint of the time value of money, Alternative B results in a faster return of invested capital.

While the payback method and the accounting rate of return method are straightforward to use, the fact that neither considers the time value of money seriously constrains both methods.

Net present value (NPV) method

Two widely used techniques that incorporate the time value of money are the net present value (NPV) method and the internal rate of return (IRR) method. The NPV method indicates whether the future cash flow stream generated by an investment will yield a positive net present value when the cash flows are discounted using the company's desired rate of return. If the discount rate represents the firm's cost of capital, and if the NPV is positive, the investment will yield a return greater than the cost of capital. Therefore, the investment should be undertaken.

One negative aspect of the NPV method is that it does not permit ranking of projects at a single discount rate. In other words, if two proposed projects both have a positive NPV using the same discount rate, one would not necessarily opt for the project with the higher NPV. Both projects would be acceptable at that rate.

EXAMPLE

Net Present Value

Tyler Company is evaluating whether or not the economics support the completion of Well X. Tyler has a target rate of return on investment of at least 10%.

Years	Cash flows	10% Present value factors	NPV	12% Present value factors	NPV
0	$(60,000)	1.000	$(60,000)	1.000	$(60,000)
1	10,000	0.909	9,090	0.893	8,930
2	10,000	0.826	8,260	0.797	7,970
3	10,000	0.751	7,510	0.712	7,120
4	10,000	0.683	6,830	0.636	6,360
5	10,000	0.621	6,210	0.567	5,670
6	10,000	0.564	5,640	0.507	5,070
7	10,000	0.513	5,130	0.452	4,520
8	10,000	0.467	4,670	0.404	4,040
9	10,000	0.424	4,240	0.361	3,610
10	10,000	0.386	3,860	0.322	3,220
NPV			$ 1,440		$ (3,490)

Note that the cash outlay is entirely paid out during the current year, and therefore the corresponding present value factor is 1.0. The NPV is computed by multiplying the cash flow for each year by the corresponding net present value factor and then summing. If the sum is positive, the project would go forward. In this example, the well has a positive NPV at 10% but a negative NPV at 12%.

Internal rate of return (IRR) method

The internal rate of return (IRR) is the discount rate that would be required in order to generate a NPV of zero. In comparing projects, the higher the IRR, the better the investment. IRR is useful because it can be compared to the firm's cost of capital, or its "hurdle rate" for new investments. Like the NPV method, the IRR method incorporates the time value of money. However, the IRR is computed in terms of a percentage rather than a monetary value.

In general there is no closed-form solution for IRR; it must be determined through an iterative process. Typically, IRR is calculated using a spreadsheet program or calculator, since the iterative process can be tedious and time consuming. If a spreadsheet program or calculator is not used, one must pick a rate of return, plug it into the IRR calculation, and see how close the NPV is to zero. Based on the initial results, the next step would involve selecting a different rate of return and repeating the process until the NPV is zero or as close to zero as possible.

One caveat regarding the IRR method is that the IRR formula assumes all future cash inflows will be reinvested at the same rate of return as the IRR. Since that assumption is often unrealistic, the IRR method should be used with caution and, as is typically the case, used in combination with other evaluation methods such as NPV, payback, etc.

The IRR for the project in the above example is 11%, i.e., at 11%, the NPV of the investment is zero. If Tyler Company has a hurdle rate of 11% or less on new investments, then Tyler would likely proceed with the investment in Well X.

Profitability index

Another present value measure that is often used to evaluate alternative E&P projects is the *profitability index (PI)*, which measures the *proportion* of the present value of dollars returned to dollars invested. The measure therefore permits comparison of the NPV and initial investment of any number of alternative projects. The basic PI formula is:

$$\text{Profitability index} = \frac{\text{NPV of future cash inflows}}{\text{Initial cash outlay}}$$

If the profitability index is greater than 1, then the project is acceptable. When comparing alternative projects, the higher the PI, the higher the rank of the project.

EXAMPLE

Profitability Index

Tyler Company is comparing three alternative investments in Field 1, Field 2, and Field 3. Tyler opts to do the analysis using a discount rate of 10%. The estimated cash flows for each investment are as follows:

Years	Field 1 Cash Flows $	Field 2 Cash Flows $	Field 3 Cash Flows $
0	$(75,000)	$(175,000)	$(475,000)
1	30,000	25,000	150,000
2	20,000	25,000	150,000
3	20,000	25,000	150,000
4	15,000	25,000	75,000
5	10,000	25,000	75,000
6	5,000	25,000	50,000
7	0	25,000	20,000
8	0	25,000	20,000
9	0	25,000	20,000
10	0	25,000	20,000

The NPVs of the future cash inflows (after the initial cash outlay) for each field using a 10% discount rate are as follows:

Years	Present Value Factors 10%	Field 1	Field 2	Field 3
1	0.909	$27,270	$ 22,725	$136,350
2	0.826	16,520	20,650	123,900
3	0.751	15,020	18,775	112,650
4	0.683	10,245	17,075	51,225
5	0.621	6,210	15,525	46,575
6	0.564	2,820	14,100	28,200
7	0.513	0	12,825	10,260
8	0.467	0	11,675	9,340
9	0.424	0	10,600	8,480
10	0.386	0	9,650	7,720
		$78,085	$153,600	$534,700

$$\text{Profitability index} = \frac{\text{NPV of future cash inflows}}{\text{Initial cash outlay}}$$

Field 1: $\text{PI} = \dfrac{\$78,085}{\$75,000} = 1.041$

Field 2: $\text{PI} = \dfrac{\$153,600}{\$175,000} = 0.878$

Field 3: $\text{PI} = \dfrac{\$534,700}{\$475,000} = 1.126$

Using the PI analysis, Field 1 and Field 3 have PI values greater than 1 using a 10% discount rate. If only one project can be selected, based on PI rankings, Tyler should choose Field 3.

COMPREHENSIVE EXAMPLE

IRR, NPV, PI

Using the cash flows for Field 1, Field 2, and Field 3, the IRR of the projects is calculated, as well as the NPV and PI of the projects using both 10% and 13% discount rates.

Years	Field 1 Cash Flows $	Field 2 Cash Flows $	Field 3 Cash Flows $
0	$(75,000)	$(175,000)	$(475,000)
1	30,000	25,000	150,000
2	20,000	25,000	150,000
3	20,000	25,000	150,000
4	15,000	25,000	75,000
5	10,000	25,000	75,000
6	5,000	25,000	50,000
7	0	25,000	20,000
8	0	25,000	20,000
9	0	25,000	20,000
10	0	25,000	20,000

IRR

	IRR
Field 1	12%
Field 2	7%
Field 3	14%

Present Value Factors

Years	Present Value Factors 10%	Present Value Factors 13%
0	1.000	1.000
1	0.909	0.885
2	0.826	0.783
3	0.751	0.693
4	0.683	0.613
5	0.621	0.543
6	0.564	0.480
7	0.513	0.425
8	0.467	0.376
9	0.424	0.333
10	0.386	0.295

Calculations

	10%			13%		
Years	Field 1 NPV	Field 2 NPV	Field 3 NPV	Field 1 NPV	Field 2 NPV	Field 3 NPV
0	$(75,000)	$(175,000)	$(475,000)	$(75,000)	$(175,000)	$(475,000)
1	27,270	22,725	136,350	26,550	22,125	132,750
2	16,520	20,650	123,900	15,660	19,575	117,450
3	15,020	18,775	112,650	13,860	17,325	103,950
4	10,245	17,075	51,225	9,195	15,325	45,975
5	6,210	15,525	46,575	5,430	13,575	40,725
6	2,820	14,100	28,200	2,400	12,000	24,000
7	0	12,825	10,260	0	10,625	8,500
8	0	11,675	9,340	0	9,400	7,520
9	0	10,600	8,480	0	8,325	6,660
10	0	9,650	7,720	0	7,375	5,900
NPV	$ 3,085	$ (21,400)	$ 59,700	$ (1,905)	$ (39,350)	$ 18,430
NPV of future cash inflows	$ 78,085	$ 153,600	$ 534,700	$ 73,095	$ 135,650	$ 493,430

Discount Factor of 10%

NPV: As can be seen from the table above, only Field 1 and Field 3 have positive NPVs at 10%.

PI: Only Field 1 and Field 3 have PIs greater than 1.

$$\text{Field 1:} \quad PI = \frac{\$78,085}{\$75,000} = 1.041$$

$$\text{Field 2:} \quad PI = \frac{\$153,600}{\$175,000} = 0.878$$

$$\text{Field 3:} \quad PI = \frac{\$534,700}{\$475,000} = 1.126$$

Discount Factor of 13%

NPV: As can be seen from the table above, only Field 3 has a positive NPV.

PI: Only Field 3 has a PI greater than 1 at 13%.

$$\text{Field 1:} \quad PI = \frac{\$73,095}{\$75,000} = 0.9746$$

$$\text{Field 2:} \quad PI = \frac{\$135,650}{\$175,000} = 0.7751$$

$$\text{Field 3:} \quad PI = \frac{\$493,430}{\$475,000} = 1.0388$$

Conclusion: Field 3 has the largest NPV and PI at either 10% or 13%. It also has the largest IRR. Based on this analysis, Tyler should rank Field 3 first. If the appropriate rate of return is 10%, Field 1 would also be a viable project. However, Field 2 would not be selected.

PROBLEMS

1. Define and discuss the following:
    tubular goods
    gathering systems
    workover operations
    shut-in payments

2. Indicate whether each of the following items is a directly attributable operating expense (D), an allocable operating expense (A), or a capital expenditure (C):

   ____ Repairs that can be traced to individual wells
   ____ Cost of pumping unit
   ____ Installation of pumping unit
   ____ Ordinary repair parts for pumping unit
   ____ Repair of pumping unit
   ____ Lease road maintenance
   ____ Wages for employees who work only on one lease
   ____ Wages for employees who designate hours worked on certain wells or leases
   ____ Wages for employees who work on multiple leases, detailed records not kept
   ____ Severance tax
   ____ District office expenses
   ____ Fuel for leases, invoices indicate particular lease
   ____ Depreciation of truck used on one lease
   ____ Depreciation of truck used on multiple leases, detailed mileage records not kept
   ____ Property taxes
   ____ Contract labor for refracturing, invoices indicate wells
   ____ Workover for the purpose of restoring production
   ____ Recompletion for the purpose of obtaining new production
   ____ Depreciation of district office facilities

3. District office expenses were $48,000 for July 2018. The district office supervised the following leases:

Lease	Number of Wells	Oil Production, bbl
A	6	5,000
B	2	1,000
C	4	2,000

   a. Record lease operating expense, assuming allocation based on the number of wells.
   b. Record lease operating expense, assuming allocation based on the barrels of oil produced.

4. Valve Oil Corporation has a waterflooding system for a reservoir in Oklahoma. Costs of operating the system for the month of August were $20,400.

   a. Give the entry to record operating expense, assuming the reservoir underlies only Lease A.
   b. Assume the reservoir underlies the following three leases:

Lease	Producing Wells	Injection Wells	Volume of Water
A	2	1	200
B	5	2	450
C	6	3	625

   Give the entry to record operating expense, assuming allocation based on the volume of water.

5. Core Petroleum incurs the following costs relative to a gathering system:

   a. Purchase and installation costs of separators and compressors: $200,000
   b. Operational costs for the system: $10,000

   Give the entry to record the above costs, assuming the gathering system serves only one lease.

6. Acceleration Oil Company purchased new tubing and casing to replace damaged tubular goods in a well. The new tubular goods were installed in a producing well. The cost of the items were as follows:

a. Tubing and casing	$80,000
b. Installation of tubular goods	15,000
c. Loading, hauling, and unloading charges	10,000

   Give any entries necessary.

7. What factors are important in determining whether to plug or complete a well?

8. Elizabeth Petroleum data in connection with Lease A are as follows:

Property cost (acquisition cost)	$ 50,000
Drilling cost (one well)	300,000
Estimated completion	500,000
Estimated selling cost per bbl	60
Estimated lifting cost per bbl	20
State severance tax	6%
Royalty interest	12.5%

   REQUIRED: Should the well be completed, assuming the following total production? Discuss your answer.

   Case A: 10,000 bbl
   Case B: 20,000 bbl
   Case C: 30,000 bbl

9. Assuming the same data as given in problem 8, was the well in each case profitable? Discuss your answer.

10. Assume the same data as given in problem 8, except the company expects the following production:

    Case A: 600 bbl per month
    Case B: 1,000 bbl per month

    REQUIRED:
    a. Determine the number of months needed for payout.
    b. Would an investment in this property be considered successful if an investor wanted a 30-month payout?

11. A saltwater disposal system is added to Lease A's gathering system at a cost of $150,000. The expense for the month of May 2016 is $15,000. Record the acquisition cost and the monthly expense assuming the following:

    Case A: The disposal system serves several wells on two different leases. There are 5 wells on Lease A and 10 wells on Lease B.

    Case B: The disposal system serves only the wells on Lease A.

12. Bore Oil Company pays the following amounts during June 2016 relating to producing leases:

Supplies for Lease A	$ 300
Fuel for Lease A	800
Labor cost for pumpers and gaugers—Lease A	2,000
Fringe benefits of pumpers and gaugers who work on several leases	400
Salaries and fringe benefits of regional supervisors	10,000
Contract labor for refracturing well of Lease A	1,000
Property tax—Lease A	1,500
Transportation cost for several leases	1,200

    Record the above transactions.

13. Choice Petroleum has the following data in connection with Lease A:

Proved property cost	$ 40,000
Drilling cost	300,000
Estimated completion cost	250,000
Estimated selling price per barrel	55
Estimated lifting cost per barrel	23
State severance tax rate	5%
Royalty interest	3/16th

    Choice Petroleum is sole owner of the working interest.

    If reserves are determined to be 20,000 barrels, is the well profitable? Would the well be profitable if reserves are 30,000 barrels? How many barrels of total production would it take to recover drilling and completion costs?

14. Optimistic Oil Corporation estimates the following costs to acquire, drill, and complete a well on Lease A:

Acquisition costs	$ 75,000
Drilling and completion costs	600,000
Selling price of oil	60/bbl
Lifting costs	25/bbl
State severance tax rate	5%
Royalty interest	20%

Would the investment be profitable if proved reserves are:

a. 20,000 barrels?

b. 30,000 barrels?

c. 40,000 barrels?

15. Burnout Oil Company has the following expenditures in August 2016:

Lease A:	Well #2—Cleaning and reacidizing formation	$10,000
Lease B:	Well #3—Testing, perforating, and completion at 10,000 feet. Current production is at 12,000 feet.	
	IDC	60,000
	Equipment from inventory	10,000

Record the above transactions.

16. Aaron Energy Corporation drilled a successful gas well. Although capable of production, the well was shut in awaiting the completion of a pipeline. Shut-in royalty payments of $1,000 per month were paid during the months of June through August. Assume the following independent cases:

a. The shut-in royalty payments were recoverable from future production. Give the entry to record payment of the shut-in royalty for June.

b. The shut-in royalty payments were not recoverable from future production. Give the entry to record payment of the shut-in royalty for June.

17. Action Oil and Gas Corporation operates the Flat Hill Basin. Along with oil, this field produces large quantities of saltwater. The saltwater is highly corrosive, and as a result, the downhole production equipment is subject to frequent replacement. During the first half of 2014, Action must replace the tubulars in wells A, B, and C. Wells X, Y, and Z (newly drilled wells) are in the process of being completed.

During April, Action receives a shipment containing 500 joints of production tubing. The invoice totals $500,000 plus $3,000 for transportation and hauling. The production tubing is installed in the following wells:

Well	Production Tubing
A	100
B	50
C	75
X	100
Y	125
Z	50

Record the purchase and installation of the production tubing.

18. Core Petroleum is considering drilling one well on either the Rago lease or the Bennett lease in Texas. Core does not have sufficient funds to drill both wells and must decide which of the two wells to drill. Information relating to each of the wells follows:

	Rago well	Bennett well
Drilling cost	$450,000	$325,000
Completion cost	500,000	525,000
Selling price per bbl	60	60
Lifting cost per bbl	24	24
State severance tax	6%	6%
Royalty interest percentage	12%	12%
Estimated monthly production	2,000 bbl	1,500 bbl

REQUIRED: Using the payback method, determine which well Core should drill.

19. Fantastic Oil Corporation is considering two alternatives for the installation of production equipment on the Panther well in the Odessa West field. Alpha costs $275,000 and Beta costs $350,000. The estimated net income generated from each is below:

Year	Net Income Alpha	Net Income Beta
1	$ 25,000	$ 50,000
2	25,000	50,000
3	25,000	50,000
4	25,000	50,000
5	25,000	50,000
6	25,000	30,000
7	25,000	30,000
8	25,000	30,000
9	25,000	30,000
10	25,000	30,000
11	25,000	15,000
12	25,000	15,000
13	25,000	15,000
14	25,000	15,000
15	25,000	15,000
Total	$375,000	$475,000

REQUIRED: Using accounting rate of return, determine which alternative Fantastic should choose.

20. Polecat Corporation is considering beginning drilling operations in three separate fields. Polecat decides to analyze these fields using a 13% discount rate. The estimated cash flows for each field are as follows:

Years	East Field Cash Flows	South Field Cash Flows	West Field Cash Flows
0	$(105,000)	$(425,000)	$(300,000)
1	40,000	100,000	50,000
2	35,000	100,000	50,000
3	30,000	100,000	50,000
4	20,000	100,000	50,000
5	15,000	90,000	50,000
6	10,000	90,000	40,000
7	5,000	90,000	40,000
8	0	40,000	40,000
9	0	40,000	40,000
10	0	20,000	40,000

REQUIRED:

a. Calculate the net present value of each field

b. Calculate the profitability index of each field.

c. Determine the internal rate of return of each field.

d. Rank the fields from best investment to worst investment.

References

1. Institute of Petroleum Accounting. 2001. *2001 PricewaterhouseCoopers Survey of U.S. Petroleum Accounting Practices.* Denton, TX: Institute of Petroleum Accounting.

2. Ibid.

9

ACCOUNTING FOR ASSET RETIREMENT OBLIGATIONS AND ASSET IMPAIRMENT[1]

For a number of years, full cost and successful efforts companies have attempted to deal with two difficult issues: accounting for the future cost of decommissioning and environmental remediation and accounting for possible impairment of long-lived assets. SEC *Reg. S-X 4-10* and *SFAS No. 19* included rather vague instructions regarding the treatment of future decommissioning and environmental remediation costs. For asset impairment, companies using the full cost method followed the ceiling test rules contained in *Reg. S-X 4-10*. However, there were no specific impairment rules applicable to long-lived assets of companies using the successful efforts method. All of that changed when the FASB issued *SFAS No. 143*, "Accounting for Asset Retirement Obligations" and *SFAS No. 144*, "Accounting for the Impairment or Disposal of Long-Lived Assets." This chapter includes a discussion of both of these significant issues.

Accounting for Asset Retirement Obligations

SFAS No. 143 applies to both successful efforts and full cost companies. Previously both *SFAS No. 19* and the full cost rules in *Reg. S-X 4-10* included the requirement that companies estimate the future cost of decommissioning (i.e., dismantlement, abandonment, and reclamation costs) related to producing properties and incorporate the estimates into the DD&A calculation. The future cost estimates were not required to be recorded in the company's financial accounts, yet they were required to be added to the amortization base for the cost center. *SFAS No. 143* amended the prior practice by requiring upfront recognition of asset retirement obligations (AROs) by capitalization of the costs to the related long-lived asset and recognition of an offsetting ARO liability.

Under *SFAS No. 143,* a company must recognize any and all obligations associated with the retirement of tangible long-lived assets. In *SFAS No. 143,* the term **retirement** is defined as the other-than-temporary removal of a long-lived asset from service and includes the sale, abandonment, or other disposal of the asset. It does not encompass the temporary idling of a long-lived asset. In addition, activities necessary to prepare a long-lived asset for an alternative use are not associated with retirement activities and are not included in the scope of the statement. Accounting for obligations associated with the maintenance of assets and the cost of replacement of assets are also not included.

Upon initial recognition of a liability for asset retirement obligations, a company must capitalize the same amount as part of the cost basis of the related long-lived asset. Since the ARO is included in the cost basis of the asset, it is subject to amortization. As such, the ARO costs are allocated to expense (through DD&A) over the useful life of the asset. Changes in the obligation that result from revisions in estimates of the amount or timing of the cash flows required to settle the future liability are recognized by increasing or decreasing the carrying value of the ARO liability and the related long-lived asset. Changes due solely to the passage of time or "accretion of the discounted liability" are recognized as an increase in the carrying value of the liability and as an expense classified as an operating item in the income statement. This expense is referred to as an **accretion expense**. Accretion of the discounted liability occurs as a consequence of the discounting process. As time passes, payment of the cash outlay draws nearer, and the present value of the future cash outlay increases. This increase is referred to as **accretion of the discount**, or sometimes as the **unwinding of the discount**.

Scope of *SFAS No. 143*

SFAS No. 143 applies to all entities that incur legally enforceable obligations related to the retirement of tangible long-lived assets. Specifically, *SFAS No. 143* applies to all legal obligations associated with retirement of tangible long-lived assets that result from acquisition, construction, development, and/or normal use of long-lived assets. These obligations include all AROs incurred any time during the life of an asset and not just during the acquisition and early operation stages. *SFAS No. 143* also applies to obligations associated with interim property retirements resulting from a legal obligation. For example, in the development of a field, a company anticipates that over time certain wells will cease to produce and will need to be plugged and abandoned (prior to the end of the life of the field). The cost of abandoning all wells, including those whose life is shorter than the field life, is within the scope of *SFAS No. 143*. The cost of workovers and maintenance is not.

An obligation that results from the improper operation of a long-lived asset is not within the scope of the standard. For example, environmental remediation liabilities that relate to pollution arising from some past act that will be corrected without regard to retirement activities are subject to the provisions of the American Institute of Certified Public Accountants (AICPA) *SOP 96-1*, "Environmental Remediation Liabilities." Similarly, a catastrophic accident occurring at a fuel storage facility caused by noncompliance with company procedures does not result from the normal operations of the facility and thus is not within *SFAS No. 143*'s scope. However, an environmental remediation liability that results from the normal operation of a long-lived asset, for example, the obligation to decontaminate a nuclear power plant site, should be accounted for under the provisions of *SFAS No. 143*.

Legally enforceable obligations

SFAS No. 143 applies to legally enforceable obligations associated with the retirement of a tangible long-lived asset. For purposes of *SFAS No. 143*, legally enforceable obligations can result from the following:

1. A government action, such as a law, statute, or ordinance
2. An agreement between entities, such as a written or oral contract
3. A promise conveyed to a third party that imposes a reasonable expectation of performance upon the promisor under the doctrine of promissory estoppel

In some cases, the determination of whether a legal obligation exists may be clear. For instance, retirement, removal, or closure obligations may be imposed by government units that have responsibility for oversight of a company's operations. In other cases, there may be an agreement between two or more parties, such as a lease or "right of use" agreement. Examples of AROs where the company is normally legally obligated include the following:

1. Decommissioning of onshore and offshore oil and gas wells and production equipment and facilities
2. Removal of pipelines
3. Removal of leasehold improvements (including legal obligations of the lessor or lessee, provided that the obligations are not included in minimum lease payments or contingent rentals as defined in *SFAS No. 13*)
4. Closure and postclosure costs of refineries
5. Removal of underground storage tanks
6. Closure and postclosure costs of certain hazardous waste storage facilities
7. Reforestation of land after cessation of operations
8. Removal of microwave towers and other electronic communication devices

In some situations, no law, statute, ordinance, or contract exists, but a company has made a promise to a third party (which may include the public at large) about its intention to perform retirement activities. In these situations, facts and circumstances need to be considered carefully, and significant judgment may be required to determine if a retirement obligation exists. Those judgments should be made within the framework of the doctrine of promissory estoppel.

The FASB refers to *Black's Law Dictionary*, seventh edition, which defines *promissory estoppel* as follows: [2]

> *The principle that a promise made without consideration may nonetheless be enforced to prevent injustice if the promisor should have reasonably expected the promisee to rely on the promise and if the promisee did actually rely on the promise to his or her detriment.*

Additionally, a legal obligation may exist even though no party has taken any formal action to enforce it. Finally, in assessing whether a legal obligation exists, a company is not permitted to forecast changes in the law or changes in the interpretation of existing laws and regulations. Companies and their legal advisors must evaluate current facts and circumstances to determine whether a legal obligation exists.

A contract may contain an option or a provision that requires one party to the contract to perform retirement activities when an asset is retired. The other party, perhaps a governmental entity, may have a history of waiving the particular provisions. Even if there is an expectation of a waiver or nonenforcement based on historical experience, the contract still imposes a legal obligation that is included in the scope of *SFAS No. 143*. For example, assume that Tyler Company enters into a 50-year agreement with the government that permits it to construct a 50-mile-long pipeline from a production site to a refinery. Under the terms of the agreement, Tyler Company is required to remove the pipeline at the end of the 50-year term or upon cessation of the use of the pipeline. Although historical experience indicates that the authorities will not actually require Tyler Company to remove the pipeline 50 years hence, Tyler Company nevertheless must accrue a liability. However, application of the expected cash flow approach (which is discussed later) in measuring the obligation could make the amount immaterial.

Obligating event

Identifying obligating events that require recognition of an asset retirement obligation is often difficult, especially in situations that involve the occurrence of a series of transactions. A company must look to the nature of the duty or responsibility to assess whether the obligating event has occurred and an asset retirement obligation should be recognized. For example, consider an oil and gas lease. Typically the terms of the lease agreement require that if drilling or production occurs, the lessor must assume the responsibility for ultimately plugging wells, dismantling equipment, and restoring the area. However, no obligation exists until wells are drilled or equipment is installed. Therefore, the drilling and equipping of wells and otherwise disturbing the environment, not the signing of the lease, constitute the obligating event and no obligation should be recorded until drilling is begun or equipment is installed. (Note that since *SFAS No. 143* applies to the retirement of tangible long-lived assets, logically ARO accrual would not be required until properties become proved.)

In another example, the construction of an offshore oil platform often creates the obligation to dismantle and remove the structure at some point in the future. The obligation to remove the facility does not change with the operation of the asset or the passage of time (although estimates of the amount of the obligation may change).

Asset recognition

When the asset retirement obligation liability is initially recognized, a corresponding increase in the carrying value of the related long-lived asset is recognized. This offsetting amount is *not* recorded in a separate asset account; rather, it is to be added to the related asset account. The capitalized asset retirement obligation, along with the historical cost of the asset, is expensed through DD&A over the asset's useful life. For example, if an offshore oil platform cost $18 million to construct, and the present value of the ARO is $6 million, the total cost of the asset is $24 million. This total amount should be amortized over the platform's useful life. (For a more complete discussion of DD&A under the successful efforts and full cost methods, see chapters 6 and 7, respectively.)

On the other hand, for companies that incur an ARO ratably (e.g., an obligation to restore land after it has been strip mined), the obligation and related asset are recognized as strip mining operations progress. *SFAS No. 143* does not preclude a company from capitalizing these costs as they are incurred each year and then recognizing them as expense in the same year.

Initial measurement—fair value

SFAS No. 143 requires that the initial measurement of an ARO liability is to be at current fair value. Companies must recognize the current fair value of the liability for an asset retirement obligation in the period in which the obligation is incurred if a reasonable estimate of fair value can be made. If a reasonable estimate of fair value cannot be made in the period in which the ARO is incurred, no liability should be recognized in that period, but disclosure of the existence of the ARO is required. The recognition of the ARO should be delayed until the period in which a reasonable estimate of fair value can be made.

The fair value of a liability for an asset retirement obligation is the amount at which the liability could be settled in a *current* transaction between willing parties, other than in a forced or liquidation transaction. Quoted market prices in active markets are the best evidence of fair value and should be used as the basis for the measurement, if available. If quoted market prices are not available, the estimate of fair value should be based on the best information available in the circumstances, including prices for similar liabilities and the results of present value or other valuation techniques.

The FASB relied heavily on *SFAC No. 7*, "Using Cash Flow and Present Value in Accounting Measurements," in developing the measurement principles in *SFAS No. 143*. *SFAC No. 7* provides a framework for using future cash flows as the basis for accounting measurements. *SFAC No. 7* indicates that (1) the objective of a present value measurement is always to determine fair value and (2) the most relevant measurement of a company's liabilities should always include the effects of the company's credit standing.

Present value techniques are often used to measure liabilities. *SFAC No. 7* describes two basic techniques: the **traditional approach** and the **expected cash flow approach**. In the traditional approach, a single set of estimated cash flows and a single interest rate (commensurate with the risk) are used to estimate fair value. In the expected cash flow approach, multiple cash flow scenarios that reflect the range of possible outcomes and a credit-adjusted risk-free rate are used to estimate fair value. Note that in the traditional approach, the associated risk is reflected in the discount rate used, whereas in the expected cash flow approach, the risk is reflected in the cash flows. The FASB recognizes that fair values will not be available for the majority of AROs and concludes that the expected cash flow approach will typically be the only appropriate technique. This is due to two factors. First, estimates of the fair value of an ARO generally will involve estimates of future cash flows that are uncertain in both timing and amount. Second, observable marketplace prices for such liabilities generally do not exist.

Credit-adjusted risk-free discount rate. When measuring an ARO using the expected cash flow approach, a company should discount the liability using a discount rate that equates to a risk-free rate for an instrument with a term equal to the estimated period until the retirement activities will be performed, adjusted to reflect the company's credit standing. This rate, which must be a company-wide rate in order to reflect the company's credit standing, is referred to in *SFAS No. 143* as a **credit-adjusted risk-free rate**.

The determination of a credit-adjusted risk-free rate is not entirely clear in *SFAS No. 143* or *SFAC No. 7*. In the United States, the risk-free rate is typically assumed to equal the rate for zero coupon U.S. Treasury instruments. Thus, a reasonable risk-free rate would appear to be the rate of a zero coupon U.S. Treasury bill with maturity dates that coincide with the expected timing of the estimated cash flows needed to satisfy the ARO. This rate would then be adjusted to reflect the obligor's credit standing. *SFAS No. 143* provides little guidance as to this aspect of the rate. However, there is general agreement that it is the rate at which the obligor could borrow to fund the obligation today. In other words, a reasonable approach might be to estimate the company's incremental borrowing rate on debt of similar maturity. The increment of that rate over the risk-free rate of the same maturity would then be the adjustment for the company's credit standing.

Expected cash flow approach. The FASB assumes that the expected cash flow approach will be used to estimate the fair value of most ARO liabilities. *SFAS No. 143* requires that the estimated cash flows reflect, to the extent possible, the marketplace's assessment of the cost and timing of the performance of the required retirement activities, i.e., the amount a third party would charge to assume the obligation. This requirement applies even if the company plans to perform the retirement activities itself. The FASB acknowledges that many AROs cannot be settled in current transactions with third parties and that many companies will perform the retirement activities themselves. Nevertheless, *SFAS No. 143* requires that the ARO initially be measured based on fair value. Therefore, if the company ultimately does perform the retirement activities and its actual cost is less than the marketplace estimate, it will recognize a gain upon the retirement of the asset. The prospect of such a gain has been the subject of considerable criticism of the Statement.

It is important to note that although *SFAS No. 143* calls for fair value measurement, amounts derived through negotiations between willing buyers and willing sellers normally are not available for AROs. In such cases, *SFAS No. 143* allows the use of a company's own assumptions about future cash flows if information about assumptions marketplace participants would use in their estimates of fair value is not available without undue cost and effort. A company's own estimates should include reasonable and supportable assumptions about all of the following:

1. The costs that a third party would incur in performing the tasks necessary to retire the asset
2. Other amounts that a third party would include in determining the price of settlement, including, for example, inflation, overhead, equipment charges, and anticipated advances in technology
3. The profit margin required by the third party
4. The extent to which the amount of a third party's costs or the timing of its costs would vary under different future scenarios, and the relative probabilities of those scenarios
5. The third party's "market risk premium," i.e., the price that a third party would demand and could expect to receive for bearing the uncertainties and unforeseeable circumstances inherent in the obligation

EXAMPLE

Expected Cash Flows

On January 1, 2012, Tyler Company completed and placed an offshore oil platform into service at a cost of $2,000,000. Legally, Tyler Company is required to dismantle and remove the platform at the end of its useful life, which is estimated to be in 10 years. Tyler Company plans to perform the decommissioning and removal work itself. However, *SFAS No. 143* requires that the fair value of the ARO be recorded. Accordingly, Tyler Company decides to estimate the fair value of the liability using the expected present value technique.

Tyler Company contacted several outside contractors that dismantle and remove offshore oil platforms to determine representative labor rate costs in the area. As the first stage in measuring the liability, Tyler Company then estimated the labor costs associated with decommissioning of its platform under a range of different scenarios and estimated the probability for each of the cash flow estimates as follows:

Cash Flow Estimate	Probability Assessment	Expected Undiscounted Cash Flow
$500,000	10%	$ 50,000
350,000	60%	210,000
400,000	30%	120,000
		$380,000

Market risk premiums. According to the FASB, the market risk premium is intended to reflect what a contractor would demand for bearing the uncertainty of a fixed price today for performing the dismantling many years in the future. The FASB has provided no additional guidance regarding the estimate of an appropriate market risk premium. This estimate will be particularly difficult in circumstances in which the retirement activities will be performed many years in the future, as is often the case. The estimate is also difficult in situations in which the company has little information about how much a contractor would charge in addition to a normal price to assume the risk that the actual costs to perform the retirement activities will change in the future.

EXAMPLE

Initial Estimation and Recording

Continuing with the previous example, assume the following:

Tyler Company estimates that contractors would estimate the cost of overhead and equipment costs by adding an additional percentage based upon labor costs. Tyler Company uses a transfer-pricing rate of 60% and has no reason to believe that its overhead and equipment rate differs from the rate that a contractor in the business of dismantling and removing offshore platforms would use.

Tyler Company determines that contractors typically add a markup on labor, allocated equipment, and overhead costs in determining their required profit margin. Tyler Company determines that 15% is representative of the profit rate that contractors in the industry generally earn to dismantle and remove offshore oil platforms.

Tyler Company believes that a contractor would typically demand a market risk premium for bearing the uncertainties and unforeseeable circumstances inherent in "locking in" today's price for a project that will not occur for 10 years. Tyler Company estimates the amount of that premium to be 5% of the estimated inflation-adjusted cash flows.

Assume that the risk-free rate of interest on January 1, 2012 is 6%, and Tyler Company adds 4% to the rate to reflect its credit standing. Therefore, the credit-adjusted risk-free rate used to compute expected present value is 10%.

A rate of inflation of 4% is expected over the next 10-year period.

Initial Measurement of the ARO Liability at January 1, 2012

	Expected Cash Flows as of 1/1/12
Expected labor costs.	$ 380,000
Allocated overhead and equipment charges (0.60 × $380,000)	228,000
Contractor's markup [0.15 × ($380,000 + $228,000)] . .	91,200
Expected cash flows before inflation adjustment	699,200
Inflation factor assuming 4% rate for 10 years	1.4802
Expected cash flows adjusted for inflation	1,034,956
Market-risk premium* (0.05 × $1,034,956)	51,748
Expected cash flows adjusted for market risk	$1,086,704
Present value (PV) using credit-adjusted risk-free rate of 10% for 10 years (PV factor = 0.385543)	$ 418,971

Entries—January 1, 2012:

Long-lived asset (platform)	2,000,000	
Cash		2,000,000
(to record the initial construction of the platform)		
Long-lived asset (platform)	418,971	
ARO liability.		418,971
(to record the initial fair value of the ARO liability)		

---

* The FASB used a 5% market risk factor in its illustrative example in *SFAS No. 143*. In practice, firms have found it difficult to determine the appropriate market risk factor that would be used in this calculation. As a consequence, firms frequently do not incorporate a market-risk premium in the calculation of the expected cash flows.

Offsetting cash inflows. The FASB has indicated that all offsetting cash inflows, including expected salvage values, should be excluded from computation of AROs and instead be included in the computation of the depreciable base of the assets.

Subsequent recognition and measurement

Companies are not required to remeasure their AROs at fair value each year following the initial recognition of AROs. Rather, companies are to recognize period-to-period changes in the liability for the asset retirement obligation resulting from (1) the passage of time (accretion), and (2) revisions to either the timing or the amount of the original estimated future cash flows.

Changes due to the passage of time. Changes in the asset retirement obligation due to the passage of time should be measured by recognizing accretion expense. The method commonly used to compute accretion expense is the **interest method of allocation**. Using the interest method of allocation, accretion expense is determined by multiplying the carrying value of the liability at the beginning of each period by a constant effective interest rate. The interest rate used should be the credit-adjusted risk-free rate applied when the liability (or component thereof) was measured initially. The rate should not change subsequent to the initial measurement.

Changes in an ARO liability resulting from the passage of time should be recognized as an increase in the carrying value of the liability and as a charge to accretion expense. The accretion expense should *not* be considered interest cost qualifying for capitalization under SFAS No. 34.

EXAMPLE

Changes Due to the Passage of Time

Continuing with the previous examples, assume that the only change in the liability balance that occurred over the subsequent 10-year period was related to the passage of time. Accretion is computed by applying the credit-adjusted risk-free rate of 10% to the beginning liability balance each year.

Accretion: Interest Method of Allocation

Date	Accretion	Liability Balance
1/1/2012		$418,971
12/31/2012	$41,897	460,868
12/31/2013	46,087	506,955
12/31/2014	50,696	557,651
12/31/2015	55,765	613,416
12/31/2016	61,342	674,758
12/31/2017	67,476	742,234
12/31/2018	74,223	816,457
12/31/2019	81,646	898,103
12/31/2020	89,810	987,913
12/31/2021	98,791	1,086,704

Chapter 9 • Accounting For Asset Retirement Obligations And Asset Impairment

The expenses to be recognized are summarized in the following table. It should be noted that *SFAS No. 19* requires that amortization of oil and gas producing assets be computed using the unit-of-production method. Tyler Company's platform would, in fact, be properly depreciated using the unit-of-production method. However, for simplicity and illustration purposes, DD&A is computed in this example using the straight-line method and assumes no salvage value.

Schedule of Expenses

Year-end	Accretion Expense	DD&A Expense for ARO
2012	$41,897	$41,897
2013	46,087	41,897
2014	50,695	41,897
2015	55,765	41,897
2016	61,342	41,897
2017	67,476	41,897
2018	74,223	41,897
2019	81,646	41,897
2020	89,810	41,897
2021	98,791	41,897

(Figures have been rounded.)

Entries—December 31, 2012—2021:

```
DD&A expense (platform)*.  . . . . . . . . . . . . .    200,000
    Accumulated DD&A .  . . . . . . . . . . . . . . . .            200,000
(to record straight-line DD&A on the platform ($2,000,000/10 yrs)


DD&A expense (ARO) .  . . . . . . . . . . . . . .     41,897
    Accumulated DD&A .  . . . . . . . . . . . . . . . .             41,897
(to record straight-line DD&A on the asset retirement cost)


Accretion expense .  . . . . . . . . . . . . . . . . . . . .   Per schedule
    ARO liability .  . . . . . . . . . . . . . . . . . . . .                 Per schedule
(to record accretion expense on the ARO liability)
```

* Note that the platform account totals $2,418,971, which includes the actual cost of the platform as well as the related ARO. For illustration purposes, in this chapter, DD&A expense is shown separately for the actual platform cost versus the ARO.

Gain or loss recognition upon settlement. One of the more controversial aspects of *SFAS No. 143* is apparent when the ARO is eventually settled. When the asset is retired and the ARO is settled, the difference between the ARO liability and the amount actually incurred is to be recognized as either a gain or loss. If the actual cost is greater than the ARO liability, a loss is recognized, and if the actual cost is less than the ARO liability, a gain is recognized. Recall that the expected cash flow estimate of the ARO's fair value requires the assumption that a third party would perform the decommissioning. If a company actually performs the work using its own labor force and equipment, the cost would presumably be less than the ARO liability, thus resulting in gain recognition.

EXAMPLE

Gain on Final Settlement

Continuing with the previous examples, assume that on December 31, 2021, Tyler Company dismantles the platform and restores the area using its own internal workforce at a cost of $1,000,000.

Entry—December 31, 2021:

ARO liability.	1,086,704	
Miscellaneous payables		1,000,000
Gain on settlement of ARO liability		86,704
(to record settlement of the ARO liability)		
Accumulated DD&A	2,418,971	
Long-lived asset (platform)		2,418,971
(to record the removal of the platform)		

Changes due to the revisions in estimates. Cash flow estimates might change, for example, due to a newly enacted law that requires greater retirement obligations or that changes the expected cash flows to settle the asset retirement obligation. Changes due to revised estimates of the amount or timing of the original undiscounted cash flows are to be recognized by increasing or decreasing the carrying value of an ARO liability and the carrying value of the related long-lived asset. *Upward revisions* in the amount of undiscounted estimated cash flows should be discounted using the credit-adjusted risk-free rate in effect at the time of the change in estimate. *Downward revisions* in the amount of undiscounted estimated cash flows should be discounted using the credit-adjusted risk-free rate that existed when the original liability or component thereof was recognized. Sometimes a company cannot identify the component of the liability (and the associated credit-adjusted risk-free rate) to which the downward revision relates as might be the case if numerous changes in estimates of future cash flows have been made. In these instances, a weighted average credit-adjusted risk-free rate may be used to discount the downward revision.

EXAMPLE

Upward Revision in Estimate

Continuing from the previous examples (other than the gain on final settlement example), assume that on December 31, 2013, Tyler Company notes that contractor labor rates have risen significantly. Tyler Company decides that it must revise its estimate of the cost to dismantle its offshore platform. Tyler Company also determines that it is appropriate to revise the probability assessments related to its decommissioning scenarios. The change in estimated labor costs results in an upward revision to the undiscounted cash flows. Since the revision is upward, Tyler Company must determine its credit-adjusted risk-free rate as of that date. The rate turns out to be 9%. Consequently, the incremental cash flows are discounted at the current rate of 9%. All of the other assumptions remain unchanged. The revised estimate of expected cash flows for labor costs is as follows:

Cash Flow Estimate	Probability Assessment	Expected Cash Flow
$336,000	10%	$ 33,600
500,000	55%	275,000
600,000	35%	210,000
	100%	$518,600

Subsequent Measurement of the ARO Liability Reflecting a Change in Labor Cost Estimate as of December 31, 2013

Incremental expected labor costs ($518,600 − $380,000).	$138,600
Allocated overhead and equipment charges (0.60 × $138,600).	83,160
Contractor's markup [0.15 × ($138,600 + $83,160)] . . .	33,264
Expected cash flows before inflation adjustment.	255,024
Inflation factor assuming 4% rate for eight years	1.3686
Expected cash flows adjusted for inflation	349,026
Market-risk premium (0.05 × $349,026)	17,451
Expected cash flows adjusted for market risk	$366,477
Present value of incremental liability using credit-adjusted risk-free rate of 9% for eight years (present value factor = 0.501866)	$183,922

Accretion: Interest Method of Allocation

Date	Accretion on Original	Accretion on Change	Total Accretion	Original Liability Balance	Increment Liability Balance	Total Liability Balance
1/1/12				$ 418,971		$ 418,971
12/31/12	$41,897		$ 41,897	460,868		460,868
12/31/13	46,087		46,087	506,955	$183,922	690,877
12/31/14	50,696*	$16,553**	67,249	557,651	200,475	758,126
12/31/15	55,765	18,043	73,808	613,416	218,518	831,934
12/31/16	61,342	19,667	81,009	674,758	238,185	912,943
12/31/17	67,476	21,437	88,913	742,234	259,622	1,001,856
12/31/18	74,223	23,366	97,589	816,457	282,988	1,099,445
12/31/19	81,646	25,469	107,115	898,103	308,457	1,206,560
12/31/20	89,810	27,761	117,571	987,913	336,218	1,324,131
12/31/21	98,791	30,259	129,050	1,086,704	366,477	1,453,181

*12/31/14—Accretion on original: 10% × $506,955

**12/31/14—Accretion on increment: 9% × $183,922

Schedule of Expenses

Year-End	Accretion Expense	DD&A Expense
2012	$ 41,897	$41,897
2013	46,087	41,897
2014	67,249	64,887*
2015	73,808	64,887
2016	81,009	64,887
2017	88,913	64,887
2018	97,589	64,887
2019	107,115	64,887
2020	117,571	64,887
2021	129,050	64,887

*[$418,971 − 2($41,897) + $183,922]/8 = $64,887

Entries:

December 31, 2013:

DD&A expense (platform)..................	200,000	
Accumulated DD&A		200,000

(to record straight-line DD&A on the platform: $2,000,000/10 yrs)

DD&A expense (platform)..................	41,897	
Accumulated DD&A		41,897

(to record straight-line DD&A on the asset retirement cost)

Accretion expense	46,087	
ARO liability.................		46,087

(to record accretion expense on the ARO liability)

Long-lived asset (platform)	183,922	
ARO liability.................		183,922

(to record the change in estimated cash flows)

December 31, 2014—2021:

DD&A expense (platform)..................	200,000	
Accumulated DD&A		200,000

(to record straight-line DD&A on the platform ($2,000,000/10 yrs)

DD&A expense (platform)..................	64,887	
Accumulated DD&A		64,887

(to record straight-line DD&A on the asset retirement cost adjusted for the change in cash flow estimate)

Accretion expense	Per schedule	
ARO liability.................		Per schedule

(to record accretion expense on the ARO liability)

On December 31, 2021, Tyler Company settles its asset retirement obligation by using an outside contractor. It incurs costs of $1,600,000, resulting in the recognition of a $146,819 loss on settlement of the obligation:

ARO liability	$1,453,181
Outside contractor	1,600,000
Loss on settlement of obligation.................	$ (146,819)

December 31, 2021:

ARO liability	1,453,181	
Loss on settlement of ARO liability	146,819	
Accounts payable (outside contractor)		1,600,000
(to record settlement of the ARO liability)		
Accumulated DD&A	2,602,893	
Long-lived asset (platform)		2,602,893
(to record the removal of the platform)		

Reassessment. *SFAS No. 143* does not provide clear guidance regarding how frequently an ARO should be reassessed to determine whether a change in the estimate of the ARO is necessary. Companies are not required to remeasure each year; however, they must have a procedure in place to identify events that may materially affect the amount of their AROs. These events are referred to as "trigger" events. A reasonable strategy is to monitor the occurrence of any trigger event that suggests that the estimated cash flows underlying the ARO liability have changed materially. If so, the cash flows should be reestimated, which may include revisions to estimated probabilities associated with different cash flow scenarios. If no trigger events have occurred, no reestimation is required (though the liability still must be accreted). For companies that report on a quarterly basis, the assessment should be updated more frequently as evidence arises that suggests that the ARO estimate may have changed by a material amount. According to *SFAS No. 143*, a change in the credit-adjusted risk-free rate is not a trigger event.

Funding and assurance provisions

Providing assurance that a firm will be able to satisfy its asset retirement obligation does not extinguish the required recognition. Even the existence of a fund or dedication of assets to satisfy the asset retirement obligation *does not* relieve the firm of the required recognition. For example, in some cases, a company may be required by a local governmental agency or as a result of contractual agreement to provide assurance that it will be able to satisfy its AROs. Such assurance may be provided through the use of surety bonds, the establishment of trust funds, letters of credit, or other third party guarantees. The ARO liability should not be reduced because of compliance with such assurance provisions, nor should the liability be considered defeased if the company remains primarily liable for the obligation. In addition, if securities or other assets are set aside for future settlement of the asset retirement obligations, those assets should not be offset against the ARO liability. However, according to *SFAS No. 143*, the existence of a fund may affect the credit-adjusted risk-free rate. It is interesting to note that since the credit-adjusted risk-free rate is the company-wide rate, presumably the FASB's suggestion that the existence of a fund might affect the credit-adjusted risk-free rate is based on the assumption that the company has relatively few AROs, but those AROs are large. Otherwise, it is unlikely that the credit-adjusted risk-free rate for a company with numerous AROs would be affected by a few AROs with funding provisions.

Conditional AROs

In implementing *SFAS No. 143*, companies have encountered a variety of situations in which the timing and/or method of settlement of retirement activities are conditional on future events or circumstances. Some long-lived assets may have an indeterminate useful life and thereby have an indeterminate settlement date for the related asset retirement obligation.

As a consequence, a variety of policies developed in practice. According to the FASB, some accountants concluded that it was appropriate to deal with uncertainty by incorporating it into the calculation of the obligation's fair value. Others did not recognize an ARO until the uncertainty ceased to exist. In 2005, the FASB issued *Interpretation No. 47*, "Accounting for Conditional Asset Retirement Obligations: An Interpretation of FASB Statement No. 143" (*FIN No. 47*). This statement was issued to add clarity and guidance in situations where sufficient information exists to estimate the fair value of an ARO, even though there is uncertainty as to the timing or method of retirement, or both.

According the FASB, situations may exist where the obligation to perform the retirement activity is unconditional even when there is uncertainty regarding the timing of the retirement or the method of settlement. In other words, while the timing or method may be conditional on a future event, the obligation to perform the retirement activity is, nevertheless, unconditional. Accordingly, *FIN No. 47* affirmed the FASB's position that recognition of a liability for the fair value of a conditional asset retirement obligation is required if the fair value of the obligation can be reasonably estimated.

In situations where a current law, regulation, or contract (including an entity's promise) requires an entity to perform asset retirement activities, the FASB has indicated that an unambiguous requirement to perform the retirement exists. This is true even if the activity can be indefinitely deferred. Since tangible long-lived assets (other than land) do not last forever, the asset retirement activity must eventually be performed. The activity is deemed to be unconditional even if the timing or method of settlement, or both, are uncertain.

FIN No. 47 indicates that when the settlement date or method of settlement has been specified by a law, contract, regulation, etc., the entity has sufficient information to apply the expected present value technique in measuring the ARO. According to *SFAS No. 143*, par. 5a:

> *Uncertainty about whether performance will be required does not defer the recognition of an asset retirement obligation because a legal obligation to stand ready to perform the retirement activities still exists, and it does not prevent the determination of a reasonable estimate of fair value because the only uncertainty is whether performance will be required.*

Since the only uncertainty in these situations is whether performance will be required, the ARO is deemed to be reasonably estimable and should be recognized.

Situations involving questions regarding whether retirement activities will be required basically involve one of two possible outcomes: retirement of the asset will either be required or will not be required. *SFAS No. 143* indicates that in situations where either outcome is equally likely, a 50% probability should be assigned to each outcome until additional information permits a more accurate assessment. In some situations, the date or method of settlement, or both, are not specified by others. In these cases, sufficient information is deemed to exist to enable application of the expected present value technique if *all* of the following are known or can be estimated:

1. The settlement date or a range of potential settlement dates
2. The method of settlement or potential methods of settlement
3. The probabilities associated with the potential settlement dates/methods

In those instances, the ARO is deemed to be reasonably estimable, and a liability should be recorded.

In implementing *SFAS No. 143*, some accountants concluded that it was not necessary to record ARO liabilities associated with tangible long-lived assets (such as refineries or plants) having indefinite lives. According to *FIN No. 47*, this approach is not acceptable. *FIN No. 47* distinguishes uncertain from indefinite and concludes that even when the life of a tangible long-lived asset is indefinite, the retirement obligation nevertheless exists.

Finally, in some cases, sufficient information may simply not be available to permit reasonable estimation of the fair value of an asset retirement obligation. In those situations, while it is not necessary to book a liability, extensive disclosure of the facts and circumstances is necessary. The entity is to monitor the particular situation and record an ARO liability as soon as a reasonable estimate can be made.

Miscellaneous

Historically, some oil and gas companies did not recognize AROs because they assumed that they would avoid the obligation by selling the facility or property before the end of its useful life. Under *SFAS No. 143*, companies are no longer permitted to assume that they will avoid recognizing AROs by selling the facility or property because they are legally obligated until such time as the property is sold and another entity assumes the obligation.

Reporting and disclosures

According to *SFAS No. 143*, a company that reports a liability for its asset retirement obligations is required to disclose the following:

1. A general description of the AROs and the associated long-lived assets.
2. The fair value of assets legally restricted for purposes of settling AROs.
3. A reconciliation of the beginning and ending aggregate carrying value of AROs showing separately the changes attributable to the following: (a) liabilities incurred in current period; (b) liabilities settled in the current period; (c) accretion expense; and (d) revisions in estimated cash flows, whenever there is a significant change in one or more of these four components during the reporting period.
4. If the fair value of an ARO cannot be reasonably estimated, disclosure of that fact and the reasons it cannot be reasonably estimated.

In what would be a significant change in practice, some industry experts believe companies should exclude future cash flows related to AROs in preparing the standardized measure of discounted future net cash flows disclosure under *SFAS No. 69*. (*SFAS No. 69* is discussed in detail in chapter 14.) The argument in favor of this approach is that it is consistent with how AROs are treated for purposes of assessing oil and gas properties for impairment in accordance with *SFAS No. 144* (see next section). The approach also is consistent with the fact that future cash flows associated with AROs will be accounted for and disclosed separately pursuant to *SFAS No. 143*. However, the standardized measure disclosure of *SFAS No. 69* is intended to reflect the future net cash flows associated with oil and gas operations. If the future cash outflows related to AROs are excluded from the disclosure, future net cash inflows associated with oil and gas operations may be overstated.

Accounting for the Impairment and Disposal of Long-Lived Assets

For many years, accountants were troubled by the possibility that the net carrying value of long-lived assets might exceed the net realizable value of the assets. Within the oil and gas industry, this issue was especially significant for full cost companies due to the fact that the full cost method permits capitalization of virtually all costs associated with mineral right acquisition, exploration, and development. Consequently, as discussed in chapter 7, the SEC included a ceiling test or impairment test in the full cost accounting rules contained in *Reg. S-X 4-10*. As assets were not as likely to be overvalued under successful efforts at the time the successful efforts rules were written, the successful efforts accounting rules included in *SFAS No. 19* did not include a formal impairment test. However, the SEC later informally required successful efforts companies to test for impairment and permanently write down those assets whose carrying values were determined to be in excess of their underlying value. The requirement that successful efforts companies assess their assets for impairment was formalized in 1995 when the FASB issued *SFAS No. 121*. In 2001, *SFAS No. 144* superseded *SFAS No. 121*, modifying the impairment rules specified in *SFAS 121*. Neither *SFAS No. 121* nor *SFAS No. 144* applies to assets accounted for using full cost. Impairment of these assets continues to fall under the SEC's ceiling test requirements.

Scope

SFAS No. 144 applies to long-lived assets that are held for use or are to be disposed of, including capital leases of lessees, long-lived assets of lessors subject to operating leases, proved oil and gas properties that are accounted for using the successful efforts method, and long-term prepaid assets. Among other items, *SFAS No. 144* does not apply to goodwill, indefinite-life intangible assets, deferred tax assets, or unproved oil and gas properties accounted for using the successful efforts method of accounting. *SFAS No. 144* (par. 3, footnote 2) indicates that oil and gas properties accounted for using the full cost method should adhere to the provisions of *Reg. S-X 4-10*. (The full cost accounting requirements of

Reg. S-X 4-10 are described in detail in chapter 7.) Unproved oil and gas properties accounted for using the successful efforts method are also excluded, since impairment provisions that apply to these properties are already in place in *SFAS No. 19*.

Asset groups

One of the more critical aspects of *SFAS No. 144* is the determination of the "asset" to be tested for impairment. Long-lived assets are to be grouped at the lowest level for which the identifiable cash flows of that asset group are largely independent of the cash flows of other groups of assets and liabilities. In oil and gas operations, oil and gas wells, equipment, and facilities are typically grouped at the field level for impairment purposes. This is because the cash flows of individual wells are typically not independent from the cash flows of other wells in the field, whereas the cash flows related to fields are often independent of the cash flows of other assets. However, more than one field could be linked together through the use of common storage, processing, or transportation facilities, or one or more fields may be linked to certain downstream transportation and refining activities. In such cases, it may be appropriate to group the fields and other facilities together into one single group for impairment purposes. When wells and facilities are connected in a manner requiring allocation of production back to the individual wells or facilities, the wells and facilities would be considered to be linked by common cash flows and would be grouped together for impairment purposes. An example of a single asset that might not qualify for grouping is a gas processing plant that receives gas from numerous fields (owned by the plant owners), as well as gas from nonowners that is processed on a contract basis. In the remainder of the chapter, either a single asset being impaired individually or assets grouped together for impairment purposes are normally referred to simply as the *asset*.

Long-lived assets to be held and used

According to *SFAS No. 144*, par. 7, **impairment** is "the condition that exists when the carrying amount of a long-lived asset (asset group) exceeds its fair value." To determine whether a long-lived asset *to be held and used* is impaired, *SFAS No. 144* requires the use of a three-step approach to recognize and measure an impairment loss. First, the Statement requires that a company test a long-lived asset for recoverability only if certain impairment indicators are present. Second, to test to see if an asset is impaired, the carrying value of the asset is compared to the undiscounted future net cash flows associated with the asset. If the carrying value exceeds the associated cash flows, the asset is impaired, and impairment must be measured and recorded. Third, the impairment loss is measured by comparing the carrying value of the asset to the fair value of the asset. Note that the amount used to test for recoverability differs from that used to measure the impairment loss. A company should recognize an asset impairment loss only if the carrying value of the long-lived asset exceeds its fair value.

Indications of impairment. A company is not required to test each long-lived asset for impairment at each balance sheet date. Rather, *SFAS No. 144* requires that a company test a long-lived asset for recoverability only if events or circumstances indicate that the asset's carrying value may not be recoverable. Examples of such events or circumstances include the following, according to *SFAS No. 144*, par. 8:

1. A significant decrease in the market price of a long-lived asset
2. A significant adverse change in the physical condition of a long-lived asset or in the extent or manner in which the asset can be used
3. Significant adverse change in the legal or business environment in which the long-lived asset is operated
4. Significant cost overruns that occur during the process of constructing or developing a long-lived asset
5. Current period or future expected operating or cash flow losses that indicate potential continued losses
6. Current expectation that it is probable that the long-lived asset will be sold or otherwise disposed of significantly prior to the end of its expected useful life

In oil and gas operations, a temporary decline in the price of oil would not be sufficient to require testing for impairment unless the decline in price was deemed to be permanent or long lasting. On the other hand, a hostile or unfavorable action by the government of a country in which a company is operating would likely indicate the need to assess the properties in that country for impairment.

Testing for recoverability. If events or circumstances indicate a long-lived asset's carrying value may not be recoverable, then the asset must be assessed for impairment by comparing its carrying value to its undiscounted expected future net cash flows. To determine the estimated net cash flows, a company should consider only the expected cash inflows less associated expected cash outflows (excluding interest expense) that are directly attributable to and are a direct result of the use and eventual disposition of the asset. *SFAS No. 144* provides additional guidance on estimating future net cash flows focusing on the following:

a. The estimation approach
b. The estimation periods
c. Types of asset-related expenditures to consider

Estimation approach. Estimated future net cash flows should be based on the remaining service potential of the asset. The service potential of an asset consists of the cash flow–generating capacity or physical output capacity of the asset. A company should incorporate its own assumptions about the use of the asset. The assumptions used in developing the estimate should be reasonable in relation to the assumptions used in developing other information used by the company for similar periods. Other information would include information used in preparing internal budgets and projections or information communicated to others. Presumably, this information would include reasonable assumptions regarding trends in future prices and costs. If proved and probable reserves are utilized in formulating the company's internal budgeting and planning, then cash flows from the recovery and sale of those reserves would be included in future cash flow projections.

Estimation periods. The cash flow estimation periods are limited to the individual asset's (or primary asset's, if an asset group) remaining estimated useful life to the company. The primary asset is the most significant component asset from which the asset group derives its cash flow–generating activities. The primary asset can be either a long-lived tangible asset being depreciated or an intangible asset being amortized.

Estimates of future cash flows used to test for recoverability of the asset should be limited to the remaining amount of time that the asset will be used by the entity. For example, many international oil and gas contracts, specifically production sharing contracts (PSCs), contain a maximum production period effectively terminating the contract after a specific number of years of production, e.g., 15 years. An oil and gas producer operating under a PSC should consider whether the contract has a maximum production period. If so, the producer's interest in the field automatically terminates at the end of the production period, even if production from the area is expected to continue beyond the maximum period. In this case, for purposes of *SFAS No. 144*, the producer can only include its share of expected future net cash flows from production of the reserves that would be recoverable prior to the end of the production period.

Types of asset-related expenditures to consider. For an asset that is complete or substantially complete, *SFAS No. 144* requires a company to base the estimates of future cash flows on the long-lived asset's existing service potential at the date it is tested for impairment. A company should include cash outflows associated with future expenditures necessary to maintain an asset's existing service potential, including those expenditures that replace the service potential of a component part of an asset (e.g., repairing an existing elevator in a building). However, *SFAS No. 144* prohibits including in the estimates of future cash flows those amounts associated with future capital expenditures that would increase the long-lived asset's service potential.

If an asset will require substantial future development expenditures in order to generate the estimated future cash inflows, the cost of the future development must be considered when projecting cash outflows. For example, consider an oil and gas field that is currently producing but will require secondary recovery, such as waterflooding, prior to reaching maximum recovery. The future cash outlay for the secondary recovery project would properly be included in projecting the future cash flows associated with the asset.

For a long-lived asset that is not yet complete and is under development, a company should base the estimates of future cash flows to test recoverability on the asset's expected service potential when development is substantially complete. The estimates should include cash flows related to all future expenditures to substantially complete the asset, including interest payments it would capitalize under *SFAS No. 34*.

One exception to the rules specifying which future expenditures to include in estimates of future net cash flows is future decommissioning costs. As discussed previously, *SFAS No. 143* requires that the estimated future cost to decommission facilities and restore the environment be estimated and capitalized as part of the cost of the related asset. Since AROs represent a future cash *outlay*, if there were no exception, including an ARO in estimates of future net cash flows would result in a lower net cash flow figure. That lower cash flow figure would be compared to a higher asset value, i.e., an asset value increased by the amount of the ARO, thus in effect double counting the ARO in calculating impairment. Consequently, the FASB concluded that for purposes of applying the impairment rules, the carrying value of the asset should include capitalized asset retirement costs. However, the future cash outflows related to the future decommissioning activities are to be *excluded* from both (a) the undiscounted net cash flows used to test the asset's recoverability, and (b) the discounted net cash flows commonly used to measure an oil and gas asset's fair value. In addition, the fair value of the asset may be based on a quoted market price that considers the costs of retirement (in other words, those costs would have been deducted in arriving at the fair value). If so, the quoted market price must be increased by the fair value of the estimated retirement costs.

Once the estimated future cash net flows are determined, the next step is to compare the results of those undiscounted estimated cash flows to the asset's carrying value. If the estimated cash flows are greater than the carrying value, *SFAS No. 144* precludes a company from recognizing an impairment loss. However, if the estimated undiscounted future cash flows are less than the asset's carrying value, a company must recognize an impairment loss.

Measuring impairment. The third and final step is to measure the amount of the impairment loss. The amount of the impairment loss a company should recognize is the amount by which the carrying value of the asset exceeds its fair value. *SFAS No. 144* defines the fair value of an asset as the amount at which the asset could be bought or sold in a current transaction between willing parties, i.e., other than in a forced or liquidation sale. Quoted market prices in active markets are the best evidence of fair value, so a company should use quoted market prices as the basis for measurement, if available. Absent quoted market prices, a company should base the measurement on the best information available, including present value techniques. For oil and gas companies, the best measurement technique is normally the discounted future net cash flows associated with the asset.

SFAS No. 144 allows two different techniques to be used in estimating discounted future net cash flows. The first is the **traditional present value approach**, where a single set of estimated cash flows and a single interest rate (commensurate with the risk) are used to estimate fair value. The second is the **expected present value approach**, where multiple cash flow scenarios with expected probabilities and a credit-adjusted risk-free rate are used to estimate fair value. Because there will typically be uncertainties in both the timing and the amount of the cash flows, the expected present value approach will often be the appropriate technique. As a result, if a company is considering alternative plans or if a company is estimating a range of possible outcomes, it should consider the likelihood of those possible outcomes. Although it is not a requirement, a company may find the probability-weighting feature of the expected present value technique useful in considering the likelihood of those possible outcomes.

The following example illustrates the application of the expected present value technique to the measurement of the impairment loss.

EXAMPLE

Expected Present Value Technique

Tyler Company, a successful efforts company, has 100% of the working interest in the Alpha Field, an offshore oil field in the Gulf of Mexico. The field constitutes a cost center and is also an asset group for purposes of testing for impairment. The net carrying value of the field is $120 million. In 2013, the government enacted a new tax. Since the new tax significantly affects the economic environment, Tyler Company must test for impairment. The following table reflects the expected cash flows from the operation of the field for the remainder of its life.

Year	Total Net Cash Flow Estimate (Market) (million $)	Probability Assessment	Expected Net Cash Flows (million $)	Undiscounted Net Cash Flows/yr (million $)	Credit-Adjusted Risk-free Rate	Present Value (million $)	Total Present Value (million $)
2014	$30.0	20%	$ 6.00				
	32.9	70%	23.03				
	29.3	10%	2.93	$ 31.96	5.1%	$30.41	
2015	$30.0	50%	$15.00				
	35.3	20%	7.06				
	15.0	30%	4.50	26.56	5.3%	23.95	
2016	$22.5	60%	$13.50				
	19.8	20%	3.96				
	8.5	20%	1.70	19.16	5.5%	16.32	
2017	$18.4	70%	$12.88				
	10.2	20%	2.04				
	9.8	10%	0.98	15.90	5.7%	12.74	
2018	$ 8.9	50%	$ 4.45				
	7.5	25%	1.88				
	6.0	25%	1.50	7.83	6.0%	5.85	
2019	$ 6.6	60%	$ 3.96				
	10.0	35%	3.50				
	16.2	5%	0.81	8.27	6.1%	5.80	
				$109.68			$95.07

Impairment Test

Net carrying value.............................	$120.00 million
Expected undiscounted cash flows	$109.68
Impairment?..................................	Yes

Measure of Impairment Loss

Net carrying value.............................	$120.00 million
Expected undiscounted cash flows	95.07
Impairment?..................................	$ 24.93 million

Since the undiscounted expected cash flows of $109.68 million are less than the $120 million carrying value, impairment must be measured and recorded. The amount of impairment is $24,930,000.

Entry

Impairment loss	24,930,000	
Allowance for impairment		24,930,000
(to record impairment on Alpha Field)		

Any impairment loss relating to an individual asset would reduce the carrying value of the long-lived asset. If an asset group is impaired, the impairment loss would be allocated to the individual long-lived assets in the group on a pro rata basis using the relative carrying values of the individual assets. However, the carrying value of any single long-lived asset may not be reduced below its fair value if the fair value is determinable without undue cost and effort.

Once an asset has been written down by recognizing impairment, the asset may not be written back up even if changes in economic circumstances indicate that the impairment has reversed. The reduced value of the asset becomes the asset's new cost basis. For depreciable assets, the new basis is the basis that would be used for the purpose of computing future DD&A.

Long-lived assets to be disposed of

SFAS No. 144 establishes accounting and reporting requirements related to long-lived assets that are to be disposed of either by sale, abandonment, or exchange for other productive assets.

Long-lived assets to be disposed of other than by sale. In some circumstances, a long-lived asset is to be disposed of in some manner other than by sale (such as by abandonment or in exchange for similar assets). In these cases, the assets continue to be classified as held and used until actually disposed of, and the impairment rules discussed for assets to be held and used apply. If the company commits to a plan of abandonment for an asset prior to the end of its previously estimated life, depreciation estimates are to be adjusted to reflect the shortened life. If an asset is to be exchanged, the related net cash flows are to be estimated based on the expected useful life of the asset assuming the exchange does not occur.

Long-lived assets to be disposed of by sale. A long-lived asset to be sold is to be classified as "held for sale" in the period in which all of the following criteria are met:

a. Management commits to a plan to sell the asset.
b. The asset is currently in a condition making it available for immediate sale.
c. An active program to locate a buyer or efforts to complete a sale are underway.
d. The sale of the asset is probable.
e. The asset is being actively marketed for sale at a reasonable price (in relation to the asset's current fair value).
f. Actions required to complete the sale make it unlikely that plans to sell the asset will change.

When a long-lived asset is classified as being held for sale, depreciation on the asset should cease, even if the asset continues in use. Such an asset should be valued on the balance sheet at the lower of its carrying value or its fair value less estimated selling costs. If the asset's carrying value is greater than its fair value less selling costs, a loss in the amount of the difference should be recognized. If the fair value subsequently increases, a gain should be recognized but not in an amount greater than the total of any losses previously recognized.

Disposal groups. *SFAS No. 144* defines a "component of an entity" to include a reporting unit, a subsidiary, or an asset group whose operations and cash flows can be clearly distinguished from the rest of the entity's. The results of operating long-lived assets that either have been disposed of through sale, abandonment, or exchange or are classified as held for sale are to be reported in the discontinued operations section of the income statement if the disposal group is a component of an entity and if the following conditions are met: (a) the operations and cash flows of the asset have or will be eliminated from ongoing operations as a result of the disposal and (b) the company will not have a significant involvement in continuing operations of the asset after the disposal transaction. For oil and gas producing companies, the field (or perhaps well or facility) is the most likely asset group and is often a "component of an entity." Thus, it would appear that normally when an oil and gas company disposes of a field or classifies the field as held for sale, the field must normally be accounted for according to *SFAS No. 144* guidelines as a discontinued operation. This means that in the period in which the field has been disposed of or is classified as held for sale, the results of operating the field, including any gain or loss on fair value adjustments, should be reported on the income statement as discontinued operations. The results of discontinued operations should be reported on the income statement separately between extraordinary items and the cumulative effect of any accounting changes.

Impairment for full cost companies

Full cost companies must apply the SEC ceiling test to their oil and gas assets versus the impairment test specified in *SFAS No. 144*. As discussed in chapter 7, the ceiling test also requires an estimate of future net cash flows. Consequently, full cost companies should also exclude future cash outflows related to an ARO that have been recorded (in accordance with *SFAS No. 143*) from the future net cash flows.

PROBLEMS

1. Define the following:

    asset retirement obligation

    retirement

    accretion

    promissory estoppel

    legally enforceable obligation

    obligating event

2. Define the following:

    impairment

    asset group

    traditional present value approach

    expected present value approach

3. Determine whether the following statements are true or false.

    a. Full cost companies do not book AROs.

    b. An oral agreement to dismantle equipment and restore the environment at the end of the productive life of a facility would result in the recording of an ARO.

    c. A change in the discount rate should result in the reestimation of AROs.

    d. A company had an oil spill resulting from a tanker running aground. The company should immediately accrue an ARO.

    e. Either gains or losses may be recognized when an ARO is settled.

    f. Companies must test all of their long-lived assets for impairment on an annual basis.

4. When are companies required to remeasure AROs, and what types of changes result in the remeasurement? How are changes recorded?

5. What is a conditional ARO? How are conditional AROs to be accounted for?

6. For purposes of booking impairment, what is an asset group? In oil and gas operations, what asset groups are most common?

7. Give examples of events or circumstances that may trigger impairment testing.

8. Lexington Company is operating a single field in Alaska in which it has a 49% working interest. The remainder of the working interest is held by another oil company. The field went onto production in early in 2012 and has an estimated life of 15 years.

    REQUIRED: Should Lexington recognize an ARO under SFAS No. 143? If so, when should the ARO be recognized?

9. Ameritec Oil and Gas Company completes construction of an offshore oil platform and places it into service on January 1, 2015. Ameritec is legally required to dismantle and remove the platform at the end of its useful life, which is estimated to be 10 years. On January 1, 2015, Ameritec recognized a liability for an asset retirement obligation and capitalized an amount for an asset retirement cost. It estimated the initial fair value of the liability using an expected present value technique. The significant assumptions used in that estimate of fair value are as follows:

    a. Labor costs are based on current marketplace wages required to hire contractors to dismantle and remove offshore oil platforms. Ameritec assigns probability assessments to a range of cash flow estimates as follows:

Cash Flow Estimate	Probability Assessment
$250,000	10%
175,000	60%
200,000	30%

b. Ameritec estimates allocated overhead and equipment charges using the rate it applies to labor costs for transfer pricing (60%). The entity has no reason to believe that its overhead and equipment rate differs from that used by contractors in the industry.

c. A contractor typically adds a markup on labor, allocated internal costs, and equipment to provide a profit margin on the job. The entity determines the profit that contractors in the industry generally earn to dismantle and remove offshore oil platforms is 15%.

d. A contractor would typically demand and receive a premium (market risk premium) for bearing the uncertainty and unforeseeable circumstances inherent in "locking in" today's price for a project that will not occur for 10 years. The entity estimates the amount of that premium to be 5% of the estimated inflation-adjusted cash flows.

e. The risk-free rate of interest on January 1, 2015 is 6%. The entity adjusts that rate by 4% to reflect the effect of its credit standing. Therefore, the credit-adjusted risk-free rate used to compute expected present value is 10%. (Round the present value factor to four decimal places.)

f. Ameritec assumes a rate of inflation of 4% over the 10-year period. (Round the factor to four decimal places.)

REQUIRED:

a. Complete the following tables:

Cash Flow Estimate	Probability Assessment	Expected Cash Flow
$250,000	10%	
175,000	60%	
200,000	30%	

Initial Measurement of the ARO Liability at January 1, 2015

	Expected Cash Flows 1/1/15
Expected labor costs	
Allocated overhead and equipment charges	
Contractor's markup	
Expected cash flows before inflation adjustment	
Inflation factor assuming 4% rate for 10 years	
Expected cash flows adjusted for inflation	
Market-risk premium	
Expected cash flows adjusted for market risk	
Present value using credit-adjusted risk-free rate of 10% for 10 years	

Accretion: Interest Method of Allocation

Year	Accretion	Liability Balance
1/1/2015		
12/31/2015		
12/31/2016		
12/31/2017		
12/31/2018		
12/31/2019		
12/31/2020		
12/31/2021		
12/31/2022		
12/31/2023		
12/31/2024		

Schedule of Expenses

Year-End	Accretion Expense	DD&A Expense
2015		
2016		
2017		
2018		
2019		
2020		
2021		
2022		
2023		
2024		

b. Prepare the journal entry that would be made on January 1, 2015 to record the asset retirement obligation.

c. Prepare the journal entries that would be made from December 31, 2015 to December 31, 2024 to record the accretion expense and the amortization expense related to the ARO.

d. On December 31, 2024, the entity settles its asset retirement obligation by using its internal workforce at a cost of $432,000. Assume no changes during the 10-year period in the cash flows used to estimate the obligation. Prepare the journal entry that would be made on December 31, 2024 to record the settlement of the asset retirement obligation.

10. Problem 10 is the same as problem 9 with respect to initial measurement of the ARO liability. Now assume that Ameritec's credit standing improves over time, causing the credit-adjusted risk-free rate to decrease by 1% to 9% at December 31, 2016.

    On December 31, 2016, Ameritec revises its estimate of labor costs and revised the probability assessments related to those labor costs. The change in labor costs results in an upward revision to the undiscounted cash flows. Consequently, the incremental cash flows are discounted at the current rate of 9%. All other assumptions remain unchanged. The revised estimate of expected cash flows for labor costs is as follows:

Cash Flow Estimate	Probability Assessment	Expected Cash Flow
$731,000	10%	$ 73,100
196,000	55%	107,800
224,000	35%	78,400
		$259,300

    REQUIRED:
    a. Complete the following tables:

    Subsequent Measurement of the ARO Liability Reflecting a Change in Labor Cost Estimate as of December 31, 2016

Incremental expected labor costs	
Allocated overhead and equipment charges	
Contractor's markup	
Expected cash flows before inflation adjustment	
Inflation factor assuming 4% rate for eight years	
Expected cash flows adjusted for inflation	
Market-risk premium	
Expected cash flows adjusted for market risk	
Present value of incremental liability using credit-adjusted risk-free rate of 9% for eight years	

Accretion: Interest Method of Allocation

Year	Accretion on Original	Accretion on Change	Total Accretion	Original Liability Balance	Increment Liability Balance	Total Liability Balance
1/1/2015						
12/31/2015						
12/31/2016						
12/31/2017						
12/31/2018						
12/31/2019						
12/31/2020						
12/31/2021						
12/31/2022						
12/31/2023						
12/31/2024						

Schedule of Expenses

Year-End	Accretion Expense	DD&A Expense
2015		
2016		
2017		
2018		
2019		
2020		
2021		
2022		
2023		
2024		

b. Prepare the journal entry that would be made on January 1, 2015 to record the asset retirement obligation.

c. Prepare the journal entries that would be made on December 31, 2015 and December 31, 2016 to record the accretion expense and the DD&A expense related to the ARO.

d. Prepare the journal entries that would be required at December 31, 2016 to record the revision in the asset retirement obligation.

e. Prepare the journal entries that would be made from December 31, 2017 to December 31, 2024 to record the accretion expense and the DD&A expense.

f. On December 31, 2024, Ameritec settles its asset retirement obligation by using an outside contractor. It incurs costs of $800,000. Prepare the journal entries that would be made on December 31, 2024 to record the settlement of the asset retirement obligation.

11. Exron Oil and Gas Company constructs a natural gas treatment facility in three phases. The first phase was completed and placed into service on December 31, 2017. The second phase was completed and placed into service on December 31, 2018, and the third phase was completed and placed into service on December 31, 2019. Exron is legally required to decommission the plant at the end of its useful life, which is estimated to be 20 years from the date that the first phase went into service. The following schedule reflects the undiscounted expected cash flows and respective credit-adjusted risk-free rates used to measure each portion of the liability through December 31, 2019. (The undiscounted cash flows below have already been adjusted for market risk premium and inflation and already incorporate labor costs, overhead, and a profit margin. They have not been discounted.)

Date	Undiscounted Expected Cash Flows	Credit-Adjusted Risk-Free Rate
12/31/17	$100,000	9.0%
12/31/18	90,000	8.0%
12/31/19	50,000	7.0%

On December 31, 2019, the entity increases by 10% its estimate of undiscounted expected cash flows that were used to measure those portions of the liability recognized on December 31, 2017 and December 31, 2018.

REQUIRED: Make the journal entries necessary to record the asset retirement obligation under *SFAS No. 143* and any DD&A expense related to the ARO and accretion expense for the following dates:

a. December 31, 2017

b. December 31, 2018

c. December 31, 2019

12. Duncan Oil Company, a successful efforts company, has capitalized costs on Property R, Property S, and Property T as of 12/31/2017 as follows:

	Property R	Property S	Property T
Net capitalized costs	$300,000	$600,000	$400,000
Expected undiscounted future net cash flows	200,000	630,000	380,000
Fair value (discounted future net cash flows)	160,000	540,000	320,000

Duncan has no other capitalized costs. The properties are located in different regions in Venezuela.

REQUIRED: If necessary, test the assets for impairment and make any necessary journal entries, assuming production costs tripled late in the year on all three properties and that production costs are not expected to decrease.

13. Ellis Company, a successful efforts company, has 100% of the working interest in a field in Texas. The field constitutes a cost center and is also an asset group for purposes of testing for impairment. In 2016, the price of oil dropped significantly; therefore, Ellis must test for impairment. The table below reflects Ellis's latest expected cash flows and risk-free rates for the remainder of the life of the field.

Year	Total Net Cash Flow Estimate (Market) (million $)	Probability Assessment	Credit-Adjusted Risk-free Rate
2017	$100.0	30%	
	90.9	60%	
	85.3	10%	6.2%
2018	$ 91.2	50%	
	88.3	20%	
	75.4	30%	6.9%
2019	$ 87.5	70%	
	91.8	20%	
	65.5	10%	7.1%
2020	$ 62.4	55%	
	50.2	20%	
	45.8	25%	6.7%
2021	$ 45.9	80%	
	55.5	10%	
	36.0	10%	5.5%
2022	$ 31.6	60%	
	21.0	20%	
	15.4	20%	6.1%

REQUIRED:

a. Assume that Ellis's carrying value for the field is $300 million. Determine whether Ellis must book impairment and, if so, record the necessary journal entry. Round the present value factors to four decimal places.

b. Assume that Ellis's carrying value for the field is $400 million. Determine whether Ellis must book impairment and, if so, record the necessary journal entry. Round the present value factors to four decimal places.

14. Frontier Company, a successful efforts company, owns a working interest in an oil field in Oklahoma. The field has been producing for a number of years and is expected to be producing for another 10 years. On January 1, 2019, the net book value of the field wells, equipment, and facilities totals $4,500,000. Frontier determines that it should book an ARO in relation to the field. The undiscounted future cash flows to settle the ARO are estimated to be $2,000,000. When discounted at a rate of 8% for 10 years, the present value is $926,390.

    Frontier's chief accountant indicates that the field should be assessed for impairment. Analysis yields the following information:

	Undiscounted	Discounted
Estimated future net cash flows **before** ARO	$5,300,000	$3,000,000
Estimated ARO cash outflows	(2,000,000)	(926,390)
Estimated future net cash flows after ARO	$3,300,000	$2,073,610
Net book value of field assets	$5,426,390	
Net book value of field assets without ARO	4,500,000	

    REQUIRED:

    a. Make the entry to record the ARO on January 1, 2019.
    b. Determine the net book value of the field after the ARO is recorded.
    c. Explain how impairment would be computed if circumstances indicated the asset may be impaired.

15. Payne Oil Corporation, a successful efforts company, has capitalized costs on three leases as of 12/31/2013 as follows:

	Lease A	Lease B	Lease C
Net capitalized costs	$ 800,000	$600,000	$900,000
Expected undiscounted future net cash flows	900,000	500,000	750,000
Fair value (discounted future net cash flows)	685,000	490,000	610,000

    Payne has no other capitalized costs. The leases are located in different counties in Oklahoma.

REQUIRED: If necessary, test the assets for impairment and make any necessary journal entries assuming taxes tripled on all three leases.

16. Baseline Petroleum, a successful efforts company, entered the oil and gas business in 2014 with the acquisition of one field. Baseline proved the field during 2014. At the end of 2014, prices were high and costs low. During 2015, Baseline continued exploration and development activities in the field, but towards the end of 2015, oil and gas prices plummeted. During 2016, due to continued low prices, Baseline suspended all exploration and development activities. Prices at the end of 2016 declined significantly once again. Data for Baseline's one field are as follows:

	2014	2015	2016
Capitalized costs before any write-downs	$ 300,000	$ 900,000	$ 900,000
Accumulated DD&A	20,000	100,000	180,000
Expected undiscounted future net cash flows	290,000	820,000	700,000
Fair value (discounted future net cash flows)	200,000	580,000	520,000

REQUIRED: Make any necessary journal entries for the above years relating to impairment.

17. The CEO of Green Oil Company, Dwight Allen, gave a speech at the World Environmental Conference. In his speech, Mr. Allen spoke about Green Oil's goal of doing no damage to the environment as a result of the company's operations and gave specific details on how the company was achieving this goal. Mr. Allen stated that no matter where the company operates and regardless of what the legal or contractual requirements, Green Oil will always leave every place it operates "better than they found it."

REQUIRED: Determine if Mr. Allen's speech could result in recognition of an ARO.

18. Solar Petroleum owns 100% of the working interest in the Bearcat Field in West Texas. The field has been producing and is expected to produce for 10 more years. Events have occurred that indicate that impairment may be necessary. Therefore, Solar must test for impairment of the field. Assume the following:

	Undiscounted	Discounted
Estimated future net cash flows before ARO	$10,600,000	$ 6,000,000
Estimated ARO cash flows	(4,000,000)	(1,852,780)
Estimated future net cash flows after ARO	$ 6,600,000	$ 4,147,220
Net book value of field assets	$10,852,780	
Net book value of field assets without ARO	9,000,000	

REQUIRED: Determine if impairment is necessary. If so, prepare the journal entries to record the impairment.

References

1. This chapter is based largely on the related chapters in Wright, Charlotte, and Rebecca Gallun. 2005. *International Petroleum Accounting.* Tulsa, OK: PennWell.

2. Garner, Bryan, and Henry Black, eds. 1999. *Black's Law Dictionary.* 7th ed. St. Paul, MN: West Group.

10

ACCOUNTING FOR REVENUE FROM OIL AND GAS SALES

Accounting for production activities was examined in chapter 8. In this chapter, accounting for revenue from the sale of oil and gas is discussed. As with production costs, revenue accounting is the same for both full cost accounting and successful efforts accounting. Before discussing revenue accounting, the process of measuring and selling oil and gas is briefly discussed. Definitions of common terminology related to the measurement and sale of oil and gas are presented in the following glossary.

DEFINITIONS

API gravity. A measure of the gravity or density of petroleum liquid in relation to water. Measurement is based on a universally accepted scale developed by the American Petroleum Institute (API) and is expressed in degrees. There is an inverse relationship between API gravity and density; the higher the density, the lower the API gravity.

Associated gas. Natural gas that overlies and is in contact with crude oil in the reservoir but not in solution with the oil. Commonly known as **gas-cap gas**. See the discussion of natural gas measurement and the accompanying illustration later in the chapter, as well as the definitions for **dissolved gas** and **nonassociated gas**.

Barrel. A unit of measure of crude oil equal to 42 U.S. gallons.

BS&W. Basic sediment and water, i.e., impurities in suspension with crude oil.

Btu. British thermal unit; the amount of heat required to raise the temperature of 1 pound of water by 1° Fahrenheit.

Casinghead gas. Natural gas that is produced with crude oil from oil wells.

City gate. System inlet of the local distribution company to which gas is delivered from a natural gas pipeline company.

Commingled gas. Gas from two or more sources combined into one stream of gas.

Condensate. A hydrocarbon liquid that condenses from a gaseous state to a liquid state when produced.

Crude oil. Unrefined liquid petroleum (hydrocarbons).

Dissolved gas. Natural gas in solution with the crude oil in the reservoir. Also referred to as **solution gas**.

Field facility. An installation designed to process natural gas from two or more leases in order to control the quality of gas sold to the market. Usually separates condensate from the natural gas and may extract other products, such as propane and butane.

Gas balancing agreement. An agreement that specifies how any gas production imbalance will be balanced between the different parties.

Gas settlement statement. A form used to record the amount of natural gas transferred to the pipeline.

Gauging. Measuring the level of oil in a tank.

Heater-treater. A vessel that heats the well fluids to separate the water, sediments, and any remaining gas from the oil in order to raise the oil to a quality acceptable for sale.

LACT unit. Lease automatic custody transfer unit that measures the flow of oil from the lease to a pipeline and determines the API gravity, BS&W, and temperature.

Local distribution company (LDC). A company that purchases gas and resells the gas to the public or to other end users.

Mcf. The standard unit for measuring natural gas—1,000 cubic feet.

MMcf. One million cubic feet of natural gas.

Natural gas. A mixture of hydrocarbons occurring in the gaseous state.

Natural gas processing plant. An installation in which the natural gas liquids are separated from the natural gas and/or the natural gas liquids are separated into liquid products, such as butane and propane.

Nonassociated gas. Natural gas in a reservoir that does not contain any oil. (See the discussion of natural gas measurement and the accompanying illustration later in the chapter.)

Posted field price. The announced or published price that some purchasers in a particular area will pay for crude oil.

Processed gas. Natural gas processed in a natural gas processing plant in order to remove the liquid hydrocarbons.

Psia. Pounds per square inch absolute. Psia is equal to pressure reading from a measurement gage plus the pressure of the atmosphere at that point.

Run ticket. A document showing the amount of oil run from a lease tank into a connecting pipeline or tank truck.

Scrubber. A vessel used to separate entrained liquids or solids from gas.

Separator. A vessel used to separate well fluids into gases and liquids.

Shrinkage. The reduction in the volume of natural gas due to the removal of impurities such as liquids, sulfur, and water vapor.

Spot sales. Short-term sale of oil or gas in the spot market, without a long-term contract.

Tank battery. Two or more tanks on a lease that can hold oil production from wells prior to sale.

Tank strapping. Measuring an oil tank so that a tank table can be prepared that shows the volume of oil in the tank at any given height.

Thief. A brass, aluminum, or glass cylinder that is lowered into an oil storage tank to obtain an oil sample.

Measurement and Sale of Oil and Natural Gas

Oil and gas must be measured either when produced or when sold. Understanding the measurement process is important to the accountant, because the amount of revenue recorded is based on information obtained during this process. The following sections discuss the measurement and sale of oil and gas.

Crude oil measurement

Fluids produced from an oil well normally contain oil, gas, and water. Before the oil can be sold, well fluids must be separated, treated, and measured. Flow lines transfer the well fluids from individual wells to a centralized separating, treating, and/or storage facility. At the centralized facility, the fluid enters separators that remove the gas from the liquid (oil and water). The liquid is then treated in heater-treater units to remove the water and other impurities from the oil. The water and other impurities, called **basic sediment and water (BS&W)**, must be below an established limit in oil sold. The limit is usually 1%.

The treated crude is then stored in a **tank battery** on the lease. Typically, a tank battery has storage capacity equal to three to seven days of production. Each tank in a tank battery is **strapped**, which means that the dimensions of a tank are determined and the volume of oil (in barrels) that a tank can hold is calculated at height intervals, usually 1/4 inch apart. The oil enters a tank through an opening at the top of the tank and is sold, i.e., delivered or run to a tank truck, barge, or pipeline, through an opening near the bottom of the tank. The oil delivered is measured by **gauging** the height of the oil in the tank both before and after delivery of the oil. A tank gauger normally determines the height of the oil in the tank in a nonautomated system by lowering a steel tape with a plumb bob on the end into the tank. When the tape touches the tank bottom, it is withdrawn, and the gauger notes the oil level on the tape.

The unit of measure of oil is a **barrel**, which is 42 U.S. gallons. Before oil is sold, the oil measurement must be converted to standard conditions, net of any BS&W. In addition, the API gravity of the oil must be determined for a standard temperature of 60° Fahrenheit (F). **Gravity**, as used here, refers to degrees of gravity on the American Petroleum Institute (API) scale and is a measure of the specific gravity or density of crude oil or condensate in comparison to water. API gravity is determined after adjusting measurements made at the actual temperature of the petroleum liquid (i.e., crude oil and condensate) to a standard temperature of 60°F. On this scale, the greater the density of the petroleum liquid (or the heavier the liquid), the lower the API gravity. The API gravity of petroleum liquid is significant in determining the sales value of the crude or condensate, because the lighter petroleum liquids (i.e., those with a higher API gravity) typically sell for a higher price than the heavier petroleum liquids (i.e., those that have a lower API gravity).

The volume of oil sold is converted to standard conditions by using the characteristics of the oil sold, i.e., the gravity, temperature, and BS&W. Thus, in addition to measuring the amount of oil delivered, the oil must also be tested to determine its characteristics. A container known as a **thief** is used to obtain samples of the oil at various levels in the tank. These samples are then analyzed to determine BS&W content, the gravity of the oil, and the temperature. In addition, the temperature of the oil in the tank is taken both before and after the run. These two temperature readings are used to correct the volume of the fluid in the tank before and after the run to a standard condition of 60°F. The temperature of the oil from the oil sample is used to correct the observed gravity at the observed temperature to the API gravity at standard temperature, i.e., 60°F. The BS&W factor is used to convert gross barrels of oil delivered to barrels net of BS&W content, at API standard.

The BS&W, gravity, temperature, the number of barrels of oil removed from a tank, and the calculations converting the barrels to standard conditions, net of BS&W, are recorded on a run ticket at the time of delivery of the oil (Fig. 10–1). The run ticket is the source document for recording the amount of oil sold.

COMPANY	
ACCT NO. _____ DIST NO.	**RUN TICKET NO.** DATE

OPERATOR	
LEASE NAME	LEASE NO.

WELL NO.		FT.	INCHES	TEMP	BBLS
	1ST MEAS.				
TANK NO.	2ND MEAS.				
	GRAVITY				
TANK SIZE	OBSERVED	TEMP.	TRUE	PRICE	
BS &W	OFF←SEAL NO.'S→ON			AMOUNT	

THIEFING FT – IN BS & WATER	BEFORE	AFTER	CALCULATIONS
REMARKS			

	DATE	GAUGER
1ST MEAS	TIME A.M. P.M.	WELL OWNER'S WITNESS
2ND MEAS	DATE	GAUGER
	TIME A.M. P.M.	WELL OWNER'S WITNESS

This ticket covers all claims for allowance. The oil represented by this ticket was received and run as property of

Fig. 10–1. Typical run ticket
Source: The run tickets presented in this chapter and in the homework problems are based on a pipeline run ticket prepared and printed by Kraftbilt, Tulsa, OK (1-800-331-7290).

When the oil is sold, the pipeline gauger, observed by the lease pumper-gauger, measures the oil removed and determines its characteristics. When the amount of oil removed from the tank and the oil's characteristics have been recorded on the run ticket, both the pipeline gauger and the lease pumper-gauger sign the ticket.

Another method of measuring oil delivered is used when the oil is delivered or transferred automatically into a pipeline. In this case, measurement of the oil is performed by a **lease automatic custody transfer (LACT) unit**. A LACT unit is an unattended unit that automatically transfers oil from the lease to the pipeline, takes samples, and collects and records production/accounting data. The data include volume measurement, temperature, and BS&W content. While the use of LACT units appears to be much more efficient than manual measurement, a substantial portion of the producing leases in the United States are not equipped with LACT units. LACT units are most effectively used in large fields or reservoirs with numerous producing wells.

If a tank battery is to collect crude oil from several wells, the wells must be connected to a gathering system. A gathering system is a network of flow lines going from the various wells to the central tank battery. Since the tank battery collects crude oil from several wells, the quantity of crude oil produced must be allocated back to the various wells. The amount of production per well is determined by using a routine productivity test. Each well is periodically put on test by diverting the flow from the production separator to a test separator that determines the production rates of oil, gas, and water of the well being tested. Once this information is obtained, each well's percentage of production can be calculated. From these percentages, actual production of the commingling facility can be allocated to each well. (See further discussion later in this chapter.)

Run ticket calculation

The data entered on a run ticket is used to determine the volume of oil released from a storage tank or flowed through a meter. The process used to complete a run ticket and determine the net volume from a tank run is described below and is illustrated with numbers from a completed run ticket (Fig. 10–2):

a. At the time a tank is installed, it is strapped and assigned a unique number, in this case number 1. This number, which is placed on the run ticket, indicates the strapping table that corresponds with the volumetric characteristics of the tank. This table will be used to convert the feet and inches measurement on the run ticket into a run volume from the tank expressed in barrels.

b. The level of oil in the tank and the temperature of the oil are measured at the beginning and at the end of the run. In this example, the beginning level was 13'0" and the ending level was 1'4". Each of these measurements, which are in feet and inches, are converted to a volume in barrels using the appropriate tank table (Table 10–1). In this example, the level of oil in the tank converted to barrels at the beginning and at the end of the run is 180.83 barrels and 18.60 barrels, respectively. The difference between these two numbers is the measured volume of the run.

Chapter 10 • Accounting for Revenue from Oil and Gas Sales

COMPANY **RUN TICKET NO. 1**						
ACCT NO. 492						
DIST NO. 14	DATE 5/1/2012					

OPERATOR: *Tyler Company*

LEASE NAME				LEASE NO.		
Ebert				5		
WELL NO.		FT.	INCHES	TEMP	BBLS	
	1ST MEAS.	13	0	80	180	83
TANK NO. 1	2ND MEAS.	01	4	75	18	60
	GRAVITY				159	44*
TANK SIZE 210	OBSERVED 32.0	TEMP. 74	TRUE 31		PRICE $125	00
BS &W 8/10%	OFF←SEAL NO.'S→ON				AMOUNT $19,930	00

THIEFING	BEFORE	AFTER	CALCULATIONS
FT – IN			*180.83 × .9910 = 179.20
BS & WATER			18.60 × .9932 = 18.47
REMARKS			160.73
			× .992
			159.44

1ST MEAS	DATE	GAUGER
	TIME A.M. P.M.	WELL OWNER'S WITNESS

2ND MEAS	DATE	GAUGER
	TIME A.M. P.M.	WELL OWNER'S WITNESS

This ticket covers all claims for allowance. The oil represented by this ticket was received and run as property of

* Net Barrels

Fig. 10–2. Typical run ticket data

Table 10–1. Strapping table

TANK NO. _____1_____ DISTRICT _____

OWNER _____ COUNTY _____

LEASE _____ STATE _____

LEASE NO. _____ STRAPPER _____

		Ft.		Ft.		Ft.		Ft.		Ft.		Ft.		Ft.
	.00	1	13.95	2	27.90	3	41.85	4	55.76	5	69.66	6	83.56	7	97.45
¼	.29	¼	14.24	¼	28.19	¼	42.14	¼	56.05	¼	69.95	¼	83.85	¼	97.74
½	.58	½	14.53	½	28.48	½	42.43	½	56.34	½	70.24	½	84.14	½	98.03
¾	.87	¾	14.82	¾	28.77	¾	42.72	¾	56.63	¾	70.53	¾	84.43	¾	98.32
1	1.16	1	15.11	1	29.06	1	43.01	1	56.92	1	70.82	1	84.72	1	98.61
¼	1.45	¼	15.40	¼	29.35	¼	43.30	¼	57.21	¼	71.11	¼	85.01	¼	98.90
½	1.74	½	15.69	½	29.64	½	43.59	½	57.50	½	71.40	½	85.30	½	99.19
¾	2.03	¾	15.98	¾	29.94	¾	43.88	¾	57.79	¾	71.69	¾	85.58	¾	99.48
2	2.32	2	16.27	2	30.23	2	44.17	2	58.08	2	71.98	2	85.87	2	99.77
¼	2.61	¼	16.56	¼	30.52	¼	44.46	¼	58.36	¼	72.27	¼	86.16	¼	100.06
½	2.90	½	16.85	½	30.81	½	44.75	½	58.65	½	72.56	½	86.45	½	100.35
¾	3.19	¾	17.15	¾	31.10	¾	45.04	¾	58.94	¾	72.85	¾	86.74	¾	100.64
3	3.48	3	17.44	3	31.39	3	45.33	3	59.23	3	73.14	3	87.03	3	100.93
¼	3.77	¼	17.73	¼	31.68	¼	45.62	¼	59.52	¼	73.43	¼	87.32	¼	101.22
½	4.06	½	18.02	½	31.97	½	45.91	½	59.81	½	73.72	½	87.61	½	101.51
¾	4.36	¾	18.31	¾	32.26	¾	46.20	¾	60.10	¾	74.00	¾	87.90	¾	101.80
4	4.65	4	18.60	4	32.55	4	46.49	4	60.39	4	74.29	4	88.19	4	102.09
¼	4.94	¼	18.89	¼	32.84	¼	46.78	¼	60.68	¼	74.58	¼	88.48	¼	102.38
½	5.23	½	19.18	½	33.13	½	47.07	½	60.97	½	74.87	½	88.77	½	102.67
¾	5.52	¾	19.47	¾	33.42	¾	47.36	¾	61.26	¾	75.16	¾	89.06	¾	102.95
5	5.81	5	19.76	5	33.71	5	47.65	5	61.55	5	75.45	5	89.35	5	103.24
¼	6.10	¼	20.05	¼	34.00	¼	47.94	¼	61.84	¼	75.74	¼	89.64	¼	103.53
½	6.39	½	20.34	½	34.30	½	48.23	½	62.13	½	76.03	½	89.93	½	103.82
¾	6.68	¾	20.63	¾	34.59	¾	48.52	¾	62.42	¾	76.32	¾	90.22	¾	104.11
6	6.97	6	20.92	6	34.88	6	48.81	6	62.71	6	76.61	6	90.51	6	104.40
¼	7.26	¼	21.21	¼	35.17	¼	49.09	¼	63.00	¼	76.90	¼	90.80	¼	104.69
½	7.55	½	21.51	½	35.46	½	49.38	½	63.29	½	77.19	½	91.09	½	104.98
¾	7.84	¾	21.80	¾	35.75	¾	49.67	¾	63.58	¾	77.48	¾	91.37	¾	105.27
7	8.13	7	22.09	7	36.04	7	49.96	7	63.87	7	77.77	7	91.66	7	105.56
¼	8.42	¼	22.38	¼	36.33	¼	50.25	¼	64.16	¼	78.06	¼	91.95	¼	105.85
½	8.72	½	22.67	½	36.62	½	50.54	½	64.45	½	78.35	½	92.24	½	106.14
¾	9.01	¾	22.96	¾	36.91	¾	50.83	¾	64.74	¾	78.64	¾	92.53	¾	106.43
8	9.30	8	23.25	8	37.20	8	51.12	8	65.03	8	78.93	8	92.82	8	106.72
¼	9.59	¼	23.54	¼	37.49	¼	51.41	¼	65.32	¼	79.22	¼	93.11	¼	107.01
½	9.88	½	23.83	½	37.78	½	51.70	½	65.61	½	79.51	½	93.40	½	107.30
¾	10.17	¾	24.12	¾	38.07	¾	51.99	¾	65.90	¾	79.79	¾	93.69	¾	107.59
9	10.46	9	24.41	9	38.36	9	52.28	9	66.19	9	80.08	9	93.98	9	107.88
¼	10.75	¼	24.70	¼	38.66	¼	52.57	¼	66.48	¼	80.37	¼	94.27	¼	108.17
½	11.04	½	24.99	½	38.95	½	52.86	½	66.77	½	80.66	½	94.56	½	108.46
¾	11.33	¾	25.28	¾	39.24	¾	53.15	¾	67.06	¾	80.95	¾	94.85	¾	108.74
10	11.62	10	25.57	10	39.53	10	53.44	10	67.34	10	81.24	10	95.14	10	109.03
¼	11.91	¼	25.87	¼	39.82	¼	53.73	¼	67.63	¼	81.53	¼	95.43	¼	109.32
½	12.20	½	26.16	½	40.11	½	54.02	½	67.92	½	81.82	½	95.72	½	109.61
¾	12.49	¾	26.45	¾	40.40	¾	54.31	¾	68.21	¾	82.11	¾	96.01	¾	109.90
11	12.78	11	26.74	11	40.69	11	54.60	11	68.50	11	82.40	11	96.30	11	110.19
¼	13.08	¼	27.03	¼	40.98	¼	54.89	¼	68.79	¼	82.69	¼	96.59	¼	110.48
½	13.37	½	27.32	½	41.27	½	55.18	½	69.08	½	82.98	½	96.88	½	110.77
¾	13.66	¾	27.61	¾	41.56	¾	55.47	¾	69.37	¾	83.27	¾	97.16	¾	111.06

c. A thief is used to collect a sample of the oil. From this sample, the temperature (74°F), the BS&W (0.8%), and the observed gravity (32° API) are determined. The volume of oil sold must be adjusted to a standard temperature of 60°F, net of any BS&W. The observed gravity must also be adjusted to a true gravity based on the observed temperature of the sample.

d. A true gravity is determined using the observed gravity (32° API) and observed temperature (74°F) of the oil during the run. The true gravity, which is the observed gravity corrected for temperature, is determined using a gravity temperature correction table given in Table 10–2. The true gravity for the oil run is 31° API.

e. The number of barrels in the tank at the beginning of the run (180.83 barrels) is adjusted to standard temperature by referring to Table 10–3. Using the true gravity and the temperature of the oil at the beginning of the run (80°F), a correction

Table 10–1. Strapping table (continued)

Ft.		Ft.		Ft.		Ft.		Ft.		Ft.		Ft.		Ft.		Ft.	
8	111.35	9	125.25	10	139.14	11	153.04	12	166.93	13	180.83	14	194.73	15	208.6		
¼	111.64	¼	125.54	¼	139.43	¼	153.33	¼	167.22	¼	181.12	¼	195.02	¼			
½	111.93	½	125.83	½	139.72	½	153.62	½	167.51	½	181.41	½	195.31	½			
¾	112.22	¾	126.11	¾	140.01	¾	153.91	¾	167.80	¾	181.70	¾	195.59	¾			
1	112.51	1	126.40	1	140.30	1	154.20	1	168.09	1	181.99	1	195.88	1			
¼	112.80	¼	126.69	¼	140.59	¼	154.49	¼	168.38	¼	182.28	¼	196.17	¼			
½	113.09	½	126.98	½	140.88	½	154.78	½	168.67	½	182.57	½	196.46	½			
¾	113.38	¾	127.27	¾	141.17	¾	155.06	¾	168.96	¾	182.86	¾	196.75	¾			
2	113.67	2	127.56	2	141.46	2	155.35	2	169.25	2	183.15	2	197.04	2			
¼	113.96	¼	127.85	¼	141.75	¼	155.64	¼	169.54	¼	183.44	¼	197.33	¼			
½	114.25	½	128.14	½	142.04	½	155.93	½	169.83	½	183.73	½	197.62	½			
¾	114.53	¾	128.43	¾	142.33	¾	156.22	¾	170.12	¾	184.01	¾	197.91	¾			
3	114.82	3	128.72	3	142.62	3	156.51	3	170.41	3	184.30	3	198.20	3			
¼	115.11	¼	129.01	¼	142.91	¼	156.80	¼	170.70	¼	184.59	¼	198.49	¼			
½	115.40	½	129.30	½	143.20	½	157.09	½	170.99	½	184.88	½	198.78	½			
¾	115.69	¾	129.59	¾	143.48	¾	157.38	¾	171.28	¾	185.17	¾	199.07	¾			
4	115.98	4	129.88	4	143.77	4	157.67	4	171.57	4	185.46	4	199.36	4			
¼	116.27	¼	130.17	¼	144.06	¼	157.96	¼	171.86	¼	185.75	¼	199.65	¼			
½	116.56	½	130.46	½	144.35	½	158.25	½	172.15	½	186.04	½	199.94	½			
¾	116.85	¾	130.75	¾	144.64	¾	158.54	¾	172.43	¾	186.33	¾	200.23	¾			
5	117.14	5	131.04	5	144.93	5	158.83	5	172.72	5	186.62	5	200.52	5			
¼	117.43	¼	131.33	¼	145.22	¼	159.12	¼	173.01	¼	186.91	¼	200.81	¼			
½	117.72	½	131.62	½	145.51	½	159.41	½	173.30	½	187.20	½	201.10	½			
¾	118.01	¾	131.90	¾	145.80	¾	159.70	¾	173.59	¾	187.49	¾	201.38	¾			
6	118.30	6	132.19	6	146.09	6	159.99	6	173.88	6	187.78	6	201.67	6			
¼	118.59	¼	132.48	¼	146.38	¼	160.28	¼	174.17	¼	188.07	¼	201.96	¼			
½	118.88	½	132.77	½	146.67	½	160.57	½	174.46	½	188.36	½	202.25	½			
¾	119.17	¾	133.06	¾	146.96	¾	160.85	¾	174.75	¾	188.65	¾	202.54	¾			
7	119.46	7	133.35	7	147.25	7	161.14	7	175.04	7	188.94	7	202.83	7			
¼	119.75	¼	133.64	¼	147.54	¼	161.43	¼	175.33	¼	189.23	¼	203.12	¼			
½	120.04	½	133.93	½	147.83	½	161.72	½	175.62	½	189.52	½	203.41	½			
¾	120.32	¾	134.22	¾	148.12	¾	162.01	¾	175.91	¾	189.80	¾	203.70	¾			
8	120.61	8	134.51	8	148.41	8	162.30	8	176.20	8	190.09	8	203.99	8			
¼	120.90	¼	134.80	¼	148.70	¼	162.59	¼	176.49	¼	190.38	¼	204.28	¼			
½	121.19	½	135.09	½	148.99	½	162.88	½	176.78	½	190.67	½	204.57	½			
¾	121.48	¾	135.38	¾	149.27	¾	163.17	¾	177.07	¾	190.96	¾	204.86	¾			
9	121.77	9	135.67	9	149.56	9	163.46	9	177.36	9	191.25	9	205.15	9			
¼	122.06	¼	135.96	¼	149.85	¼	163.75	¼	177.65	¼	191.54	¼	205.44	¼			
½	122.35	½	136.25	½	150.14	½	164.04	½	177.94	½	191.83	½	205.73	½			
¾	122.64	¾	136.54	¾	150.43	¾	164.33	¾	178.22	¾	192.12	¾	206.02	¾			
10	122.93	10	136.83	10	150.72	10	164.62	10	178.51	10	192.41	10	206.31	10			
¼	123.22	¼	137.12	¼	151.01	¼	164.91	¼	178.80	¼	192.70	¼	206.60	¼			
½	123.51	½	137.41	½	151.30	½	165.20	½	179.09	½	192.99	½	206.89	½			
¾	123.80	¾	137.69	¾	151.59	¾	165.49	¾	179.38	¾	193.28	¾	207.17	¾			
11	124.09	11	137.98	11	151.88	11	165.78	11	179.67	11	193.57	11	207.46	11			
¼	124.38	¼	138.27	¼	152.17	¼	166.07	¼	179.96	¼	193.86	¼	207.75	¼			
½	124.67	½	138.56	½	152.46	½	166.36	½	180.25	½	194.15	½	208.04	½			
¾	124.96	¾	138.85	¾	152.75	¾	166.64	¾	180.54	¾	194.44	¾	208.33	¾			

¼" INCREMENTS
144 X .29068
240 X .28968
720 X .28950
208.62720

Computed by
R. K. McFARLAND

factor of 0.9910 is obtained from the table. The number of barrels in the tank at the end of the run is also corrected for temperature. The second correction factor for the number of barrels in the tank at the end of the run, also taken from Table 10–3 using the true gravity and the temperature of the oil at the end of the run (75°F), is 0.9932. The factors from the table are multiplied by the volumes of oil before and after the run to get the number of barrels adjusted to standard temperature, 179.20 and 18.47, respectively. The adjusted number of barrels at the beginning of the run and at the end of the run are then subtracted to get gross measured volume, 160.73 barrels.

f. The measured volume, now corrected for gravity and temperature, is multiplied by the correction factor for BS&W (1.00 – 0.008) to determine the net barrels (159.44 barrels). The net barrels are multiplied by the selling price per barrel to compute gross revenue from the run.

Table 10–2. Generalized crude oils API correction to 60°F

TEMP F	30.0	30.5	31.0	31.5	32.0	32.5	33.0	33.5	34.0	34.5	35.0	TEMP F
				API GRAVITY AT OBSERVED TEMPERATURE								
				CORRESPONDING API GRAVITY AT 60 F								
60.0	30.0	30.5	31.0	31.5	32.0	32.5	33.0	33.5	34.0	34.5	35.0	60.0
60.5	30.0	30.5	31.0	31.5	32.0	32.5	33.0	33.5	34.0	34.5	35.0	60.5
61.0	29.9	30.4	30.9	31.4	31.9	32.4	32.9	33.4	33.9	34.4	34.9	61.0
61.5	29.9	30.4	30.9	31.4	31.9	32.4	32.9	33.4	33.9	34.4	34.9	61.5
62.0	29.9	30.4	30.9	31.4	31.9	32.4	32.9	33.3	33.8	34.3	34.8	62.0
62.5	29.8	30.3	30.8	31.3	31.8	32.3	32.8	33.3	33.8	34.3	34.8	62.5
63.0	29.8	30.3	30.8	31.3	31.8	32.3	32.8	33.3	33.8	34.3	34.7	63.0
63.5	29.8	30.3	30.8	31.3	31.7	32.2	32.7	33.2	33.7	34.2	34.7	63.5
64.0	29.7	30.2	30.7	31.2	31.7	32.2	32.7	33.2	33.7	34.2	34.7	64.0
64.5	29.7	30.2	30.7	31.2	31.7	32.2	32.7	33.2	33.7	34.2	34.7	64.5
65.0	29.7	30.2	30.6	31.1	31.6	32.1	32.6	33.1	33.6	34.1	34.6	65.0
65.5	29.6	30.1	30.6	31.1	31.6	32.1	32.6	33.1	33.6	34.1	34.6	65.5
66.0	29.6	30.1	30.6	31.1	31.6	32.1	32.6	33.1	33.5	34.0	34.5	66.0
66.5	29.5	30.0	30.5	31.0	31.5	32.0	32.5	33.0	33.5	34.0	34.5	66.5
67.0	29.5	30.0	30.5	31.0	31.5	32.0	32.5	33.0	33.5	34.0	34.5	67.0
67.5	29.5	30.0	30.5	30.9	31.4	31.9	32.4	32.9	33.4	33.9	34.4	67.5
68.0	29.4	29.9	30.4	30.9	31.4	31.9	32.4	32.9	33.4	33.9	34.4	68.0
68.5	29.4	29.9	30.4	30.9	31.4	31.9	32.4	32.9	33.4	33.9	34.4	68.5
69.0	29.4	29.9	30.4	30.8	31.3	31.8	32.3	32.8	33.3	33.8	34.3	69.0
69.5	29.3	29.8	30.3	30.8	31.3	31.8	32.3	32.8	33.3	33.8	34.3	69.5
70.0	29.3	29.8	30.3	30.8	31.3	31.8	32.3	32.8	33.2	33.7	34.2	70.0
70.5	29.3	29.8	30.3	30.7	31.2	31.7	32.2	32.7	33.2	33.7	34.2	70.5
71.0	29.2	29.7	30.2	30.7	31.2	31.7	32.2	32.7	33.1	33.6	34.1	71.0
71.5	29.2	29.7	30.2	30.7	31.2	31.6	32.1	32.6	33.1	33.6	34.1	71.5
72.0	29.2	29.7	30.2	30.6	31.1	31.6	32.1	32.6	33.1	33.6	34.1	72.0
72.5	29.1	29.6	30.1	30.6	31.1	31.6	32.1	32.6	33.1	33.6	34.1	72.5
73.0	29.1	29.6	30.1	30.6	31.1	31.6	32.0	32.5	33.0	33.5	34.0	73.0
73.5	29.1	29.6	30.0	30.5	31.0	31.5	32.0	32.5	33.0	33.4	33.9	73.5
74.0	29.0	29.5	30.0	30.5	31.0	31.5	32.0	32.5	32.9	33.4	33.9	74.0
74.5	29.0	29.5	30.0	30.5	31.0	31.5	32.0	32.4	32.9	33.4	33.9	74.5
75.0	29.0	29.5	29.9	30.4	30.9	31.4	31.9	32.4	32.9	33.4	33.9	75.0

* DENOTES EXTRAPOLATED VALUE   API GRAVITY = 30.0 TO 35.0

Table 10–3. Generalized crude oils volume correction to 60°F

TEMP. F	30.0	30.5	31.0	31.5	32.0	32.5	33.0	33.5	34.0	34.5	35.0	TEMP F
					API GRAVITY AT 60 F							
					FACTOR FOR CORRECTING VOLUME TO 60 F							
75.0	0.9933	0.9933	0.9932	0.9932	0.9931	0.9931	0.9931	0.9930	0.9930	0.9929	0.9929	75.0
75.5	0.9931	0.9930	0.9930	0.9930	0.9929	0.9929	0.9928	0.9928	0.9927	0.9927	0.9926	75.5
76.0	0.9929	0.9928	0.9928	0.9927	0.9927	0.9926	0.9926	0.9925	0.9925	0.9925	0.9924	76.0
76.5	0.9926	0.9926	0.9925	0.9925	0.9925	0.9924	0.9924	0.9923	0.9923	0.9922	0.9922	76.5
77.0	0.9924	0.9924	0.9923	0.9923	0.9922	0.9922	0.9921	0.9921	0.9920	0.9920	0.9919	77.0
77.5	0.9922	0.9921	0.9921	0.9920	0.9920	0.9919	0.9919	0.9918	0.9918	0.9917	0.9917	77.5
78.0	0.9920	0.9919	0.9919	0.9918	0.9918	0.9917	0.9917	0.9916	0.9916	0.9915	0.9915	78.0
78.5	0.9917	0.9917	0.9916	0.9916	0.9915	0.9915	0.9914	0.9914	0.9913	0.9913	0.9912	78.5
79.0	0.9915	0.9915	0.9914	0.9914	0.9913	0.9913	0.9912	0.9911	0.9911	0.9910	0.9910	79.0
79.5	0.9913	0.9912	0.9912	0.9911	0.9911	0.9910	0.9910	0.9909	0.9909	0.9908	0.9907	79.5
80.0	0.9911	0.9910	0.9910	0.9909	0.9908	0.9908	0.9907	0.9907	0.9906	0.9906	0.9905	80.0
80.5	0.9908	0.9908	0.9907	0.9907	0.9906	0.9906	0.9905	0.9904	0.9904	0.9903	0.9903	80.5
81.0	0.9906	0.9906	0.9905	0.9904	0.9904	0.9903	0.9903	0.9902	0.9902	0.9901	0.9900	81.0
81.5	0.9904	0.9903	0.9903	0.9902	0.9902	0.9901	0.9900	0.9900	0.9899	0.9899	0.9898	81.5
82.0	0.9902	0.9901	0.9901	0.9900	0.9899	0.9899	0.9898	0.9897	0.9897	0.9896	0.9896	82.0
82.5	0.9900	0.9899	0.9898	0.9898	0.9897	0.9896	0.9896	0.9895	0.9894	0.9894	0.9893	82.5
83.0	0.9897	0.9897	0.9896	0.9895	0.9895	0.9894	0.9893	0.9893	0.9892	0.9891	0.9891	83.0
83.5	0.9895	0.9894	0.9894	0.9893	0.9892	0.9892	0.9891	0.9890	0.9890	0.9889	0.9888	83.5
84.0	0.9893	0.9892	0.9891	0.9891	0.9890	0.9889	0.9889	0.9888	0.9887	0.9887	0.9886	84.0
84.5	0.9891	0.9890	0.9889	0.9889	0.9888	0.9887	0.9886	0.9886	0.9885	0.9884	0.9884	84.5
85.0	0.9888	0.9888	0.9887	0.9886	0.9886	0.9885	0.9884	0.9884	0.9883	0.9882	0.9881	85.0
85.5	0.9886	0.9885	0.9885	0.9884	0.9883	0.9883	0.9882	0.9881	0.9880	0.9880	0.9879	85.5
86.0	0.9884	0.9883	0.9882	0.9882	0.9881	0.9880	0.9879	0.9879	0.9878	0.9877	0.9877	86.0
86.5	0.9882	0.9881	0.9880	0.9879	0.9879	0.9878	0.9877	0.9876	0.9876	0.9875	0.9874	86.5
87.0	0.9879	0.9879	0.9878	0.9877	0.9876	0.9876	0.9875	0.9874	0.9873	0.9873	0.9872	87.0
87.5	0.9877	0.9876	0.9876	0.9875	0.9874	0.9873	0.9873	0.9872	0.9871	0.9870	0.9869	87.5
88.0	0.9875	0.9874	0.9873	0.9873	0.9872	0.9871	0.9870	0.9869	0.9869	0.9868	0.9867	88.0
88.5	0.9873	0.9872	0.9871	0.9870	0.9869	0.9869	0.9868	0.9867	0.9866	0.9865	0.9865	88.5
89.0	0.9870	0.9870	0.9869	0.9868	0.9867	0.9866	0.9866	0.9865	0.9864	0.9863	0.9862	89.0
89.5	0.9868	0.9867	0.9867	0.9866	0.9865	0.9864	0.9863	0.9862	0.9862	0.9861	0.9860	89.5
90.0	0.9866	0.9865	0.9864	0.9863	0.9863	0.9862	0.9861	0.9860	0.9859	0.9858	0.9857	90.0

* DENOTES EXTRAPOLATED VALUE   API GRAVITY = 30.0 TO 35.0

Crude oil sales

Crude oil marketing has changed dramatically over the past two decades. Historically, almost all crude oil sales transactions were governed by fixed contracts. Since prices had been quite stable, these contracts rarely had provisions that called for the price to be reexamined. The oil markets were essentially seller markets, since the sellers controlled the supply side of the business. In the early 1980s, demand declined and supply increased. Purchasers became disenchanted with these fixed price contracts, since with rising supply, they believed that they were consistently overpaying. Purchasers turned instead to the spot market. Today, crude oil is usually sold under some form of contractual arrangement; however, these contracts are typically linked to prices in the spot market or the futures market. These two markets play a key role in the determination of crude oil sales prices.

A spot arrangement involves an agreement to buy or sell one shipment of oil under a price that is agreed upon at the time of the arrangement with no long-term agreement, i.e., at a current market price. The spot market balances supply and demand. When one company has excess crude oil supply, it offers some of its excess for sale. Conversely, if it needs additional oil, it may opt to purchase oil on a short-term basis. Collectively, prices in spot markets are seen as a signal of the balance between supply and demand. When supply is low, prices rise, and when there is excess oil supply, prices fall.

There are "spot markets" for different regions of the world (e.g., U.S. Gulf Coast, Rotterdam/Northwest Europe, Singapore/South East Asia, etc.). The spot price reflects a base price for crude of a particular quality. This base price is then adjusted for varying quality crudes available in the same geographical region. West Texas Intermediate (WTI) normally serves as the base price for crude oil sold into the U.S. Gulf Coast region. Similarly, North Sea Brent Blend is typically the base for the Northwest Europe market, and Dubai crude is tied to the spot price for Singapore/South East Asian markets.

Key indicators of quality are API gravity (or density) and sulfur content. As discussed earlier, heavier crude oils, i.e., those with lower API gravity and/or crudes with higher sulfur content, sell for less than lighter, low sulfur crudes. Crude oil with a sulfur content less than 0.5% is referred to as "sweet," while crude oil with sulfur content higher than 0.5% is referred to as "sour." Since sour crude oil requires more refining and processing than sweet crude, it sells for a lower price.

While spot markets involve the current sales of physical barrels of oil, crude oil is also sold on the futures markets, i.e., the New York Mercantile Exchange (NYMEX) or the International Petroleum Exchange in London. In a futures contract, a seller promises to deliver a given quantity of crude at a specified place, price, and time in the future. A buyer, seeking a guaranteed quantity at a guaranteed price and location, agrees to purchase the crude offered by the seller. In addition to physical actual crude oil sales, the futures market is also used by speculators seeking to buy and sell the **option** to purchase crude oil in the future. Speculators buy the option to purchase crude at an agreed-upon price for future delivery under the hope that the market price will rise. If the market price does indeed rise, they can sell their right and profit from the difference between the price they locked into and the price in effect as the delivery date nears.

In the United States, some domestically produced crude oil is still sold at a posted price, i.e., the published price that some purchasers in an area will pay for crude oil. Named for the sheet that was literally posted in a producing field, posted prices are established by the buyers, usually refiners. Posted prices generally apply to a particular quality of crude oil available for delivery in a particular area. For example, WTI refers to low sulfur, light crude oil priced for delivery in the U.S. Gulf Coast region. North Sea Brent Blend is a light, low sulfur blend of oil produced from 15 separate fields in the North Sea.

The postings also indicate the factors by which the price should be adjusted (higher or lower) for oil that varies from a specified API gravity. A common pricing benchmark for crude oil gravity is 40°API. For example, assume that the posted price for WTI is $150/bbl and the price adjustment is $0.08/degree higher for WTI crude oil with an API gravity higher than 40°, and $0.10/degree lower for WTI crude oil with an API gravity below 40°. If the crude oil sold is WTI with an actual gravity of 25°, the price would be lower than $150/bbl. Since the crude oil is 15° below 40°API gravity (40° – 25°), the actual price for that delivery would be $148.50 [$150 – (15 × $0.10)]. The price may also be adjusted depending on the actual location of the crude oil and the cost to transport the oil to the purchaser's refinery. In decades past, posted prices remained relatively stable even while spot prices fluctuated. Today, they quickly reflect market conditions.

Natural gas measurement

Gas, unlike oil, is not stored on site after production but is instead transferred immediately to a pipeline or gathering system. (Figure 10–3 diagrams the flow of gas from the wellhead to the point of consumption.) Gas measurement is more complicated than oil measurement because of different state and federal measurement standards and because of the physical characteristics of natural gas. Gas volume varies significantly with changes in pressure and temperature. Therefore, it is especially important that measured volumes of gas be converted to volumes at a standard set of conditions. Generally, standard conditions are a temperature of 60°F and a pressure at 14.65 psia. Federal standard conditions are a temperature of 60°F, with pressure at 14.73 psia for onshore leases and 15.025 psia for offshore leases.

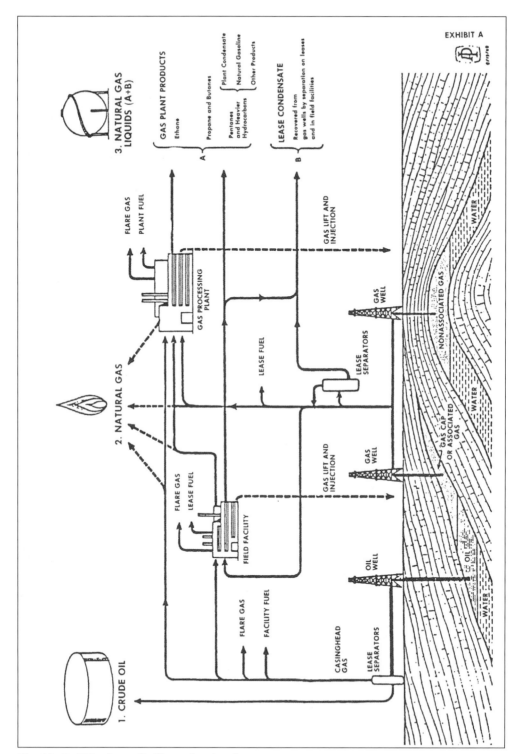

Fig. 10–3. API schematic flow of natural gas
Source: Council of Petroleum Accountants Societies, Inc. 1993. Accounting Guideline (AG)-15. April. "Gas Accounting Manual." p. I-4. Copyright 1993 by the Council of Petroleum Accountants Societies, Inc. (COPAS). Reprinted with permission of the American Petroleum Institute and COPAS. COPAS documents are available for purchase at www.COPAS.org.

Gas volume is measured in thousand cubic feet (Mcf), adjusted to standard conditions. Several instruments are used for measuring natural gas. One of the most important and widely used instruments is the orifice meter. An orifice meter records the flow of gas through pressure differentials. The data are recorded and then converted to standard cubic feet by using conversion factors based upon a number of variables. Such variables include the size of the line, the size of the orifice, the flowing temperature, the flowing pressure, the specific gravity of the gas, and the differential pressure.

Other measuring devices are used, such as mass flowmeters (normally at refineries), turbine meters, positive-displacement meters, and electronic flow computers. The latter are used in connection with an orifice or turbine, and give instantaneous and continuous metered volumes. For further details concerning these methods, see *COPAS AG -15*, "Gas Accounting Manual."[1]

Sales of natural gas (whether processed or nonprocessed) are normally reported on a gas settlement statement. The form of the statement varies by purchaser, but certain necessary information should be included. Such information includes purchaser identification, lease identity, producer identification, quantity delivered from lease, month of production, Btu factor, gross or net value due the lease, lease production taxes, value due the producer, and pressure base(s). This information is necessary whether the sale is directly from the lease or from a central facility. If a central facility is used, gas allocation to individual wells must be made.

Natural gas sales

Prior to 1985, most sales of natural gas were made using long-term contracts, often for terms of 10 years or longer. In contrast, most gas sales today are made on the spot market on a short-term basis. **Spot sales** of gas involve the sale of gas at current market prices by the producer to pipelines or their affiliates, which then transport the gas to the end user or local distribution company. Spot sales of gas also include **direct sales** that are made by the producer to the end user or local distribution company. In direct sales, the gas is transported by a third-party pipeline company, which instead of owning the gas merely transports it, charging a transportation fee. Spot sales also include sales to **marketers**, or middlemen who buy and resell gas. Brokers may also be used. **Brokers**, instead of taking title to the gas, find a buyer for the gas and charge a brokerage fee for their services. In 1990, similar to crude oil, natural gas futures started trading on the NYMEX.

There are three types of natural gas prices: wellhead prices, spot prices, and futures prices. In the United States, these prices are typically based on the price at the Henry Hub. The Henry Hub is in Louisiana and is the point of intersection of nine interstate and four intrastate pipelines. In natural gas contracts, the Henry Hub is typically used as the physical contract delivery point. The price at the Henry Hub is also typically used as the price benchmark for spot trades of natural gas.

Standard Division Order

Due to the fact that U.S. laws allow private ownership of oil and gas reserves, leases located in the United States frequently have numerous interest owners. It is the operator's responsibility to assure that all owners are identified and compensated. The legal document setting up each interest owner's share of production is called a **division order**. A division order is a contract between the lessors (RI owners), lessees, and the purchaser of the oil or gas. A division order must be executed in order for the operator to sell the production, collect the revenue from the sale of the oil and gas, and pay the correct owners their correct share of production or the proceeds from production. Thus, a division order must generally be prepared and signed by all interest owners prior to the sale of oil or gas from a particular lease. The information from the division order is typically set up in a division order database in the operator's accounting system so that the appropriate accounting entries can be made and payment distributed to the appropriate parties. As shown in Figure 10–4, exhibit A, which is part of the division order, is prepared, showing the interest owners' names, addresses, and interests owned, i.e., ownership decimals.

The ownership decimals in a division order are based on the following:

- The lease contract for royalty owners
- The operating agreement or unit agreement for working interest owners
- The assignment document for overriding royalty interest owners

DIVISION ORDER

To: Date: *November 20, 2012*

Property Number: *5474-3456-3*   Effective Date: *Date of First Sale*
Property Name: *Lomax Heirs*
Operator: *Tyler Oil Company*
County and State: *Harris County, Texas*

Property
Description: [PROPERTY LEGAL DESCRIPTION]
Production: __X__ Oil __X__ Gas _____ Other:_____

Owner Name and Address:	OWNER NUMBER
	Type of Interest:
	Decimal Interest:
See Exhibit A	

The undersigned certifies the ownership of their decimal interest in production or proceeds as described above payable by ___*Tyler Oil Company*___ (Payor).
                        (Company Name)

Payor shall be notified, in writing, of any change in ownership, decimal interest, or payment address. All such changes shall be effective the first day of the month following receipt of such notice.

Payor is authorized to withhold payment pending resolution of a title dispute or adverse claim asserted regarding the interest in production claimed herein by the undersigned. The undersigned agrees to indemnify and reimburse Payor any amount attributable to an interest to which the undersigned is not entitled.

Payor may accrue proceeds until the total amount equals *$25.00*, or pay ____ whichever occurs first, or as required by applicable state statute.

This Division Order does not amend any lease or operating agreement between the undersigned and the lessee or operator or any other contracts for the purchase of oil or gas.

In addition to the terms and conditions of this Division Order, the undersigned and Payor may have certain statutory rights under the laws of the state in which the property is located.

Special Clauses:

Owner(s) Signature(s):   _____   _____
Owner(s) Tax I.D. Number(s):   _____   _____

Owner Daytime Telephone #:   _____   _____
Owner FAX Telephone #:   _____   _____

> Federal Law requires you to furnish your Social Security or Taxpayer Identification Number. Failure to comply will result in 31% tax withholding and will not be refundable by Payor

Fig. 10–4. Division order

Exhibit A

Property Number: *5474-3456-3*
Property Name: *Lomax Heirs*
Operator: *Tyler Oil Company*
County and State: *Harris County, Texas*

Effective Date: *Date of First Sale*

Property Description: [PROPERTY LEGAL DESCRIPTION]

Production: __*X*__ Oil __*X*__ Gas _____ Other: _____

OWNER NUMBER	OWNER NAME	OWNER ADDRESS	TYPE OF INTEREST	DECIMAL INTEREST
243662	Tyler Oil Company		WI	0.365625
345927	Big John Oil Company		WI	0.162500
298732	Wildcat Oil Company		WI	0.121875
675233	Mabel Lomax		RI	0.093750
723122	E.B. Lomax		RI	0.031250
723123	Bruce Lomax		RI	0.031250
723123	Kay Jerin Lomax		RI	0.031250
823345	James Larkin		ORI	0.081250
823345	Janice Larkin		ORI	0.040625
823346	Estelle Bradt		ORI	0.040625

Source: National Association of Division Order Analysts (NADOA) Model Form Division Order (Adopted 9/95)

Fig. 10–4. Division order (continued)

DETERMINATION OF REVENUE

Revenue from the sale of oil or gas produced from a property must be divided among all owners of an economic interest in that property according to the ownership interests shown on the division order. In international operations, the division of revenue between owners may differ from the division in the United States. Revenue splits in international operations are discussed in chapter 15. The following example illustrates the computation of the share of revenue for all the economic interests in a property in the United States. In contrast to revenue, operating costs are divided only among the working interest owners. Since most properties will, at a minimum, have a royalty interest, the working interest's percentage of revenue is usually not the same as the working interest's percentage of costs. The following example also illustrates what portion of the costs each interest bears.

EXAMPLE

Revenue and Cost Allocation to Ownership Interests

a. Mr. Z owns the mineral rights in some land in Texas. He leases this land to Tyler Company, reserving a 1/8 royalty. Each owner's share of the first year's gross production of 40,000 barrels is as follows.

Royalty Interest, Mr. Z:	1/8 RI
	1/8 × 40,000 = 5,000 bbl
	(0% of the costs)
Working Interest, Tyler Company:	100% of the WI (7/8 revenue interest)
	7/8 × 40,000 = 35,000 bbl
	(100% WI = 100% of the costs)

b. Assume that on January 1 of year two, Tyler Company assigned to Intex Co. a 1/5 ORI (overriding royalty interest), and that production in year two was 80,000 barrels. Note that an interest created out of the working interest is stated in terms of that working interest owner's revenue share at the time the interest was created. Consequently, the ORI's share of revenue is 1/5 × 7/8.

Royalty Interest, Mr. Z:	1/8 RI
	1/8 × 80,000 = 10,000 bbl
	(0% of the costs)
ORI, Intex Co.:	1/5 × WI share of revenue = 1/5 × 7/8 ORI
	1/5 × 7/8 × 80,000 = 14,000 bbl
	(0% of the costs)
100% WI, Tyler Company:	Revenue interest: 7/8 × 4/5*
	7/8 × 4/5 × 80,000 = 56,000 bbl
	(100% WI = 100% of the costs)

c. Assume that on January 1 of year three, Tyler Company assigned to Joint Company a working interest of 25% and that the production is again 80,000 barrels.

Royalty Interest, Mr. Z:	1/8 RI
	$1/8 \times 80{,}000 = 10{,}000$ bbl
	(0% of the costs)
ORI, Intex Co.:	$1/5 \times 7/8$ ORI
	$1/5 \times 7/8 \times 80{,}000 = 14{,}000$ bbl
	(0% of the costs)
75% WI, Tyler Company:	Revenue interest: $7/8 \times 4/5 \times 75\%$
	$7/8 \times 4/5 \times 75\% \times 80{,}000 = 42{,}000$ bbl
	(75% WI = 75% of the costs)
25% WI, Joint Company:	Revenue interest: $7/8 \times 4/5 \times 25\%$
	$7/8 \times 4/5 \times 25\% \times 80{,}000 = 14{,}000$ bbl
	(25% WI = 25% of the costs)

---

* The working interest share of revenue is what is left after paying all the nonworking interests. To compute the amount of the working interest's share of revenue, start with the original whole interest (1) and subtract the royalty interest (1/8) and the ORI ($1/5 \times 7/8$) as follows:

WI share of revenue = $1 - 1/8 - 1/5 \times 7/8$

The working interest share of revenue can then be reduced as follows:

$$\begin{aligned} \text{WI} &= 7/8 - 1/5 \times 7/8 \\ &= 7/8(1 - 1/5) \quad \text{(factoring out the 7/8)} \\ &= 7/8 \times 4/5 \end{aligned}$$

Note that as opposed to the DD&A calculations shown in chapters 6 and 7, total production is used in the distribution of revenue versus only the working interest's share of production.

In the example above, ownership interests were shown as fractions because in the authors' opinion, fractions more effectively illustrate how the owners' interests are computed. In a division order, however, the interests are stated as decimals. In determining division order decimals, fractional ownership is converted to decimal ownership. The decimal calculation is comparable with the preceding example where fractional ownership is shown. Either approach may be used in determining the gross revenue applicable to each owner.

EXAMPLE

Decimal Ownership

Tyler Company has production on a lease with the following ownership interest:

Royalty interest:	3/16 RI
ORI:	1/5 of WI
Production payment:	20% of WI, created after the ORI was created
Working interest:	
75% WI	Tyler Company
25% WI	Baker Company

Originally, Tyler Company had 100% of the working interest. Later an ORI and then a production payment interest were carved out. Several years after the ORI and production payment interest were created, Tyler conveyed 25% of its working interest to Baker Company. Based on these facts, ownership interests as shown on a division order are as follows:

Total interest	1.000000	
Royalty interest, 3/16	0.187500	
Remainder		0.812500
ORI, 1/5 × 0.8125	0.162500	
Remainder		0.650000
Production payment, 0.20 × 0.65	0.130000	
Remainder		0.520000
Tyler Company, 0.75 × 0.52	0.390000	
Baker Company, 0.25 × 0.52	0.130000	

The division order would show the owners' interests as follows:

Royalty interest	0.187500
ORI	0.162500
Production payment	0.130000
Tyler Company (75% WI)	0.390000
Baker Company (25% WI)	0.130000
	1.000000

Each owner's portion of the total revenue is determined by multiplying the decimal amounts by total revenue. For example, if total revenue were $56,000, the royalty owner would receive $56,000 × 0.187500, or $10,500.

UNITIZATIONS

As discussed in chapter 1, two or more properties are often combined for a more efficient, economical operation. The parties may voluntarily agree to unitize, or alternatively, the regulatory authority overseeing the area may require that the parties unitize. Typically, when properties are unitized, a unitization agreement is negotiated among all of the parties. In most cases, the unitization agreement effectively supercedes the lease agreement. In order to be effective, the agreement to unitize normally requires approval of a previously agreed-upon percentage of *all* of the interest parties (i.e., 85% of all working interest, royalty interest, and ORI owners in the unitized area). If approved, all of the interests in all of the unitized properties are redetermined based upon whatever sharing factors are stipulated in the unitization agreement.

Sharing factors, called participation factors, typically are based on such elements as acreage contributed, pay sand or sand thickness contributed, or net recoverable barrels of oil in place contributed. The redetermined interests are used to share revenues and costs. All of the working interest owners typically share a common obligation to all of the royalty owners. The following simple example illustrates this process.

EXAMPLE

Unitization—Determination of Interest

Assume that the following three leases are being combined to form the South End Unit:

Lease 1

Working interest:	100% Tyler Company
Royalty interest:	1/16 Mable Smith
	1/16 James Jones
Acreage contributed:	100 acres
Estimated net recoverable bbl:	50,000 bbl

Lease 2

Working interest:	50% Texas Company
	50% ArkTex Company
Royalty interest:	1/8 Fred Luna
Acreage contributed:	500 acres
Estimated net recoverable bbl:	100,000 bbl

Lease 3

Working interest:	100% Frank Company
Royalty interest:	1/6 Anna Bell
Acreage contributed:	200 acres
Estimated net recoverable bbl:	10,000 bbl

Participation Factors Based on Acreage Contributed

	Working Interest	Royalty Interest
Tyler Company	100% (100/800) = 0.125	
Texas Company	50% (500/800) = 0.3125	
ArkTex Company	50% (500/800) = 0.3125	
Frank Company	100% (200/800) = 0.25	
Mable Smith		1/16 (100/800) = 0.0078
James Jones		1/16 (100/800) = 0.0078
Fred Luna		1/8 (500/800) = 0.0781
Anna Bell		1/6 (200/800) = 0.0417

Note: Recall that the working interest is stated in terms of how the costs are to be shared, while the royalty interest is stated in terms of its share of gross revenue. In order to calculate the working interest owners' share of gross revenue, the working interest owners' participation factors would be multiplied by 1 minus the royalty on the property that they contributed. For example, for Tyler Company, the working interest after the unitization is 0.125, so Tyler will be responsible for 12.5% of the costs on the unit. Tyler's share of revenue is 10.94% (0.125 × 7/8 = 0.1094).

Participation Factors Based on Estimated Net Recoverable Barrels Contributed

	Working Interest	Royalty Interest
Tyler Company	100% (50/160) = 0.3125	
Texas Company	50% (100/160) = 0.3125	
ArkTex Company	50% (100/160) = 0.3125	
Frank Company	100% (10/160) = 0.0625	
Mable Smith		1/16 (50/160) = 0.0195
James Jones		1/16 (50/160) = 0.0195
Fred Luna		1/8 (100/160) = 0.0781
Anna Bell		1/6 (10/160) = 0.0104

Participation factors may also be based on multiple elements that are combined in a weighted average calculation to redetermine interests. This calculation is illustrated in the example below.

EXAMPLE

Participation Factors Based on Multiple Elements

Using the information from the previous example, assume that the parties agree to weight acreage contributed by a 45% factor and net recoverable barrels by 55%.

Participation Factors Based on Acreage Contributed (45%) and Estimated Net Recoverable Barrels Contributed (55%)

Working Interest:

Tyler Company	100% [(100/800 × .45) + (50/160 × .55)] = 0.2281
Texas Company	50% [(500/800 × .45) + (100/160 × .55)] = 0.3125
ArkTex Company	50% [(500/800 × .45) + (100/160 × .55)] = 0.3125
Frank Company	100% [(200/800 × .45) + (10/160 × .55)] = 0.1469

Royalty Interest:

Mable Smith	1/16 [(100/800 × .45) + (50/160 × .55)] = 0.0143
James Jones	1/16 [(100/800 × .45) + (50/160 × .55)] = 0.0143
Fred Luna	1/8 [(500/800 × .45) + (100/160 × .55)] = 0.0781
Anna Bell	1/6 [(200/800 × .45) + (10/160 × .55)] = 0.0245

Several working interest owners with properties producing from the same formation may wish to develop their properties based on an overall plan for the area. Rather than actually unitizing, the parties may enter into a **cooperative unit**. In a cooperative unit, the working interests share in the total costs and production from the unit. The royalty owners, however, receive a royalty based on the production from their own leases.

OIL AND GAS REVENUES

Who is responsible for paying severance taxes? Who is responsible for paying the royalty owners and other owners of economic interests their pro rata share of lease revenue? The answers depend on legal requirements and contractual provisions.

Frequently, the purchaser assumes the responsibility of collecting and paying severance taxes on both oil and gas production. In doing so, the purchaser is acting as a conduit, i.e., the purchaser collects the taxes owed by the economic interest owners of the property by simply deducting the taxes owed from revenue due the economic interest owners. The purchaser then sends the severance tax amount to the state.

In the past, the purchaser of oil has frequently assumed the obligation of paying to each owner of an economic interest that owner's share of revenue, less applicable severance taxes.

If the purchaser does not assume this responsibility, the lease operator normally distributes the applicable amounts to the other economic interest owners. The purchaser of gas usually remits 100% of the proceeds, less severance taxes, to the lease operator. The lease operator ordinarily then assumes the responsibility of paying the other economic interest owners their share of revenue, net of severance tax.

Sometimes the purchaser of either oil or gas does not assume either responsibility, i.e., the responsibility of paying severance taxes and each economic interest owner his or her share of revenue net of taxes. In this case, the purchaser remits 100% of the proceeds to the operator. The operator ordinarily then assumes both responsibilities. For simplicity, the purchaser in most of the examples in this chapter assumes either both responsibilities or neither responsibility.

RECORDING OIL REVENUE

Gross revenue from oil sales is determined by multiplying the net amount of oil delivered by the price for that particular grade of oil. The net amount of oil delivered is recorded on a run ticket, which serves as the source document for determining the quantity of oil sold.

Once the sales price has been determined and the oil sold, the accounting entries are made. The following example illustrates recording revenue from the sale of oil for both the working interest owner and the royalty interest owner. In this example, as well as in the remaining examples in the chapter, it is assumed that the severance tax, as is the usual case, is a specified percentage of the selling price.

In part a of the first example, an analysis is presented showing each economic interest owner's portion of gross revenue, severance tax, and revenue net of tax. This analysis provides the amounts to be debited or credited in the revenue entries made by the various interest owners.

In part b, it is assumed that the purchaser distributes severance tax to the state and remits the net revenue amounts to the various interest owners. Therefore, the working interest owner records a receivable for its share of revenue after tax due from the purchaser. The working interest owner also records its share of severance tax expense and credits oil sales (revenue) for its share of gross revenue. The amounts in the analysis presented in part a or the equivalent calculations given in parenthesis after each debit and credit may be used to obtain the amounts to debit and credit.

In part c of the example, it is assumed that the purchaser remits 100% of the proceeds to the working interest owner. The working interest owner remits the severance tax owed by all the economic interest owners to the state. The working interest owner also remits the revenue net of taxes to all economic interest owners—in this case, just the royalty interest owner—for their share of revenue net of tax. Again, either the analysis prepared in part a or the calculations given in parentheses may be used to determine the amounts to be debited or credited. In this case, the working interest owner records the gross revenue amount as a receivable and, as in part b, records its own share of tax expense and revenue. A liability to the state for production taxes is recognized by the working interest owner for both its share of production taxes and the royalty interest owner's share. The working interest owner also records a payable to the royalty interest owner for revenue due the royalty interest owner net of production tax.

Note that regardless of who bears the responsibility for paying the severance tax and paying the economic interest owners their share of revenue, i.e., part b compared to part c, the working interest owner (or any other economic interest owner) records the same amount of revenue and expenses.

Part d gives the entry for the royalty interest owner. The entry for the royalty interest owner would be the same under any situation because the royalty interest owner always receives only his share of revenue after tax, i.e., his portion of revenue net of severance tax. The royalty owner records his net receivable, tax expense, and revenue.

EXAMPLE

Oil Revenue

Tyler Company owns 100% of the working interest in the Gilbert Lease in Texas. Gilbert Jones owns a 1/8 royalty interest in the lease. During February, 4,000 (gross) barrels of oil (after correction for temperature, gravity, and BS&W) were produced and sold. Assume the price for oil is $64/bbl and the production tax in Texas is 5%.

a. Analysis:

Revenue Interest	a Gross Revenue	b Severance Tax	c Net of Severance Tax
	(4,000 × $64 × interest)	(5% × a)	(a – b)
WI: 7/8	$224,000	$11,200	$212,800
RI: 1/8	32,000	1,600	30,400
	$256,000	$12,800	$243,200

b. The purchaser assumes the responsibility of distributing severance taxes and royalty income.

Entry by Tyler Company (100% WI)—Amounts may be taken from the above analysis or may be calculated as shown below:

A/R—purchaser (4,000 × $64 × 0.95 × 7/8)   212,800
Production tax expense (4,000 × $64 × 0.05 × 7/8) . . .   11,200
    Oil sales (4,000 × $64 × 7/8).   224,000

c. Tyler Company assumes the responsibility and receives 100% of the proceeds.

   Entry by Tyler Company (100% WI)—Amounts may be taken from the analysis in part a above or from the calculations in parentheses.

A/R—purchaser (WI + RI) (4,000 × $64)	256,000	
Production tax expense (WI) (4,000 × $64 × 0.05 × 7/8)	11,200	
Production tax payable (WI + RI) (4,000 × $64 × 0.05)		12,800
Royalty payable (net of severance tax) (4,000 × $64 × 0.95 × 1/8)		30,400
Oil sales (WI) (4,000 × $64 × 7/8)		224,000

d. The royalty interest owner's entry under either part b or part c above.

   Entry by Gilbert Jones (RI)

A/R (4,000 × $64 × 0.95 × 1/8)	30,400	
Production tax expense (4,000 × $64 × 0.05 × 1/8)	1,600	
Oil sales (4,000 × $64 × 1/8)		32,000

The following example illustrates a situation in which the oil is sold to a refinery owned 100% by the working interest owner. The working interest owner's entry to record the sale in this situation differs only from the previous example in that an intracompany account receivable and an intracompany revenue account must be recognized. The purchaser, as distinguished from the producer, i.e., the working interest owner, differentiates between intracompany oil purchases and outside oil purchases.

EXAMPLE

Intracompany Sales

Assume the same facts as in the above example except that Tyler Company sold the oil to a refinery owned 100% by Tyler Company.

Entry for producer:

A/R—intracompany (WI + RI)	256,000	
Production tax expense	11,200	
Production tax payable		12,800
Royalty payable (net of taxes)		30,400
Oil sales—intracompany (WI)		224,000

Entry for purchaser:

Crude oil purchases—intracompany (4,000 × $64 × 7/8).	224,000	
Crude oil purchases—outside (RI) (4,000 × $64 × 1/8).	32,000	
A/P—intracompany		256,000

EXAMPLE

Multiple Nonworking Interests

Tyler Company has production on a lease with the following ownership interests:

Royalty Interest	1/8
Overriding Royalty Interest (ORI)	1/5 of total WI
Working Interest	Tyler Company (75% WI)
	Ewing Company (25% WI)

During February, 4,000 (gross) barrels of oil (after correction for temperature, gravity, and BS&W) were produced and sold. Assume the price for oil is $64/bbl, and the production tax in Texas is 5%.

a. Analysis:

Interest Type	Share of Revenue	a Gross Revenue (4,000 × $64 × interest)	b Severance Tax (5% × a)	c Net of Severance Tax (a – b)
WI	Tyler = (7/8 × 4/5) 75%	$134,400	$ 6,720	$127,680
WI	Ewing = (7/8 × 4/5) 25%	44,800	2,240	42,560
ORI	(1/5 × 7/8)	44,800	2,240	42,560
RI	(1/8)	32,000	1,600	30,400
Total	(8/8)	$256,000	$12,800	$243,200

b. The purchaser assumes the responsibility of distributing severance taxes and royalty income.

   Entry by Tyler, the operator—Amounts may be taken from the above analysis or may be calculated as shown below:

A/R—purchaser (Tyler) (4,000 × $64 × 0.95 × 7/8 × 4/5 × 75%)	127,680	
Production tax expense (Tyler) (4,000 × $64 × 0.05 × 7/8 × 4/5 × 75%).	6,720	
Oil sales (Tyler) (4,000 × $64 × 7/8 × 4/5 × 75%) . .		134,400

c. Tyler Company assumes the responsibility and receives 100% of the proceeds.

   Entry by Tyler—Amounts may be taken from the analysis in part a above or from the calculations in parentheses.

A/R—purchaser (WI + RI + ORI) (4,000 × $64)	256,000	
Production tax expense (WI) (4,000 × $64 × 0.05 × 7/8 × 4/5 × 75%)	6,720	
Production tax payable (WI + RI + ORI) (4,000 × $64 × 0.05)		12,800
Royalty payable (4,000 × $64 × 0.95 × 1/8)		30,400
ORI payable (4,000 × $64 × 0.95 × 1/5 × 7/8). . . .		42,560
Ewing payable (4,000 × $64 × 0.95 × 7/8 × 4/5 × 25%)		42,560
Oil sales (Tyler) (4,000 × $64 × 7/8 × 4/5 × 75%) . .		134,400

d. The royalty interest owner's entry under either part b or part c above.

A/R (4,000 × $64 × 0.95 × 1/8)	30,400	
Production tax expense (4,000 × $64 × 0.05 × 1/8). . . .	1,600	
Oil sales (4,000 × $64 × 1/8)		32,000

e. The overriding royalty interest owner's entry under either part b or c above.

   Entry by ORI

A/R (4,000 × $64 × 0.95 × 1/5 × 7/8)	42,560	
Production tax expense (4,000 × $64 × 0.05 × 1/5 × 7/8).	2,240	
Oil sales (4,000 × $64 × 1/5 × 7/8)		44,800

f. Ewing's entry under either part b or part c above.

Entry by joint working interest owner
A/R (4,000 × $64 × 0.95 × 7/8 × 4/5 × 25%)	42,560	
Production tax expense (4,000 × $64 × 0.05 × 7/8 × 4/5 × 25%).	2,240	
Oil sales (4,000 × $64 × 7/8 × 4/5 × 25%)		44,800

Frequently, a portion of the oil or gas produced on a lease is used on that lease as fuel for generators, boilers, pumps, or other equipment. When oil or gas is used on the lease from which it was produced, the financial accounting impact is typically neutral. Neither revenue nor expense is recognized because any revenue or expense would normally offset. Royalty interest owners generally do not receive royalty payments on oil or gas produced and used on the same lease because most lease agreements have a "free fuel" clause. The payment of severance tax on oil or gas produced and used on the same lease depends upon state law.

Since operators typically operate many leases in close proximity to one another, it is possible that oil or gas produced on one lease might be used on a nearby lease with differing ownership. In that case, revenue, royalty, and severance taxes would be recorded by the producing lease, while operating expense is recorded for the lease using the product. In the following example, oil produced on Lease A is transferred to Lease B to be used on Lease B in lease operations. The working interest ownership in both leases is the same, but the royalty ownership is different. The working interest owner treats this transaction as a sale of the oil from Lease A and as an operating expense on Lease B. The value assigned to the oil would normally correspond to the price at which the oil produced in the area could have been sold.

EXAMPLE

Oil Used Off Lease by Operator

Tyler Company operates Leases A and B. The company uses 10 barrels of oil obtained from Lease A as fuel on Lease B. Assume the oil is priced on both leases at $80/bbl, that production taxes are 5%, and that the royalty interest on Lease A is a 1/8 royalty.

Analysis:

Revenue Interest	a Gross Revenue	b Severance Tax	c Net of Severance Tax
	(10 × $80)	(5% × a)	(a − b)
WI (7/8)	$700	$35	$665
RI (1/8)	100	5	95
Total (8/8)	$800	$40	$760

Entry to record transfer of 10 barrels of oil

Lease operating expense—Lease B (10 × $80)	800	
Production tax expense—Lease A (10 × $80 × 5% × 7/8)	35	
Production tax payable (WI + RI) (10 × $80 × 5%)		40
Royalty payable—Lease A (10 × $80 × 95% × 1/8)		95
Oil sales—Lease A (10 × $80 × 7/8)		700

CRUDE OIL EXCHANGES

Producing companies often exchange crude oil in order to operate more efficiently. For example, crude oil production in one area may be exchanged for crude in another location closer to the purchaser so that the oil does not have to be physically transported. Alternately, future production could be exchanged for an immediate delivery of crude oil in order to meet refinery demands.

Crude oil exchanges are often tracked and accounted for using volumetric information. Under the **inventory method**, the number of barrels received in an exchange is debited to a volume-based physical inventory account with a corresponding credit to exchange inventory. The reverse entry would be made when the oil received is paid back with a future delivery.

A common alternate method of recording exchanges is based on the monetary value of the oil exchanged. The **gross purchase and sales method** treats each exchange receipt as a purchase and each exchange delivery as a sale. Corresponding entries are made to accounts payable and accounts receivable, respectively.

Regardless of which basic method is used, differentials will usually exist, because the monetary values of the barrels exchanged will normally not be equal due to differences in quality, crude type, or gravity. These differentials are settled in cash and are recorded as a revenue by the company receiving the payment and as an expense by the company making the payment.

Recording Gas Revenue

In the domestic market and many foreign markets, the volume of gas sold is measured in thousand cubic feet (Mcf). The selling price of gas, however, is usually based on the heating value of the gas, as measured in million British thermal units (MMBtu). (A Btu is the quantity of energy necessary to raise the temperature of 1 pound of water by 1° Fahrenheit.) Because of this difference in measurement units, it is important to know how to convert gas measurements to MMBtu and to Mcf. (Gas volumes are determined by metering the gas. The heating value, on the other hand, is determined through sampling and analyzing the gas in the pipeline.)

EXAMPLE

MMBtus versus Mcfs

Tyler Company sold 500 Mcf of gas with a heat content of 1.040 MMBtu/Mcf. The selling price of the gas was $8.40 per MMBtu.

To determine MMBtus for gas sold:

500 Mcf × 1.040 MMBtu/Mcf = <u>520 MMBtu</u>
(Note that the units of Mcfs cancel, giving a result in units of MMBtus.)

To determine total sales price:

520 MMBtu × $8.40/MMBtu = <u>$4,368</u>
(Note that the units of MMBtus cancel, giving a result in units of dollars.)

To determine unit sales price per Mcf:

$8.40/MMBtu × 1.040 MMBtu/Mcf = <u>$8.736 per Mcf</u>
(Note that the units of MMBtus cancel, giving a result in units of dollars per Mcf.)

The temperature standard for gas is typically 60°F, but the pressure standard varies. Both the volume of gas in Mcf and the heating content of the gas in MMBtu/Mcf must be at the same standard pressure (psia). If they are not at the same standard pressure, one of the measurements must be converted to the pressure of the other measurement. The formulas for converting from one pressure to another are as follows:

Converting Volume:

$$\text{Mcf @ current psia} \times \frac{\text{current psia}}{\text{desired psia}} = \text{Mcf @ desired psia}$$

The calculation above is based on the ideal gas law.

Converting Btu:

$$\text{MMBtu/Mcf @ current psia} \times \frac{\text{desired psia}}{\text{current psia}} = \text{MMBtu/Mcf @ desired psia}$$

In the calculation above, the desired psia is divided by the current psia rather than the current psia being divided by the desired psia, because the calculation involves MMBtu *per* Mcf.

EXAMPLE

Pressure Base

Tyler Company sold 500 Mcf of gas at 14.65 psia with a heat content of 1.040 MMBtu/Mcf at 14.73 psia. The selling price of the gas was $8.00 per MMBtu.

To convert Mcf @ 14.65 psia to a standard pressure of 14.73 psia:

$$500 \text{ Mcf} \times \frac{14.65}{14.73} = 497.28445 \text{ Mcf @ 14.73 psia}$$

To determine MMBtus for gas sold:

497.28445 Mcf × 1.040 MMBtu/Mcf = <u>517.1758 MMBtu</u>

Alternatively, to convert MMBtu/Mcf @ 14.73 psia to a standard pressure of 14.65 psia:

$$1.040 \text{ MMbtu/Mcf} \times \frac{14.65}{14.73} = 1.03435 \text{ MMBtu/Mcf @ 14.65 psia}$$

To determine MMBtus for gas sold:

500 Mcf × 1.03435 MMBtu/Mcf = <u>517.175 MMBtu</u>

To determine sales price of gas sold:

517.175 MMBtu × $8.00/MMBtu = <u>$4,137.40</u>

Although most sales contracts are written based on MMBtus, most accounting records and internal reports are based on gas volumes and prices in Mcfs. Therefore, any subsequent examples illustrating gas sales utilize Mcfs versus MMBtus.

Once the price of natural gas is determined and a gas settlement statement is received from the purchaser, the accounting entries are made. Accounting for gas revenue is similar to accounting for oil revenue except for the following:

a. No gas is stored on the lease prior to sale.

b. The measurement of gas is more complex than the measurement of oil.

c. Distribution of the proceeds to the economic interest owners is usually handled by the operator rather than the purchaser.

An illustration of gas revenue accounting follows.

EXAMPLE

Gas Revenue

Tyler Company owns a 100% working interest in a lease on which there is a 1/8 royalty interest. During one month, Tyler produced and sold 300 (gross) Mcf of gas at $10.00/Mcf. Assume production taxes are 5%.

a. **Analysis:**

	Revenue Interest	a Gross Revenue	b Severance Tax	c Net of Severance Tax
		(300 × $10 × interest)	(5% × a)	(a – b)
WI	7/8	$2,625	$131.25	$2,493.75
RI	1/8	375	18.75	356.25
Total	8/8	$3,000	$150.00	$2,850.00

b. The operator assumes the responsibility of distributing royalty, and the purchaser assumes the responsibility of paying severance taxes.

Entry by Tyler Company

A/R (300 × $10 × 95%)	2,850.00	
Production tax expense (300 × $10 × 0.05 × 7/8)	131.25	
Royalty payable (300 × $10 × 0.95 × 1/8).		356.25
Gas sales (300 × $10 × 7/8)		2,625.00

c. The operator assumes the responsibility of distributing royalty income and severance taxes.

Entry by Tyler Company

A/R (300 × $10)	3,000.00	
Production tax expense (300 × $10 × 0.05 × 7/8)	131.25	
Production tax payable (WI + RI) (300 × $10 × 5%)		150.00
Royalty payable (300 × $10 × 0.95 × 1/8).		356.25
Gas sales (300 × $10 × 7/8)		2,625.00

d. The royalty interest owner's entry under either part b or part c above:

Entry by RI

A/R (300 × $10 × 0.95 × 1/8)	356.25	
Severance tax expense (300 × $10 × 0.05 × 1/8)	18.75	
Gas sales (300 × $10 × 1/8)		375.00

Gas, like oil, is also often used for lease operation. The gas may be used as fuel for operating lease equipment such as pumping units, heater-treaters, and compressors or as fuel for the lessor's residence or other activities. Such gas is measured by metering if the quantities involved are substantial. Otherwise, the volume used is normally estimated based on engineering estimates of fuel used by the various pieces of equipment. The value assigned to the gas is generally equivalent to the value at which the gas could have been sold. If gas produced on one lease is used on another lease owned by the same company, the accounting entry is similar to the entry in the previous example. This entry also is similar to the oil revenue entry for the same situation.

EXAMPLE

Gas Used Off Lease by Operator

Tyler Company owns 100% of the working interests on Lease A and Lease B. Tyler Company used 300 Mcf of gas produced on Lease A and valued at $10/Mcf to operate lease equipment on Lease B. Assume production taxes were 5% and a 1/8 royalty interest on Lease A.

Entry—Analysis is the same as in the previous example.

Lease operating expense—Lease B	3,000.00	
Production tax expense—Lease A (300 × $10 × 0.05 × 7/8)	131.25	
Production tax payable (RI + WI)(300 × $10 × 0.05)		150.00
Royalty payable—Lease A (300 × $10 × 0.95 × 1/8)		356.25
Gas sales—Lease A (300 × $10 × 7/8)		2,625.00

Gas may also be used in lease operations for gas injection or gas lift. In a **gas injection** operation, gas is injected into a formation to maintain pressure or for secondary recovery. Reproduced injected gas cannot usually be distinguished from the original formation gas. Therefore, an assumption must be made concerning the recovery of gas, i.e., whether the produced gas is original formation gas or injected gas (see *COPAS AG-15* for further details). **Gas-lift gas** is injected into the wellbore to help lift the oil to the surface. Gas-lift gas, unlike injected gas, returns immediately to the mouth of the well without entering the reservoir. Normally the sales price for recovered gas-lift gas is lower, because when it is reproduced, its pressure has diminished. Gas going into a pipeline must either be at a prescribed pressure or at the same pressure as the gas already in the pipeline. Thus if the pressure of the recovered gas-lift gas is too low, compression is necessary.

EXAMPLE

Gas Injection or Gas Lift

If Tyler Company had used the gas in the preceding example for injection or for gas lift, the entry to record the use of the gas would have been the same as described above. When the gas is recovered and sold, the entry would be the following (assuming 100% recovery and the same selling price):

Entry

A/R—purchaser	3,000	
Lease operating expense—Lease B		3,000

If gas is produced and used on the same lease, royalty payments and production taxes may not have to be paid. The liability for royalty payments and production taxes would be determined from the lease agreement and state laws. However, most leases provide that gas may be used on the same lease free from the obligation of payment of royalty. If state laws do not levy production taxes on such gas, then an accounting entry would not have to be made to record the production and use of gas on the same lease. If an entry is made, it would be as follows:

Entry

Lease operating expense.	XXX	
Gas sales		XXX

If the gas is used for injection or gas lift on the same lease from which it was produced, usually no accounting entry is made at the time of production and use. Also, there usually is no production tax or royalties payable on such gas. When the gas is recovered and sold, the regular accounting entry would be made to record the sale of the gas.

Vented or flared gas

Under normal circumstances, gas flaring is prohibited because of environmental considerations. Generally, permission to vent or flare gas must be obtained from the appropriate regulatory agency. A request to flare or vent gas may arise when the gas is of low quality or low pressure or when sale of the gas is not economically feasible (i.e., an oil well with no gas pipelines available). If venting or flaring is allowed, normally no value is assigned to vented or flared gas, and revenue accounts are not credited. In most states, this type of gas is not subject to severance or production taxes. Vented or flared gas is not usually measured, but the volume is estimated.

Nonprocessed natural gas

Gas may be of marketable quality when produced and, therefore, not need any processing. In this case, the gas may be sold directly to a gas pipeline, local distribution company, or end user. (All of the gas sales examples in this chapter apply to nonprocessed natural gas that is of marketable quality when produced.)

Natural gas processing

Often gas is not of marketable quality when produced because of impurities and liquids. Since gas may not usually be disposed of by flaring, the gas must be processed to make it marketable. Such processing involves removing impurities and saleable liquids and compressing the gas to meet pipeline pressure specifications. Typically the gas is either sold to a processing plant or is processed by a plant for a fee. In either case, accounting entries to record the sale are similar to the accounting discussed above. (For more information, see *COPAS AG-15.*)

Stored natural gas

The demand for gas is highly seasonal. As a result, if wells are to produce at capacity during the low demand summer months, the excess gas must be placed in underground storage facilities for use in the higher demand winter months. Royalties usually must be paid to royalty interest owners when the gas is severed from the lease, even if the gas is being stored by the producer for sale at a later time.

Take-or-pay provision

Some older gas sales/purchase contracts, most often those entered into with a pipeline, contain a take-or-pay clause. This clause requires the purchaser to take a specified minimum volume of gas each month. If the minimum volume of gas is not taken, the purchaser must pay for the deficiency gas, even though the gas is not taken. However, the purchaser generally may offset the deficiency payment against any future purchases in excess of the specified volume. The producer records the deficiency payment as a deferred credit until make-up gas is taken by the purchaser, as follows:

Entry
Cash XXX
    Deferred credit XXX

When the purchaser buys more than the specified minimum amount of gas from the producer, the producer recognizes revenue and reduces the deferred credit. If the price of gas has increased since the deficiency arose, most contracts require the purchaser to pay an additional amount.

EXAMPLE

Take-or-Pay

Tyler Company has a contract with Southwest Pipeline in which Southwest has agreed to purchase 20,000 Mcf of gas per month at $12/Mcf. Southwest must pay for the gas, even if it chooses not to take delivery of the full amount. In August, Southwest took delivery of 16,000 Mcf produced from a lease with a 1/8 royalty and in which Tyler owns 100% of the working interest. Production taxes are 5%. Tyler assumes the responsibility of distributing royalty income and paying production taxes.

Entries

To record the sale of 16,000 Mcf:

Cash (16,000 × $12)	192,000	
Production tax expense (16,000 × $12 × 5% × 7/8) . . .	8,400	
Royalty payable (16,000 × $12 × 1/8 × 0.95)		22,800
Production tax payable (16,000 × $12 × 5%)		9,600
Gas sales (16,000 × $12 × 7/8)		168,000

To record the deferred credit:

Cash (4,000 × $12).	48,000	
Deferred revenue.		48,000

In September, Southwest took delivery of 24,000 Mcf.

Entries

To record the current month's sale:

Cash (20,000 × $12)	240,000	
Production tax expense (20,000 × $12 × 5% × 7/8) . . .	10,500	
Royalty payable (20,000 × $12 × 1/8 × 0.95)		28,500
Production tax payable (20,000 × $12 × 5%)		12,000
Gas sales (20,000 × $12 × 7/8)		210,000

To record the makeup volume of gas:

Deferred revenue (4,000 × $12)	48,000	
Production tax expense (4,000 × $12 × 5% × 7/8)	2,100	
Royalty payable (4,000 × $12 × 1/8 × 0.95).		5,700
Production tax payable (4,000 × $12 × 5%)		2,400
Gas sales (4,000 × $12 × 7/8)		42,000

Very few contracts containing a take-or-pay provision still exist. Moreover, contracts containing such provisions are not currently being written. (For more information, see *COPAS AG-15*.)

Timing of Revenue Recognition

Theoretically, revenue from oil and gas sales should be recognized when the sale is made rather than when the oil or gas is produced. Legally, title normally passes when physical control of the oil or gas is transferred from one party to another. Therefore, theoretically, revenue should be recognized at the point of delivery.

Revenue from crude oil

In practice, revenue from crude oil sales is typically recognized either when the oil is produced or when the oil is sold. When oil revenue is recognized at the time of sale, the revenue is recognized based on pipeline runs or run tickets. Revenue recognition at the time of the sale is the basis used for all the previous examples in this chapter.

EXAMPLE

Revenue Recognition at Time of Sale

Tyler Company produces 3,000 (gross) barrels in the month of October. The sales price is $77. Assume the purchaser of the oil will pay the royalty owner and severance taxes. Assume that Tyler owns 100% of the working interest, the royalty interest is 1/8, and the state severance tax is 8%.

Entry at time of sale

A/R—purchaser (3,000 × $77 × 0.92 × 7/8)	185,955	
Production tax expense (3,000 × $77 × 0.08 × 7/8) . . .	16,170	
Oil sales (3,000 × $77 × 7/8).		202,125

A company may recognize revenue at the point of sale and disregard inventory in lease tanks. If that is the case, no entry is made at the time of production. (See chapter 8 for more discussion of oil and gas inventory.)

If, on the other hand, oil revenue is recognized when the oil is produced, the revenue is determined by multiplying the number of barrels produced by the expected sales price. Inventory is also then recognized and carried in an inventory account on the books of the producer until the crude oil is sold.

EXAMPLE

Revenue Recognition at Time of Production

Using the same data as in the previous example, the entry to record the production and sale of the oil would be the following.

a. Entry at time of production is as follows:

    Entry
    Crude oil inventory (3,000 × $77 × 7/8)   202,125
      Oil sales   202,125

b. The production is sold in November. Entry at the time of the sale is as follows:

    Entry
    A/R—purchaser (3,000 × $77 × 0.92 × 7/8)   185,955
    Production tax expense (3,000 × $77 × 0.08 × 7/8) . . .   16,170
      Crude oil inventory   202,125

Any price difference between the expected price at the time of production and actual sales price at the time of sale would be treated as an adjustment to oil revenue.

Revenue from natural gas

Since gas cannot be stored on the lease, nonprocessed gas is generally not produced unless there is a buyer for the gas. Traditionally under long-term contracts, the delivery point for nonprocessed gas sales was at the wellhead, and therefore gas sales were recognized essentially concurrently with production, based on pipeline runs. (All gas revenue examples in this chapter are on this basis.) With spot market sales, title of the gas often passes at an extended delivery point, i.e., at *city gate* rather than at wellhead. In these cases, the pipeline is transporting the gas to a delivery point versus purchasing it at the wellhead. Accounting for gas sold in the spot market is complicated by the extended delivery points. The extended delivery points result in transportation charges, pipeline losses, and the need to recognize gas in the pipeline as inventory.

Several methods of accounting for spot sales are currently used. The example below illustrates two methods. The first is the traditional accounting whereby the gas is sold to a pipeline at the wellhead. The second method is for a spot sale in which the gas is sold to an industrial plant, and the pipeline only transports the gas instead of purchasing the gas.

EXAMPLE

Gas Revenue Recognition

Delivery Point Wellhead (pipeline is purchaser)

Tyler Company produces and sells 20,000 (gross) Mcf of gas to a pipeline for $9/Mcf. Tyler owns a 100% WI in the lease, which has a 1/5 royalty interest. Ignore severance taxes.

Entry to record sale

A/R (20,000 × $9)	180,000	
Royalty payable (20,000 × $9 × 1/5)		36,000
Gas sales (20,000 × $9 × 4/5)		144,000

Delivery Point Gas Inlet at Municipal Gas Company (pipeline transports gas)

Tyler Company also produces 30,000 Mcf of gas from the same lease to be transported by a pipeline to the local municipal gas company. The gas is sold at $10/Mcf. The transportation charge is $0.40/Mcf. Tyler elects not to pass any of the transportation charge on to the royalty owner.

Entry to record gas delivered to pipeline

Inventory (30,000 × $10)	300,000	
Royalty payable (30,000 × $10 × 1/5)		60,000
Deferred revenue (30,000 × $10 × 4/5)		240,000

Entry to record delivery of gas to municipal gas company

A/R (30,000 × $10)	300,000	
Deferred revenue	240,000	
Transportation expense (30,000 × $0.40)	12,000	
A/P transporter (30,000 × $0.40)		12,000
Inventory		300,000
Gas sales (30,000 × $10 × 4/5)		240,000

Some small producers defer recording both oil and gas revenue until a check and accompanying statement are received from the purchaser. This practice, referred to as **recording revenue at the time of settlement**, does not adhere to the principles of accrual accounting. However, as long as production and prices are relatively stable from accounting period to accounting period, it does not result in a material distortion of revenue, and therefore its use is allowed.

Revenue Reporting to Interest Owners

Royalty owners and other interest owners are normally paid monthly, based on their revenue percentage as specified in the division order. The payments are made by either the operator or the purchaser, depending upon which entity has assumed that responsibility. The statement is typically a check-stub type that contains the information needed by the interest owners for their tax records and for other record-keeping purposes. Information provided on the monthly check stub includes the number of barrels or Mcfs sold, the average price, the owner's decimal ownership, and the applicable state severance tax withheld.

Additional Topics

The following sections discuss miscellaneous revenue accounting topics.

Gas imbalances

One of the more challenging issues in revenue accounting is gas imbalances. Imbalances can be of two types: producer gas imbalances or pipeline gas imbalances. **Producer gas imbalances** occur when there are two or more working interest owners in a single property, and one working interest owner takes more than his proportionate share of the gas produced from the property. **Pipeline gas imbalances** occur between a producer and a pipeline company. For example, a producer agrees to sell a quantity of gas to an end user. The end user typically has no means to store the excess gas that results when the actual demanded quantity is less than the contracted quantity. The end user also typically has no reserve to use when the actual demanded quantity is greater than the contracted quantity. In these cases, the buyer takes what it needs from the pipeline. If the take is more or less than the amount of gas that the producer put into the pipeline, a pipeline imbalance results.

Producer gas imbalances

Producer gas imbalances occur for a variety of reasons, but with the advent of spot sales and direct sales, the imbalances have become more complex. Currently, each working interest owner on a property may market its gas to one or more purchasers, which includes marketers, brokers, end users, and other producers. This creates multiple gas purchasers. In some instances it is not possible to deliver the gas simultaneously to all purchasers. Thus one purchaser will take all the production for a period of time, after which another purchaser will take all the production, and so forth. Only the working interest owner who sold the gas to the purchaser currently taking the gas is paid by that purchaser for the gas received. Thus, at any point in time, an imbalance situation exists.

This imbalance is magnified due to the fact that natural gas wells rarely produce at a steady rate. As a result, the volumes delivered to the purchasers are seldom at exactly the average quoted production rate. Imbalances may also occur because some of the working interest owners on a property may be unwilling or unable, because of market conditions, lack of connection to wellhead, etc., to market their gas currently. The working interest owners on a property usually sign a gas balancing agreement, which specifies the rights and duties of each party when a gas imbalance situation exists. The American Association of Professional Landmen (AAPL) publishes a model form gas balancing agreement (AAPL Form 610-E). This model form is widely used in practice today. Additionally, a detailed discussion of a variety of industry practices appears in *COPAS AG-22*, "Producer Gas Imbalances."[2]

In a producer gas imbalance situation, the operator typically maintains records or reports in which allocated and delivered volumes are tracked. This enables the operator to request adjustments of future deliveries of gas to each purchaser because of deliveries being over or short in the past.

The industry has developed several methods of dealing with gas imbalances, called **production balancing**. Under one method, an "in-kind" balancing, the underproduced party may give notice to take up to an agreed-upon quantity of the overproduced party's share of gas until balanced. Some agreements also allow the operator to periodically require a cash balancing. If balancing is not possible, e.g., the reservoir has been depleted, the operating or gas balancing agreement may provide for a cash settlement under both the sales method and entitlement method as described in the following.

For financial accounting purposes, there are two methods of recognizing revenues—the sales method and the entitlement method. Both are generally accepted accounting methods. Using the **sales method**, the working interest owners recognize their actual sales of gas regardless of the amount of production they are entitled to for the period. The sales method assumes that any production sold by a working interest owner comes from that party's share of the total reserves in place. Thus, whatever quantity is sold in any given period is revenue for that party. No receivables, payables, or unearned revenue are recorded unless a working interest owner's aggregate sales from the property exceed its share of the total reserves in place. If such a situation arises, the parties would likely choose to cash balance, or in some instances, the overdelivered partner might choose to negotiate to buy out the underdelivered party's share.

Using the **entitlement method**, each owner recognizes revenue based on its ownership share of total gas produced during the period, regardless of which owner actually sells and receives payment for the gas. Thus, the overdelivered owner would recognize unearned revenue for the amount received in excess of its entitled share. The underdelivered owner would recognize a receivable and revenue for the amount of its share of production sold by another owner for which it was not paid. While both the sales and entitlement methods are widely used, the sales method is the more popular of the two.

Companies must determine their reserves for the property in a manner that is consistent with their revenue recognition method. For example, a company using the sales method would reduce its estimated reserves for the quantity of gas it actually sells during the period. A company using the entitlement method would reduce its reserves by the amount of the share of production it was entitled to, regardless of whether the quantity it actually sold was more or less than the entitled quantity.

The following example illustrates an in-kind production balancing arrangement and the sales and entitlement methods of recording revenues.

EXAMPLE

Producer Gas Imbalances

Shepherd Field is jointly owned by Tyler Company (60% WI), which acts as field operator, and Garza Company (40% WI). There is a 1/8 royalty, which is shared proportionally by Tyler and Garza. The two working interest owners agree that Tyler's purchaser will take gas produced in April, and Garza's purchaser will take gas produced in May. Gas allocations will be equalized in June. Ignore severance taxes.

Gross production and gas prices were as follows:

	Production	Price
April	300,000 Mcf	$5.00/Mcf
May	280,000 Mcf	$5.00/Mcf
June	200,000 Mcf	$5.00/Mcf

Each working interest owner receives payment only for gas delivered to his purchaser(s). The following Gas Balance Report prepared by Tyler Company summarizes the production deliveries and equalization of gas for April through June:

	April	May	Cumulative	June	Cumulative Total
Gross Production:	300,000	280,000	580,000	200,000	780,000
Allocated shares based on WI% (including royalty):					
Tyler Company (60%)	180,000	168,000	348,000	120,000	468,000
Garza Company (40%)	120,000	112,000	232,000	80,000	312,000
	300,000	280,000	580,000	200,000	780,000
Deliveries taken by:					
Tyler Company's Purchaser	300,000	—	300,000	168,000	468,000
Garza Company's Purchaser	—	280,000	280,000	32,000	312,000
	300,000	280,000	580,000	200,000	780,000
Over/(Under)delivered:					
Tyler Company	120,000	(168,000)	(48,000)	48,000	0
Garza Company	(120,000)	168,000	48,000	(48,000)	0

Sales Method

The journal entries made by each company for the three-month period are shown below, assuming that both companies use the sales method:

APRIL
Tyler Company

Accounts receivable (300,000 × $5)............	1,500,000	
Royalty payable (300,000 × $5 × 1/8).........		187,500
Gas revenue (300,000 × $5 × 7/8)...........		1,312,500

Garza Company
No entry

MAY
Tyler Company
No entry

Garza Company

Accounts receivable (280,000 × $5)............	1,400,000	
Royalty payable (280,000 × $5 × 1/8).........		175,000
Gas revenue (280,000 × $5 × 7/8)...........		1,225,000

JUNE
Tyler Company

Accounts receivable (168,000 × $5)............	840,000	
Royalty payable (168,000 × $5 × 1/8).........		105,000
Gas revenue (168,000 × $5 × 7/8)...........		735,000

Garza Company

Accounts receivable (32,000 × $5).............	160,000	
Royalty payable (32,000 × $5 × 1/8)		20,000
Gas revenue (32,000 × $5 × 7/8)		140,000

Entitlement Method

The journal entries made by each company for the three-month period are shown below, assuming that both companies use the entitlement method for both revenue and royalty.

APRIL

Tyler Company

Accounts receivable (300,000 × $5)	1,500,000	
Royalty payable (180,000 × $5 × 1/8).		112,500
Unearned revenue—overdelivered (120,000 × $5) .		600,000
Gas revenue (180,000 × $5 × 7/8).		787,500

Garza Company

Accounts receivable—underdelivered (120,000 × $5) .	600,000	
Royalty payable (120,000 $5 × 1/8).		75,000
Gas revenue (120,000 × $5 × 7/8).		525,000

MAY

Tyler Company

Unearned revenue—overdelivered (120,000 × $5) . . .	600,000	
Accounts receivable—underdelivered (48,000* × $5) .	240,000	
Royalty payable (168,000 × $5 × 1/8).		105,000
Gas revenue (168,000 × $5 × 7/8).		735,000

* Tyler's share of production to date based on its working interest percentage is 348,000 Mcf. The actual amount taken to date by Tyler's purchaser is 300,000 Mcf.

Garza Company

Accounts receivable (280,000 × $5)	1,400,000	
Royalty payable (112,000 × $5 × 1/8)		70,000
Accounts receivable—underdelivered (120,000 × $5)		600,000
Unearned revenue—overdelivered (48,000* × $5)		240,000
Gas revenue (112,000 × $5 × 7/8)		490,000

* Garza's share of production to date based on its working interest percentage is 232,000 Mcf. The actual amount taken to date by Garza's purchaser is 280,000 Mcf.

JUNE

Tyler Company

Accounts receivable (168,000* × $5)	840,000	
Royalty payable (120,000 × $5 × 1/8)		75,000
Accounts receivable—underdelivered		240,000
Gas revenue (120,000 × $5 × 7/8)		525,000

*The 168,000 barrels equals Tyler's 60% share of June's 200,000 barrels of production, plus the 48,000 underdelivered barrels.

Garza Company

Accounts receivable (32,000* × $5)	160,000	
Unearned revenue—overdelivered	240,000	
Royalty payable (80,000 × $5 × 1/8)		50,000
Gas revenue (80,000 × $5 × 7/8)		350,000

*The 32,000 barrels equals Garza's 40% share of June's 200,000 barrels of production less the 48,000 overdelivered barrels.

In addition to recognizing revenue on gas sales, companies must also pay royalties and severance taxes. Lease agreements typically do not specify whether the sales or entitlement method is to be used in the calculation of royalties. However, most companies—even those using the entitlement method for financial accounting purposes—use the sales method for determining royalties. Most states require the sales method for determination of severance taxes.

Pipeline gas imbalances

In order for the various interest owners in a lease to market their own gas, there must be communication between each working interest owner (producer), the operator, the pipeline company, and each producer's customer. The operator must advise each producer of the anticipated gas production from each well during the next month. Each producer negotiates with its purchaser for the quantity of gas the purchaser will require during the next month. Then through a process of **nomination** and **confirmation,** the operator agrees to put a quantity of gas into the pipeline on behalf of the producer, and the purchaser agrees to take that quantity out of the pipeline during the next month.

The actual amount of gas production rarely equals the total of the confirmed gas nominations. The parties involved should execute an agreement that includes a method to determine the following:

    a. The amount of actual gas production (in Mcfs) to be allocated to each party
    b. Each producer's resulting overdeliveries and underdeliveries to the pipeline

There are two methods commonly used to determine the imbalance position of the parties. These are allocation of production volumes based on **confirmed nominations** and allocation based on **entitlement**. These two methods are similar to the sales and entitlement methods illustrated and discussed in the producer gas imbalance situation. Regardless of which method is used, a monthly production volume allocation statement is prepared.

Financial accounting for pipeline gas imbalances involves recording overdeliveries and underdeliveries to the pipeline company as receivables or payables. Once overdeliveries and underdeliveries are determined through allocation, the issue of valuing the receivables and payables must be resolved. Since each producer is selling its gas to its own various customers, multiple gas sales prices will exist. Which price should be used to value the imbalance? The FASB Emerging Issues Task Force (EITF) and the SEC have studied this issue. They generally agree that in valuing the imbalances, receivables should be valued at the lowest of the possible sales prices, and payables valued at the highest (see *EITF Issue No. 90-22*).

The following example demonstrates accounting for pipeline imbalances.

EXAMPLE

Pipeline Gas Imbalances

Lehman Field is jointly owned by Tyler Company, which acts as field operator, and Conroe Company. Each company owns a 50% WI. Tyler estimates that gross gas production during August will be 50,000 Mcf. Tyler Company makes confirmed nominations of 40,000 Mcf, and Conroe Company makes confirmed nominations of 10,000 Mcf to a pipeline company. Actual August production totals 80,000 Mcf. Assume that the appropriate price is $6/Mcf. The pipeline company has paid Tyler and Conroe for the confirmed nominations only. Therefore, Tyler and Conroe must recognize receivables for the gas sold in excess of the confirmed nominations.

Allocation using the confirmed nomination method and entitlement method would be the following:

Confirmed Nomination Method

	Tyler Company	Conroe Company
Actual production	80,000 Mcf	80,000 Mcf
Nomination ratio	× 40/50 = 80%	× 10/50 = 20%
Allocated actual production	64,000 Mcf	16,000 Mcf
Less: confirmed nomination	(40,000) Mcf	(10,000) Mcf
Imbalance receivable from pipeline	24,000 Mcf	6,000 Mcf
$6.00/Mcf	× $ 6	× $ 6
Pipeline imbalance receivable	$144,000	$36,000
Allocated actual production	64,000 Mcf	16,000 Mcf
$6.00/Mcf	× $ 6	× $ 6
Revenue recognized	$384,000	$96,000

Entitlement Method

	Tyler Company	Conroe Company
Actual production	80,000 Mcf	80,000 Mcf
Ownership interest	× 0.50	× 0.50
Allocated actual production	40,000 Mcf	40,000 Mcf
Less: confirmed nomination	(40,000) Mcf	(10,000) Mcf
Imbalance receivable from pipeline	0 Mcf	30,000 Mcf
$6.00/Mcf	× $ 6	× $ 6
Pipeline imbalance receivable	$ 0	$180,000
Allocated actual production	40,000 Mcf	40,000 Mcf
$6.00/Mcf	× $ 6	× $ 6
Revenue recognized	$240,000	$240,000

The conditions that result in pipeline gas imbalances also result in producer gas imbalances. For example, in the confirmed nomination method illustration above, Tyler Company sold more and Conroe Company sold less than their respective 50% share of gas production. Thus, in addition to accounting for the pipeline gas imbalance, it would be necessary for the companies to also account for the resulting producer gas imbalances.

Allocation of oil and gas

Production from different wells or leases is often commingled and measured at the point of delivery. Total production sold may need to be allocated back or reconciled to the individual well or lease production in order to comply with tax requirements and management needs or because of different ownership of the properties involved. The method for determining production quantities from individual wells depends upon whether the well is metered individually. Each gas well/completion is usually metered individually, as is generally required by state law. Oil wells are not required to be metered individually. As a result, oil wells usually are not metered individually but instead are put on test periodically to determine or estimate the total amount of oil, gas, and water produced from the well.

A well **put on test** means that the stream of fluid from the well is diverted from the production separator to a test separator. The test separator measures the amount of oil, gas, and water produced from the well for 24 hours. The test results are used to estimate the production from that well for a period of time—usually a month. When all wells have been tested, the test results are used to allocate total production to the various wells or properties. This allocation method is similar to the relative sales value method used for many other accounting purposes.

The following example illustrates a one-stage allocation of oil in which production from one lease is allocated back or reconciled to the individual well production, based on a 24-hour test rate of production for the individual wells.

EXAMPLE

Allocation Back to Well

Assume the following data for Lease A:

Well	24-hr Test (bbl)	Days Produced	Total Production Sold
1	20	30	2,900—measured at
2	35	26	point of delivery*
3	50	24	

* Commingled production from three wells

Allocation

Well	24-hr Test (bbl)	Days Produced	Theoretical Production*		Ratio**		Actual Production
1	20	30	600	×	1.07	=	642
2	35	26	910	×	1.07	=	974
3	50	24	1,200	×	1.07	=	1,284
			2,710	×	1.07	=	2,900

* 24-hr test × days produced = theoretical production
** 2,900/2,710 = 1.07

An equivalent computational approach more commonly used in accounting when allocating a lump sum, referred to as the *relative sales method*, is as follows:

Well 1: $2{,}900 \times \dfrac{600}{600 + 910 + 1{,}200} = 642$ bbl

Well 2: $2{,}900 \times \dfrac{910}{600 + 910 + 1{,}200} = 974$ bbl

Well 3: $2{,}900 \times \dfrac{1{,}200}{600 + 910 + 1{,}200} = 1{,}284$ bbl

The same basic principles of allocation apply when production from more than one lease has been commingled. This usually occurs when one central processing system has been installed as the central delivery point for two or more leases. The processing system will consist of the equipment necessary for processing the oil and gas from the wells. Included would typically be a separator, heater-treater, gas scrubber (to remove liquids entrained with the gas), test separator, tanks, and possibly a LACT unit.

When the production from more than one lease is being commingled, generally the following apply:

- The production from all the leases will be measured at the central tank battery.
- The production from each lease will be measured.
- The production from each well will be measured or estimated.

In this case, at least a two-stage allocation would be involved, starting with the disposition point of the oil or gas and flowing toward the source of the oil or gas. The following example illustrates a two-stage allocation of oil in which production from two leases is delivered to a central tank battery.

EXAMPLE

Two-Stage Allocation

The production from each well on Lease A and Lease B is estimated based on a 24-hour test. The production from each well on each lease is then commingled, and the commingled flow is measured before leaving the individual lease. The production from both leases is commingled and delivered to a central tank battery. Assume the following data for Lease A:

Well	24-hr Test (bbl)	Days Produced	Theoretical Production
1	20	30	600
2	35	26	910
3	50	24	1,200
			2,710

Assume the following data for Lease B:

Well	24-hr Test (bbl)	Days Produced	Theoretical Production
1	25	28	700
2	40	29	1,160
			1,860

Measured production is 2,450 barrels from Lease A and 1,663 barrels from Lease B, for a total of 4,113 barrels. After treatment at the central tank battery, 4,000 barrels are sold.

Allocation

Step 1: Allocate sales from tank battery to each lease.

Ratio of sales to total production from leases = 4,000/4,113 = 0.9725

Lease	Measured Production		Ratio		Allocated Sales
A	2,450	×	0.9725	=	2,383
B	1,663	×	0.9725	=	1,617
	4,113				4,000

Step 2: Allocate sales from each lease to individual wells.

Lease A: Ratio of sales allocated to Lease A to total theoretical production from wells on Lease A = 2,383/2,710 = 0.8793.

Well	24-hr Test (bbl)	Days Produced	Theoretical Production		Ratio		Actual Production
1	20	30	600	×	0.8793	=	528
2	35	26	910	×	0.8793	=	800
3	50	24	1,200	×	0.8793	=	1,055
			2,710				2,383

Lease B: Ratio of sales allocated to Lease B to total theoretical production from wells on Lease B = 1,617/1,860 = 0.8694.

Well	24-hr Test (bbl)	Days Produced	Theoretical Production		Ratio		Actual Production
1	25	28	700	×	0.8694	=	609
2	40	29	1,160	×	0.8694	=	1,008
			1,860				1,617

The exact method of allocation depends upon the points at which the production is measured or metered.

The allocation procedure may be complicated when the quality of the oil varies. The value of a barrel of oil is partially dependent on the quality of the oil, specifically the API gravity of the oil, which can vary from lease to lease. Consequently, when significant differences in gravity between the various leases exist, the allocation procedure illustrated above must be modified to determine sales revenue due each lease from the quantities of oil produced and sold from each lease. (For further discussion, see *COPAS AG-6*, "Oil Accounting Manual."[3])

While conceptually the allocation of gas is the same as allocation of oil, the actual allocation process is much more complicated. (For further discussion, see *COPAS AG-15*.)

Minimum royalty—an advance revenue to royalty owners

Minimum royalties usually arise from an agreement in which the lessee agrees to pay a stated minimum amount of money to the royalty owner whenever the actual production revenue to which the royalty owner is entitled is less than the minimum amount. This payment may be paid prior to production and/or during the production phase. Minimum royalty payments may or may not be recoverable from future production. If nonrecoverable, the payments usually are expensed. If recoverable, the payments should be held in a suspense account until recovered. If the payments have not been recovered by the time the lease terminates, they should be expensed.

EXAMPLE

Minimum Royalty

a. During 2011 before production was established, Tyler Company paid minimum royalty payments of $100 a month for five months. The minimum royalty payments were recoverable out of future royalty payments. The entry to record the minimum royalty payments paid would be as follows:

    Entry

Minimum royalty suspense (total)	500	
Cash		500

b. Production was established late in 2011. The royalty payment payable from the first month of production was $250. The entry to record payment of the royalty would be as follows:

    Entry

Royalty payable	250	
Minimum royalty suspense		150
Cash		100

    Note that the royalty owner is still entitled to a minimum of $100/month, and thus only $150 of the suspense may be recovered. The royalty payable account originated in the entry to record revenue.

c. The royalty payment from the second month of production was $400. The entry to record the payment of the royalty would be as follows:

Entry

Royalty payable	400	
Minimum royalty suspense		300
Cash		100

d. Disaster struck and the lease was abandoned. The entry necessary related to the minimum royalty would be as follows:

Entry

Expense	50	
Minimum royalty suspense		50

e. Instead, assume in part a above that the minimum royalty payments were not recoverable. The entry to record the minimum royalty payments would be as follows:

Entry

Minimum royalty expense (total)	500	
Cash		500

Oil and gas revenue accounting is a complicated topic. This chapter discusses only the fundamental points. For more guidance on some of these issues, see *COPAS AG-15* and *AG-6*.[4,5]

PROBLEMS

1. Why is the temperature of oil or gas important when measuring the volume of oil or gas sold?

2. Define the following:

    BS&W

    gravity

    tank battery

    LACT unit

    Mcf

    minimum royalty

    MMBtu

3. Define the following:

    run ticket

    settlement statement

    thief

    gauging

    strapping

    Define the following and explain their relevance to oil and gas revenue accounting:

    Henry Hub

    West Texas Intermediate

    North Sea Brent Blend

    Dubai crude

4. Mr. Lomax owns the mineral rights in some land in Texas. He leases the land to Frank Oil Company, reserving a 1/5 royalty. During 2011, Frank Oil Company makes the following assignments:

    a. To Mr. Jones, an ORI of 1/7 of Frank's interest

    b. To Ms. Wilson, a production payment interest of 30,000 barrels of oil to be paid out of 1/4 of the working interest's share of production (i.e., Ms. Wilson gets 1/4 of this production until she receives a total of 30,000 barrels)

    c. To Mr. Smith, a joint working interest of 1/3 after giving consideration to all the above assignments.

    REQUIRED: Assuming gross production is 45,000 barrels, calculate each owner's share.

5. Blow Out Oil Company owns a 100% WI in Lease A. Lease A is burdened with a 1/6 royalty. During the month of June, a total of 12,000 barrels of oil was produced and sold. Assume the selling price of the oil was $72/bbl and the production tax was 5%.

    REQUIRED: Give the entry required to record the sale of the oil for each of the following:

    a. The purchaser assumes responsibility for distributing taxes and royalty income.

    b. Blow Out Oil Company assumes the responsibility.

6. Prospect Oil Company sold a total of 1,000 Mcf of gas at $9.00/Mcf. The lease provides a 1/5 RI. The working interest owner receives 100% of the revenues (net of 5% severance tax) and then distributes the amount due to the royalty interest owner.

    REQUIRED: Give the entry by Prospect to record the sale of the gas.

7. Jackson Oil Company used 100 Mcf of gas obtained from Lease A and valued at $8.40/Mcf for gas injection on Lease B. Assume production taxes are 5% and the royalty on Lease A is a 1/6 RI.

a. Give the entry necessary to record the transfer of the gas.

b. Give the entry assuming 100% of the gas is recovered and sold at $8.40/Mcf.

8. Gamma Oil Company sold or used the gas produced on Lease A during January as follows:

   a. 300 Mcf used as fuel to operate lease equipment
   b. 800 Mcf sold to R Company at $12/Mcf

   Assume a 1/7 RI and a production tax of 5%, and assume that the lease agreement has a free fuel clause, but production taxes still have to be paid according to state law. The selling price of gas is currently $12/Mcf.

   REQUIRED: Give the entries necessary to record the gas sold or used, assuming the operator distributes taxes and royalty.

9. Big John Oil Company purchased 100 barrels of oil from JD Operator. The gross value of the oil was $5,000. The severance tax rate was 4%. Give the entry to record revenue for JD, assuming Big John disbursed the royalty and remitted all taxes, and assuming a division order as follows:

Property No. 35	Interest
JD Company	0.875
Diane Royalty Owner	0.125

10. Jayhawk Oil Company has a working interest in a property. In addition to the 1/5 royalty, Jayhawk agreed to pay the royalty owner a minimum royalty of $400/month. Gas production on the lease began in the third month after the lease contract was signed. Total sales revenue during the third month and the next two months was $6,000 each month. The severance tax rate was 10%. Give the revenue and minimum royalty entries for the first five months, assuming Jayhawk takes the responsibility of distributing taxes and royalty, and also assuming the following:

    a. The minimum royalty payments were not recoverable.
    b. The minimum royalty payments were recoverable.

11. Complete the run ticket on the following page and give the entry to record the sale of the oil at $75/bbl assuming a severance tax rate of 5% and a 1/5 RI. Use the tables given in the chapter. The tank number is 1, the observed temperature is 73°F, and the observed API gravity is 33°. The first tank measurement at the beginning of the run is 14'4" at a temperature of 77°F, and the second tank measurement at the end of the run is 1'6" at 78°F. The BS&W content is 0.002.

12. Information from a run ticket shows that 1,000 net barrels of oil with an API gravity of 36° were sold. The selling price is based on a contract price of $66/bbl, adjusted downward 4¢ for each degree of gravity less than 40°. Compute the selling price for the 1,000 barrels.

ACCT NO.	COMPANY **RUN TICKET NO.**				
DIST NO.	DATE				
OPERATOR					
LEASE NAME				LEASE NO.	
WELL NO.		FT.	INCHES	TEMP	BBLS
	1ST MEAS.				
TANK NO.	2ND MEAS.				
	GRAVITY				
TANK SIZE	OBSERVED	TEMP.	TRUE	PRICE	
BS &W	OFF←SEAL NO.'S→ON			AMOUNT	

THIEFING	BEFORE	AFTER	CALCULATIONS
FT – IN			
BS & WATER			
REMARKS			

1ST MEAS
DATE	GAUGER
TIME A.M. P.M.	WELL OWNER'S WITNESS

2ND MEAS
DATE	GAUGER
TIME A.M. P.M.	WELL OWNER'S WITNESS

This ticket covers all claims for allowance. The oil represented by this ticket was received and run as property of

Source: The run tickets presented in this chapter and in the homework answers are based on a pipeline run ticket prepared and printed by Kraftbilt, Tulsa, Oklahoma, 1-800-331-7290.

13. Determine the barrels of production allocated to each well given the following data:

Well	24-hr Test, bbl	Days Produced	Total Production Sold
1	40	28	3,300 barrels—measured at point of delivery (production from three wells commingled)
2	50	30	
3	35	24	

    Round the ratio to three decimal places and round barrels to the nearest whole barrel.

14. Cameron Oil Company produced a total of 2,000 barrels of oil in June 2012. The expected selling price was $60/bbl. The purchaser pays the severance taxes and the royalty interest owner and remits the remainder to Cameron Oil. The royalty interest is 1/5, and the severance tax rate is 10%.

    a. Prepare the entry for Cameron Oil Company and the RI owner, assuming the oil was sold in June 2012 at the expected selling price.

    b. Prepare entries for Cameron Oil Company, assuming that revenue is recognized as produced based on the expected selling price and that the oil produced in June was sold in July 2012 for $66/bbl.

15. Foundation Oil Company owns a working interest in the Carpenter Lease in Texas. The lease is burdened with a 3/16 royalty interest. During February, a total of 3,000 barrels of oil is sold at $66/bbl to a refinery owned 100% by Foundation Oil. Assume the severance tax rate is 5% in Texas.

    REQUIRED: Prepare entries for the producer and the refinery purchaser.

16. Cameron Oil Company operates Leases X and Y. Cameron Oil transfers 50 barrels of oil from Lease X to Lease Y to be used as fuel on Lease Y. The current spot oil price is $66/bbl and the severance tax rate is 5%. Cameron owns 100% working interests in Lease X and Lease Y. Lease X has a 1/8 royalty interest, and Lease Y a 3/16 royalty interest.

    REQUIRED: repare an entry for the transfer of the oil.

17. Stephens Oil Company produces a total of 2,000 barrels of oil in June that is sold in July. The posted field price and the actual selling price is $66/bbl. The severance tax rate is 5%. The purchaser of the oil will pay the severance tax to the state and also will pay the royalty interest owner. The royalty interest is 1/8.

    REQUIRED: Prepare entries assuming Stephens Oil recognizes revenue (a) at time of sale, and (b) at time of production based on the posted field price.

18. Gusher Oil Company has the following transactions in 2011:

   a. Minimum royalty payments of $200/month are paid during the months of January through March. The minimum royalty payments are recoverable from future royalty payments.

   b. Production was sold in April 2011, and the royalty payable in April is $300.

   c. The royalty payment in May is $500.

   d. The well quit producing in June, and the lease was abandoned.

   REQUIRED:

   1. Prepare entries for the above transactions.
   2. Prepare entries assuming the royalty payments are not recoverable.

19. Ms. Kyle owns some mineral rights in Texas that she leases to Seagull Oil Company, reserving a 1/8 royalty interest. During 2011, Seagull Oil made the following assignments:

   a. To Mr. Hall, an ORI of 1/6

   b. To Mr. Clark, a production payment interest of 10,000 barrels of oil to be paid out of 1/5 of the working interest's share of production

   c. To Ms. Wilson, a joint working interest of 40% after giving consideration to the above assignments

   REQUIRED:

   a. Prepare the decimals to be used in the division order.
   b. Assuming production of 12,000 (gross) barrels of oil, calculate the number of barrels each owner would receive.

20. Joyner Oil Company sells 10,000 (gross) Mcf of gas at $9/Mcf. The lease provides for a 1/6 RI, and the working interest owner has distributed an ORI of 1/10. The severance tax rate is 7%. Record the entries for the royalty interest owner, ORI owner, and working interest owner, assuming:

   a. The purchaser assumes responsibility for distributing taxes and royalty income.
   b. Joyner Oil assumes the responsibility for distributing taxes and royalty income.

21. Jayhawk Oil Company's production for Lease A and Lease B is gathered into a common system and sold. Total sales for the month are 6,562 barrels. Assume the following data for Lease A and Lease B:

	Well	24-hour Test, bbl	Days Produced
Lease A:	1	40	24
	2	30	30
	3	50	28
Lease B:	1	50	30
	2	80	24

Measured production is 3,300 barrels from Lease A and 3,500 barrels from Lease B.

REQUIRED:

a. Allocate production to each lease.

b. Allocate the amounts per lease determined in part a to the wells. Round the ratios to four decimal places.

22. Kyle Company's production from each well on Lease C and Lease D is estimated based on a 24-hour test. Oil produced from each well on each lease is commingled and measured before leaving each lease. The oil produced from each lease is then commingled and delivered to a central tank battery. Assume the following data for Lease C:

Well	24-hr Test, bbl	Days Produced
1	30	28
2	40	30

Assume the following data for Lease D:

Well	24-hr Test, bbl	Days Produced
1	20	27
2	30	26
3	35	30

Measured production is 2,000 barrels from Lease C and 2,280 barrels from Lease D. After treatment at the central tank battery, 4,100 barrels are sold.

REQUIRED: Allocate the 4,100 barrels sold to each lease and then to each well in a two-stage allocation. Round the ratios to three decimal places.

23. Cameron Company (70% WI) and Garcia Company (30% WI) own a joint working interest in the Dowling Field. There is a 1/8 royalty owner. The 1/8 royalty is shared proportionally by Cameron and Garcia. Cameron and Garcia agree that Cameron's purchaser will take March's gas production and Garcia's purchaser will take April's production. Gas allocations will be equalized in May. Ignore severance taxes.

Gross production and gas prices were as follows:

	Production	Price
March	300,000 Mcf	$10/Mcf
April	200,000 Mcf	10/Mcf
May	220,000 Mcf	9/Mcf

Each working interest owner receives payment only for gas delivered to his purchaser.

REQUIRED:

a. Prepare a gas balance report.

b. Prepare entries for the three-month period for both parties assuming both companies use the entitlement method for both revenue and royalty.

24. Young Company has a 100% WI in some property in Texas. The property is burdened with a 1/8 royalty. Young produces and sells a total of 130,000 Mcf of gas from the property during July. Of the 130,000 Mcf, 50,000 Mcf of gas is sold to a pipeline for $10.00/Mcf. Young sells the remaining 80,000 Mcf of gas to the local gas company for $10.00/Mcf. The 80,000 Mcf of gas, which must be transported to the local gas company, will be transported by Isaac Pipeline Company. Transportation charges are $0.26 per Mcf.

REQUIRED: Ignore severance taxes. Prepare entries to record the sales by Young Company.

25. Two leases in far West Texas are being combined to form the West End Unit.

    Lease 1

WI: 100%	Quinn Company
RI: 1/7	Sam Sofer
1/7	John Hill
Acreage contributed:	300 acres
Estimated net recoverable bbl:	150,000 bbl

    Lease 2

WI: 75%	Core Company
25%	Format Company
RI: 1/8	E.B. Lomax
Acreage contributed:	200 acres
Estimated net recoverable bbl:	250,000 bbl

REQUIRED:

a. Determine the participation factors for each party, assuming the participation factors are based on the acreage contributed.

b. Determine the participation factors for each party, assuming the participation factors are based on the net recoverable barrels contributed.

c. Determine the participation factors for each party, assuming the participation factors are based on the acreage contributed being weighted 40% and the net recoverable barrels being weighted 60%.

26. Bruno Oil Company has production on a lease in Louisiana with the following ownership interest:

RI —1/5 RI
ORI— 1/16 of 4/5 of gross production
Joint WI— Lomax Company (40%) and Bruno Company (60%)

During April, 10,000 (gross) barrels of oil (after correction for temperature, gravity, and BS&W) were produced and sold. Assume the price for oil is $60.00/bbl, and the severance tax rate in Louisiana is 5%.

REQUIRED:

a. Prepare the journal entry for Bruno Oil to record the sale of the oil, given that the purchaser assumes the responsibility of distributing severance taxes and royalty income.

b. Prepare the journal entry for Bruno Oil to record the sale of the oil, given that Bruno Company assumes the responsibility and receives 100% of the proceeds.

c. Prepare the journal entry for the royalty interest owner to record the sale of the oil.

d. Prepare the journal entry for the overriding royalty interest owner to record the sale of the oil.

e. Prepare the journal entry for Lomax to record the sale of the oil.

27. Belmont Company, a successful efforts company located in California, sold 2,500 (gross) Mcf of gas with a heat content of 1.030 MMBtu/Mcf. The selling price of the gas was $8.00/MMBtu.

REQUIRED:

a. Determine the MMBtus for the gas sold.

b. Determine the total sales price.

c. Determine the unit sales price per Mcf.

28. During September 2010, Gamma Oil Company sold 2,000 Mcf of gas at 14.65 psia with a heat content of 1.030 MMBtu/Mcf at 14.73 psia. The selling price of the gas was $8.80/MMBtu.

REQUIRED:

a. Convert Mcf to a standard pressure of 14.73 psia and determine the MMBtus for the gas sold.

b. Convert MMBtu/Mcf to a standard pressure of 14.65 psia and determine the MMBtus for the gas sold.

c. Determine the sales price of the gas sold.

29. Bruno Field is jointly owned by Ramsey Company (70% WI), which acts as field operator, and Garza Company (30% WI). There is a 1/6 royalty. The 1/6 royalty is shared proportionally by Ramsey and Garza. The two working interest owners have agreed that Ramsey's purchaser will take gas produced in July, and Garza's purchaser will take gas produced in August. Gas allocations will be equalized in September. Assume each working interest owner receives payment only for gas delivered to his purchaser(s). Ignore severance taxes.

Gross production and gas prices were as follows:

	Production	Price
July	100,000 Mcf	$8.00/Mcf
August	120,000 Mcf	8.00/Mcf
September	190,000 Mcf	8.00/Mcf

REQUIRED:

a. Prepare the Gas Balance Report for Ramsey Company to summarize the production deliveries and equalization of gas for July through September.

b. Prepare the journal entries for each company during the three-month period, assuming that both companies use the sales method for both revenue and royalty.

c. Prepare the journal entries for each company during the three-month period, assuming that both companies use the entitlement method for both revenue and royalty.

30. Foundation Field, located in East Texas, is jointly owned by Bryant Company (60% WI) and Joyner Company (40% WI). Bryant, which is the operator, estimates that gross gas production during July will be 40,000 Mcf. Bryant Company makes confirmed nominations of 30,000 Mcf, and Joyner Company makes confirmed nominations of 10,000 Mcf to a pipeline company. Actual July production totals 50,000 Mcf. Assume that the appropriate price is $10.80/Mcf.

REQUIRED: Determine the number of Mcfs of actual production to be allocated to each party, along with each producer's resulting overdeliveries and underdeliveries to the pipeline, using the following:

a. The confirmed nominations method
b. The entitlement method

REFERENCES

1. Council of Petroleum Accountants Societies, Inc. 1993. *Accounting Guideline (AG)-15.* "Gas Accounting Manual." Denver, CO: COPAS.

2. Council of Petroleum Accountants Societies, Inc. 2001. *Accounting Guideline (AG)-22.* "Producer Gas Imbalances." Denver, CO: COPAS.

3. Council of Petroleum Accountants Societies, Inc. 1988. *Accounting Guideline (AG)-6*, "Oil Accounting Manual." Denver, CO: COPAS.

4. COPAS. 1993. *Accounting Guideline (AG)-15.*

5. COPAS. 1988. *Accounting Guideline (AG)-6.*

11

BASIC OIL AND GAS TAX ACCOUNTING

Income tax laws often differ substantially from financial accounting standards and procedures. The objective of financial accounting is to provide information that is relevant to those making investment decisions. In contrast, income tax accounting focuses on determining the amount of income taxes that companies must pay according to specific statutes and regulations. Income tax rules related to the capitalization (deferral) versus deduction (expensing) of an expenditure can differ dramatically from the financial accounting treatment of the expenditure. Similarly, the determination of gross income and taxable income for tax purposes can differ dramatically from the determination of revenue and net income for financial accounting purposes.

In this book, the focus has been on financial accounting, specifically the successful efforts and full cost methods. The statutes and regulations governing income taxation contain no special provisions for companies using the successful efforts versus full cost method for financial accounting. Accordingly, the decision to capitalize or expense a particular cost for financial accounting (successful efforts versus full cost) has no impact on whether the same item of cost is capitalized (deferred) or expensed (deducted) in determining taxable income for income tax purposes.

A company that operates in the United States is responsible for determining its taxable income in accordance with U.S. tax laws and paying any taxes due to the U.S. government. A company operating in an international location must comply with the tax laws of the local jurisdiction as well as with the tax laws of its home country. For example, a U.S. company operating in Thailand must pay any income taxes due to the Thai government in accordance with the income tax laws in Thailand. In addition, the company must determine the amount of U.S. income tax by applying U.S. tax statutes and regulations to those same operations. Consistently, a foreign company operating in the United States must pay U.S. income taxes on its profits generated in the United States. Additionally, it must determine the tax consequences of its U.S. operations under the tax laws prevailing in its home country. Tax treaties frequently exist between governments that provide for tax credits designed to avoid double taxation on the same profits. Tax treaties are typically quite complicated and beyond the scope of topics covered in this chapter.

To further complicate matters, in the United States, some provisions differ depending on whether the operation in question is located domestically or internationally. For example, U.S. income tax statutes prescribe a calculation for use in determining depreciation deductions on equipment used in U.S. domestic petroleum operations and a different calculation for computing depreciation on equipment used in international petroleum operations. Thus, in preparing its income tax returns, a U.S. company operating both domestically and internationally must comply with foreign income tax laws, U.S. income tax laws on domestic operations, and U.S. income tax laws on foreign operations.

This chapter introduces basic oil and gas tax accounting, which is applicable to all entities, large or small, integrated or independent, full cost or successful efforts. Exceptions relating to independent producers, such as percentage depletion, and differences applying only to integrated companies, such as the 30% holdback of IDC, are noted and examined. Transactions that are discussed include those involving nondrilling costs, acquisition costs, drilling costs, completion costs, operational costs, and the disposition of capitalized costs. Discussion of the numerous complex issues, e.g., foreign tax treaties and alternative minimum tax, are not included.

This chapter is intended to provide only a basic overview of taxation for oil and gas exploration and production activities. It is not intended to provide the reader with an authoritative source of reference relating to the various tax issues that are discussed.

Lessee's Transactions

The following sections focus on income taxation issues relating to lessee's transactions. The lessor's transactions are discussed later in the chapter.

Nondrilling costs

The nondrilling costs discussed in this section include: geological and geophysical (G&G) costs, delay rentals, ad valorem taxes, legal costs, dry-hole contributions, bottom-hole contributions, and shut-in royalties.

Generally, the first costs incurred in an oil and gas operation are G&G costs. G&G costs are considered inherently capital in nature and must be added to the tax basis of property acquired or retained. As a first step, a reconnaissance G&G survey is often made on a broad area called a **project area**. The survey is done to locate portions of the project area having the greatest potential for containing oil and gas. *Rev. Rul. 77-188* indicates the following:

Each separable noncontiguous portion of the original project area in which such a specific geological feature is identified is a separate 'area of interest.' The original project area is subdivided into as many small projects as there are areas of interest located and identified within the original project area.

Areas or tracts of land are considered to be contiguous if they share a common border. (Tracts that touch at a corner are adjacent, not contiguous.)

All costs of the project area are initially allocated to a single project. If interest areas warranting further evaluation are identified, the initial reconnaissance costs of the project area are allocated equally to the areas of interest regardless of their relative sizes. If only one area of interest is identified, all reconnaissance costs are allocated to that area of interest. If no areas of interest are identified, the reconnaissance costs are deductible.

After areas of interest are identified, any costs incurred to further evaluate an area are capitalized to that area of interest. If further evaluation results in property interests being acquired within an area of interest, the allocated reconnaissance costs and the cost of detail work performed within the area of interest are allocated to any property taken on a comparative acreage basis. There is no deduction if only a part of an area of interest is acquired. If the detailed survey does not result in property acquisition, the accumulated costs may be written off in the year in which the company decides that it is not interested in pursuing property acquisition in that area.

EXAMPLE

G&G Costs

Tyler Company incurs G&G costs of $180,000 for a reconnaissance survey of project area 1102. The survey indicates three areas of interest. Detailed surveys are made on the areas of interest at the following costs:

	Number of Acres	Cost
Interest Area A	640 acres	$260,000
Interest Area B	335 acres	200,000
Interest Area C	640 acres	350,000

After evaluating the surveys, Tyler Company acquired the following acreage:

Interest Area A	One lease of 320 acres
Interest Area B	One lease of 160 acres
Interest Area C	No leases acquired

Allocation of G&G Cost in Determination of Tax Basis per Acre

	Area A	Area B	Area C
Allocation of reconnaissance costs ($180,000/3)	$ 60,000	$ 60,000	$ 60,000
Detailed survey costs	260,000	200,000	350,000
Tax basis of area of interest	$320,000	$260,000	$410,000
Acreage acquired (acres)	320	160	0
Tax loss—immediately deductible			$410,000
Tax basis of property	$320,000	$260,000	0
Tax basis per acre	$ 1,000	$ 1,625	0

Instead, assume that the 320 acres acquired in Interest Area A included the following leases:

	Acres	WI	Net Acres
Lease 1:	120 acres	100% WI	120
Lease 2:	120 acres	25% WI	30
Lease 3:	80 acres	25% WI	20
Total net acreage			170

Tax Basis per Acre for the Leases Acquired in Area A

$320,000/170 acres = $1,882.35

Allocation of Costs of Interest Area A

Lease 1:	120 acres × $1,882.35	=	$225,882
Lease 2:	30 acres × $1,882.35	=	56,471
Lease 3:	20 acres × $1,882.35	=	37,647
Total			$320,000

Shooting rights, which are the rights to access a property, should be accounted for in the same way as G&G costs. Consequently, the cost of shooting rights should ultimately be capitalized if a lease is obtained. If no lease is obtained, these rights become deductible as indicated above.

A typical lease contract requires the lessee to commence drilling operations within one year. If the lessee does not commence such operations within one year, a delay rental payment must be made. The tax law allows the lessee to either capitalize delay rentals or deduct them in the period incurred. The election must be made on a year-by-year, lease-by-lease basis. Except for companies with significant net operating losses, delay rental payments are almost universally expensed as incurred. For the lessor, any delay rentals payments received must be treated as ordinary income not subject to depletion.

Ad valorem taxes paid by the lessee on his working interest in an unproved property are ordinary and necessary business expenses, and are deductible in the period incurred. Ad valorem taxes paid by the lessee for the lessor can generally be recovered from future royalty payments to the lessor, and are neither an expense nor revenue to the lessee. If the ad valorem tax amount is recoverable, a receivable is established for the payment.

Legal costs incurred to defend or examine the title to a property are considered capital expenditures and should be capitalized as part of the basis of the underlying property. Maintenance costs of lease and land records, however, appear to be ordinary and necessary business expenses.

Bottom-hole contributions for successful wells are considered to be acquisition costs or retention costs and are capitalized as incurred. These costs are deemed to be payment for information relating to the acquisition of adjoining acreage or retention of an interest. As such, they are capitalized and added to the basis of the underlying property. Bottom-hole contributions for dry holes and dry-hole contributions also should be capitalized.

Shut-in royalty payments may have three basic characteristics:

- The payments are not recoverable from future production.
- Failure to make payments terminates the lease.
- The payments are recoverable from future production.

When shut-in payments have one of the first two characteristics above, the payments are considered to be delay rentals and are accounted for as described previously. Those shut-in payments recoverable from future production are treated as a receivable until recovered from production.

Acquisition costs

Acquisition costs paid by the lessee should be capitalized and added to the basis of the property acquired. The major acquisition cost is the lease bonus paid by the lessee to the lessor. Other acquisition costs that should be capitalized include broker fees, legal fees, filing fees, and the cost of title examinations. Acquisition costs are recovered through cost depletion or percentage depletion. Certain other costs, although not typical acquisition costs, are treated as acquisition costs and are capitalized and depleted along with normal acquisition costs. These costs include capitalized G&G, shooting rights, test-well contributions, and capitalized delay rental payments, if any. The lessor treats any bonus received as ordinary income.

EXAMPLE

Acquisition Costs

Tyler Company incurred $10,000 of G&G reconnaissance costs in North Texas that resulted in one area of interest being identified. Detailed survey costs totaling $15,000 were subsequently incurred in the area, resulting in two leases being acquired. Tyler paid a lease bonus of $50/acre to Sam Jones in return for a 150 acre lease, and $45/acre to Manuel Ramirez for a 250 acre lease. Legal fees, landmen commissions, filing fees, and title examination fees were incurred in the amount of $2,000 on the Jones lease and $1,500 on the Ramirez lease.

The Tax Basis of Each Lease

	Jones Lease	Ramirez Lease
Capitalized G&G = $10,000 + $15,000		
Acre-by-acre allocation:		
$25,000/400 acres = $62.50/acre		
$62.50 × 150	$ 9,375	
$62.50 × 250		$15,625
Lease bonus:		
$50 × 150	7,500	
$45 × 250		11,250
Other fees	2,000	1,500
Tax basis	$18,875	$28,375

In some instances, a producing property is subleased (i.e., the current working interest owner conveys the working interest in the property to another party and retains a nonoperating interest). When subleasing occurs, the **grantee** or **sublessee** (i.e., the new working interest owner) of the producing property must apportion the consideration paid between the leasehold and equipment. The equipment received should be allocated its fair market value, with the remainder being allocated to the leasehold.

Producing property that is subleased generally provides for a retained overriding royalty interest by the **grantor** or **lessee** (i.e., the original working interest owner). Therefore, when a producing property is subleased to another party by the lessee, the consideration received by the grantor (lessee) is considered as payment for equipment. When the cash consideration received is greater than the depreciable basis of the equipment, the excess is considered as a lease bonus received, i.e., it is treated as ordinary income. When the cash consideration received is less than the basis of the equipment, the excess equipment basis should be transferred to the depletable basis of the interest retained and depleted along with that interest.

EXAMPLE

Sublease

Tyler Company owns a producing lease with leasehold costs of $5,000, accumulated depletion of $1,000, well equipment of $300,000, and accumulated depreciation of $75,000. The lease is subleased to Sam Jones for $400,000, and Tyler retains a 3/8 ORI in the property. At the date of sublease, the equipment had a fair market value of $325,000.

Tax Treatment of the Sublease by Tyler Company

Tax basis of property:
Cost	$ 5,000
Less: Accumulated depletion	(1,000)
Basis	$ 4,000

Tax basis of equipment:
Cost	$300,000
Less: Accumulated depreciation	(75,000)
Basis	$225,000

Since the cash received ($400,000) is greater than the depreciable basis of the equipment ($225,000), the excess ($175,000) is treated as ordinary income as shown below.

Payment for equipment:
Sales price	$400,000
Less: Basis	(225,000)
Taxable income	$175,000

Tyler's basis of the proved property (versus equipment) is transferred to the basis of the retained ORI.

Basis of retained ORI	$ 4,000

Tax Treatment of the Sublease by Sam Jones

The fair market value (FMV) of the equipment, $325,000, is allocated to equipment and the remainder, $75,000, to the leasehold.

Tax basis of equipment (FMV)	$325,000
Tax basis of leasehold	75,000
Total payment	$400,000

Drilling operations

As will be seen, the tax treatment of costs related to drilling operations depends upon whether the producer is an independent producer or an integrated producer. An **integrated producer** is an entity that has refining or retailing activities in addition to producing activities. An **independent producer** may also have some refining or retailing activities. However, in order to be considered as an independent producer for tax purposes, a company's refining or retailing activities must be below a specified maximum amount. A refiner is a company (i.e., a taxpayer) whose refinery runs exceed 75,000 bbl/day (based on average daily runs rather than actual) for that tax year. Average daily refinery runs are computed by dividing total refinery runs for the tax year by the total number of days in the tax year. (This maximum was raised from 50,000 bbl/day to 75,000 bbl/day in 2005.) A retailer is defined as being a taxpayer that sells more than $5,000,000 annually of oil, gas, or any resulting product through retail outlets connected to or controlled by the taxpayer. For more information, see IRS Code Sec. 613A(d).

During drilling operations, the most important consideration is the proper separation of drilling costs between (1) intangible drilling and development costs (IDC), and (2) lease and well equipment. This importance is due to the ability, if the proper election is made, to expense as incurred the majority of IDC for an integrated producer or 100% of IDC for an independent producer. Lease and well equipment, in contrast, must be capitalized and recovered through depreciation. Two separate elections may be involved in expensing or capitalizing IDC. If an election is made in the first tax return filed by the taxpayer to expense IDC, then all IDC incurred by the taxpayer may be expensed in the year incurred, except for integrated producers, who must capitalize 30% of IDC incurred on productive wells. Even if a taxpayer fails to elect to expense IDC, a second election may be made to currently expense IDC of dry holes.

The 30% capitalized by integrated producers must be amortized over 60 months, beginning with the month in which the costs are paid or incurred. The total amount amortized for the year is deducted currently by an integrated producer in addition to the remaining 70% of the related IDC expenditures. Note that 30% of the IDC for a well in progress must also be capitalized and amortized. If the well is determined to be dry, any remaining unamortized IDC would be expensed in the year in which the well is plugged and abandoned.

Equipment costs, which must be capitalized, generally include all tangible property costs incurred both before and after the Christmas tree, plus intangible costs incurred after the Christmas tree. Intangible costs incurred prior to and including the installation of the Christmas tree are IDC amounts.

IDC is defined as any expenditure that in itself does not have a salvage value and is "incident to and necessary for the drilling of wells and the preparation of wells for the production of oil and gas" [*U.S. Treas. Reg. Section 1.612-4(a)*]. Costs incurred relating to the following activities are considered to be IDC:

a. Agreements and negotiations in obtaining an operator for the well
b. Agreements and negotiations with drilling contractors for bids on drilling
c. Survey and seismic work for locating a wellsite
d. Constructing a road to the drilling location

e. Dirt work on location—cellar, pits, and drilling pack
f. Transporting the rig to location and rig-up costs
g. Water, fuel, and other items necessary for drilling wells
h. Setting deadmen (anchors in the ground used to stabilize the drilling rig)
i. Technical services rendered during the drilling activities by engineers, geologists, and fluid technicians
j. Logging and drillstem test services
k. Swabbing, fracturing, and acidizing
l. Rental equipment for oil storage during testing
m. Removing rig from location, trucking, bulldozers, and labor
n. Dirt work for cleanup of drillsite
o. Cementing and installing surface casing
p. Transportation of casing and tubing from supply point
q. Perforating casing and electrical logging
r. Saltwater, freshwater, and gas injection wells drilled solely for pressurization or flooding of producing zone (intangible costs only)
s. Water supply wells if the water is to be used for drilling or secondary recovery (intangible costs only)

As shown by this list, IDC is incurred for wages, fuel, repairs, hauling, supplies, etc., in preparing to drill wells (clearing ground, making roads, digging mud pits, etc.). IDC is also incurred in drilling, shooting, and cleaning wells, in constructing derricks, and in completing the well by installing casing and the Christmas tree.

Equipment costs

Lease and well equipment installed or constructed in connection with the drilling, completion, and production of a well must be capitalized. These costs normally are recovered through depreciation. Some typical lease and well equipment costs follow:

a. Surface casing—even though permanently cemented and unsalvageable
b. Well casing
c. Tubing
d. Transportation of casing and tubing from manufacturer to supply point
e. Stabilizers, guide shoes, centralizers, and other downhole equipment
f. Wellhead—Christmas tree
g. Saltwater disposal equipment and necessary pipelines, including cost of drilling disposal wells
h. Pump jack, treaters, separators
i. Recycling equipment, including necessary flow lines
j. Dirt-moving necessary for the tank battery and operation roads

k. Digging, refilling, and backhoe work for installing flow lines from the well to the tank battery

l. Installation and labor costs for the tank battery, flow lines, pump jacks, separators, and similar items

m. Construction of a turnaround pad at the tank battery, with additional overflow pits

The following example illustrates the tax treatment of IDC and equipment costs:

EXAMPLE

IDC versus Equipment

Tyler Company incurred the following costs in connection with the drilling and completion of a well during the first three months of 2010. The company had elected to expense IDC as incurred. The IDC costs are indicated by an X in the column headed IDC, and the equipment costs are indicated by an X in the column headed "Eq." Tyler's IDC deduction and the basis of Tyler's equipment are shown below, assuming the following:

a. Tyler is an independent producer.
b. Tyler is an integrated company.

IDC	Eq.		
X		Legal and negotiating costs for hiring an operator and drilling contractor	$ 10,000
X		G&G costs in selection of the wellsite	15,000
X		Construction of roads to the drilling site	12,000
X		Water well and fuel costs during drilling	9,000
X		Drilling contractor costs on a footage and day-rate basis	316,000
X		Costs incurred in connection with technical services (engineers, geologists, logging, swabbing, testing)	75,000
X		Dirt work costs for cleaning up the drillsite	2,000
X		Cement for casing	2,000
	X	Well casing	20,000
X		Transportation of the well casing from the supply point	500
	X	Equipment necessary for completing the well (including wellhead equipment, treaters, separators, flow lines) . . .	200,000
X		Labor costs for installing the casing and wellhead equipment	4,000
	X	Labor to install the treaters, separators, and flow line . . .	12,000

a. **If Tyler is an independent producer**

The deduction for IDC is equal to the sum of all IDC costs (as itemized above).

IDC deduction = $10,000 + $15,000 + $12,000 + $9,000 + $316,000 + $75,000 + $2,000 + $2,000 + $500 + $4,000 = $445,500

Basis of lease and well equipment = ($20,000 + $200,000 + $12,000) = $232,000

b. If Tyler Company is an integrated producer

IDC deduction = 70% of current IDC expenditures (70% x $445,500) = $311,850

Capitalized IDC of (30% x $445,500) = $133,650 deductible over the next 60 months

Basis of lease and well equipment = ($20,000 + $200,000 + $12,000) = $232,000

The capitalized IDC is amortized on a straight-line basis over 60 months at a rate of $2,227.50 per month ($133,650/60), beginning with the month in which the IDC was incurred or paid. In the event that the well is determined to be dry, any remaining unamortized IDC would be expensed immediately.

The following is a more detailed example of the required treatment for IDC incurred by an integrated producer.

EXAMPLE

Capitalized IDC of Integrated Producer

Tyler Company, an integrated producer, incurs the following IDC costs for the years indicated. All IDC costs marked with an asterisk (*) relate to dry-hole IDC. Assume no IDC was incurred prior to 2011.

Date Incurred	Amount	Date Incurred	Amount	Date Incurred	Amount
5/11/11	$100,000*	3/14/12	$300,000	6/12/13	$180,000*
9/11/11	200,000*	7/22/12	360,000	10/14/13	400,000

Amount of IDC Deductible in Each of the Following Years

2011

IDC incurred in 2011:
$100,000 × 100% = $100,000
200,000 × 100% = 200,000
$300,000

2012

IDC incurred in 2012:	$300,000 × 70% =	$210,000
	300,000 × 30% × 10/60 =	15,000
	360,000 × 70% =	252,000
	360,000 × 30% × 6/60 =	10,800
IDC incurred in 2011:	100% expensed in 2011 =	0
		$487,800

2013

IDC incurred in 2013:	$180,000 × 100% =	$180,000
	400,000 × 70% =	280,000
	400,000 × 30% × 3/60 =	6,000
IDC incurred in 2012:	300,000 × 30% × 12/60 =	18,000
	360,000 × 30% × 12/60 =	21,600
IDC incurred in 2011:	100% expensed in 2011 =	0
		$505,600

Assume 2013 was the last year in which IDC was incurred. The amount of IDC deductible in 2018 would be computed as follows:

2018

All IDC incurred prior to 2013 would be fully expensed by 2018.

IDC incurred in 2013: 100% of $180,000 expensed in 2013

$400,000 × 70% expensed in 2013

$400,000 × 30% × 9/60 = $18,000 expensed in 2018

The number of months of amortization per year for the capitalized IDC relating to the $400,000 can be diagramed as follows:

Year:	2013	2014	2015	2016	2017	2018	Total months amortized
# of Months:	3	12	12	12	12	9	60

Production operations

Revenue arising from the production and sale of oil and gas products is ordinary income. The portion of revenue from production paid to the royalty interest owner is sometimes referred to as a royalty expense. This payment to the royalty interest owner is actually a division of revenue and is not included in the gross income of the working interest owner. The royalty is taxable income to the royalty owner.

Ordinary and necessary business expenses incurred in lifting the oil and gas and treating it for sale are deductible as necessary and reasonable operating expenditures. Examples of these costs are pumpers' and gaugers' salaries, the cost of travel to and from the lease, repairs, minor workovers, fuel costs, maintenance costs, depreciation, and depletion.

Depreciation and depletion, which are two important expenses connected with lifting and treating the oil and gas, are discussed in the following sections.

Depletion. Leasehold costs are recovered through cost depletion or possibly through percentage depletion. (Percentage depletion, which is available only to independent producers and royalty owners, and only if larger than cost depletion, is discussed later in this chapter.) Leasehold costs include such costs as lease bonuses, title examinations, and legal fees in drafting contracts. There are other capitalized costs that are treated as leasehold costs and depleted. These costs include G&G costs, test-well contributions, elected capitalized IDC (versus the 30% IDC required to be capitalized on productive wells by integrated producers), and capitalized delay rentals, if any. Cost depletion is based on activities connected with a particular property. Generally, a property is defined as either a single mineral interest or multiple mineral interests that are obtained at the same time and under the same lease contract. Operating and nonoperating interests are classified as separate property units.

Cost depletion is computed as follows:

$$B \times \frac{Y}{Z + Y}$$

where:

B = basis at end of period (unrecovered costs at year-end)

Z = reserves at year-end

Y = barrels or Mcfs sold during the year

Reserves at year-end for the property unit are not limited to proved reserves. In a number of cases, the IRS has also required that taxpayers include probable and even prospective reserves, depending on the particular situation. Note that the cost depletion formula is very similar to the unit-of-production formula used for computing financial DD&A under successful efforts and full cost. However, the tax calculation requires the use of barrels or Mcfs *sold* rather than barrels or Mcfs *produced*.

Depreciation. Depreciable property placed in service after December 31, 1986 is recovered using the depreciation methods prescribed by the Tax Reform Act of 1986 and the 1993 Revenue Reconciliation Act. Depreciable property consists of tangible assets, including the intangible costs of installing such assets, if after the Christmas tree. For the assets commonly used in oil and gas exploration activities, the law currently provides for a 7-year life for tangible personal property and a 39-year life for nonresidential real property.

Lease and well equipment of a producer is typically seven-year property, which includes the following: lease equipment, well equipment, exploration equipment, offshore drilling equipment and platforms, facilities to store oil and gas, etc. Lease and well equipment is depreciated over a seven-year period. The calculation is not straight-line, but rather a 200% declining balance method referred to as **Modified Accelerated Cost Recovery System (MACRS).** (A 200% declining balance method is equivalent to 2 times the straight-line rate applied to the declining balance. The straight-line rate for seven-year property would be 1/7.) Under this method, the 200% declining balance rate is used early in the life of the asset until such time as the rate yields an amount that is lower than the straight-line rate. At that point it is appropriate to switch to the straight-line rate.

A half-year convention modified by a mid-quarter convention is specified. Under the half-year convention, assets are considered to have been placed in service at midyear regardless of when the particular assets were actually placed in service. However, if more than 40% of the assets were placed in service during the last quarter of the year, the mid-quarter convention applies. Under the mid-quarter convention, assets purchased during a quarter are considered to have been placed in service at mid-quarter, regardless of when the assets were actually placed in service during that quarter. As a result of the half-year convention, the recovery life in effect spans 8 years for lease and well equipment. Lease and well equipment of a producer placed in service in a foreign location is typically depreciated over 12 to 14 years.

The MACRS rates for U.S. domestic lease and well equipment are as follows:

Year	Rate
1	14.29%
2	24.49%
3	17.49%
4	12.49%
5	8.93%
6	8.92%
7	4.46%

EXAMPLE

Depreciation of Seven-Year Property

On May 2, 2012, Tyler Company placed in service lease and well equipment costing $20,000.

Year	Depreciation	Year	Depreciation
2012	$2,857 (1)	2016	$1,785 SL (5)
2013	4,898 (2)	2017	1,785 SL
2014	3,499 (3)	2018	1,785 SL
2015	2,499 (4)	2019	892 SL

Calculations:
(1) ($20,000 × 1/7 × 2) × ½ year *or* $20,000 × 0.1429* = $2,857
(2) ($20,000 − $2,857) × 1/7 × 2 *or* $20,000 × 0.2449 = $4,898
(3) ($20,000 − $2,857 − $4,898) × 1/7 × 2 *or* $20,000 × 0.1749* = $3,499
(4) ($20,000 − $2,857 − $4,898 − $3,499) × 1/7 × 2 *or* $20,000 × 0.1249* = $2,499
(5) $6,247/3.5 years (straight line) = $1,785

* Minor difference due to rounding.

The taxpayer may elect to depreciate a class of property using straight-line depreciation. This election is irrevocable and applies to all property in that class placed in service during that taxable year. The taxpayer may also elect to exclude a property from the specified IRS depreciation provisions and depreciate the property using the unit-of-production method if the taxpayer can determine the oil and gas reserves applicable to the property.

Nonresidential real property, i.e., commercial real estate property, includes those equipment items permanently attached to the land or having the characteristic of a building, such as a gas processing plant. Under the 1993 Revenue Reconciliation Act, nonresidential real property must be depreciated over 39 years using the mid-month convention. Under the mid-month convention, property placed in service (or disposed of) during any month is treated as being placed in service (or disposed of) on the mid-point of the month.

Salvage value is ignored for assets being depreciated using both double declining balance and straight-line depreciation. The following example illustrates depreciation for both 7-year and 39-year property and depletion.

EXAMPLE

Depreciation and Depletion

Tyler Company, an independent producer, owns the working interest in a lease in Nueces County, Texas. The following information concerning the lease is for the year ended December 31, 2012. Assume Tyler's cost depletion is larger than percentage depletion, and therefore Tyler's depletion expense is based on cost depletion. Also assume IDC was expensed in the year incurred. All reserve, production, and sales data apply only to Tyler Company.

Well equipment cost (end of second full year of life)..	$100,000
Leasehold cost and capitalized G&G.	18,000
Capitalized test-well contribution.	2,000
Building (18-year life, placed in service 1/1/11).	135,000
Accumulated depreciation, well equipment (1/1/12)..	14,286
Accumulated depreciation, building (1/1/12).	3,317
Accumulated depletion (1/1/12).	1,200
Percentage depletion for 2012 (assumed).	1,600
Reserves 12/31/12	36,000 bbl
Production for the year	5,000 bbl
Sales for the year.	4,000 bbl

Computation of Depreciation for 2012

Well equipment: Year 1: [($100,000 × 1/7) × 2] ½ yr. = $14,286
Year 2: [($100,000 – $14,286) × 1/7] × 2 = $24,490

Building: Year 1: $135,000/39 × 11.5/12 = $3,317 first year*
Year 2: $135,000/39 = $3,462 second year

* Since the building was placed in service in January 2011, and given that the mid-month convention applies, depreciation for only 11.5 months out of 12 months would be recognized in the first year.

Computation of Cost Depletion for 2012

$$(\$18{,}000 + \$2{,}000 - \$1{,}200) \times \frac{4{,}000}{36{,}000 + 4{,}000} = \$1{,}880$$

Losses from unproductive property

For the lessee, losses from unproductive property may be taken for tax purposes in the following situations:

- Abandonment of unproved property
- Abandonment of wells and equipment when the well or reservoir is depleted, but the lease is not abandoned
- Abandonment of leasehold, wells and equipment, and facilities when the reservoir is depleted, and the lease is abandoned
- Drilling exploratory or developmental dry holes

In general, losses may be deducted in the year in which the property is deemed worthless. Property is deemed worthless if the title is relinquished or if the property is abandoned. Abandonment or title relinquishment is considered to be a closed and completed transaction, thereby proving worthlessness.

Determination of worthlessness may also be made by examining identifiable events. The most common identifiable events that indicate worthlessness are the following:

- Depletion of oil and gas
- Drilling dry hole(s) on leased property
- Drilling dry hole(s) on adjacent property

Once a property is deemed worthless, the deductible amount is determined as follows:

- *Abandonment of unproved property.* The deductible loss amount is the acquisition cost of the property, including other capitalized costs such as G&G costs and test-well contributions.
- *Abandonment of wells and equipment when a well or reservoir is depleted but the lease is not abandoned.* The amount deductible is the undepreciated value of the equipment (i.e., book value), plus the unamortized portion of any capitalized IDC, less salvage value. However, if the company uses a system whereby the depreciable assets are combined into one main account, referred to as **mass asset accounting**, no deduction is usually allowed for the abandonment.
- *Abandonment of leasehold, wells and equipment, and facilities when the reservoir is depleted and the lease is abandoned.* The deductible amount is the total unrecovered cost of the equipment, leasehold, and IDC, less salvage value.
- *Drilling exploratory and developmental dry holes.* The deductible amount for an independent or integrated producer is the total of IDC (not yet deducted) and equipment, less salvage value.

The following two examples present abandonment transactions for an independent producer and an integrated producer.

EXAMPLE

Independent Producer Abandonment

Transactions for the Tyler Company, an independent producer, in 2013 follow:

a. Abandoned unproved Lease A—acquisition costs totaling $4,000.
b. Abandoned Wells 1 and 2 on the Jones lease. The Jones lease was not abandoned because Well 3 is producing in another formation.

Costs Incurred on the Jones Lease

	Jones Lease	Well 1	Well 2	Well 3
Acquisition costs	$ 5,000			
IDC (deducted prior to 2013 in year drilled)		$320,000	$400,000	$380,000
Equipment costs		160,000	180,000	200,000
Accumulated depletion	4,000			
Accumulated depreciation.		120,000	150,000	90,000

The salvageable equipment located on Well 1 and Well 2 was transferred to a warehouse owned by Tyler Company. The FMV of the transferred equipment was $50,000. Assume each well was depreciated individually.

c. Use the same facts as item b, except that Well 3 and the lease were also abandoned.
d. A dry hole costing $275,000 was drilled on unproved Lease B. The IDC cost was $240,000 and equipment was $35,000. (No equipment was salvaged.)

Tyler Company's tax losses for each of these situations are determined below.

Tax Losses

a. Lease A—abandonment loss $\underline{\$\ \ 4,000}$
    Remaining basis of Lease A $\underline{\$\ \ \ \ \ \ 0}$

b. Cost of equipment on Well 1 $160,000
    Less: Accumulated depreciation $\underline{(120,000)}$
    Equipment basis Well 1 $ 40,000

    Cost of equipment on Well 2 180,000
    Less: Accumulated depreciation $\underline{(150,000)}$
    Equipment basis Well 2 30,000

    Less: Salvage value of equipment $\underline{(50,000)}$
    Abandonment loss on wells $\underline{\$\ 20,000}$

c. Cost of equipment on Well 1 $160,000
    Less: Accumulated depreciation $\underline{(120,000)}$
    Equipment basis Well 1 $ 40,000

    Cost of equipment on Well 2 180,000
    Less: Accumulated depreciation $\underline{(150,000)}$
    Equipment basis Well 2 30,000

    Cost of equipment on Well 3 200,000
    Less: Accumulated depreciation $\underline{(90,000)}$
    Equipment basis Well 3 110,000

    Less: Salvage value of equipment (50,000)
    Abandonment loss on wells $\overline{130,000}$

    Leasehold 5,000
    Less: Accumulated depletion $\underline{(4,000)}$
    Abandonment loss on leasehold $\underline{1,000}$

    Total abandonment loss $\underline{\$131,000}$

d. IDC (deductible as drilled) $240,000
    Nonsalvageable equipment $\underline{35,000}$
    Total tax loss on dry hole $\underline{\$275,000}$
    (assuming the lease itself is retained)

EXAMPLE

Integrated Producer Abandonment

Transactions for the Tyler Company, an integrated producer, in 2013 are as follows:

a. Abandoned unproved Lease A—acquisition costs totaling $4,000.
b. Abandoned Wells 1 and 2 on the Jones lease. The Jones lease was not abandoned because Well 3 is producing in another formation.

Costs Incurred on the Jones Lease

	Jones Lease	Well 1	Well 2	Well 3
Acquisition costs	$5,000			
IDC (deducted in year drilled).		$320,000	$400,000	$380,000
Capitalized IDC (30%)		137,000	171,000	163,000
Equipment costs		160,000	180,000	200,000
Accumulated depletion	4,000			
Accumulated depreciation.		120,000	150,000	90,000
Accumulated IDC amortization (assumed)		137,000	57,000	41,000

The salvageable equipment located on Well 1 and Well 2 was transferred to a warehouse owned by Tyler Company. The fair market value of the transferred equipment was $50,000. Assume each well was depreciated individually.

c. Use the same facts as in item b, except that Well 3 and the lease were also abandoned.
d. A dry hole costing $275,000 was drilled on unproved Lease B. The IDC cost was $240,000, and equipment was $35,000. (No equipment was salvaged.)

Tyler Company's tax losses for each of these situations are determined below:

Tax Losses

a. Lease A—abandonment loss $ 4,000
   Remaining basis of Lease A $ 0

b.
Cost of equipment on Well 1	$ 160,000	
Less: Accumulated depreciation	(120,000)	
Equipment basis Well 1		$ 40,000
Capitalized IDC	137,000	
Less: Accumulated IDC	(137,000)	
Net IDC remaining		0
Cost of equipment on Well 2	180,000	
Less: Accumulated depreciation	(150,000)	
Equipment basis Well 2		30,000
Capitalized IDC	171,000	
Less: Accumulated IDC	(57,000)	
Net IDC remaining		114,000
Less: Salvage value of equipment		(50,000)
Net abandonment loss on equipment and IDC		$134,000

c.
Cost of equipment on Well 1	$ 160,000	
Less: Accumulated depreciation	(120,000)	
Equipment basis Well 1		$ 40,000
Capitalized IDC	137,000	
Less: Accumulated IDC	(137,000)	
Net IDC remaining		0
Cost of equipment on Well 2	180,000	
Less: Accumulated depreciation	(150,000)	
Equipment basis Well 2		30,000
Capitalized IDC	171,000	
Less: Accumulated IDC	(57,000)	
Net IDC remaining		114,000
Cost of equipment on Well 3	200,000	
Less: Accumulated depreciation	(90,000)	
Equipment basis Well 3		110,000
Capitalized IDC	163,000	
Less: Accumulated IDC	(41,000)	
Net IDC remaining		122,000
Less: Salvage value of equipment		(50,000)
Net abandonment loss on equipment and IDC		366,000
Leasehold	5,000	
Less: Accumulated depletion	(4,000)	
Abandonment loss on leasehold		1,000
Total abandonment loss		$367,000

d. As well is drilled:

IDC deduction	$240,000 × 70% =	$168,000
IDC capitalized	$240,000 × 30% =	$ 72,000
Equipment capitalized		$ 35,000

When well is determined to be dry*:

IDC (deduct remaining capitalized IDC)	$ 72,000	
Nonsalvageable equipment	35,000	
Total tax loss on dry hole		$107,000
(assuming the lease itself is retained)		

Total deductions relating to well
($168,000 + $107,000): $275,000

* Capitalized IDC is amortized over 60 months beginning with the month incurred. This calculation assumes the well was drilled and determined to be dry all within one month.

Note that for income tax purposes, the designation of a well as exploratory or development has no significance. The only considerations are whether the costs are IDC versus equipment, whether the well is dry or successful, and whether the producer is an independent or integrated producer.

The following comprehensive example illustrates tax accounting for both independent and integrated producers.

COMPREHENSIVE EXAMPLE

During 2011

Tyler Company, an integrated company, has decided to survey a project area in Deaf Smith County, Texas. A G&G team was hired to perform a reconnaissance survey of the eastern half of the county for $30,000. After the survey was complete and a thorough study was made of the seismic information, three interest areas were identified and labeled areas A, B, and C. The G&G team made a detailed survey of each of the interest areas. Interest Area A's survey, covering 640 acres, cost $24,000; Interest Area B's survey, covering 320 acres, cost $15,000; Interest Area C's survey, covering 160 acres, cost $10,000. After a detailed evaluation, Tyler Company acquired the following leases:

Interest Area A: None

Interest Area B: Richards—320 acres; 100% WI; 1/8 RI; bonus, $75/acre; other acquisition costs, $2,500; delay rental, $2/acre

Interest Area C: Rylander—80 acres; 100% WI; 3/16 RI; bonus, $50/acre; other costs, $1,500; delay rental, $1/acre

Tax Effects of Transactions Occurring during 2011

Basis of Areas of Interest

	Area A (640 acres)	Area B (320 acres)	Area C (160 acres)
Allocation of reconnaissance costs ($30,000/3)	$10,000	$10,000	$10,000
Detailed survey costs	24,000	15,000	10,000
Capitalized basis of areas	$34,000	$25,000	$20,000

Disposition of Capitalized Area of Interest Costs

Area A

No interest, therefore, entire basis of $34,000 is currently deductible.

Area B

Acquired Richards Lease (320 acres). Costs capitalized to Area B added to basis of Richards Lease:

Capitalized costs of Area B	$25,000
Bonus (320 × $75)	24,000
Other acquisition costs	2,500
Basis of Richards Lease	$51,500

Area C

Acquired Rylander Lease (80 acres). Costs capitalized to Area C added to basis of Rylander Lease:

Capitalized costs of Area C	$20,000
Bonus (80 × $50)	4,000
Other acquisition costs	1,500
Basis of Rylander Lease	$25,500

During 2012

Tyler Company made the following payments in 2012:

Delay rental—Richards lease, $640, paid March 2012
Dry-hole contribution of $5,000 in connection with the Richards lease
Bottom-hole contribution in connection with the Rylander lease, $10,000 (well was dry)
Drilling costs of Rylander No. 1:

IDC (assume incurred September 16)	$302,000
Equipment costs (assume incurred during September)	212,000

The well was completed in late September 2012, and 1,600 total gross barrels of oil were produced and sold during the year. Estimated total gross reserves at year-end were 30,400 barrels. The oil was sold for $60/bbl in 2012. Lifting costs were $25/bbl. The lessee distributes the royalty to Rylander. The severance tax rate is 7%. Because Tyler Company is an integrated producer, the company must use cost depletion.

Tyler Company's taxable income and expenses in 2012 and the tax basis of these two properties are determined below.

Tax Effects of Transactions Occurring during 2012

Gross income (1,600 bbl × 13/16 × $60)		$ 78,000
Deductions for both leases:		
Delay rental	$ 640	
IDC (70% × $302,000)	211,400	
Amortization of capitalized IDC (see below*)	6,040	
Severance taxes ($78,000 × 7%)	5,460	
Lifting costs (1,600 bbl × $25)	40,000	
Depletion (see below**)	1,775	
Depreciation (see below***)	30,286	
Total deductions		$295,601

*IDC:

Capitalized IDC—Rylander No. 1 ($302,000 × 0.30)		$90,600
2012 write-off ($90,600 × 4/60)		(6,040)
Remaining balance to be written over next 56 months		$84,560

**Depletion:	Richards	Rylander
1/1/12 Basis	$51,500	$25,500
Test-well contributions	5,000	10,000
Balance subject to depletion	$56,500	$35,500
2012 Depletion	0	(1,775)
12/31/12 Basis	$56,500	$33,725

$$\frac{1,600 \times 13/16}{30,400 \times 13/16 + 1,600 \times 13/16} \times \$35,500 = \underline{\$1,775}$$

***Depreciation:

Capitalized cost of tangible equipment	$212,000
2012 write-off ($212,000 × 1/7 × 2 × 1/2)	(30,286)
Undepreciated balance	$181,714

If Tyler had been an independent producer, all of the income and expenses would have been identical with the exception of IDC and possibly depletion expense. If Tyler had been an independent producer, the entire $302,000 of IDC would have been deductible during 2012, and depletion expense would have been the larger of cost versus percentage depletion.

Table 11–1 presents a comparison of capitalizing or expensing selected expenditures under successful efforts, full cost, and tax accounting. The abbreviations used are SE = successful efforts, FC = full cost, E = Expense, C = Capitalize, and NA = Not Applicable. The tax treatments of expenditures by the lessor, which are included in the table below, are discussed in more detail later in the chapter.

Chapter 11 • Basic Oil And Gas Tax Accounting 441

Table 11–1. Summary of accounting treatment of costs—successful efforts vs. full cost vs. tax

Item	SE	FC	Tax	
			Lessor	Lessee
1. Broad G&G studies—an area of interest is identified	E	C	NA	C
2. Broad G&G studies—an area of interest is not identified	E	C	NA	E
3. Seismic studies on 2,000 acres—land is leased	E	C	NA	C
4. Seismic studies on 2,000 acres—no land is leased	E	C	NA	E
5. Bottom-hole contribution—well successful	E	C	NA	C
6. Dry-hole contribution	E	C	NA	C
7. Bonus	C	C	Income	C
8. Delay rental payments	E	C	Income	E (C)
9. Minor workover	E	E	NA	E
10. Water injection well (intangible costs)	C	C	NA	E(C)/IDC*
11. Option to lease	C	C	Income	C
12. Water disposal well (intangible costs)	C	C	NA	C
13. Successful exploratory drilling a. Intangible costs b. Tangible costs	 C C	 C C	 NA NA	 E (C)* C
14. Exploratory drilling—dry hole a. Intangible costs b. Tangible costs (net of salvage)	 E E	 C C	 NA NA	 E (C)+ E
15. Successful development drilling a. Intangible costs b. Tangible costs	 C C	 C C	 NA NA	 E (C)* C
16. Development drilling—dry hole a. Intangible costs b. Tangible costs (net of salvage)	 C C	 C C	 NA NA	 E (C)+ E

			Tax	
Item	SE	FC	Lessor	Lessee
17. Deepening a development well to unexplored depths—dry hole				
a. Intangible costs	E	C	NA	E(C)/IDC⁺
b. Tangible costs (net of salvage)	E	C	NA	E
18. Abandonment of well on producing lease (well had been depreciated separately for tax purposes)	§	§	NA	E
19. Abandonment of lease (no other nearby leases)	E	§	NA	E
20. Production costs	E	E	NA	E
21. Water supply well (intangible costs)				
a. Water to be used for development drilling	C	C	NA	E(C) / IDC*
b. Water to be used for secondary recovery	C	C	NA	E(C) / IDC*
22. Exploratory stratigraphic test well—dry (intangible costs)	E	C	NA	E(C)/IDC⁺

* Depends upon election to expense and whether independent or integrated producer.
⁺ Depends upon election to expense.
§ Do not recognize a loss; instead, charge accumulated DD&A or abandoned costs.

The next example compares the computation of DD&A under tax accounting, successful efforts accounting, and full cost accounting:

EXAMPLE

DD&A under Three Accounting Methods

Tyler Company began operations in September 2010. By December 2012, Tyler had acquired only two properties, both located in the United States and both considered significant. Costs incurred on those properties during the first years of operations are given below net of accumulated DD&A. During the first years of operations, Tyler drilled two dry holes and one successful well on Lease A and one dry hole on Lease B. No equipment was salvaged from the dry holes. Production on Lease A began on January 1, 2011. As a result of the dry hole on Lease B, Tyler impaired Lease B $10,000 on December 31, 2011, for successful efforts accounting purposes. Early in August 2012, Tyler incurred $30,000 of IDC on Lease A (in wells-in-progress—IDC at 12/31/12) and $150,000 on Lease B (in wells-in-progress—IDC at 12/31/12). All reserve, production, and sales data given below apply only to Tyler Company.

Costs Incurred 9/10/10 through 12/31/12

	Lease A	Lease B
Related G&G costs	$ 50,000	$ 80,000
Bonus	60,000	50,000
Impairment (successful efforts)		(10,000)
Delay rentals, cumulative	0	5,000
Wells-in-progress—IDC (at 12/31/12)	30,000	150,000
Dry exploratory wells—IDC	200,000	250,000
Dry exploratory wells—L&WE	70,000	75,000
Successful exploratory well—IDC	300,000	—
Successful exploratory well—L&WE	100,000	—
Dry development wells—IDC	350,000	—
Dry development wells—L&WE	120,000	—
Other data:		
Salvage value of wells	45,000	—
Future development costs	400,000	—
Production	10,000 bbl	—
Sales	9,000 bbl	—
Proved reserves, 12/31/12	500,000 bbl	—
Proved developed reserves, 12/31/12	100,000 bbl	—

Tyler also placed in service on 1/1/11 a building that cost $108,000 and has a life of 20 years with no salvage value. The building houses the corporate headquarters that supports oil and gas operations in the United States and non–oil and gas operations in Canada. The operations conducted in the building are general in nature and are not directly attributable to any specific exploration, development, or production activity. Since the building is not directly related to exploration, development, or production and supports activities in more than one cost center, it is depreciated using straight-line depreciation for financial accounting.

a. DD&A for 2012 for financial accounting purposes assuming Tyler uses successful efforts accounting is computed below.

Proved Property, Lease A Only

$$\$60,000 \times \frac{10,000}{500,000 + 10,000} = \underline{\$1,176}$$

Wells and Equipment to Be Amortized, Lease A

Total wells = $300,000 + $100,000 + $350,000 + $120,000 – $45,000 = $\underline{\$825,000}$

$$\$825,000 \times \frac{10,000}{100,000 + 10,000} = \underline{\$75,000}$$

Depreciation Building (supporting general and administrative functions only)

$$\frac{\$108,000 - \$0}{20} = \underline{\$5,400}$$

b. DD&A for 2012 for financial accounting purposes assuming Tyler uses full cost accounting and assuming inclusion of all possible costs in the amortization base is shown below. The revenue method is ignored.

Total costs to amortize = all costs from both Lease A (minus salvage) and Lease B (impairment not included because Lease B would not have been impaired due to its inclusion in amortization) and plus future development costs
= $1,235,000 (Lease A) + $610,000 (Lease B) + $400,000 (development costs)
= $\underline{\$2,245,000}$

$$\$2,245,000 \times \frac{10,000}{500,000 + 10,000} = \underline{\$44,020}$$

Depreciation Building (supporting general and administrative functions only)

$$\frac{\$108,000 - \$0}{20} = \underline{\$5,400}$$

c. DD&A for 2012 for tax reporting purposes assuming that Tyler is an independent producer is shown below. Percentage depletion is ignored. Assume proved reserves are used to compute cost depletion.

Depletion, Lease A Only

$$(\$50,000 + \$60,000) \times \frac{9,000}{500,000 + 9,000} = \underline{\$1,945}$$

Depreciation in Second Year

L&WE from successful exploratory well (assume $100,000 is gross cost):

$$(\$100,000 - \$14,286^*) \times 1/7 \times 2 = \underline{\$24,490}$$

*Depreciation for the first year: $100,000 \times 1/7 \times 2 \times 1/2 = \underline{\$14,286}$

Building (assume $108,000 is gross cost):

$108,000/39 = \underline{\$2,769/\text{year}}$ (except first and last years)

d. DD&A for 2012 for tax reporting purposes assuming Tyler is an integrated producer is shown below.

Depletion and depreciation would be unchanged. However, 30% of all the IDC for successful wells and wells in progress must be capitalized and amortized over 60 months.

Total IDC Incurred for Successful Wells and Wells in Progress

Prior to 2012 (but within 60 months)	$ 300,000
During 2012	180,000
	$ 480,000

Total IDC Capitalized

Prior to 2012	$300,000 × 30% =	$ 90,000
During 2012	180,000 × 30% =	54,000
		$144,000

Total IDC Amortized during 2012

$90,000/60 months × 12 months =	$ 18,000
$54,000/60 months × 5 months =	4,500
	$ 22,500

Percentage depletion

Percentage depletion is one of the two methods used for income tax purposes in determining the depletion amount to be deducted in arriving at taxable income. Percentage depletion, when allowed and if greater than cost depletion, must be taken in lieu of cost depletion. Generally, percentage depletion is allowed only for royalty interest owners and independent producers of oil and gas. Companies with integrated functional areas, i.e., exploring, producing, refining, and distributing, are not allowed to take percentage depletion.

Percentage depletion is computed by multiplying the gross revenues from a property by a percentage established by federal statutes. The depletion percentage is currently 15% for oil and gas produced from nonmarginal wells. (Note that the percentage depletion rate for regulated natural gas and natural gas sold under a fixed-price contract is 22%.) The amount of oil or gas subject to depletion cannot exceed an average daily production of 1,000 barrels of oil or 6,000 Mcf of gas. The amount of percentage depletion allowed cannot be greater than 100% of net income from a particular property. Further, percentage depletion may not exceed 65% of the taxpayer's taxable income before depletion from all sources. Amounts disallowed as a result of the 65% limitation may be carried over.

Percentage depletion allowed is deducted from the capitalized leasehold costs and elected capitalized IDC (if any). Percentage depletion may be taken even if all capitalized costs have been recovered. In contrast, cost depletion is limited to the amount of capitalized costs.

Prior to October 12, 1990, a transferee of a proved oil or gas property was denied percentage depletion. The Revenue Act of 1990 repealed this prohibition on percentage depletion for transferees. Therefore, a person receiving a proved property can take percentage depletion if the other percentage depletion rules are met.

EXAMPLE

Percentage Depletion

Tyler Company is an independent producer. In 2014, the company had sales of 23,360 barrels from Lease A (less than 1,000 bbl/day). The average selling price for the oil was $125/bbl. Net income was $985,000 from Lease A, and taxable income was the same amount.

Computation of Percentage Depletion

	Lease A
Gross sales (23,360 bbl × $125)	$2,920,000
Tentative % depletion (gross revenue × rate: $2,920,000 × 0.15)	438,000
Net income limitation (100% × NI)	985,000
Tentative allowable deduction	438,000
Taxable income limitation (0.65 × taxable income: 0.65 × $985,000)	640,250
Percentage depletion	$ 438,000*

*This amount must be compared with cost depletion—the greater amount would be the actual depletion expense taken.

In the above example, only one lease was used to illustrate percentage depletion. When more than one lease is involved, the computation of the net income and the application of the depletable quantity limitation and the 65% limitation become complicated. Allocation of various expenses must be made between leases before net income can be determined. In addition, the maximum percentage depletion allowable under the 65% tax income limitation must be allocated and possibly reallocated between properties. The following example briefly illustrates the 100% and 65% limitations. (For a more detailed discussion of these rules, see *Oil and Gas: Federal Income Taxation* by Patrick A. Hennessee.[1])

EXAMPLE

Percentage Depletion: 100% and 65% Limitations

Tyler Company owns three properties (classified as individual properties for tax purposes) at year-end 2011. Data for the three properties are presented below:

	Properties			
	A	B	C	Total
Total gross revenues subject to % depletion	$400,000	$500,000	$300,000	$1,200,000
Net (loss) income before depletion—2011	50,000	100,000	(30,000)	120,000
Calculation of Allowable Percentage Depletion				
Gross revenues × 15%	$ 60,000	$ 75,000	$ 45,000	$ 180,000
100% Net income limitation	50,000	100,000	(30,000)	120,000
Allowable % depletion subject to 100% limitation	50,000	75,000	0	125,000
Total taxable income before depletion x 65% (i.e., $120,000 × 0.65)				78,000
Percentage depletion (compare $125,000 with $78,000)				$ 78,000

Property

The term **property** is an extremely important concept in the U.S. income tax statutes relating to oil and gas operations. Almost all of the tax statutes are applied on a property-by-property basis. This includes the following areas of oil and gas tax accounting:

- G&G costs capitalized and allocated to a property unit
- Capitalized IDC
- Cost and percentage depletion
- Gain or loss on disposal of property
- Recapture of IDC as ordinary income
- Tax preference items

According to *U.S. Treas. Reg. Section 1.614-1(b)*, property means "each separate interest owned by the taxpayer in each mineral deposit in each tract or parcel of land." Each economic interest is generally regarded as a separate property, unless it was acquired at the same time, is the same kind of interest, is from the same assignor, and is contiguous. For example, offshore tracts or properties that are bid on separately are considered to be separate properties, even if the tracts are contiguous. **Separate tract or parcel** relates to the physical scope of land. If two parcels of land are not contiguous, then even if they are purchased at the same time from the same assignor, two separate properties are created.

Interests generally are classified as operating (working interest) or nonoperating (royalty or ORI, production payment, or net profits interest). Nonoperating interests may, with the consent of the commissioner, be aggregated.

Property differentiation can be very important in the recapture of IDC and depletion. As will be seen in the next section, separation of properties into small property units may be advantageous for IDC recapture.

Recapture of IDC and depletion

All IDC that has been expensed and any depletion expense that reduced the basis of the property must be recaptured as ordinary income versus a capital gain to the extent a property is sold at a gain. (As there is currently no difference between the capital gains rate versus the ordinary income rate for a business, this section does not have any tax implications for a business unless the business has a capital loss.) Specifically, the taxpayer is required to recapture as ordinary income the lower of the following amounts:

- The total of any IDC expensed as incurred, IDC expensed through amortization, and any depletion expense that reduced the adjusted basis of the property
- The amount of the gain on the sale

EXAMPLE

Recapturable IDC and Depletion Given

Mineral deposit A has recapturable IDC and depletion of $400,000, and mineral deposit B has recapturable IDC of $200,000. Both properties together are sold for $1,000,000. The selling price is allocated $100,000 to A and $900,000 to B. Both deposits have a zero basis.

Assuming the same property: The property has a zero basis, therefore the gain on the sale is $1,000,000. Ordinary income is the lower of $600,000 versus the $1,000,000 gain:

Section 1231 gain (capital gain)	$400,000
Ordinary income	600,000

Assuming separate properties:

Property A—Ordinary income is the lower of $400,000 versus the $100,000 gain:

Section 1231 gain	$ 0
Ordinary income	100,000

Property B—Ordinary income is the lower of $200,000 versus the $900,000 gain:

Section 1231 gain	$700,000
Ordinary income	200,000

The following example illustrates the computation of the amount of IDC and depletion to recapture as ordinary income for both an integrated and independent company.

EXAMPLE

Recapturable IDC and Depletion Not Given

On January 1, 2011, Tyler Company bought an undeveloped lease for $200,000. Early in January 2011, Tyler Company incurred IDC of $450,000. Reserves of 75,000 barrels were discovered, and 30,000 barrels were produced and sold. Gross income from production was $1,000,000. On January 3, 2012, Tyler Company sold the property for $420,000. All reserve, production, and sales data apply only to Tyler Company.

a. The income tax effects of the sale assuming an independent producer and percentage depletion of 15% are determined below. The 100% and 65% limitations are ignored.

Cost Depletion

$$\$200,000 \times \frac{30,000}{75,000} = \underline{\$80,000}$$

Percentage Depletion

15% × $1,000,000 = $\underline{\$150,000}$

Depletion expense for 2011 is $\underline{\$150,000}$

Amount of Recapturable IDC and Depletion

$150,000 (depletion) + $450,000 (IDC) = $\underline{\$600,000}$

The tax basis of the property is $50,000 ($200,000 − $150,000); therefore, the gain on the sale is $\underline{\$370,000}$ ($420,000 − $50,000). Ordinary income is the lower of $600,000 versus $370,000. Therefore, the entire gain on the sale is ordinary income.

b. The income tax effects for the sale are determined below assuming the same facts as shown in the above example, except that Tyler Company is an integrated producer. (An integrated producer cannot take percentage depletion, and 30% of any IDC must be capitalized and amortized over 60 months.)

Actual Cost Depletion Expense

$$\$200,000 \times \frac{30,000}{75,000} = \underline{\$80,000}$$

IDC Expensed to Date

$$\$450,000 \times 0.70 = \$315,000$$

$$(\$450,000 \times 0.30) \times 12/60 = \underline{27,000}$$

$$\underline{\$342,000}$$

Amount of Recapturable IDC and Depletion

$80,000 (depletion) + $342,000 (IDC) = $\underline{\$422,000}$

Amount of Gain

Basis in property

Leasehold		$200,000
Less: Cost depletion		(80,000)
Leasehold basis		120,000
Plus: Capitalized IDC ($135,000 – $27,000)		108,000
		$228,000

Gain	=	$420,000	–	$228,000	=	$192,000
		(Selling price)		(Basis in property)		

Ordinary income is the lower of $422,000 versus $192,000. Therefore, the entire gain on the sale is ordinary income.

The recapture provision illustrated by the preceding example applies to the disposition of all properties placed in service after December 31, 1986.

The example is unrealistically simple because the property was owned for only one year. If the property had been owned for more than one year, the amount of recapturable IDC would have had to be calculated for each year the property was owned. In addition, in the above example, equipment was ignored. Generally in a transaction where IDC must be recaptured, equipment would also be sold. The selling price should be allocated between the equipment, the leasehold, and the IDC, based on relative fair market values. In practice, the equipment is treated as being sold at net book value, which results in any gain being allocated to the leasehold and IDC.

Lessor's Transactions

The previous discussion has centered upon the determination of taxable income by the lessee. Attention is now turned to the income tax accounting for transactions involving the lessor.

Acquisition costs

The lessor should capitalize any acquisition costs incurred prior to leasing the property to an oil company and deplete the costs capitalized. If the mineral rights were obtained separately from the surface rights, the amount capitalized should be the acquisition cost of the mineral rights. If the property were purchased in fee, the purchase price should be allocated between the mineral and surface rights.

Provided the lessor is a royalty owner—or an independent producer—either cost depletion or percentage depletion may be taken on acquisition cost amounts, with the actual depletion expense being the larger of the two. Percentage depletion, unlike cost depletion, is allowed even after the cost has been recovered. The lessor may, in many situations, have zero or a small dollar amount of acquisition costs. Therefore, cost depletion is not a viable option for many lessors.

EXAMPLE

Lessor's Acquisition Costs

Mr. Ethan purchased some mineral rights for $40,000. Ethan later leased the property and received a 1/5 royalty. The year-end total gross reserve estimate is 45,000 barrels, and total gross sales during the year were 5,000 barrels.

The tax treatment for the acquisition cost of the mineral rights and cost depletion for Ethan are determined below. Percentage depletion is ignored.

Tax Treatment

Depletion calculation:

Royalty interest sales:	$1/5 \times 5,000$	= 1,000 bbl
Royalty interest in reserves:	$1/5 \times 45,000$	= 9,000 bbl

Cost depletion:

$$\frac{1,000}{9,000 + 1,000} \times \$40,000 = \underline{\$4,000}$$

Mineral rights basis:

Tax basis of property at beginning of year	$40,000
Less: Cost depletion	(4,000)
Tax basis of property at end of year	$36,000

Revenue

Revenue received by a lessor as his share of production is a royalty and is treated as ordinary income. Percentage depletion is calculated by taking 15% of the revenue from a property. For a royalty interest owner, this translates into 15% of the royalty from a property. Lease bonuses paid to the lessor are also considered to be in the nature of a royalty amount paid in advance—an advance royalty. However, even though considered a royalty, effective August 16, 1986, advance royalty payments, lease bonuses, or any other amounts payable without regard to production from a property no longer qualify for percentage depletion by royalty owners or independent producers.

Delay rentals, ad valorem taxes paid by the lessee for the lessor, minimum and shut-in royalties, and options to lease are all considered to be ordinary income to the lessor. Delay rentals received and ad valorem taxes paid by the lessee are not considered royalties and therefore cannot be reduced by percentage depletion. Minimum and shut-in royalties are advance royalties; however, they also are not subject to percentage depletion. Percentage depletion can be taken only on royalties from actual production.

EXAMPLE

Lessor's Revenue

Mabel Lomax received the following benefits in 2011 in connection with Lease A, from which no oil or gas was produced during 2011:

Received delay rental.	$2,000
Received notice that lessee paid Lomax's ad valorem taxes in the amount of.	1,500
Received shut-in royalty for well drilled in 2011	5,000
Royalties from actual production	0

Income tax treatment of transactions

Ordinary income not subject to percentage depletion

Delay rental	$2,000
Ad valorem taxes paid by lessee	1,500
Shut-in royalty	5,000
Total	$8,500
Income subject to percentage depletion	$ 0
Depletion deduction	$ 0

PROBLEMS

1. Big Tree Petroleum incurs G&G costs of $21,000 for Project Area 12. Three areas of interest are identified. Detailed G&G is conducted on the areas of interest at the following costs:

Interest Area	Cost
1	$55,000
2	60,000
3	70,000

As a result of the detailed G&G studies, the following leases were acquired:

Interest Area	Lease	Acres	Interest
1	A	1,000	100% of WI
	B	3,000	50% of WI
	C	2,000	25% of WI
2	D	5,000	100% of WI
3	None	None	None

Determine the tax basis of any assets and the amount of any tax deductions.

2. Coyote Oil Company incurred intangible costs during 20XA related to the following:

	Amount	Date Incurred
Dry exploratory well	$100,000	May 1
Dry development well	200,000	May 1
Successful exploratory well	250,000	May 1
Water injection well	270,000	Aug 1
Deepening a well (successful)	80,000	Aug 1
Workover, minor	50,000	Aug 1

   a. Assuming Coyote is an independent producer, how much IDC can it deduct for 20XA?
   b. How much IDC could Coyote deduct as an integrated producer?

3. On January 1, 20XA, Rain Oil Corporation bought a developed lease for $300,000. During 20XA, Rain Oil Corporation incurred $600,000 of IDC on a successful well. Reserves of 400,000 barrels were discovered, and 100,000 barrels were produced and sold. Gross income from production was $2,000,000. On January 4, 20XB, the company sold the property for $700,000. All reserve, production, and sales data apply only to Rain Oil Corporation.

   a. Determine the amount of ordinary income and the amount of any capital gain assuming Rain Oil is an integrated producer and that part of the IDC is capitalized as required. Assume that all of the IDC was incurred on April 1, 20XA.

   b. Determine the amount of ordinary income and the amount of any capital gain assuming an independent producer, percentage depletion of 15%, and that all the IDC was expensed as incurred. Ignore the 100% and 65% limitations.

4. During 20XA, Core Petroleum incurred G&G costs of $20,000 for Project Area 15. Two areas of interest were identified. Detailed seismic studies were conducted on the areas of interests at the following costs:

Interest Area	Cost
1	$60,000
2	40,000

   As a result of the detailed seismic studies, the following leases were obtained:

Interest Area	Lease	Acres	Interest	Bonus Paid
1	None	—	—	—
2	A	2,000	20% of WI	$50/acre
	B	4,000	100% of WI	$30/acre

During 20XB, Core Petroleum made the following payments:

	Lease A	Lease B
Delay rental, paid March 20XB	$10,000	
Bottom-hole contribution (well was successful)		$25,000
Dry-hole contribution	20,000	
Lessor's ad valorem taxes (nonrecoverable)	2,000	
Drilling costs:		
IDC (assume incurred January 1, 20XB)		300,000
Equipment		80,000

The well on Lease B was completed early in January 20XC and was successful. Core Petroleum's share of production from the well was 10,000 barrels of oil. All 10,000 barrels of oil were sold during 20XC. Core's share of estimated reserves at year-end was 300,000 barrels. The selling price of the oil was $60/bbl, and lifting costs were $200,000. Lease A was abandoned in March 20XC, and Lease B was abandoned early in January 20XD. No oil was produced during 20XD.

a. Determine the tax effects for the above transactions in each year, assuming Core is an independent producer. Ignore percentage depletion, but remember DD&A.

b. Determine any tax effects that would be different if Core were an integrated producer rather than an independent producer.

5. Augusta Oil Corporation, an independent producer, began operations in June 20XA. During the first 2½ years of operation, Augusta acquired only two U.S. properties, which were noncontiguous. Costs incurred on those properties during those 2½ years are given below, *net of accumulated DD&A*. Augusta drilled three dry holes and one successful well. No equipment was salvaged from the dry holes. Production on the successful well started on January 1, 20XB. During 20XC, Augusta incurred total IDC of $100,000 on Lease A (in wells-in-progress—IDC at 12/31/20XC) and $200,000 on Lease B (in wells-in-progress—IDC at 12/31/20XC). All 20XC IDC on both leases was incurred early in May. All reserve, production, and sales data below apply only to Augusta Oil Corporation.

Costs incurred 6/20XA through 12/20XC:

	Lease A	Lease B
Related G&G costs	$20,000	$40,000
Bonus	30,000	50,000
Delay rentals, cumulative	10,000	5,000
Wells-in-progress—IDC (at 12/31/20XC)	100,000	200,000
Dry exploratory wells—IDC	250,000	300,000
Dry exploratory wells—L&WE	50,000	75,000
Successful exploratory well—IDC	400,000	—
Successful exploratory well—L&WE	150,000	—

	Lease A	Lease B
Dry development wells—IDC	200,000	—
Dry development wells—L&WE	100,000	—
Other data		
Future development costs	300,000	
Production	20,000 bbl	—
Sales	18,000 bbl	—
Proved reserves, 12/31/20XC	600,000 bbl	—
Proved developed reserves, 12/31/20XC	200,000 bbl	—

Augusta also placed in service on 1/1/20XB a building that cost $117,000 and has a life of 25 years, with a salvage value of $7,000. The building houses the corporate headquarters that supports oil and gas operations in the United States and non–oil and gas operations in Mexico. The operations conducted in the building are general in nature and are not directly attributable to any specific exploration, development, or production activity. The building is not directly related to exploration, development, or production and supports activities in more than one cost center. As such, it is depreciated using straight-line depreciation for financial accounting.

   a. Compute DD&A for 20XC for the following accounting methods assuming that Augusta is an independent producer:

      1) Successful efforts.

      2) Full cost, assuming inclusion of all possible costs in the amortization base. Ignore the revenue method.

      3) Tax: ignore percentage depletion, use proved reserves to compute cost depletion, and assume that the cost of the equipment and the building are gross costs.

   b. Calculate DD&A for tax, assuming instead that Augusta is an integrated producer.

6. How does the formula for tax depletion differ from the formula for successful efforts or full cost depletion?

7. How does the tax depletion of leasehold costs differ from successful efforts or full cost depletion? Include in your answer a discussion of the reserves used (proved reserves or proved developed reserves), which costs are amortized, and the cost center used (i.e., are property costs amortized separately or by some type of grouping?). Assume for full cost all possible costs are amortized.

8. How does amortization of drilling costs differ under the three methods? Include in your answer a discussion of the method used, which reserves are used, if any, and which drilling costs are amortized—dry versus successful, exploratory well versus development well, completed versus uncompleted, and IDC versus L&WE. Also include the cost center used—i.e., are drilling costs amortized separately or by some grouping? Assume for full cost all possible costs are amortized.

9. How does amortization of lease equipment such as storage tanks and separators differ under the three methods? Include in your answer a discussion of the method used, including which reserves are used, if any, and the cost center used.

10. Cowboy Oil Company, an integrated producer, has an unproved property with acquisition and capitalized G&G costs of $35,000. Cowboy also has a proved property with the following costs:

	Lease	Well 1	Well 2	Well 3	Total
Acquisition costs	$40,000				$ 40,000
Capitalized IDC costs		$ 50,000	$280,000	$170,000	500,000
Equipment costs		100,000	500,000	300,000	900,000
Accumulated depletion	(8,000)				(8,000)
Accumulated amortization		(13,000)	(110,000)	(57,000)	(180,000)
Accumulated depreciation		(25,000)	(200,000)	(100,000)	(325,000)
Net	$32,000	$112,000	$470,000	$313,000	$927,000

Determine the amount of the tax loss in each of the following situations:

a. Cowboy drilled a dry hole on the unproved property costing $250,000 for IDC and $60,000 for equipment. Equipment worth $10,000 was salvaged.

b. As a result of the dry hole, Cowboy decided to abandon the unproved property.

c. Cowboy abandoned Well 1 on the proved property. Wells 2 and 3 are still producing. Assume that the wells had been depreciated separately.

d. Assume that instead of the circumstances in part c, Cowboy abandoned the entire lease, and that equipment worth $27,000 was salvaged.

11. Indicate which items are to be capitalized (C), expensed (E), and part capitalized and part expensed (C/E) for successful efforts, full cost, and tax accounting. Assume the maximum tax deductions are taken.

Item	Successful Efforts	Full Cost	Tax Lessee
Broad G&G studies—an area of interest is identified			
Broad G&G studies—an area of interest is not identified			
Seismic studies on 2,000 acres—land is leased			
Seismic studies on 2,000 acres—no land is leased			
Bottom-hole contribution—well successful			
Dry-hole contribution			
Bonus			
Delay rental payments			
Minor workover			
Water injection well (intangible costs)			

Item	Successful Efforts	Full Cost	Tax Lessee
Option to lease			
Water disposal well (intangible costs)			

Wells-in-progress—proved property
    a. Intangible costs
    b. Tangible costs			

Wells-in-progress—unproved property
    a. Intangible costs
    b. Tangible costs			

Successful exploratory drilling
    a. Intangible costs
    b. Tangible costs			

Exploratory drilling—dry hole
    a. Intangible costs
    b. Tangible costs (net of salvage)			

Successful development drilling
    a. Intangible costs
    b. Tangible costs			

Development drilling—dry hole
    a. Intangible costs
    b. Tangible costs (net of salvage)			

Deepening a development well to unexplored depths—dry hole
    a. Intangible costs
    b. Tangible costs (net of salvage)			

Abandonment of well on producing lease (well had been depreciated separately for tax purposes)			
Abandonment of lease (no other nearby leases)			
Production costs			

Water supply well (intangible costs)
    a. Water to be used for development drilling
    b. Water to be used for secondary recovery			

Exploratory stratigraphic test well—dry (intangible costs)			

12. On March 1, 20XA, Chuck Larson purchases mineral rights (MR) for $30,000. On June 1, 20XA, he leases the mineral rights to Grey Wolf Oil Company, retaining a 1/5 royalty interest (RI). Grey Wolf Oil Company pays Larson a lease bonus of $10,000. On June 1, 20XB, a delay rental of $1,000 is received by Larson. Oil royalties of $8,000 are paid to Larson in 20XC. Reserves at 12/31/20XC are 20,000 barrels, and production and sales for the year are 3,000 barrels. The reserve, production, and sales data apply only to Chuck Larson.

    Determine the tax basis of any assets owned by Chuck Larson and the amount of any tax revenues reported and any tax deductions taken by Chuck Larson in each of the three years.

13. Harper Oil Corporation paid the following amounts in 20XD:

Shut-in royalty payments (not recoverable)	$1,200
Shut-in royalty payments (failure to make payments terminates lease)	4,000
Shut-in royalty payments (recoverable from future production)	2,400

    Determine the tax basis of any assets and the amount of any tax deductions.

14. Garage Oil Company, an independent producer, has the following account balances at 1/1/20XA:

Unproved property—Lease A	$ 10,000
Equipment—Lease B	150,000
Acquisition costs—Lease B	8,000
Accumulated depletion	5,000
Accumulated depreciation	110,000

    Determine the amount of the tax loss on the following dates:

    a. On March 1, 20XA, the unproved property is abandoned.

    b. On April 2, 20XA, Lease B is abandoned with salvageable equipment in the amount of $12,000.

15. Tiger Energy has the following information:

Leasehold costs	$ 6,000
Accumulated depletion.	1,000
L&WE	200,000
Accumulated depreciation.	75,000

The lease is subleased to Phil Oil Corporation for $300,000, and Tiger retains an 1/16 ORI. At the date of the sublease, the FMV of the equipment is 180,000.

REQUIRED: Determine the tax basis of Tiger's and Phil's assets and the amount of any tax revenue.

16. During 20XB, Sunflower Energy incurs the following costs relating to Lease A, a producing property:

Supplies for Lease A	$ 300
Labor cost for pumpers and gaugers—Lease A	800
Ad valorem tax on Lease A	1,600
Maintenance cost of lease records	1,500
Refracturing cost—Lease A (workover)	2,000
Transportation of personnel to wells—Lease A	800

REQUIRED: Determine the tax basis of any assets and the amount of any tax deductions.

17. Tiger Energy, an independent producer, has average production from Lease A of 100 bbl/ day in 20XA from Lease A. The average selling price of oil in 20XA is $58/ bbl. Net income from Lease A in 20XA is $820,000, and taxable income of the company is $2,000,000.

    Compute percentage depletion.

18. Define the following:

    a. IDC

    b. elected capitalized IDC

    c. sublease

19. Sauer Oil Corporation, an integrated producer, incurs IDC costs in the following years as indicated. The IDC amounts marked with an asterisk (*) relate to dry-hole IDC.

Date Incurred	Amount	Date Incurred	Amount	Date Incurred	Amount	Date Incurred	Amount
3/12/XA	$200,000	4/10/XB	$280,000*	2/1/XC	$600,000	3/4/XD	$240,000
7/18/XA	300,000*	11/2/XB	320,000*	12/2/XC	150,000	9/14/XD	180,000*

REQUIRED: Compute the amount that may be deducted for IDC in the years 20XA, 20XB, 20XC, and 20XD.

20. Green Oil Corporation owns and operates four oil and gas properties that are classified for tax purposes as four separate properties. Data for the four properties are presented below:

Properties	1	2	3	4
Total gross revenues subject to % depletion	$200,000	$100,000	$60,000	$300,000
Net (loss) income before depletion	40,000	10,000	(5,000)	70,000

REQUIRED: Compute the amount of percentage depletion that could be deducted on Green's tax return.

21. Tiger Energy owns only one lease in the United States, Lease Q. The following information for Lease Q, which is burdened with a 1/6 royalty, is as of 12/31/20XD. All reserve, production, and sales data apply only to Tiger Energy.

Acquisition costs	$ 40,000
Test-well contributions	12,000
G&G, direct	70,000
Wells-in-progress—IDC (incurred 6/24/20XD)	80,000
Wells-in-progress—L&WE	6,000
Wells—L&WE (end of 2nd year)	250,000
Wells—IDC (incurred in 20XB, and all productive)	600,000
Future development costs	400,000
Proved reserves, 12/31/20XD	300,000
Proved developed reserves, 12/31/20XD	100,000
Production	18,000
Sales	15,000

Additional data: Tiger also placed in service on 8/1/20XB a building that cost $140,000 and has a life of 10 years, with a salvage value of $9,000. The building houses the corporate headquarters that supports oil and gas operations in the United States and non–oil and gas operations in Canada. The operations conducted in the building are general in nature and are not directly attributable to any specific exploration, development, or production activity. Since the building is not directly related to exploration, development, or production and supports activities in more than one cost center, it is depreciated using straight-line depreciation for financial accounting.

Compute DD&A for 20XD for the following accounting methods.

a. Tax: assuming Tiger is an independent producer (ignore percentage depletion) and accumulated depletion is $10,000. Use proved reserves to compute cost depletion.

b. Tax: assuming Tiger is an integrated producer and accumulated depletion is $10,000. Use proved reserves to compute cost depletion.

c. Successful efforts: assuming accumulated DD&A—proved properties is $5,000; accumulated DD&A—wells is $100,000.

d. Full cost: assuming exclusion of all possible costs from the amortization base. Assume accumulated DD&A is $400,000.

REFERENCES

1. Hennessee, Patrick A. 2006. *Oil and Gas: Federal Income Taxation.* Chicago: CCH.

12

JOINT INTEREST ACCOUNTING

Joint Operations

Oil and gas exploration and production operations are typically high-risk, high-cost activities. In order to spread the cost and risk, oil and gas properties—especially those requiring high capital expenditures—typically are jointly owned by two or more working interest owners. In some instances, companies jointly acquire the working interest from the outset. In other cases, the joint operation may not come about until later. For example, if a number of companies own working interests in several small leases in an area, it may make economic sense for the companies to jointly operate all of the leases as one joint property. In other instances, where leases in an area have produced essentially all of the reserves that are economically recoverable from primary recovery, secondary or tertiary recovery may provide a means to produce additional reserves. In many instances, secondary or tertiary production may only be possible if the leases are unitized and operated as a single property. Finally, due to conservation and economic conditions, the local state or federal government may require the companies to operate the properties jointly.

If separate properties are combined before reserves are discovered, the combination of the properties is typically referred to as a **pooling**. If separate properties are combined after the development of all or some of the properties, the combination of the properties is typically referred to as a **unitization**. Poolings and unitizations are discussed in more detail in chapters 1 and 13. Joint working interests may also come about as a result of any number of different types of sharing arrangements. These types of arrangements are discussed in chapter 13.

Joint operations may be undertaken in three legal forms:
- Joint venture of undivided interests
- Legal partnership
- Jointly owned corporation

Joint ventures of undivided interests are by far the most common form of joint operation. In an undivided interest, the parties share the interest in an entire lease. For example, a company having a 50% undivided interest in a 640 acre lease owns a 50% interest in the entire 640 acres. This differs from a situation where a company owns a 100% interest in 320 acres that were carved out of a 640 acre lease. The former is an undivided interest, and the latter is a divided interest.

In a joint venture of undivided interests, companies may acquire an undivided interest in a property when the property is initially acquired. Alternatively, an undivided interest may be created at a later date if the companies pool, unitize, or otherwise join their properties together in such a manner that all of the parties have an undivided interest in the entire property. Companies having divided interests before a unitization or pooling will have an undivided interest after the properties are pooled or unitized. Additionally, a company may also acquire an undivided interest in a property through any number of different types of sharing arrangements (chapter 13).

Legal partnerships are much less common than joint ventures. When companies join together to form a legal partnership, they normally do so under the laws of a particular state. Typically, companies join together in a partnership formed for the exploration and production of a particular project. Legal partnerships are frequently utilized to achieve certain income tax or legal objectives.

Jointly owned corporations are very rare domestically. Internationally, most operations are conducted as joint ventures; however, there may be some instances where jointly owned corporations are formed. The laws of certain foreign countries may call for oil and gas operations to be carried out only by locally incorporated companies. In order to comply with the local laws, two or more companies may find it necessary to set up a company in that country. The companies establishing the foreign company usually own all of the stock in the new corporation.

Joint venture contracts

There are several types of contracts that may be encountered in joint operations. Examples of typical contracts include unit agreements, pooling agreements, and exploration agreements. The most commonly encountered agreement is the joint operating agreement (JOA). This agreement exists among working interest owners who own undivided interests in a joint property. Through negotiation, the working interest owners designate one of the owners—typically the one with the largest interest percentage—as the operator. The other working interest owners are designated as nonoperators. The operator manages the day-to-day operations of the property, but all the working interest owners participate in the property by voting on any major decisions affecting the property. The operator, as normally required by the operating agreement, generally pays all of the costs of exploration, development, and operation of the joint interest property, and then sends a joint interest billing to the nonoperators for their proportionate share of the costs. Alternatively, the operator may estimate the amount of cash that will be required to operate the property each month and request a cash advance from the nonoperators for their portion of the expected cash outlays. The operator then pays the respective vendors. At the end of the month, the operator sends a statement to the nonoperators detailing how the cash was actually spent.

The operator may act for the nonoperators in marketing the oil and gas, or depending on the particular situation, each of the working interest owners may contract individually to market its own share of production. If the operator markets the oil or gas, the revenue is generally distributed to the working interest owners at the end of each month, according to their respective percentage ownership interests. The royalties are paid to the royalty owners. If each working interest owner separately markets its respective share of oil and gas, then each working interest owner is responsible for paying the royalty on the quantity of oil and gas sold by that particular working interest owner during the month.

THE JOINT OPERATING AGREEMENT

When a joint interest situation is created, the parties involved, i.e., the operator and nonoperators, generally execute an operating agreement. In domestic operations, the typical form used for the JOA is the *AAPL Form 610*, which is available from the American Association of Professional Landmen (AAPL) based in Fort Worth, Texas (Fig. 12–1).

Fig. 12–1. Joint operating agreement
Source: AAPL Joint Operating Agreement. Copyright 1989. All rights reserved. American Association of Professional Landmen. Fort Worth, TX. Approved form A.A.P.L. No. 610, 1989.

A.A.P.L. FORM 610 - 1989
MODEL FORM OPERATING AGREEMENT

OPERATING AGREEMENT

DATED

_____ · _____ ,
year

OPERATOR _____

CONTRACT AREA _____

_____

_____

_____

_____

COUNTY OR PARISH OF _____ , STATE OF _____

COPYRIGHT 1989 — ALL RIGHTS RESERVED
AMERICAN ASSOCIATION OF PETROLEUM
LANDMEN, 4100 FOSSIL CREEK BLVD.
FORT WORTH, TEXAS, 76137, APPROVED FORM.
A.A.P.L. NO. 610 - 1989

TABLE OF CONTENTS

Article	Title	Page
I.	DEFINITIONS	1
II.	EXHIBITS	1
III.	INTERESTS OF PARTIES	2
	A. OIL AND GAS INTERESTS:	2
	B. INTERESTS OF PARTIES IN COSTS AND PRODUCTION:	2
	C. SUBSEQUENTLY CREATED INTERESTS:	2
IV.	TITLES	2
	A. TITLE EXAMINATION:	2
	B. LOSS OR FAILURE OF TITLE:	3
	1. Failure of Title	3
	2. Loss by Non-Payment or Erroneous Payment of Amount Due	3
	3. Other Losses	3
	4. Curing Title	3
V.	OPERATOR	4
	A. DESIGNATION AND RESPONSIBILITIES OF OPERATOR:	4
	B. RESIGNATION OR REMOVAL OF OPERATOR AND SELECTION OF SUCCESSOR:	4
	1. Resignation or Removal of Operator	4
	2. Selection of Successor Operator	4
	3. Effect of Bankruptcy	4
	C. EMPLOYEES AND CONTRACTORS:	4
	D. RIGHTS AND DUTIES OF OPERATOR:	4
	1. Competitive Rates and Use of Affiliates	4
	2. Discharge of Joint Account Obligations	4
	3. Protection from Liens	4
	4. Custody of Funds	5
	5. Access to Contract Area and Records	5
	6. Filing and Furnishing Governmental Reports	5
	7. Drilling and Testing Operations	5
	8. Cost Estimates	5
	9. Insurance	5
VI.	DRILLING AND DEVELOPMENT	5
	A. INITIAL WELL:	5
	B. SUBSEQUENT OPERATIONS:	5
	1. Proposed Operations	5
	2. Operations by Less Than All Parties	6
	3. Stand-By Costs	7
	4. Deepening	8
	5. Sidetracking	8
	6. Order of Preference of Operations	8
	7. Conformity to Spacing Pattern	9
	8. Paying Wells	9
	C. COMPLETION OF WELLS; REWORKING AND PLUGGING BACK:	9
	1. Completion	9
	2. Rework, Recomplete or Plug Back	9
	D. OTHER OPERATIONS:	9
	E. ABANDONMENT OF WELLS:	9
	1. Abandonment of Dry Holes	9
	2. Abandonment of Wells That Have Produced	10
	3. Abandonment of Non-Consent Operations	10
	F. TERMINATION OF OPERATIONS:	10
	G. TAKING PRODUCTION IN KIND	10
	(Option 1) Gas Balancing Agreement	10
	(Option 2) No Gas Balancing Agreement	11
VII.	EXPENDITURES AND LIABILITY OF PARTIES	11
	A. LIABILITY OF PARTIES:	11
	B. LIENS AND SECURITY INTERESTS:	11
	C. ADVANCES:	12
	D. DEFAULTS AND REMEDIES:	12
	1. Suspension of Rights	13
	2. Suit for Damages	13
	3. Deemed Non-Consent	13
	4. Advance Payment	13
	5. Costs and Attorneys' Fees	13
	E. RENTALS, SHUT-IN WELL PAYMENTS AND MINIMUM ROYALTIES:	13
	F. TAXES:	13
VIII.	ACQUISITION, MAINTENANCE OR TRANSFER OF INTEREST	14
	A. SURRENDER OF LEASES:	14
	B. RENEWAL OR EXTENSION OF LEASES:	14
	C. ACREAGE OR CASH CONTRIBUTIONS:	14

TABLE OF CONTENTS

  D. ASSIGNMENT; MAINTENANCE OF UNIFORM INTEREST: 15
  E. WAIVER OF RIGHTS TO PARTITION: 15
  F. PREFERENTIAL RIGHT TO PURCHASE: 15
 IX. **INTERNAL REVENUE CODE ELECTION** 15
 X. **CLAIMS AND LAWSUITS** 15
 XI. **FORCE MAJEURE** 16
 XII. **NOTICES** 16
 XIII. **TERM OF AGREEMENT** 16
 XIV. **COMPLIANCE WITH LAWS AND REGULATIONS** 16
  A. LAWS, REGULATIONS AND ORDERS: 16
  B. GOVERNING LAW: 16
  C. REGULATORY AGENCIES: 16
 XV. **MISCELLANEOUS** 17
  A. EXECUTION: 17
  B. SUCCESSORS AND ASSIGNS: 17
  C. COUNTERPARTS: 17
  D. SEVERABILITY: 17
 XVI. **OTHER PROVISIONS** 17

OPERATING AGREEMENT

THIS AGREEMENT, entered into by and between _____ hereinafter designated and referred to as "Operator," and the signatory party or parties other than Operator, sometimes hereinafter referred to individually as "Non-Operator," and collectively as "Non-Operators."

WITNESSETH:

WHEREAS, the parties to this agreement are owners of Oil and Gas Leases and/or Oil and Gas Interests in the land identified in Exhibit "A," and the parties hereto have reached an agreement to explore and develop these Leases and/or Oil and Gas Interests for the production of Oil and Gas to the extent and as hereinafter provided,

NOW, THEREFORE, it is agreed as follows:

ARTICLE I.
DEFINITIONS

As used in this agreement, the following words and terms shall have the meanings here ascribed to them:

A. The term "AFE" shall mean an Authority for Expenditure prepared by a party to this agreement for the purpose of estimating the costs to be incurred in conducting an operation hereunder.

B. The term "Completion" or "Complete" shall mean a single operation intended to complete a well as a producer of Oil and Gas in one or more Zones, including, but not limited to, the setting of production casing, perforating, well stimulation and production testing conducted in such operation.

C. The term "Contract Area" shall mean all of the lands, Oil and Gas Leases and/or Oil and Gas Interests intended to be developed and operated for Oil and Gas purposes under this agreement. Such lands, Oil and Gas Leases and Oil and Gas Interests are described in Exhibit "A."

D. The term "Deepen" shall mean a single operation whereby a well is drilled to an objective Zone below the deepest Zone in which the well was previously drilled, or below the Deepest Zone proposed in the associated AFE, whichever is the lesser.

E. The terms "Drilling Party" and "Consenting Party" shall mean a party who agrees to join in and pay its share of the cost of any operation conducted under the provisions of this agreement.

F. The term "Drilling Unit" shall mean the area fixed for the drilling of one well by order or rule of any state or federal body having authority. If a Drilling Unit is not fixed by any such rule or order, a Drilling Unit shall be the drilling unit as established by the pattern of drilling in the Contract Area unless fixed by express agreement of the Drilling Parties.

G. The term "Drillsite" shall mean the Oil and Gas Lease or Oil and Gas Interest on which a proposed well is to be located.

H. The term "Initial Well" shall mean the well required to be drilled by the parties hereto as provided in Article VI.A.

I. The term "Non-Consent Well" shall mean a well in which less than all parties have conducted an operation as provided in Article VI.B.2.

J. The terms "Non-Drilling Party" and "Non-Consenting Party" shall mean a party who elects not to participate in a proposed operation.

K. The term "Oil and Gas" shall mean oil, gas, casinghead gas, gas condensate, and/or all other liquid or gaseous hydrocarbons and other marketable substances produced therewith, unless an intent to limit the inclusiveness of this term is specifically stated.

L. The term "Oil and Gas Interests" or "Interests" shall mean unleased fee and mineral interests in Oil and Gas in tracts of land lying within the Contract Area which are owned by parties to this agreement.

M. The terms "Oil and Gas Lease," "Lease" and "Leasehold" shall mean the oil and gas leases or interests therein covering tracts of land lying within the Contract Area which are owned by the parties to this agreement.

N. The term "Plug Back" shall mean a single operation whereby a deeper Zone is abandoned in order to attempt a Completion in a shallower Zone.

O. The term "Recompletion" or "Recomplete" shall mean an operation whereby a Completion in one Zone is abandoned in order to attempt a Completion in a different Zone within the existing wellbore.

P. The term "Rework" shall mean an operation conducted in the wellbore of a well after it is Completed to secure, restore, or improve production in a Zone which is currently open to production in the wellbore. Such operations include, but are not limited to, well stimulation operations but exclude any routine repair or maintenance work or drilling, Sidetracking, Deepening, Completing, Recompleting, or Plugging Back of a well.

Q. The term "Sidetrack" shall mean the directional control and intentional deviation of a well from vertical so as to change the bottom hole location unless done to straighten the hole or to drill around junk in the hole to overcome other mechanical difficulties.

R. The term "Zone" shall mean a stratum of earth containing or thought to contain a common accumulation of Oil and Gas separately producible from any other common accumulation of Oil and Gas.

Unless the context otherwise clearly indicates, words used in the singular include the plural, the word "person" includes natural and artificial persons, the plural includes the singular, and any gender includes the masculine, feminine, and neuter.

ARTICLE II.
EXHIBITS

The following exhibits, as indicated below and attached hereto, are incorporated in and made a part hereof:

_____ A. Exhibit "A," shall include the following information:
    (1) Description of lands subject to this agreement,
    (2) Restrictions, if any, as to depths, formations, or substances,
    (3) Parties to agreement with addresses and telephone numbers for notice purposes,
    (4) Percentages or fractional interests of parties to this agreement,
    (5) Oil and Gas Leases and/or Oil and Gas Interests subject to this agreement,
    (6) Burdens on production.
_____ B. Exhibit "B," Form of Lease.
_____ C. Exhibit "C," Accounting Procedure.
    D. Exhibit "D," Insurance.
_____ E. Exhibit "E," Gas Balancing Agreement.
_____ F. Exhibit "F," Non-Discrimination and Certification of Non-Segregated Facilities.
_____ G. Exhibit "G," Tax Partnership.
    H. Other: _____

If any provision of any exhibit, except Exhibits "E," "F" and "G," is inconsistent with any provision contained in the body of this agreement, the provisions in the body of this agreement shall prevail.

ARTICLE III.
INTERESTS OF PARTIES

A. Oil and Gas Interests:

If any party owns an Oil and Gas Interest in the Contract Area, that Interest shall be treated for all purposes of this agreement and during the term hereof as if it were covered by the form of Oil and Gas Lease attached hereto as Exhibit "B," and the owner thereof shall be deemed to own both royalty interest in such lease and the interest of the lessee thereunder.

B. Interests of Parties in Costs and Production:

Unless changed by other provisions, all costs and liabilities incurred in operations under this agreement shall be borne and paid, and all equipment and materials acquired in operations on the Contract Area shall be owned, by the parties as their interests are set forth in Exhibit "A." In the same manner, the parties shall also own all production of Oil and Gas from the Contract Area subject, however, to the payment of royalties and other burdens on production as described hereafter.

Regardless of which party has contributed any Oil and Gas Lease or Oil and Gas Interest on which royalty or other burdens may be payable and except as otherwise expressly provided in this agreement, each party shall pay or deliver, or cause to be paid or delivered, all burdens on its share of the production from the Contract Area up to, but not in excess of, _____ and shall indemnify, defend and hold the other parties free from any liability therefor. Except as otherwise expressly provided in this agreement, if any party has contributed hereto any Lease or Interest which is burdened with any royalty, overriding royalty, production payment or other burden on production in excess of the amounts stipulated above, such party so burdened shall assume and alone bear all such excess obligations and shall indemnify, defend and hold the other parties hereto harmless from any and all claims attributable to such excess burden. However, so long as the Drilling Unit for the productive Zone(s) is identical with the Contract Area, each party shall pay or deliver, or cause to be paid or delivered, all burdens on production from the Contract Area due under the terms of the Oil and Gas Lease(s) which such party has contributed to this agreement, and shall indemnify, defend and hold the other parties free from any liability therefor.

No party shall ever be responsible, on a price basis higher than the price received by such party, to any other party's lessor or royalty owner, and if such other party's lessor or royalty owner should demand and receive settlement on a higher price basis, the party contributing the affected Lease shall bear the additional royalty burden attributable to such higher price.

Nothing contained in this Article III.B. shall be deemed an assignment or cross-assignment of interests covered hereby, and in the event two or more parties contribute to this agreement jointly owned Leases, the parties' undivided interests in said Leaseholds shall be deemed separate leasehold interests for the purposes of this agreement.

C. Subsequently Created Interests:

If any party has contributed hereto a Lease or Interest that is burdened with an assignment of production given as security for the payment of money, or if, after the date of this agreement, any party creates an overriding royalty, production payment, net profits interest, assignment of production or other burden payable out of production attributable to its working interest hereunder, such burden shall be deemed a "Subsequently Created Interest." Further, if any party has contributed hereto a Lease or Interest burdened with an overriding royalty, production payment, net profits interest, or other burden payable out of production created prior to the date of this agreement, and such burden is not shown on Exhibit "A," such burden also shall be deemed a Subsequently Created Interest to the extent such burden causes the burdens on such party's Lease or Interest to exceed the amount stipulated in Article III.B. above.

The party whose interest is burdened with the Subsequently Created Interest (the "Burdened Party") shall assume and alone bear, pay and discharge the Subsequently Created Interest and shall indemnify, defend and hold harmless the other parties from and against any liability therefor. Further, if the Burdened Party fails to pay, when due, its share of expenses chargeable hereunder, all provisions of Article VII.B. shall be enforceable against the Subsequently Created Interest in the same manner as they are enforceable against the working interest of the Burdened Party. If the Burdened Party is required under this agreement to assign or relinquish to any other party, or parties, all or a portion of its working interest and/or the production attributable thereto, said other party, or parties, shall receive said assignment and/or production free and clear of said Subsequently Created Interest, and the Burdened Party shall indemnify, defend and hold harmless said other party, or parties, from any and all claims and demands for payment asserted by owners of the Subsequently Created Interest.

ARTICLE IV.
TITLES

A. Title Examination:

Title examination shall be made on the Drillsite of any proposed well prior to commencement of drilling operations and, if a majority in interest of the Drilling Parties so request or Operator so elects, title examination shall be made on the entire Drilling Unit, or maximum anticipated Drilling Unit, of the well. The opinion will include the ownership of the working interest, minerals, royalty, overriding royalty and production payments under the applicable Leases. Each party contributing Leases and/or Oil and Gas Interests to be included in the Drillsite or Drilling Unit, if appropriate, shall furnish to Operator all abstracts (including federal lease status reports), title opinions, title papers and curative material in its possession free of charge. All such information not in the possession of or made available to Operator by the parties, but necessary for the examination of the title, shall be obtained by Operator. Operator shall cause title to be examined by attorneys on its staff or by outside attorneys. Copies of all title opinions shall be furnished to each Drilling Party. Costs incurred by Operator in procuring abstracts, fees paid outside attorneys for title examination (including preliminary, supplemental, shut-in royalty opinions and division order title opinions) and other direct charges as provided in Exhibit "C" shall be borne by the Drilling Parties in the proportion that the interest of each Drilling Party bears to the total interest of all Drilling Parties as such interests appear in Exhibit "A." Operator shall make no charge for services rendered by its staff attorneys or other personnel in the performance of the above functions.

Each party shall be responsible for securing curative matter and pooling amendments or agreements required in connection with Leases or Oil and Gas Interests contributed by such party. Operator shall be responsible for the preparation and recording of pooling designations or declarations and communitization agreements as well as the conduct of hearings before governmental agencies for the securing of spacing or pooling orders or any other orders necessary or appropriate to the conduct of operations hereunder. This shall not prevent any party from appearing on its own behalf at such hearings. Costs incurred by Operator, including fees paid to outside attorneys, which are associated with hearings before governmental agencies, and which costs are necessary and proper for the activities contemplated under this agreement, shall be direct charges to the joint account and shall not be covered by the administrative overhead charges as provided in Exhibit "C."

Operator shall make no charge for services rendered by its staff attorneys or other personnel in the performance of the above functions.

No well shall be drilled on the Contract Area until after (1) the title to the Drillsite or Drilling Unit, if appropriate, has been examined as above provided, and (2) the title has been approved by the examining attorney or title has been accepted by all of the Drilling Parties in such well.

B. Loss or Failure of Title:

1. <u>Failure of Title:</u> Should any Oil and Gas Interest or Oil and Gas Lease be lost through failure of title, which results in a reduction of interest from that shown on Exhibit "A," the party credited with contributing the affected Lease or Interest (including, if applicable, a successor in interest to such party) shall have ninety (90) days from final determination of title failure to acquire a new lease or other instrument curing the entirety of the title failure, which acquisition will not be subject to Article VIII.B., and failing to do so, this agreement, nevertheless, shall continue in force as to all remaining Oil and Gas Leases and Interests; and,

(a) The party credited with contributing the Oil and Gas Lease or Interest affected by the title failure (including, if applicable, a successor in interest to such party) shall bear alone the entire loss and it shall not be entitled to recover from Operator or the other parties any development or operating costs which it may have previously paid or incurred, but there shall be no additional liability on its part to the other parties hereto by reason of such title failure;

(b) There shall be no retroactive adjustment of expenses incurred or revenues received from the operation of the Lease or Interest which has failed, but the interests of the parties contained on Exhibit "A" shall be revised on an acreage basis, as of the time it is determined finally that title failure has occurred, so that the interest of the party whose Lease or Interest is affected by the title failure will thereafter be reduced in the Contract Area by the amount of the Lease or Interest failed;

(c) If the proportionate interest of the other parties hereto in any producing well previously drilled on the Contract Area is increased by reason of the title failure, the party who bore the costs incurred in connection with such well attributable to the Lease or Interest which has failed shall receive the proceeds attributable to the increase in such interest (less costs and burdens attributable thereto) until it has been reimbursed for unrecovered costs paid by it in connection with such well attributable to such failed Lease or Interest;

(d) Should any person not a party to this agreement, who is determined to be the owner of any Lease or Interest which has failed, pay in any manner any part of the cost of operation, development, or equipment, such amount shall be paid to the party or parties who bore the costs which are so refunded;

(e) Any liability to account to a person not a party to this agreement for prior production of Oil and Gas which arises by reason of title failure shall be borne severally by each party (including a predecessor to a current party) who received production for which such accounting is required based on the amount of such production received, and each such party shall severally indemnify, defend and hold harmless all other parties hereto for any such liability to account;

(f) No charge shall be made to the joint account for legal expenses, fees or salaries in connection with the defense of the Lease or Interest claimed to have failed, but if the party contributing such Lease or Interest hereto elects to defend its title it shall bear all expenses in connection therewith; and

(g) If any party is given credit on Exhibit "A" to a Lease or Interest which is limited solely to ownership of an interest in the wellbore of any well or wells and the production therefrom, such party's absence of interest in the remainder of the Contract Area shall be considered a Failure of Title as to such remaining Contract Area unless that absence of interest is reflected on Exhibit "A."

2. <u>Loss by Non-Payment or Erroneous Payment of Amount Due:</u> If, through mistake or oversight, any rental, shut-in well payment, minimum royalty or royalty payment, or other payment necessary to maintain all or a portion of an Oil and Gas Lease or Interest is not paid or is erroneously paid, and as a result a Lease or Interest terminates, there shall be no monetary liability against the party who failed to make such payment. Unless the party who failed to make the required payment secures a new Lease or Interest covering the same interest within ninety (90) days from the discovery of the failure to make proper payment, which acquisition will not be subject to Article VIII.B., the interests of the parties reflected on Exhibit "A" shall be revised on an acreage basis, effective as of the date of termination of the Lease or Interest involved, and the party who failed to make proper payment will no longer be credited with an interest in the Contract Area on account of ownership of the Lease or Interest which has terminated. If the party who failed to make the required payment shall not have been fully reimbursed, at the time of the loss, from the proceeds of the sale of Oil and Gas attributable to the lost Lease or Interest, calculated on an acreage basis, for the development and operating costs previously paid on account of such Lease or Interest, it shall be reimbursed for unrecovered actual costs previously paid by it (but not for its share of the cost of any dry hole previously drilled or wells previously abandoned) from so much of the following as is necessary to effect reimbursement:

(a) Proceeds of Oil and Gas produced prior to termination of the Lease or Interest, less operating expenses and lease burdens chargeable hereunder to the person who failed to make payment, previously accrued to the credit of the lost Lease or Interest, on an acreage basis, up to the amount of unrecovered costs;

(b) Proceeds of Oil and Gas, less operating expenses and lease burdens chargeable hereunder to the person who failed to make payment, up to the amount of unrecovered costs attributable to that portion of Oil and Gas thereafter produced and marketed (excluding production from any wells thereafter drilled) which, in the absence of such Lease or Interest termination, would be attributable to the lost Lease or Interest on an acreage basis and which as a result of such Lease or Interest termination is credited to other parties, the proceeds of said portion of the Oil and Gas to be contributed by the other parties in proportion to their respective interests reflected on Exhibit "A"; and,

(c) Any monies, up to the amount of unrecovered costs, that may be paid by any party who is, or becomes, the owner of the Lease or Interest lost, for the privilege of participating in the Contract Area or becoming a party to this agreement.

3. <u>Other Losses:</u> All losses of Leases or Interests committed to this agreement, other than those set forth in Articles IV.B.1. and IV.B.2. above, shall be joint losses and shall be borne by all parties in proportion to their interests shown on Exhibit "A". This shall include but not be limited to the loss of any Lease or Interest through failure to develop or because express or implied covenants have not been performed (other than performance which requires only the payment of money), and the loss of any Lease by expiration at the end of its primary term if it is not renewed or extended. There shall be no readjustment of interests in the remaining portion of the Contract Area on account of any joint loss.

4. <u>Curing Title:</u> In the event of a Failure of Title under Article IV.B.1. or a loss of title under Article IV.B.2. above, any Lease or Interest acquired by any party hereto (other than the party whose interest has failed or was lost) during the ninety (90) day period provided by Article IV.B.1. and Article IV.B.2. above covering all or a portion of the interest that has failed or was lost shall be offered at cost to the party whose interest has failed or was lost, and the provisions of Article VIII.B. shall not apply to such acquisition.

ARTICLE V.
OPERATOR

A. Designation and Responsibilities of Operator:

_____ shall be the Operator of the Contract Area, and shall conduct and direct and have full control of all operations on the Contract Area as permitted and required by, and within the limits of this agreement. In its performance of services hereunder for the Non-Operators, Operator shall be an independent contractor not subject to the control or direction of the Non-Operators except as to the type of operation to be undertaken in accordance with the election procedures contained in this agreement. Operator shall not be deemed, or hold itself out as, the agent of the Non-Operators with authority to bind them to any obligation or liability assumed or incurred by Operator as to any third party. Operator shall conduct its activities under this agreement as a reasonable prudent operator, in a good and workmanlike manner, with due diligence and dispatch, in accordance with good oilfield practice, and in compliance with applicable law and regulation, but in no event shall it have any liability as Operator to the other parties for losses sustained or liabilities incurred except such as may result from gross negligence or willful misconduct.

B. Resignation or Removal of Operator and Selection of Successor:

1. Resignation or Removal of Operator: Operator may resign at any time by giving written notice thereof to Non-Operators. If Operator terminates its legal existence, no longer owns an interest hereunder in the Contract Area, or is no longer capable of serving as Operator, Operator shall be deemed to have resigned without any action by Non-Operators, except the selection of a successor. Operator may be removed only for good cause by the affirmative vote of Non-Operators owning a majority interest based on ownership as shown on Exhibit "A" remaining after excluding the voting interest of Operator; such vote shall not be deemed effective until a written notice has been delivered to the Operator by a Non-Operator detailing the alleged default and Operator has failed to cure the default within thirty (30) days from its receipt of the notice or, if the default concerns an operation then being conducted, within forty-eight (48) hours of its receipt of the notice. For purposes hereof, "good cause" shall mean not only gross negligence or willful misconduct but also the material breach of or inability to meet the standards of operation contained in Article V.A. or material failure or inability to perform its obligations under this agreement.

Subject to Article VII.D.1., such resignation or removal shall not become effective until 7:00 o'clock A.M. on the first day of the calendar month following the expiration of ninety (90) days after the giving of notice of resignation by Operator or action by the Non-Operators to remove Operator, unless a successor Operator has been selected and assumes the duties of Operator at an earlier date. Operator, after effective date of resignation or removal, shall be bound by the terms hereof as a Non-Operator. A change of a corporate name or structure of Operator or transfer of Operator's interest to any single subsidiary, parent or successor corporation shall not be the basis for removal of Operator.

2. Selection of Successor Operator: Upon the resignation or removal of Operator under any provision of this agreement, a successor Operator shall be selected by the parties. The successor Operator shall be selected from the parties owning an interest in the Contract Area at the time such successor Operator is selected. The successor Operator shall be selected by the affirmative vote of two (2) or more parties owning a majority interest based on ownership as shown on Exhibit "A"; provided, however, if an Operator which has been removed or is deemed to have resigned fails to vote or votes only to succeed itself, the successor Operator shall be selected by the affirmative vote of the party or parties owning a majority interest based on ownership as shown on Exhibit "A" remaining after excluding the voting interest of the Operator that was removed or resigned. The former Operator shall promptly deliver to the successor Operator all records and data relating to the operations conducted by the former Operator to the extent such records and data are not already in the possession of the successor operator. Any cost of obtaining or copying the former Operator's records and data shall be charged to the joint account.

3. Effect of Bankruptcy: If Operator becomes insolvent, bankrupt or is placed in receivership, it shall be deemed to have resigned without any action by Non-Operators, except the selection of a successor. If a petition for relief under the federal bankruptcy laws is filed by or against Operator, and the removal of Operator is prevented by the federal bankruptcy court, all Non-Operators and Operator shall comprise an interim operating committee to serve until Operator has elected to reject or assume this agreement pursuant to the Bankruptcy Code, and an election to reject this agreement by Operator as a debtor in possession, or by a trustee in bankruptcy, shall be deemed a resignation as Operator without any action by Non-Operators, except the selection of a successor. During the period of time the operating committee controls operations, all actions shall require the approval of two (2) or more parties owning a majority interest based on ownership as shown on Exhibit "A." In the event there are only two (2) parties to this agreement, during the period of time the operating committee controls operations, a third party acceptable to Operator, Non-Operator and the federal bankruptcy court shall be selected as a member of the operating committee, and all actions shall require the approval of two (2) members of the operating committee without regard for their interest in the Contract Area based on Exhibit "A."

C. Employees and Contractors:

The number of employees or contractors used by Operator in conducting operations hereunder, their selection, and the hours of labor and the compensation for services performed shall be determined by Operator, and all such employees or contractors shall be the employees or contractors of Operator.

D. Rights and Duties of Operator:

1. Competitive Rates and Use of Affiliates: All wells drilled on the Contract Area shall be drilled on a competitive contract basis at the usual rates prevailing in the area. If it so desires, Operator may employ its own tools and equipment in the drilling of wells, but its charges therefor shall not exceed the prevailing rates in the area and the rate of such charges shall be agreed upon by the parties in writing before drilling operations are commenced, and such work shall be performed by Operator under the same terms and conditions as are customary and usual in the area in contracts of independent contractors who are doing work of a similar nature. All work performed or materials supplied by affiliates or related parties of Operator shall be performed or supplied at competitive rates, pursuant to written agreement, and in accordance with customs and standards prevailing in the industry.

2. Discharge of Joint Account Obligations: Except as herein otherwise specifically provided, Operator shall promptly pay and discharge expenses incurred in the development and operation of the Contract Area pursuant to this agreement and shall charge each of the parties hereto with their respective proportionate shares upon the expense basis provided in Exhibit "C." Operator shall keep an accurate record of the joint account hereunder, showing expenses incurred and charges and credits made and received.

3. Protection from Liens: Operator shall pay, or cause to be paid, as and when they become due and payable, all accounts of contractors and suppliers and wages and salaries for services rendered or performed, and for materials supplied on, to or in respect of the Contract Area or any operations for the joint account thereof, and shall keep the Contract Area free from

liens and encumbrances resulting therefrom except for those resulting from a bona fide dispute as to services rendered or materials supplied.

4. <u>Custody of Funds:</u> Operator shall hold for the account of the Non-Operators any funds of the Non-Operators advanced or paid to the Operator, either for the conduct of operations hereunder or as a result of the sale of production from the Contract Area, and such funds shall remain the funds of the Non-Operators on whose account they are advanced or paid until used for their intended purpose or otherwise delivered to the Non-Operators or applied toward the payment of debts as provided in Article VII.B. Nothing in this paragraph shall be construed to establish a fiduciary relationship between Operator and Non-Operators for any purpose other than to account for Non-Operator funds as herein specifically provided. Nothing in this paragraph shall require the maintenance by Operator of separate accounts for the funds of Non-Operators unless the parties otherwise specifically agree.

5. <u>Access to Contract Area and Records:</u> Operator shall, except as otherwise provided herein, permit each Non-Operator or its duly authorized representative, at the Non-Operator's sole risk and cost, full and free access at all reasonable times to all operations of every kind and character being conducted for the joint account on the Contract Area and to the records of operations conducted thereon or production therefrom, including Operator's books and records relating thereto. Such access rights shall not be exercised in a manner interfering with Operator's conduct of an operation hereunder and shall not obligate Operator to furnish any geologic or geophysical data of an interpretive nature unless the cost of preparation of such interpretive data was charged to the joint account. Operator will furnish to each Non-Operator upon request copies of any and all reports and information obtained by Operator in connection with production and related items, including, without limitation, meter and chart reports, production purchaser statements, run tickets and monthly gauge reports, but excluding purchase contracts and pricing information to the extent not applicable to the production of the Non-Operator seeking the information. Any audit of Operator's records relating to amounts expended and the appropriateness of such expenditures shall be conducted in accordance with the audit protocol specified in Exhibit "C."

6. <u>Filing and Furnishing Governmental Reports:</u> Operator will file, and upon written request promptly furnish copies to each requesting Non-Operator not in default of its payment obligations, all operational notices, reports or applications required to be filed by local, State, Federal or Indian agencies or authorities having jurisdiction over operations hereunder. Each Non-Operator shall provide to Operator on a timely basis all information necessary to Operator to make such filings.

7. <u>Drilling and Testing Operations:</u> The following provisions shall apply to each well drilled hereunder, including but not limited to the Initial Well:

(a) Operator will promptly advise Non-Operators of the date on which the well is spudded, or the date on which drilling operations are commenced.

(b) Operator will send to Non-Operators such reports, test results and notices regarding the progress of operations on the well as the Non-Operators shall reasonably request, including, but not limited to, daily drilling reports, completion reports, and well logs.

(c) Operator shall adequately test all Zones encountered which may reasonably be expected to be capable of producing Oil and Gas in paying quantities as a result of examination of the electric log or any other logs or cores or tests conducted hereunder.

8. <u>Cost Estimates.</u> Upon request of any Consenting Party, Operator shall furnish estimates of current and cumulative costs incurred for the joint account at reasonable intervals during the conduct of any operation pursuant to this agreement. Operator shall not be held liable for errors in such estimates so long as the estimates are made in good faith.

9. <u>Insurance:</u> At all times while operations are conducted hereunder, Operator shall comply with the workers compensation law of the state where the operations are being conducted; provided, however, that Operator may be a self-insurer for liability under said compensation laws in which event the only charge that shall be made to the joint account shall be as provided in Exhibit "C." Operator shall also carry or provide insurance for the benefit of the joint account of the parties as outlined in Exhibit "D" attached hereto and made a part hereof. Operator shall require all contractors engaged in work on or for the Contract Area to comply with the workers compensation law of the state where the operations are being conducted and to maintain such other insurance as Operator may require.

In the event automobile liability insurance is specified in said Exhibit "D," or subsequently receives the approval of the parties, no direct charge shall be made by Operator for premiums paid for such insurance for Operator's automotive equipment.

ARTICLE VI.
DRILLING AND DEVELOPMENT

A. Initial Well:

On or before the _____ day of _____ , _____ , Operator shall commence the drilling of the Initial Well at the following location:

and shall thereafter continue the drilling of the well with due diligence to

The drilling of the Initial Well and the participation therein by all parties is obligatory, subject to Article VI.C.1. as to participation in Completion operations and Article VI.F. as to termination of operations and Article XI as to occurrence of force majeure.

B. Subsequent Operations:

1. <u>Proposed Operations:</u> If any party hereto should desire to drill any well on the Contract Area other than the Initial Well, or if any party should desire to Rework, Sidetrack, Deepen, Recomplete or Plug Back a dry hole or a well no longer capable of producing in paying quantities in which such party has not otherwise relinquished its interest in the proposed objective Zone under this agreement, the party desiring to drill, Rework, Sidetrack, Deepen, Recomplete or Plug Back such a well shall give written notice of the proposed operation to the parties who have not otherwise relinquished their interest in such objective Zone

under this agreement and to all other parties in the case of a proposal for Sidetracking or Deepening, specifying the work to be performed, the location, proposed depth, objective Zone and the estimated cost of the operation. The parties to whom such a notice is delivered shall have thirty (30) days after receipt of the notice within which to notify the party proposing to do the work whether they elect to participate in the cost of the proposed operation. If a drilling rig is on location, notice of a proposal to Rework, Sidetrack, Recomplete, Plug Back or Deepen may be given by telephone and the response period shall be limited to forty-eight (48) hours, exclusive of Saturday, Sunday and legal holidays. Failure of a party to whom such notice is delivered to reply within the period above fixed shall constitute an election by that party not to participate in the cost of the proposed operation. Any proposal by a party to conduct an operation conflicting with the operation initially proposed shall be delivered to all parties within the time and in the manner provided in Article VI.B.6.

If all parties to whom such notice is delivered elect to participate in such a proposed operation, the parties shall be contractually committed to participate therein provided such operations are commenced within the time period hereafter set forth, and Operator shall, no later than ninety (90) days after expiration of the notice period of thirty (30) days (or as promptly as practicable after the expiration of the forty-eight (48) hour period when a drilling rig is on location, as the case may be), actually commence the proposed operation and thereafter complete it with due diligence at the risk and expense of the parties participating therein; provided, however, said commencement date may be extended upon written notice of same by Operator to the other parties, for a period of up to thirty (30) additional days if, in the sole opinion of Operator, such additional time is reasonably necessary to obtain permits from governmental authorities, surface rights (including rights-of-way) or appropriate drilling equipment, or to complete title examination or curative matter required for title approval or acceptance. If the actual operation has not been commenced within the time provided (including any extension thereof as specifically permitted herein or in the force majeure provisions of Article XI) and if any party hereto still desires to conduct said operation, written notice proposing same must be resubmitted to the other parties in accordance herewith as if no prior proposal had been made. Those parties that did not participate in the drilling of a well for which a proposal to Deepen or Sidetrack is made hereunder shall, if such parties desire to participate in the proposed Deepening or Sidetracking operation, reimburse the Drilling Parties in accordance with Article VI.B.4. in the event of a Deepening operation and in accordance with Article VI.B.5. in the event of a Sidetracking operation.

2. Operations by Less Than All Parties:

(a) Determination of Participation. If any party to whom such notice is delivered as provided in Article VI.B.1. or VI.C.1. (Option No. 2) elects not to participate in the proposed operation, then, in order to be entitled to the benefits of this Article, the party or parties giving the notice and such other parties as shall elect to participate in the operation shall, no later than ninety (90) days after the expiration of the notice period of thirty (30) days (or as promptly as practicable after the expiration of the forty-eight (48) hour period when a drilling rig is on location, as the case may be) actually commence the proposed operation and complete it with due diligence. Operator shall perform all work for the account of the Consenting Parties; provided, however, if no drilling rig or other equipment is on location, and if Operator is a Non-Consenting Party, the Consenting Parties shall either: (i) request Operator to perform the work required by such proposed operation for the account of the Consenting Parties, or (ii) designate one of the Consenting Parties as Operator to perform such work. The rights and duties granted to and imposed upon the Operator under this agreement are granted to and imposed upon the party designated as Operator for an operation in which the original Operator is a Non-Consenting Party. Consenting Parties, when conducting operations on the Contract Area pursuant to this Article VI.B.2., shall comply with all terms and conditions of this agreement.

If less than all parties approve any proposed operation, the proposing party, immediately after the expiration of the applicable notice period, shall advise all Parties of the total interest of the parties approving such operation and its recommendation as to whether the Consenting Parties should proceed with the operation as proposed. Each Consenting Party, within forty-eight (48) hours (exclusive of Saturday, Sunday and legal holidays) after delivery of such notice, shall advise the proposing party of its desire to (i) limit participation to such party's interest as shown on Exhibit "A" or (ii) carry only its proportionate part (determined by dividing such party's interest in the Contract Area by the interests of all Consenting Parties in the Contract Area) of Non-Consenting Parties' interests, or (iii) carry its proportionate part (determined as provided in (ii)) of Non-Consenting Parties' interests together with all or a portion of its proportionate part of any Non-Consenting Parties' interests that any Consenting Party did not elect to take. Any interest of Non-Consenting Parties that is not carried by a Consenting Party shall be deemed to be carried by the party proposing the operation if such party does not withdraw its proposal. Failure to advise the proposing party within the time required shall be deemed an election under (i) . In the event a drilling rig is on location, notice may be given by telephone, and the time permitted for such a response shall not exceed a total of forty-eight (48) hours (exclusive of Saturday, Sunday and legal holidays). The proposing party, at its election, may withdraw such proposal if there is less than 100% participation and shall notify all parties of such decision within ten (10) days, or within twenty-four (24) hours if a drilling rig is on location, following expiration of the applicable response period. If 100% subscription to the proposed operation is obtained, the proposing party shall promptly notify the Consenting Parties of their proportionate interests in the operation and the party serving as Operator shall commence such operation within the period provided in Article VI.B.1., subject to the same extension right as provided therein.

(b) Relinquishment of Interest for Non-Participation. The entire cost and risk of conducting such operations shall be borne by the Consenting Parties in the proportions they have elected to bear same under the terms of the preceding paragraph. Consenting Parties shall keep the leasehold estates involved in such operations free and clear of all liens and encumbrances of every kind created by or arising from the operations of the Consenting Parties. If such an operation results in a dry hole, then subject to Articles VI.B.6. and VI.E.3., the Consenting Parties shall plug and abandon the well and restore the surface location at their sole cost, risk and expense; provided, however, that those Non-Consenting Parties that participated in the drilling, Deepening or Sidetracking of the well shall remain liable for, and shall pay, their proportionate shares of the cost of plugging and abandoning the well and restoring the surface location insofar only as those costs were not increased by the subsequent operations of the Consenting Parties. If any well drilled, Reworked, Sidetracked, Deepened, Recompleted or Plugged Back under the provisions of this Article results in a well capable of producing Oil and/or Gas in paying quantities, the Consenting Parties shall Complete and equip the well to produce at their sole cost and risk, and the well shall then be turned over to Operator (if the Operator did not conduct the operation) and shall be operated by it at the expense and for the account of the Consenting Parties. Upon commencement of operations for the drilling, Reworking, Sidetracking, Recompleting, Deepening or Plugging Back of any such well by Consenting Parties in accordance with the provisions of this Article, each Non-Consenting Party shall be deemed to have relinquished to Consenting Parties, and the Consenting Parties shall own and be entitled to receive, in proportion to their respective interests, all of such Non-Consenting Party's interest in the well and share of production therefrom or, in the case of a Reworking, Sidetracking,

Deepening, Recompleting or Plugging Back, or a Completion pursuant to Article VI.C.1. Option No. 2, all of such Non-Consenting Party's interest in the production obtained from the operation in which the Non-Consenting Party did not elect to participate. Such relinquishment shall be effective until the proceeds of the sale of such share, calculated at the well, or market value thereof if such share is not sold (after deducting applicable ad valorem, production, severance, and excise taxes, royalty, overriding royalty and other interests not excepted by Article III.C. payable out of or measured by the production from such well accruing with respect to such interest until it reverts), shall equal the total of the following:

(i) _____ % of each such Non-Consenting Party's share of the cost of any newly acquired surface equipment beyond the wellhead connections (including but not limited to stock tanks, separators, treaters, pumping equipment and piping), plus 100% of each such Non-Consenting Party's share of the cost of operation of the well commencing with first production and continuing until each such Non-Consenting Party's relinquished interest shall revert to it under other provisions of this Article, it being agreed that each Non-Consenting Party's share of such costs and equipment will be that interest which would have been chargeable to such Non-Consenting Party had it participated in the well from the beginning of the operations; and

(ii) _____ % of (a) that portion of the costs and expenses of drilling, Reworking, Sidetracking, Deepening, Plugging Back, testing, Completing, and Recompleting, after deducting any cash contributions received under Article VIII.C., and of (b) that portion of the cost of newly acquired equipment in the well (to and including the wellhead connections), which would have been chargeable to such Non-Consenting Party if it had participated therein.

Notwithstanding anything to the contrary in this Article VI.B., if the well does not reach the deepest objective Zone described in the notice proposing the well for reasons other than the encountering of granite or practically impenetrable substance or other condition in the hole rendering further operations impracticable, Operator shall give notice thereof to each Non-Consenting Party who submitted or voted for an alternative proposal under Article VI.B.6. to drill the well to a shallower Zone than the deepest objective Zone proposed in the notice under which the well was drilled, and each such Non-Consenting Party shall have the option to participate in the initial proposed Completion of the well by paying its share of the cost of drilling the well to its actual depth, calculated in the manner provided in Article VI.B.4. (a). If any such Non-Consenting Party does not elect to participate in the first Completion proposed for such well, the relinquishment provisions of this Article VI.B.2. (b) shall apply to such party's interest.

(c) <u>Reworking, Recompleting or Plugging Back.</u> An election not to participate in the drilling, Sidetracking or Deepening of a well shall be deemed an election not to participate in any Reworking or Plugging Back operation proposed in such a well, or portion thereof, to which the initial non-consent election applied that is conducted at any time prior to full recovery by the Consenting Parties of the Non-Consenting Party's recoupment amount. Similarly, an election not to participate in the Completing or Recompleting of a well shall be deemed an election not to participate in any Reworking operation proposed in such a well, or portion thereof, to which the initial non-consent election applied that is conducted at any time prior to full recovery by the Consenting Parties of the Non-Consenting Party's recoupment amount. Any such Reworking, Recompleting or Plugging Back operation conducted during the recoupment period shall be deemed part of the cost of operation of said well and there shall be added to the sums to be recouped by the Consenting Parties _____ % of that portion of the costs of the Reworking, Recompleting or Plugging Back operation which would have been chargeable to such Non-Consenting Party had it participated therein. If such a Reworking, Recompleting or Plugging Back operation is proposed during such recoupment period, the provisions of this Article VI.B. shall be applicable as between said Consenting Parties in said well.

(d) <u>Recoupment Matters.</u> During the period of time Consenting Parties are entitled to receive Non-Consenting Party's share of production, or the proceeds therefrom, Consenting Parties shall be responsible for the payment of all ad valorem, production, severance, excise, gathering and other taxes, and all royalty, overriding royalty and other burdens applicable to Non-Consenting Party's share of production not excepted by Article III.C.

In the case of any Reworking, Sidetracking, Plugging Back, Recompleting or Deepening operation, the Consenting Parties shall be permitted to use, free of cost, all casing, tubing and other equipment in the well, but the ownership of all such equipment shall remain unchanged; and upon abandonment of a well after such Reworking, Sidetracking, Plugging Back, Recompleting or Deepening, the Consenting Parties shall account for all such equipment to the owners thereof, with each party receiving its proportionate part in kind or in value, less cost of salvage.

Within ninety (90) days after the completion of any operation under this Article, the party conducting the operations for the Consenting Parties shall furnish each Non-Consenting Party with an inventory of the equipment in and connected to the well, and an itemized statement of the cost of drilling, Sidetracking, Deepening, Plugging Back, testing, Completing, Recompleting, and equipping the well for production; or, at its option, the operating party, in lieu of an itemized statement of such costs of operation, may submit a detailed statement of monthly billings. Each month thereafter, during the time the Consenting Parties are being reimbursed as provided above, the party conducting the operations for the Consenting Parties shall furnish the Non-Consenting Parties with an itemized statement of all costs and liabilities incurred in the operation of the well, together with a statement of the quantity of Oil and Gas produced from it and the amount of proceeds realized from the sale of the well's working interest production during the preceding month. In determining the quantity of Oil and Gas produced during any month, Consenting Parties shall use industry accepted methods such as but not limited to metering or periodic well tests. Any amount realized from the sale or other disposition of equipment newly acquired in connection with any such operation which would have been owned by a Non-Consenting Party had it participated therein shall be credited against the total unreturned costs of the work done and of the equipment purchased in determining when the interest of such Non-Consenting Party shall revert to it as above provided; and if there is a credit balance, it shall be paid to such Non-Consenting Party.

If and when the Consenting Parties recover from a Non-Consenting Party's relinquished interest the amounts provided for above, the relinquished interests of such Non-Consenting Party shall automatically revert to it as of 7:00 a.m. on the day following the day on which such recoupment occurs, and, from and after such reversion, such Non-Consenting Party shall own the same interest in such well, the material and equipment in or pertaining thereto, and the production therefrom as such Non-Consenting Party would have been entitled to had it participated in the drilling, Sidetracking, Reworking, Deepening, Recompleting or Plugging Back of said well. Thereafter, such Non-Consenting Party shall be charged with and shall pay its proportionate part of the further costs of the operation of said well in accordance with the terms of this agreement and Exhibit "C" attached hereto.

3. <u>Stand-By Costs:</u> When a well which has been drilled or Deepened has reached its authorized depth and all tests have been completed and the results thereof furnished to the parties, or when operations on the well have been otherwise terminated pursuant to Article VI.F., stand-by costs incurred pending response to a party's notice proposing a Reworking,

Sidetracking, Deepening, Recompleting, Plugging Back or Completing operation in such a well (including the period required under Article VI.B.6. to resolve competing proposals) shall be charged and borne as part of the drilling or Deepening operation just completed. Stand-by costs subsequent to all parties responding, or expiration of the response time permitted, whichever first occurs, and prior to agreement as to the participating interests of all Consenting Parties pursuant to the terms of the second grammatical paragraph of Article VI.B.2. (a), shall be charged to and borne as part of the proposed operation, but if the proposal is subsequently withdrawn because of insufficient participation, such stand-by costs shall be allocated between the Consenting Parties in the proportion each Consenting Party's interest as shown on Exhibit "A" bears to the total interest as shown on Exhibit "A" of all Consenting Parties.

In the event that notice for a Sidetracking operation is given while the drilling rig to be utilized is on location, any party may request and receive up to five (5) additional days after expiration of the forty-eight hour response period specified in Article VI.B.1. within which to respond by paying for all stand-by costs and other costs incurred during such extended response period; Operator may require such party to pay the estimated stand-by time in advance as a condition to extending the response period. If more than one party elects to take such additional time to respond to the notice, standby costs shall be allocated between the parties taking additional time to respond on a day-to-day basis in the proportion each electing party's interest as shown on Exhibit "A" bears to the total interest as shown on Exhibit "A" of all the electing parties.

4. Deepening: If less than all the parties elect to participate in a drilling, Sidetracking, or Deepening operation proposed pursuant to Article VI.B.1., the interest relinquished by the Non-Consenting Parties to the Consenting Parties under Article VI.B.2. shall relate only and be limited to the lesser of (i) the total depth actually drilled or (ii) the objective depth or Zone of which the parties were given notice under Article VI.B.1. ("Initial Objective"). Such well shall not be Deepened beyond the Initial Objective without first complying with this Article to afford the Non-Consenting Parties the opportunity to participate in the Deepening operation.

In the event any Consenting Party desires to drill or Deepen a Non-Consent Well to a depth below the Initial Objective, such party shall give notice thereof, complying with the requirements of Article VI.B.1., to all parties (including Non-Consenting Parties). Thereupon, Articles VI.B.1. and 2. shall apply and all parties receiving such notice shall have the right to participate or not participate in the Deepening of such well pursuant to said Articles VI.B.1. and 2. If a Deepening operation is approved pursuant to such provisions, and if any Non-Consenting Party elects to participate in the Deepening operation, such Non-Consenting party shall pay or make reimbursement (as the case may be) of the following costs and expenses:

(a) If the proposal to Deepen is made prior to the Completion of such well as a well capable of producing in paying quantities, such Non-Consenting Party shall pay (or reimburse Consenting Parties for, as the case may be) that share of costs and expenses incurred in connection with the drilling of said well from the surface to the Initial Objective which Non-Consenting Party would have paid had such Non-Consenting Party agreed to participate therein, plus the Non-Consenting Party's share of the cost of Deepening and of participating in any further operations on the well in accordance with the other provisions of this Agreement; provided, however, all costs for testing and Completion or attempted Completion of the well incurred by Consenting Parties prior to the point of actual operations to Deepen beyond the Initial Objective shall be for the sole account of Consenting Parties.

(b) If the proposal is made for a Non-Consent Well that has been previously Completed as a well capable of producing in paying quantities, but is no longer capable of producing in paying quantities, such Non-Consenting Party shall pay (or reimburse Consenting Parties for, as the case may be) its proportionate share of all costs of drilling, Completing, and equipping said well from the surface to the Initial Objective, calculated in the manner provided in paragraph (a) above, less those costs recouped by the Consenting Parties from the sale of production from the well. The Non-Consenting Party shall also pay its proportionate share of all costs of re-entering said well. The Non-Consenting Parties' proportionate part (based on the percentage of such well Non-Consenting Party would have owned had it previously participated in such Non-Consent Well) of the costs of salvable materials and equipment remaining in the hole and salvable surface equipment used in connection with such well shall be determined in accordance with Exhibit "C." If the Consenting Parties have recouped the cost of drilling, Completing, and equipping the well at the time such Deepening operation is conducted, then a Non-Consenting Party may participate in the Deepening of the well with no payment for costs incurred prior to re-entering the well for Deepening.

The foregoing shall not imply a right of any Consenting Party to propose any Deepening for a Non-Consent Well prior to the drilling of such well to its Initial Objective without the consent of the other Consenting Parties as provided in Article VI.F.

5. Sidetracking: Any party having the right to participate in a proposed Sidetracking operation that does not own an interest in the affected wellbore at the time of the notice shall, upon electing to participate, tender to the wellbore owners its proportionate share (equal to its interest in the Sidetracking operation) of the value of that portion of the existing wellbore to be utilized as follows:

(a) If the proposal is for Sidetracking an existing dry hole, reimbursement shall be on the basis of the actual costs incurred in the initial drilling of the well down to the depth at which the Sidetracking operation is initiated.

(b) If the proposal is for Sidetracking a well which has previously produced, reimbursement shall be on the basis of such party's proportionate share of drilling and equipping costs incurred in the initial drilling of the well down to the depth at which the Sidetracking operation is conducted, calculated in the manner described in Article VI.B.4(b) above. Such party's proportionate share of the cost of the well's salvable materials and equipment down to the depth at which the Sidetracking operation is initiated shall be determined in accordance with the provisions of Exhibit "C."

6. Order of Preference of Operations. Except as otherwise specifically provided in this agreement, if any party desires to propose the conduct of an operation that conflicts with a proposal that has been made by a party under this Article VI, such party shall have fifteen (15) days from delivery of the initial proposal, in the case of a proposal to drill a well or to perform an operation on a well where no drilling rig is on location, or twenty-four (24) hours, exclusive of Saturday, Sunday and legal holidays, from delivery of the initial proposal, if a drilling rig is on location for the well on which such operation is to be conducted, to deliver to all parties entitled to participate in the proposed operation such party's alternative proposal, such alternate proposal to contain the same information required to be included in the initial proposal. Each party receiving such proposals shall elect by delivery of notice to Operator within five (5) days after expiration of the proposal period, or within twenty-four (24) hours (exclusive of Saturday, Sunday and legal holidays) if a drilling rig is on location for the well that is the subject of the proposals, to participate in one of the competing proposals. Any party not electing within the time required shall be deemed not to have voted. The proposal receiving the vote of parties owning the largest aggregate percentage interest of the parties voting shall have priority over all other competing proposals; in the case of a tie vote, the

initial proposal shall prevail. Operator shall deliver notice of such result to all parties entitled to participate in the operation within five (5) days after expiration of the election period (or within twenty-four (24) hours, exclusive of Saturday, Sunday and legal holidays, if a drilling rig is on location). Each party shall then have two (2) days (or twenty-four (24) hours if a rig is on location) from receipt of such notice to elect by delivery of notice to Operator to participate in such operation or to relinquish interest in the affected well pursuant to the provisions of Article VI.B.2.; failure by a party to deliver notice within such period shall be deemed an election not to participate in the prevailing proposal.

   7. Conformity to Spacing Pattern. Notwithstanding the provisions of this Article VI.B.2., it is agreed that no wells shall be proposed to be drilled to or Completed in or produced from a Zone from which a well located elsewhere on the Contract Area is producing, unless such well conforms to the then-existing well spacing pattern for such Zone.

   8. Paying Wells. No party shall conduct any Reworking, Deepening, Plugging Back, Completion, Recompletion, or Sidetracking operation under this agreement with respect to any well then capable of producing in paying quantities except with the consent of all parties that have not relinquished interests in the well at the time of such operation.

C. Completion of Wells; Reworking and Plugging Back:

   1. Completion: Without the consent of all parties, no well shall be drilled, Deepened or Sidetracked, except any well drilled, Deepened or Sidetracked pursuant to the provisions of Article VI.B.2. of this agreement. Consent to the drilling, Deepening or Sidetracking shall include:

   ☐ Option No. 1: All necessary expenditures for the drilling, Deepening or Sidetracking, testing, Completing and equipping of the well, including necessary tankage and/or surface facilities.

   ☐ Option No. 2: All necessary expenditures for the drilling, Deepening or Sidetracking and testing of the well. When such well has reached its authorized depth, and all logs, cores and other tests have been completed, and the results thereof furnished to the parties, Operator shall give immediate notice to the Non-Operators having the right to participate in a Completion attempt whether or not Operator recommends attempting to Complete the well, together with Operator's AFE for Completion costs if not previously provided. The parties receiving such notice shall have forty-eight (48) hours (exclusive of Saturday, Sunday and legal holidays) in which to elect by delivery of notice to Operator to participate in a recommended Completion attempt or to make a Completion proposal with an accompanying AFE. Operator shall deliver any such Completion proposal, or any Completion proposal conflicting with Operator's proposal, to the other parties entitled to participate in such Completion in accordance with the procedures specified in Article VI.B.6. Election to participate in a Completion attempt shall include consent to all necessary expenditures for the Completing and equipping of such well, including necessary tankage and/or surface facilities but excluding any stimulation operation not contained on the Completion AFE. Failure of any party receiving such notice to reply within the period above fixed shall constitute an election by that party not to participate in the cost of the Completion attempt; provided, that Article VI.B.6. shall control in the case of conflicting Completion proposals. If one or more, but less than all of the parties, elect to attempt a Completion, the provisions of Article VI.B.2. hereof (the phrase "Reworking, Sidetracking, Deepening, Recompleting or Plugging Back" as contained in Article VI.B.2. shall be deemed to include "Completing") shall apply to the operations thereafter conducted by less than all parties; provided, however, that Article VI.B.2 shall apply separately to each separate Completion or Recompletion attempt undertaken hereunder, and an election to become a Non-Consenting Party as to one Completion or Recompletion attempt shall not prevent a party from becoming a Consenting Party in subsequent Completion or Recompletion attempts regardless whether the Consenting Parties as to earlier Completions or Recompletions have recouped their costs pursuant to Article VI.B.2.; provided further, that any recoupment of costs by a Consenting Party shall be made solely from the production attributable to the Zone in which the Completion attempt is made. Election by a previous Non-Consenting Party to participate in a subsequent Completion or Recompletion attempt shall require such party to pay its proportionate share of the cost of salvable materials and equipment installed in the well pursuant to the previous Completion or Recompletion attempt, insofar and only insofar as such materials and equipment benefit the Zone in which such party participates in a Completion attempt.

   2. Rework, Recomplete or Plug Back: No well shall be Reworked, Recompleted or Plugged Back except a well Reworked, Recompleted, or Plugged Back pursuant to the provisions of Article VI.B.2. of this agreement. Consent to the Reworking, Recompleting or Plugging Back of a well shall include all necessary expenditures in conducting such operations and Completing and equipping of said well, including necessary tankage and/or surface facilities.

D. Other Operations:

   Operator shall not undertake any single project reasonably estimated to require an expenditure in excess of _____ _____ Dollars ($ _____) except in connection with the drilling, Sidetracking, Reworking, Deepening, Completing, Recompleting or Plugging Back of a well that has been previously authorized by or pursuant to this agreement; provided, however, that, in case of explosion, fire, flood or other sudden emergency, whether of the same or different nature, Operator may take such steps and incur such expenses as in its opinion are required to deal with the emergency to safeguard life and property but Operator, as promptly as possible, shall report the emergency to the other parties. If Operator prepares an AFE for its own use, Operator shall furnish any Non-Operator so requesting an information copy thereof for any single project costing in excess of _____ Dollars ($ _____). Any party who has not relinquished its interest in a well shall have the right to propose that Operator perform repair work or undertake the installation of artificial lift equipment or ancillary production facilities such as salt water disposal wells or to conduct additional work with respect to a well drilled hereunder or other similar project (but not including the installation of gathering lines or other transportation or marketing facilities, the installation of which shall be governed by separate agreement between the parties) reasonably estimated to require an expenditure in excess of the amount first set forth above in this Article VI.D. (except in connection with an operation required to be proposed under Articles VI.B.1. or VI.C.1. Option No. 2, which shall be governed exclusively by those Articles). Operator shall deliver such proposal to all parties entitled to participate therein. If within thirty (30) days thereof Operator secures the written consent of any party or parties owning at least _____ % of the interests of the parties entitled to participate in such operation, each party having the right to participate in such project shall be bound by the terms of such proposal and shall be obligated to pay its proportionate share of the costs of the proposed project as if it had consented to such project pursuant to the terms of the proposal.

E. Abandonment of Wells:

   1. Abandonment of Dry Holes: Except for any well drilled or Deepened pursuant to Article VI.B.2., any well which has been drilled or Deepened under the terms of this agreement and is proposed to be completed as a dry hole shall not be

plugged and abandoned without the consent of all parties. Should Operator, after diligent effort, be unable to contact any party, or should any party fail to reply within forty-eight (48) hours (exclusive of Saturday, Sunday and legal holidays) after delivery of notice of the proposal to plug and abandon such well, such party shall be deemed to have consented to the proposed abandonment. All such wells shall be plugged and abandoned in accordance with applicable regulations and at the cost, risk and expense of the parties who participated in the cost of drilling or Deepening such well. Any party who objects to plugging and abandoning such well by notice delivered to Operator within forty-eight (48) hours (exclusive of Saturday, Sunday and legal holidays) after delivery of notice of the proposed plugging shall take over the well as of the end of such forty-eight (48) hour notice period and conduct further operations in search of Oil and/or Gas subject to the provisions of Article VI.B.; failure of such party to provide proof reasonably satisfactory to Operator of its financial capability to conduct such operations or to take over the well within such period or thereafter to conduct operations on such well or plug and abandon such well shall entitle Operator to retain or take possession of the well and plug and abandon the well. The party taking over the well shall indemnify Operator (if Operator is an abandoning party) and the other abandoning parties against liability for any further operations conducted on such well except for the costs of plugging and abandoning the well and restoring the surface, for which the abandoning parties shall remain proportionately liable.

 2. Abandonment of Wells That Have Produced: Except for any well in which a Non-Consent operation has been conducted hereunder for which the Consenting Parties have not been fully reimbursed as herein provided, any well which has been completed as a producer shall not be plugged and abandoned without the consent of all parties. If all parties consent to such abandonment, the well shall be plugged and abandoned in accordance with applicable regulations and at the cost, risk and expense of all the parties hereto. Failure of a party to reply within sixty (60) days of delivery of notice of proposed abandonment shall be deemed an election to consent to the proposal. If, within sixty (60) days after delivery of notice of the proposed abandonment of any well, all parties do not agree to the abandonment of such well, those wishing to continue its operation from the Zone then open to production shall be obligated to take over the well as of the expiration of the applicable notice period and shall indemnify Operator (if Operator is an abandoning party) and the other abandoning parties against liability for any further operations on the well conducted by such parties. Failure of such party or parties to provide proof reasonably satisfactory to Operator of their financial capability to conduct such operations or to take over the well within the required period or thereafter to conduct operations on such well shall entitle Operator to retain or take possession of such well and plug and abandon the well.

 Parties taking over a well as provided herein shall tender to each of the other parties its proportionate share of the value of the well's salvable material and equipment, determined in accordance with the provisions of Exhibit "C," less the estimated cost of salvaging and the estimated cost of plugging and abandoning and restoring the surface; provided, however, that in the event the estimated plugging and abandoning and surface restoration costs and the estimated cost of salvaging are higher than the value of the well's salvable material and equipment, each of the abandoning parties shall tender to the parties continuing operations their proportionate shares of the estimated excess cost. Each abandoning party shall assign to the non-abandoning parties, without warranty, express or implied, as to title or as to quantity, or fitness for use of the equipment and material, all of its interest in the wellbore of the well and related equipment, together with its interest in the Leasehold insofar and only insofar as such Leasehold covers the right to obtain production from that wellbore in the Zone then open to production. If the interest of the abandoning party is or includes an Oil and Gas Interest, such party shall execute and deliver to the non-abandoning party or parties an oil and gas lease, limited to the wellbore and the Zone then open to production, for a term of one (1) year and so long thereafter as Oil and/or Gas is produced from the Zone covered thereby, such lease to be on the form attached as Exhibit "B." The assignments or leases so limited shall encompass the Drilling Unit upon which the well is located. The payments by, and the assignments or leases to, the assignees shall be in a ratio based upon the relationship of their respective percentage of participation in the Contract Area to the aggregate of the percentages of participation in the Contract Area of all assignees. There shall be no readjustment of interests in the remaining portions of the Contract Area.

 Thereafter, abandoning parties shall have no further responsibility, liability, or interest in the operation of or production from the well in the Zone then open other than the royalties retained in any lease made under the terms of this Article. Upon request, Operator shall continue to operate the assigned well for the account of the non-abandoning parties at the rates and charges contemplated by this agreement, plus any additional cost and charges which may arise as the result of the separate ownership of the assigned well. Upon proposed abandonment of the producing Zone assigned or leased, the assignor or lessor shall then have the option to repurchase its prior interest in the well (using the same valuation formula) and participate in further operations therein subject to the provisions hereof.

 3. Abandonment of Non-Consent Operations: The provisions of Article VI.E.1. or VI.E.2. above shall be applicable as between Consenting Parties in the event of the proposed abandonment of any well excepted from said Articles; provided, however, no well shall be permanently plugged and abandoned unless and until all parties having the right to conduct further operations therein have been notified of the proposed abandonment and afforded the opportunity to elect to take over the well in accordance with the provisions of this Article VI.E.; and provided further, that Non-Consenting Parties who own an interest in a portion of the well shall pay their proportionate shares of abandonment and surface restoration costs for such well as provided in Article VI.B.2.(b).

F. Termination of Operations:

 Upon the commencement of an operation for the drilling, Reworking, Sidetracking, Plugging Back, Deepening, testing, Completion or plugging of a well, including but not limited to the Initial Well, such operation shall not be terminated without consent of parties bearing _____ % of the costs of such operation; provided, however, that in the event granite or other practically impenetrable substance or condition in the hole is encountered which renders further operations impractical, Operator may discontinue operations and give notice of such condition in the manner provided in Article VI.B.1, and the provisions of Article VI.B. or VI.E. shall thereafter apply to such operation, as appropriate.

G. Taking Production in Kind:

 ☐ Option No. 1: Gas Balancing Agreement Attached

  Each party shall take in kind or separately dispose of its proportionate share of all Oil and Gas produced from the Contract Area, exclusive of production which may be used in development and producing operations and in preparing and treating Oil and Gas for marketing purposes and production unavoidably lost. Any extra expenditure incurred in the taking in kind or separate disposition by any party of its proportionate share of the production shall be borne by such party. Any party taking its share of production in kind shall be required to pay for only its proportionate share of such part of Operator's surface facilities which it uses.

  Each party shall execute such division orders and contracts as may be necessary for the sale of its interest in production from the Contract Area, and, except as provided in Article VII.B., shall be entitled to receive payment directly from the purchaser thereof for its share of all production.

If any party fails to make the arrangements necessary to take in kind or separately dispose of its proportionate share of the Oil produced from the Contract Area, Operator shall have the right, subject to the revocation at will by the party owning it, but not the obligation, to purchase such Oil or sell it to others at any time and from time to time, for the account of the non-taking party. Any such purchase or sale by Operator may be terminated by Operator upon at least ten (10) days written notice to the owner of said production and shall be subject always to the right of the owner of the production upon at least ten (10) days written notice to Operator to exercise at any time its right to take in kind, or separately dispose of, its share of all Oil not previously delivered to a purchaser. Any purchase or sale by Operator of any other party's share of Oil shall be only for such reasonable periods of time as are consistent with the minimum needs of the industry under the particular circumstances, but in no event for a period in excess of one (1) year.

Any such sale by Operator shall be in a manner commercially reasonable under the circumstances but Operator shall have no duty to share any existing market or to obtain a price equal to that received under any existing market. The sale or delivery by Operator of a non-taking party's share of Oil under the terms of any existing contract of Operator shall not give the non-taking party any interest in or make the non-taking party a party to said contract. No purchase shall be made by Operator without first giving the non-taking party at least ten (10) days written notice of such intended purchase and the price to be paid or the pricing basis to be used.

All parties shall give timely written notice to Operator of their Gas marketing arrangements for the following month, excluding price, and shall notify Operator immediately in the event of a change in such arrangements. Operator shall maintain records of all marketing arrangements, and of volumes actually sold or transported, which records shall be made available to Non-Operators upon reasonable request.

In the event one or more parties' separate disposition of its share of the Gas causes split-stream deliveries to separate pipelines and/or deliveries which on a day-to-day basis for any reason are not exactly equal to a party's respective proportionate share of total Gas sales to be allocated to it, the balancing or accounting between the parties shall be in accordance with any Gas balancing agreement between the parties hereto, whether such an agreement is attached as Exhibit "E" or is a separate agreement. Operator shall give notice to all parties of the first sales of Gas from any well under this agreement.

☐ **Option No. 2: No Gas Balancing Agreement:**

Each party shall take in kind or separately dispose of its proportionate share of all Oil and Gas produced from the Contract Area, exclusive of production which may be used in development and producing operations and in preparing and treating Oil and Gas for marketing purposes and production unavoidably lost. Any extra expenditure incurred in the taking in kind or separate disposition by any party of its proportionate share of the production shall be borne by such party. Any party taking its share of production in kind shall be required to pay for only its proportionate share of such part of Operator's surface facilities which it uses.

Each party shall execute such division orders and contracts as may be necessary for the sale of its interest in production from the Contract Area, and, except as provided in Article VII.B., shall be entitled to receive payment directly from the purchaser thereof for its share of all production.

If any party fails to make the arrangements necessary to take in kind or separately dispose of its proportionate share of the Oil and/or Gas produced from the Contract Area, Operator shall have the right, subject to the revocation at will by the party owning it, but not the obligation, to purchase such Oil and/or Gas or sell it to others at any time and from time to time, for the account of the non-taking party. Any such purchase or sale by Operator may be terminated by Operator upon at least ten (10) days written notice to the owner of said production and shall be subject always to the right of the owner of the production upon at least ten (10) days written notice to Operator to exercise its right to take in kind, or separately dispose of, its share of all Oil and/or Gas not previously delivered to a purchaser; provided, however, that the effective date of any such revocation may be deferred at Operator's election for a period not to exceed ninety (90) days if Operator has committed such production to a purchase contract having a term extending beyond such ten (10) -day period. Any purchase or sale by Operator of any other party's share of Oil and/or Gas shall be only for such reasonable periods of time as are consistent with the minimum needs of the industry under the particular circumstances, but in no event for a period in excess of one (1) year.

Any such sale by Operator shall be in a manner commercially reasonable under the circumstances, but Operator shall have no duty to share any existing market or transportation arrangement or to obtain a price or transportation fee equal to that received under any existing market or transportation arrangement. The sale or delivery by Operator of a non-taking party's share of production under the terms of any existing contract of Operator shall not give the non-taking party any interest in or make the non-taking party a party to said contract. No purchase of Oil and Gas and no sale of Gas shall be made by Operator without first giving the non-taking party ten days written notice of such intended purchase or sale and the price to be paid or the pricing basis to be used. Operator shall give notice to all parties of the first sale of Gas from any well under this Agreement.

All parties shall give timely written notice to Operator of their Gas marketing arrangements for the following month, excluding price, and shall notify Operator immediately in the event of a change in such arrangements. Operator shall maintain records of all marketing arrangements, and of volumes actually sold or transported, which records shall be made available to Non-Operators upon reasonable request.

ARTICLE VII.
EXPENDITURES AND LIABILITY OF PARTIES

A. Liability of Parties:

The liability of the parties shall be several, not joint or collective. Each party shall be responsible only for its obligations, and shall be liable only for its proportionate share of the costs of developing and operating the Contract Area. Accordingly, the liens granted among the parties in Article VII.B. are given to secure only the debts of each severally, and no party shall have any liability to third parties hereunder to satisfy the default of any other party in the payment of any expense or obligation hereunder. It is not the intention of the parties to create, nor shall this agreement be construed as creating, a mining or other partnership, joint venture, agency relationship or association, or to render the parties liable as partners, co-venturers, or principals. In their relations with each other under this agreement, the parties shall not be considered fiduciaries or to have established a confidential relationship but rather shall be free to act on an arm's-length basis in accordance with their own respective self-interest, subject, however, to the obligation of the parties to act in good faith in their dealings with each other with respect to activities hereunder.

B. Liens and Security Interests:

Each party grants to the other parties hereto a lien upon any interest it now owns or hereafter acquires in Oil and Gas Leases and Oil and Gas Interests in the Contract Area, and a security interest and/or purchase money security interest in any interest it now owns or hereafter acquires in the personal property and fixtures on or used or obtained for use in connection therewith, to secure performance of all of its obligations under this agreement including but not limited to payment of expense, interest and fees, the proper disbursement of all monies paid hereunder, the assignment or relinquishment of interest in Oil and Gas Leases as required hereunder, and the proper performance of operations hereunder. Such lien and security interest granted by each party hereto shall include such party's leasehold interests, working interests, operating rights, and royalty and overriding royalty interests in the Contract Area now owned or hereafter acquired and in lands pooled or unitized therewith or otherwise becoming subject to this agreement, the Oil and Gas when extracted therefrom and equipment situated thereon or used or obtained for use in connection therewith (including, without limitation, all wells, tools, and tubular goods), and accounts (including, without limitation, accounts arising from gas imbalances or from the sale of Oil and/or Gas at the wellhead), contract rights, inventory and general intangibles relating thereto or arising therefrom, and all proceeds and products of the foregoing.

To perfect the lien and security agreement provided herein, each party hereto shall execute and acknowledge the recording supplement and/or any financing statement prepared and submitted by any party hereto in conjunction herewith or at any time following execution hereof, and Operator is authorized to file this agreement or the recording supplement executed herewith as a lien or mortgage in the applicable real estate records and as a financing statement with the proper officer under the Uniform Commercial Code in the state in which the Contract Area is situated and such other states as Operator shall deem appropriate to perfect the security interest granted hereunder. Any party may file this agreement, the recording supplement executed herewith, or such other documents as it deems necessary as a lien or mortgage in the applicable real estate records and/or a financing statement with the proper officer under the Uniform Commercial Code.

Each party represents and warrants to the other parties hereto that the lien and security interest granted by such party to the other parties shall be a first and prior lien, and each party hereby agrees to maintain the priority of said lien and security interest against all persons acquiring an interest in Oil and Gas Leases and Interests covered by this agreement by, through or under such party. All parties acquiring an interest in Oil and Gas Leases and Oil and Gas Interests covered by this agreement, whether by assignment, merger, mortgage, operation of law, or otherwise, shall be deemed to have taken subject to the lien and security interest granted by this Article VII.B. as to all obligations attributable to such interest hereunder whether or not such obligations arise before or after such interest is acquired.

To the extent that parties have a security interest under the Uniform Commercial Code of the state in which the Contract Area is situated, they shall be entitled to exercise the rights and remedies of a secured party under the Code. The bringing of a suit and the obtaining of judgment by a party for the secured indebtedness shall not be deemed an election of remedies or otherwise affect the lien rights or security interest as security for the payment thereof. In addition, upon default by any party in the payment of its share of expenses, interests or fees, or upon the improper use of funds by the Operator, the other parties shall have the right, without prejudice to other rights or remedies, to collect from the purchaser the proceeds from the sale of such defaulting party's share of Oil and Gas until the amount owed by such party, plus interest as provided in "Exhibit C," has been received, and shall have the right to offset the amount owed against the proceeds from the sale of such defaulting party's share of Oil and Gas. All purchasers of production may rely on a notification of default from the non-defaulting party or parties stating the amount due as a result of the default, and all parties waive any recourse available against purchasers for releasing production proceeds as provided in this paragraph.

If any party fails to pay its share of cost within one hundred twenty (120) days after rendition of a statement therefor by Operator, the non-defaulting parties, including Operator, shall, upon request by Operator, pay the unpaid amount in the proportion that the interest of each such party bears to the interest of all such parties. The amount paid by each party so paying its share of the unpaid amount shall be secured by the liens and security rights described in Article VII.B., and each paying party may independently pursue any remedy available hereunder or otherwise.

If any party does not perform all of its obligations hereunder, and the failure to perform subjects such party to foreclosure or execution proceedings pursuant to the provisions of this agreement, to the extent allowed by governing law, the defaulting party waives any available right of redemption from and after the date of judgment, any required valuation or appraisement of the mortgaged or secured property prior to sale, any available right to stay execution or to require a marshalling of assets and any required bond in the event a receiver is appointed. In addition, to the extent permitted by applicable law, each party hereby grants to the other parties a power of sale as to any property that is subject to the lien and security rights granted hereunder, such power to be exercised in the manner provided by applicable law or otherwise in a commercially reasonable manner and upon reasonable notice.

Each party agrees that the other parties shall be entitled to utilize the provisions of Oil and Gas lien law or other lien law of any state in which the Contract Area is situated to enforce the obligations of each party hereunder. Without limiting the generality of the foregoing, to the extent permitted by applicable law, Non-Operators agree that Operator may invoke or utilize the mechanics' or materialmen's lien law of the state in which the Contract Area is situated in order to sercure the payment to Operator of any sum due hereunder for services performed or materials supplied by Operator.

C. Advances:

Operator, at its election, shall have the right from time to time to demand and receive from one or more of the other parties payment in advance of their respective shares of the estimated amount of the expense to be incurred in operations hereunder during the next succeeding month, which right may be exercised only by submission to each such party of an itemized statement of such estimated expense, together with an invoice for its share thereof. Each such statement and invoice for the payment in advance of estimated expense shall be submitted on or before the 20th day of the next preceding month. Each party shall pay to Operator its proportionate share of such estimate within fifteen (15) days after such estimate and invoice is received. If any party fails to pay its share of said estimate within said time, the amount due shall bear interest as provided in Exhibit "C" until paid. Proper adjustment shall be made monthly between advances and actual expense to the end that each party shall bear and pay its proportionate share of actual expenses incurred, and no more.

D. Defaults and Remedies:

If any party fails to discharge any financial obligation under this agreement, including without limitation the failure to make any advance under the preceding Article VII.C. or any other provision of this agreement, within the period required for such payment hereunder, then in addition to the remedies provided in Article VII.B. or elsewhere in this agreement, the remedies specified below shall be applicable. For purposes of this Article VII.D., all notices and elections shall be delivered only by Operator, except that Operator shall deliver any such notice and election requested by a non-defaulting Non-Operator, and when Operator is the party in default, the applicable notices and elections can be delivered by any Non-Operator. Election of any one or more of the following remedies shall not preclude the subsequent use of any other remedy specified below or otherwise available to a non-defaulting party.

1. <u>Suspension of Rights:</u> Any party may deliver to the party in default a Notice of Default, which shall specify the default, specify the action to be taken to cure the default, and specify that failure to take such action will result in the exercise of one or more of the remedies provided in this Article. If the default is not cured within thirty (30) days of the delivery of such Notice of Default, all of the rights of the defaulting party granted by this agreement may upon notice be suspended until the default is cured, without prejudice to the right of the non-defaulting party or parties to continue to enforce the obligations of the defaulting party previously accrued or thereafter accruing under this agreement. If Operator is the party in default, the Non-Operators shall have in addition the right, by vote of Non-Operators owning a majority in interest in the Contract Area after excluding the voting interest of Operator, to appoint a new Operator effective immediately. The rights of a defaulting party that may be suspended hereunder at the election of the non-defaulting parties shall include, without limitation, the right to receive information as to any operation conducted hereunder during the period of such default, the right to elect to participate in an operation proposed under Article VI.B. of this agreement, the right to participate in an operation being conducted under this agreement even if the party has previously elected to participate in such operation, and the right to receive proceeds of production from any well subject to this agreement.

2. <u>Suit for Damages:</u> Non-defaulting parties or Operator for the benefit of non-defaulting parties may sue (at joint account expense) to collect the amounts in default, plus interest accruing on the amounts recovered from the date of default until the date of collection at the rate specified in Exhibit "C" attached hereto. Nothing herein shall prevent any party from suing any defaulting party to collect consequential damages accruing to such party as a result of the default.

3. <u>Deemed Non-Consent:</u> The non-defaulting party may deliver a written Notice of Non-Consent Election to the defaulting party at any time after the expiration of the thirty-day cure period following delivery of the Notice of Default, in which event if the billing is for the drilling of a new well or the Plugging Back, Sidetracking, Reworking or Deepening of a well which is to be or has been plugged as a dry hole, or for the Completion or Recompletion of any well, the defaulting party will be conclusively deemed to have elected not to participate in the operation and to be a Non-Consenting Party with respect thereto under Article VI.B. or VI.C., as the case may be, to the extent of the costs unpaid by such party, notwithstanding any election to participate theretofore made. If election is made to proceed under this provision, then the non-defaulting parties may not elect to sue for the unpaid amount pursuant to Article VII.D.2.

Until the delivery of such Notice of Non-Consent Election to the defaulting party, such party shall have the right to cure its default by paying its unpaid share of costs plus interest at the rate set forth in Exhibit "C," provided, however, such payment shall not prejudice the rights of the non-defaulting parties to pursue remedies for damages incurred by the non-defaulting parties as a result of the default. Any interest relinquished pursuant to this Article VII.D.3. shall be offered to the non-defaulting parties in proportion to their interests, and the non-defaulting parties electing to participate in the ownership of such interest shall be required to contribute their shares of the defaulted amount upon their election to participate therein.

4. <u>Advance Payment:</u> If a default is not cured within thirty (30) days of the delivery of a Notice of Default, Operator, or Non-Operators if Operator is the defaulting party, may thereafter require advance payment from the defaulting party of such defaulting party's anticipated share of any item of expense for which Operator, or Non-Operators, as the case may be, would be entitled to reimbursement under any provision of this agreement, whether or not such expense was the subject of the previous default. Such right includes, but is not limited to, the right to require advance payment for the estimated costs of drilling a well or Completion of a well as to which an election to participate in drilling or Completion has been made. If the defaulting party fails to pay the required advance payment, the non-defaulting parties may pursue any of the remedies provided in this Article VII.D. or any other default remedy provided elsewhere in this agreement. Any excess of funds advanced remaining when the operation is completed and all costs have been paid shall be promptly returned to the advancing party.

5. <u>Costs and Attorneys' Fees.</u> In the event any party is required to bring legal proceedings to enforce any financial obligation of a party hereunder, the prevailing party in such action shall be entitled to recover all court costs, costs of collection, and a reasonable attorney's fee, which the lien provided for herein shall also secure.

E. Rentals, Shut-in Well Payments and Minimum Royalties:

Rentals, shut-in well payments and minimum royalties which may be required under the terms of any lease shall be paid by the party or parties who subjected such lease to this agreement at its or their expense. In the event two or more parties own and have contributed interests in the same lease to this agreement, such parties may designate one of such parties to make said payments for and on behalf of all such parties. Any party may request, and shall be entitled to receive, proper evidence of all such payments. In the event of failure to make proper payment of any rental, shut-in well payment or minimum royalty through mistake or oversight where such payment is required to continue the lease in force, any loss which results from such non-payment shall be borne in accordance with the provisions of Article IV.B.2.

Operator shall notify Non-Operators of the anticipated completion of a shut-in well, or the shutting in or return to production of a producing well, at least five (5) days (excluding Saturday, Sunday and legal holidays) prior to taking such action, or at the earliest opportunity permitted by circumstances, but assumes no liability for failure to do so. In the event of failure by Operator to so notify Non-Operators, the loss of any lease contributed hereto by Non-Operators for failure to make timely payments of any shut-in well payment shall be borne jointly by the parties hereto under the provisions of Article IV.B.3.

F. Taxes:

Beginning with the first calendar year after the effective date hereof, Operator shall render for ad valorem taxation all property subject to this agreement which by law should be rendered for such taxes, and it shall pay all such taxes assessed thereon before they become delinquent. Prior to the rendition date, each Non-Operator shall furnish Operator information as to burdens (to include, but not be limited to, royalties, overriding royalties and production payments) on Leases and Oil and Gas Interests contributed by such Non-Operator. If the assessed valuation of any Lease is reduced by reason of its being subject to outstanding excess royalties, overriding royalties or production payments, the reduction in ad valorem taxes resulting therefrom shall inure to the benefit of the owner or owners of such Lease, and Operator shall adjust the charge to such owner or owners so as to reflect the benefit of such reduction. If the ad valorem taxes are based in whole or in part upon separate valuations of each party's working interest, then notwithstanding anything to the contrary herein, charges to the joint account shall be made and paid by the parties hereto in accordance with the tax value generated by each party's working interest. Operator shall bill the other parties for their proportionate shares of all tax payments in the manner provided in Exhibit "C."

If Operator considers any tax assessment improper, Operator may, at its discretion, protest within the time and manner prescribed by law, and prosecute the protest to a final determination, unless all parties agree to abandon the protest prior to final determination. During the pendency of administrative or judicial proceedings, Operator may elect to pay, under protest, all such taxes and any interest and penalty. When any such protested assessment shall have been finally determined, Operator shall pay the tax for the joint account, together with any interest and penalty accrued, and the total cost shall then be assessed against the parties, and be paid by them, as provided in Exhibit "C."

Each party shall pay or cause to be paid all production, severance, excise, gathering and other taxes imposed upon or with respect to the production or handling of such party's share of Oil and Gas produced under the terms of this agreement.

ARTICLE VIII.
ACQUISITION, MAINTENANCE OR TRANSFER OF INTEREST

A. Surrender of Leases:

The Leases covered by this agreement, insofar as they embrace acreage in the Contract Area, shall not be surrendered in whole or in part unless all parties consent thereto.

However, should any party desire to surrender its interest in any Lease or in any portion thereof, such party shall give written notice of the proposed surrender to all parties, and the parties to whom such notice is delivered shall have thirty (30) days after delivery of the notice within which to notify the party proposing the surrender whether they elect to consent thereto. Failure of a party to whom such notice is delivered to reply within said 30-day period shall constitute a consent to the surrender of the Leases described in the notice. If all parties do not agree or consent thereto, the party desiring to surrender shall assign, without express or implied warranty of title, all of its interest in such Lease, or portion thereof, and any well, material and equipment which may be located thereon and any rights in production thereafter secured, to the parties not consenting to such surrender. If the interest of the assigning party is or includes an Oil and Gas Interest, the assigning party shall execute and deliver to the party or parties not consenting to such surrender an oil and gas lease covering such Oil and Gas Interest for a term of one (1) year and so long thereafter as Oil and/or Gas is produced from the land covered thereby, such lease to be on the form attached hereto as Exhibit "B." Upon such assignment or lease, the assigning party shall be relieved from all obligations thereafter accruing, but not theretofore accrued, with respect to the interest assigned or leased and the operation of any well attributable thereto, and the assigning party shall have no further interest in the assigned or leased premises and its equipment and production other than the royalties retained in any lease made under the terms of this Article. The party assignee or lessee shall pay to the party assignor or lessor the reasonable salvage value of the latter's interest in any well's salvable materials and equipment attributable to the assigned or leased acreage. The value of all salvable materials and equipment shall be determined in accordance with the provisions of Exhibit "C," less the estimated cost of salvaging and the estimated cost of plugging and abandoning and restoring the surface. If such value is less than such costs, then the party assignor or lessor shall pay to the party assignee or lessee the amount of such deficit. If the assignment or lease is in favor of more than one party, the interest shall be shared by such parties in the proportions that the interest of each bears to the total interest of all such parties. If the interest of the parties to whom the assignment is to be made varies according to depth, then the interest assigned shall similarly reflect such variances.

Any assignment, lease or surrender made under this provision shall not reduce or change the assignor's, lessor's or surrendering party's interest as it was immediately before the assignment, lease or surrender in the balance of the Contract Area; and the acreage assigned, leased or surrendered, and subsequent operations thereon, shall not thereafter be subject to the terms and provisions of this agreement but shall be deemed subject to an Operating Agreement in the form of this agreement.

B. Renewal or Extension of Leases:

If any party secures a renewal or replacement of an Oil and Gas Lease or Interest subject to this agreement, then all other parties shall be notified promptly upon such acquisition or, in the case of a replacement Lease taken before expiration of an existing Lease, promptly upon expiration of the existing Lease. The parties notified shall have the right for a period of thirty (30) days following delivery of such notice in which to elect to participate in the ownership of the renewal or replacement Lease, insofar as such Lease affects lands within the Contract Area, by paying to the party who acquired it their proportionate shares of the acquisition cost allocated to that part of such Lease within the Contract Area, which shall be in proportion to the interests held at that time by the parties in the Contract Area. Each party who participates in the purchase of a renewal or replacement Lease shall be given an assignment of its proportionate interest therein by the acquiring party.

If some, but less than all, of the parties elect to participate in the purchase of a renewal or replacement Lease, it shall be owned by the parties who elect to participate therein, in a ratio based upon the relationship of their respective percentage of participation in the Contract Area to the aggregate of the percentages of participation in the Contract Area of all parties participating in the purchase of such renewal or replacement Lease. The acquisition of a renewal or replacement Lease by any or all of the parties hereto shall not cause a readjustment of the interests of the parties stated in Exhibit "A," but any renewal or replacement Lease in which less than all parties elect to participate shall not be subject to this agreement but shall be deemed subject to a separate Operating Agreement in the form of this agreement.

If the interests of the parties in the Contract Area vary according to depth, then their right to participate proportionately in renewal or replacement Leases and their right to receive an assignment of interest shall also reflect such depth variances.

The provisions of this Article shall apply to renewal or replacement Leases whether they are for the entire interest covered by the expiring Lease or cover only a portion of its area or an interest therein. Any renewal or replacement Lease taken before the expiration of its predecessor Lease, or taken or contracted for or becoming effective within six (6) months after the expiration of the existing Lease, shall be subject to this provision so long as this agreement is in effect at the time of such acquisition or at the time the renewal or replacement Lease becomes effective; but any Lease taken or contracted for more than six (6) months after the expiration of an existing Lease shall not be deemed a renewal or replacement Lease and shall not be subject to the provisions of this agreement.

The provisions in this Article shall also be applicable to extensions of Oil and Gas Leases.

C. Acreage or Cash Contributions:

While this agreement is in force, if any party contracts for a contribution of cash towards the drilling of a well or any other operation on the Contract Area, such contribution shall be paid to the party who conducted the drilling or other operation and shall be applied by it against the cost of such drilling or other operation. If the contribution be in the form of acreage, the party to whom the contribution is made shall promptly tender an assignment of the acreage, without warranty of title, to the Drilling Parties in the proportions said Drilling Parties shared the cost of drilling the well. Such acreage shall become a separate Contract Area and, to the extent possible, be governed by provisions identical to this agreement. Each party shall promptly notify all other parties of any acreage or cash contributions it may obtain in support of any well or any other operation on the Contract Area. The above provisions shall also be applicable to optional rights to earn acreage outside the Contract Area which are in support of well drilled inside the Contract Area.

If any party contracts for any consideration relating to disposition of such party's share of substances produced hereunder, such consideration shall not be deemed a contribution as contemplated in this Article VIII.C.

D. Assignment; Maintenance of Uniform Interest:

For the purpose of maintaining uniformity of ownership in the Contract Area in the Oil and Gas Leases, Oil and Gas Interests, wells, equipment and production covered by this agreement no party shall sell, encumber, transfer or make other disposition of its interest in the Oil and Gas Leases and Oil and Gas Interests embraced within the Contract Area or in wells, equipment and production unless such disposition covers either:

1. the entire interest of the party in all Oil and Gas Leases, Oil and Gas Interests, wells, equipment and production; or

2. an equal undivided percent of the party's present interest in all Oil and Gas Leases, Oil and Gas Interests, wells, equipment and production in the Contract Area.

Every sale, encumbrance, transfer or other disposition made by any party shall be made expressly subject to this agreement and shall be made without prejudice to the right of the other parties, and any transferee of an ownership interest in any Oil and Gas Lease or Interest shall be deemed a party to this agreement as to the interest conveyed from and after the effective date of the transfer of ownership; provided, however, that the other parties shall not be required to recognize any such sale, encumbrance, transfer or other disposition for any purpose hereunder until thirty (30) days after they have received a copy of the instrument of transfer or other satisfactory evidence thereof in writing from the transferor or transferee. No assignment or other disposition of interest by a party shall relieve such party of obligations previously incurred by such party hereunder with respect to the interest transferred, including without limitation the obligation of a party to pay all costs attributable to an operation conducted hereunder in which such party has agreed to participate prior to making such assignment, and the lien and security interest granted by Article VII.B. shall continue to burden the interest transferred to secure payment of any such obligations.

If, at any time the interest of any party is divided among and owned by four or more co-owners, Operator, at its discretion, may require such co-owners to appoint a single trustee or agent with full authority to receive notices, approve expenditures, receive billings for and approve and pay such party's share of the joint expenses, and to deal generally with, and with power to bind, the co-owners of such party's interest within the scope of the operations embraced in this agreement; however, all such co-owners shall have the right to enter into and execute all contracts or agreements for the disposition of their respective shares of the Oil and Gas produced from the Contract Area and they shall have the right to receive, separately, payment of the sale proceeds thereof.

E. Waiver of Rights to Partition:

If permitted by the laws of the state or states in which the property covered hereby is located, each party hereto owning an undivided interest in the Contract Area waives any and all rights it may have to partition and have set aside to it in severalty its undivided interest therein.

F. Preferential Right to Purchase:

☐ (Optional; Check if applicable.)

Should any party desire to sell all or any part of its interests under this agreement, or its rights and interests in the Contract Area, it shall promptly give written notice to the other parties, with full information concerning its proposed disposition, which shall include the name and address of the prospective transferee (who must be ready, willing and able to purchase), the purchase price, a legal description sufficient to identify the property, and all other terms of the offer. The other parties shall then have an optional prior right, for a period of ten (10) days after the notice is delivered, to purchase for the stated consideration on the same terms and conditions the interest which the other party proposes to sell; and, if this optional right is exercised, the purchasing parties shall share the purchased interest in the proportions that the interest of each bears to the total interest of all purchasing parties. However, there shall be no preferential right to purchase in those cases where any party wishes to mortgage its interests, or to transfer title to its interests to its mortgagee in lieu of or pursuant to foreclosure of a mortgage of its interests, or to dispose of its interests by merger, reorganization, consolidation, or by sale of all or substantially all of its Oil and Gas assets to any party, or by transfer of its interests to a subsidiary or parent company or to a subsidiary of a parent company, or to any company in which such party owns a majority of the stock.

ARTICLE IX.
INTERNAL REVENUE CODE ELECTION

If, for federal income tax purposes, this agreement and the operations hereunder are regarded as a partnership, and if the parties have not otherwise agreed to form a tax partnership pursuant to Exhibit "G" or other agreement between them, each party thereby affected elects to be excluded from the application of all of the provisions of Subchapter "K," Chapter 1, Subtitle "A," of the Internal Revenue Code of 1986, as amended ("Code"), as permitted and authorized by Section 761 of the Code and the regulations promulgated thereunder. Operator is authorized and directed to execute on behalf of each party hereby affected such evidence of this election as may be required by the Secretary of the Treasury of the United States or the Federal Internal Revenue Service, including specifically, but not by way of limitation, all of the returns, statements, and the data required by Treasury Regulations §1.761. Should there be any requirement that each party hereby affected give further evidence of this election, each such party shall execute such documents and furnish such other evidence as may be required by the Federal Internal Revenue Service or as may be necessary to evidence this election. No such party shall give any notices or take any other action inconsistent with the election made hereby. If any present or future income tax laws of the state or states in which the Contract Area is located or any future income tax laws of the United States contain provisions similar to those in Subchapter "K," Chapter 1, Subtitle "A," of the Code, under which an election similar to that provided by Section 761 of the Code is permitted, each party hereby affected shall make such election as may be permitted or required by such laws. In making the foregoing election, each such party states that the income derived by such party from operations hereunder can be adequately determined without the computation of partnership taxable income.

ARTICLE X.
CLAIMS AND LAWSUITS

Operator may settle any single uninsured third party damage claim or suit arising from operations hereunder if the expenditure does not exceed _____ Dollars ($ _____) and if the payment is in complete settlement of such claim or suit. If the amount required for settlement exceeds the above amount, the parties hereto shall assume and take over the further handling of the claim or suit, unless such authority is delegated to Operator. All costs and expenses of handling, settling, or otherwise discharging such claim or suit shall be at the joint expense of the parties participating in the operation from which the claim or suit arises. If a claim is made against any party or if any party is sued on account of any matter arising from operations hereunder over which such individual has no control because of the rights given Operator by this agreement, such **party** shall immediately notify all other parties, and the claim or suite shall be treated as any other claim or suit involving operations hereunder.

ARTICLE XI.
FORCE MAJEURE

If any party is rendered unable, wholly or in part, by force majeure to carry out its obligations under this agreement, other than the obligation to indemnify or make money payments or furnish security, that party shall give to all other parties prompt written notice of the force majeure with reasonably full particulars concerning it; thereupon, the obligations of the party giving the notice, so far as they are affected by the force majeure, shall be suspended during, but no longer than, the continuance of the force majeure. The term "force majeure," as here employed, shall mean an act of God, strike, lockout, or other industrial disturbance, act of the public enemy, war, blockade, public riot, lightning, fire, storm, flood or other act of nature, explosion, governmental action, governmental delay, restraint or inaction, unavailability of equipment, and any other cause, whether of the kind specifically enumerated above or otherwise, which is not reasonably within the control of the party claiming suspension.

The affected party shall use all reasonable diligence to remove the force majeure situation as quickly as practicable. The requirement that any force majeure shall be remedied with all reasonable dispatch shall not require the settlement of strikes, lockouts, or other labor difficulty by the party involved, contrary to its wishes; how all such difficulties shall be handled shall be entirely within the discretion of the party concerned.

ARTICLE XII.
NOTICES

All notices authorized or required between the parties by any of the provisions of this agreement, unless otherwise specifically provided, shall be in writing and delivered in person or by United States mail, courier service, telegram, telex, telecopier or any other form of facsimile, postage or charges prepaid, and addressed to such parties at the addresses listed on Exhibit "A." All telephone or oral notices permitted by this agreement shall be confirmed immediately thereafter by written notice. The originating notice given under any provision hereof shall be deemed delivered only when received by the party to whom such notice is directed, and the time for such party to deliver any notice in response thereto shall run from the date the originating notice is received. "Receipt" for purposes of this agreement with respect to written notice delivered hereunder shall be actual delivery of the notice to the address of the party to be notified specified in accordance with this agreement, or to the telecopy, facsimile or telex machine of such party. The second or any responsive notice shall be deemed delivered when deposited in the United States mail or at the office of the courier or telegraph service, or upon transmittal by telex, telecopy or facsimile, or when personally delivered to the party to be notified, provided, that when response is required within 24 or 48 hours, such response shall be given orally or by telephone, telex, telecopy or other facsimile within such period. Each party shall have the right to change its address at any time, and from time to time, by giving written notice thereof to all other parties. If a party is not available to receive notice orally or by telephone when a party attempts to deliver a notice required to be delivered within 24 or 48 hours, the notice may be delivered in writing by any other method specified herein and shall be deemed delivered in the same manner provided above for any responsive notice.

ARTICLE XIII.
TERM OF AGREEMENT

This agreement shall remain in full force and effect as to the Oil and Gas Leases and/or Oil and Gas Interests subject hereto for the period of time selected below; provided, however, no party hereto shall ever be construed as having any right, title or interest in or to any Lease or Oil and Gas Interest contributed by any other party beyond the term of this agreement.

☐ Option No. 1: So long as any of the Oil and Gas Leases subject to this agreement remain or are continued in force as to any part of the Contract Area, whether by production, extension, renewal or otherwise.

☐ Option No. 2: In the event the well described in Article VI.A., or any subsequent well drilled under any provision of this agreement, results in the Completion of a well as a well capable of production of Oil and/or Gas in paying quantities, this agreement shall continue in force so long as any such well is capable of production, and for an additional period of _____ days thereafter; provided, however, if, prior to the expiration of such additional period, one or more of the parties hereto are engaged in drilling, Reworking, Deepening, Sidetracking, Plugging Back, testing or attempting to Complete or Re-complete a well or wells hereunder, this agreement shall continue in force until such operations have been completed and if production results therefrom, this agreement shall continue in force as provided herein. In the event the well described in Article VI.A., or any subsequent well drilled hereunder, results in a dry hole, and no other well is capable of producing Oil and/or Gas from the Contract Area, this agreement shall terminate unless drilling, Deepening, Sidetracking, Completing, Re-completing, Plugging Back or Reworking operations are commenced within _____ days from the date of abandonment of said well. "Abandonment" for such purposes shall mean either (i) a decision by all parties not to conduct any further operations on the well or (ii) the elapse of 180 days from the conduct of any operations on the well, whichever first occurs.

The termination of this agreement shall not relieve any party hereto from any expense, liability or other obligation or any remedy therefor which has accrued or attached prior to the date of such termination.

Upon termination of this agreement and the satisfaction of all obligations hereunder, in the event a memorandum of this Operating Agreement has been filed of record, Operator is authorized to file of record in all necessary recording offices a notice of termination, and each party hereto agrees to execute such a notice of termination as to Operator's interest, upon request of Operator, if Operator has satisfied all its financial obligations.

ARTICLE XIV.
COMPLIANCE WITH LAWS AND REGULATIONS

A. Laws, Regulations and Orders:

This agreement shall be subject to the applicable laws of the state in which the Contract Area is located, to the valid rules, regulations, and orders of any duly constituted regulatory body of said state; and to all other applicable federal, state, and local laws, ordinances, rules, regulations and orders.

B. Governing Law:

This agreement and all matters pertaining hereto, including but not limited to matters of performance, non-performance, breach, remedies, procedures, rights, duties, and interpretation or construction, shall be governed and determined by the law of the state in which the Contract Area is located. If the Contract Area is in two or more states, the law of the state of _____ shall govern.

C. Regulatory Agencies:

Nothing herein contained shall grant, or be construed to grant, Operator the right or authority to waive or release any rights, privileges, or obligations which Non-Operators may have under federal or state laws or under rules, regulations or orders promulgated under such laws in reference to oil, gas and mineral operations, including the location, operation, or production of wells, on tracts offsetting or adjacent to the Contract Area.

With respect to the operations hereunder, Non-Operators agree to release Operator from any and all losses, damages, injuries, claims and causes of action arising out of, incident to or resulting directly or indirectly from Operator's interpretation or application of rules, rulings, regulations or orders of the Department of Energy or Federal Energy Regulatory Commission or predecessor or successor agencies to the extent such interpretation or application was made in good faith and does not constitute gross negligence. Each Non-Operator further agrees to reimburse Operator for such Non-Operator's share of production or any refund, fine, levy or other governmental sanction that Operator may be required to pay as a result of such an incorrect interpretation or application, together with interest and penalties thereon owing by Operator as a result of such incorrect interpretation or application.

ARTICLE XV.
MISCELLANEOUS

A. Execution:

This agreement shall be binding upon each Non-Operator when this agreement or a counterpart thereof has been executed by such Non-Operator and Operator notwithstanding that this agreement is not then or thereafter executed by all of the parties to which it is tendered or which are listed on Exhibit "A" as owning an interest in the Contract Area or which own, in fact, an interest in the Contract Area. Operator may, however, by written notice to all Non-Operators who have become bound by this agreement as aforesaid, given at any time prior to the actual spud date of the Initial Well but in no event later than five days prior to the date specified in Article VI.A. for commencement of the Initial Well, terminate this agreement if Operator in its sole discretion determines that there is insufficient participation to justify commencement of drilling operations. In the event of such a termination by Operator, all further obligations of the parties hereunder shall cease as of such termination. In the event any Non-Operator has advanced or prepaid any share of drilling or other costs hereunder, all sums so advanced shall be returned to such Non-Operator without interest. In the event Operator proceeds with drilling operations for the Initial Well without the execution hereof by all persons listed on Exhibit "A" as having a current working interest in such well, Operator shall indemnify Non-Operators with respect to all costs incurred for the Initial Well which would have been charged to such person under this agreement if such person had executed the same and Operator shall receive all revenues which would have been received by such person under this agreement if such person had executed the same.

B. Successors and Assigns:

This agreement shall be binding upon and shall inure to the benefit of the parties hereto and their respective heirs, devisees, legal representatives, successors and assigns, and the terms hereof shall be deemed to run with the Leases or Interests included within the Contract Area.

C. Counterparts:

This instrument may be executed in any number of counterparts, each of which shall be considered an original for all purposes.

D. Severability:

For the purposes of assuming or rejecting this agreement as an executory contract pursuant to federal bankruptcy laws, this agreement shall not be severable, but rather must be assumed or rejected in its entirety, and the failure of any party to this agreement to comply with all of its financial obligations provided herein shall be a material default.

ARTICLE XVI.
OTHER PROVISIONS

IN WITNESS WHEREOF, this agreement shall be effective as of the _____ day of _____,
_____.
   year

ATTEST OR WITNESS: **OPERATOR**

_____   By _____
_____   Type or print name

                                           Title _____
                                           Date _____
                                           Tax ID or S.S. No. _____

                          NON-OPERATORS

_____   By _____
_____   Type or print name

                                           Title _____
                                           Date _____
                                           Tax ID or S.S. No. _____

_____   By _____
_____   Type or print name

                                           Title _____
                                           Date _____
                                           Tax ID or S.S. No. _____

_____   By _____
_____   Type or print name

                                           Title _____
                                           Date _____
                                           Tax ID or S.S. No. _____

ACKNOWLEDGMENTS

Note: The following forms of acknowledgment are the short forms approved by the Uniform Law on Notarial Acts. The validity and effect of these forms in any state will depend upon the statutes of that state.

Individual acknowledgment:

State of _____)

) ss.

County of _____)

    This instrument was acknowledged before me on

_____ by _____

(Seal, if any)

                                Title (and Rank) _____

                                My commission expires: _____

Acknowledgment in representative capacity:

State of _____)

) ss.

County of _____)

    This instrument was acknowledged before me on

_____ by _____ as

_____ of _____

(Seal, if any)

                                Title (and Rank) _____

                                My commission expires: _____

The JOA spells out how the property is to be operated. One important part of the JOA is the accounting procedure. Both the JOA itself and the accounting procedure are of significant importance to the accountant. The main provisions of a typical JOA and accounting procedure are discussed in the following paragraphs. It is important to remember that, as with any negotiated contract, the parties are free to include any provisions and language they believe to be important. Therefore, the discussion that follows relates to the typical case. It is always important to thoroughly analyze any contract to be sure the specific provisions the parties have agreed upon are understood.

The JOA delineates the duties and responsibilities of the operator and nonoperators. The JOA typically covers all phases (i.e., exploration, development, and production) of operation of the joint property. The major subsections that may be included in a standard JOA are as follows:

1. **Definitions.** Defines basic terms used in the agreement. Examples include operator, nonoperator, contract area, authorization for expenditure (AFE), oil and gas lease, drillsite, etc.

2. **Exhibits.** Contains a list of all of the exhibits that form the appendices of the agreement. For example, "Exhibit A" contains the legal description of the properties, the parties to the agreement, and the percentage or fractional ownership interest of each owner. "Exhibit C" is the accounting procedure. Other exhibits that may be included are exhibits containing the gas balancing agreement, information about insurance, and information about the lease agreement.

3. **Interests of parties.** Defines how specific revenues, costs, and liabilities will be distributed according to the ownership interests specified in the exhibits.

4. **Titles.** Describes title examination requirements for drillsites, how the costs of the title process will be distributed, and how loss of title would be handled.

5. **Operator.** Designates the operator and the general rights and duties of the operator. Describes the process for the resignation or removal of the operator and the process for appointment of a successor. All records and reports—including the accounting records—to be maintained by the operator and provided to the nonoperators and governmental units are listed.

6. **Drilling and development.** Specifies procedures to be followed in drilling and development activities and termination operations. Includes the specific date and location for the drilling of the initial well. Participation in the initial well is typically obligatory for all parties. Describes the procedures that must be followed in undertaking subsequent drilling and operations, including a section providing for "operations by less than all parties." This section provides the process by which carried working interests are created and allows penalties in nonconsent situations related to drilling, deepening, rework, and abandonment of wells. (Nonconsent situations are discussed in more detail later in the chapter.)

7. **Expenditures and liabilities.** Specifies that any liabilities are to be several and not joint (i.e., each party is individually responsible only for its own obligations and liabilities, and only its own proportionate share of development and operating costs). Gives the operator the right to demand and receive cash advances from the other working interest owners. Spells out the remedies that exist in the event that any owner fails to discharge its financial obligations related to the property. Provides for the payment of shut-in royalties, minimum royalties, delay rentals, and ad valorem taxes.

8. **Acquisition, maintenance, or transfer of interest.** Specifies procedures to follow when surrendering or renewing a lease or assigning interests. Also gives preferential rights of purchase to other working interest owners when one working interest owner wishes to sell part or all of its interest.

9. **Internal revenue code election.** States whether the joint venture is to be operated or be taxed as a partnership.

10. **Claims and lawsuits.** Defines the procedures to be followed in case of legal action relating to operations of the venture, and authorizes the operator to settle claims for uninsured damages up to a maximum amount.

11. **Force majeure.** Provides that if any party is unable to meet its nonmonetary obligations, such as the operator's obligation to proceed with drilling activities, because of circumstances beyond that party's control—act of God, war, fire, etc.—the party's obligation will be temporarily suspended.

12. **Notices.** States that all notices between parties will be in writing.

13. **Terms of agreement.** States that the agreement will remain in effect so long as the underlying lease is in effect whether through production, extension, renewal, or otherwise.

14. **Compliance with laws and regulations.** States that the agreement is subject to all applicable laws, and states that the operator cannot waive or release nonoperators from any rights, privileges, or obligations arising from governmental regulation.

15. **Miscellaneous.** Other provisions, including those concerning successors, severability, and the signature page.

In summary, the operator normally manages all developing, operating, recordkeeping, and administrative activities pertaining to the joint property. For major expenditures, such as drilling a well, the operator is generally required to obtain written authority from the nonoperators via an AFE. (A sample AFE appears in appendix A.) This requirement provides a limit on expenditures without express written authority from the nonoperators. Periodic reports that describe the activities performed by the operator must generally be given to the nonoperators. The operator normally is required to pay all costs of developing and operating the joint property and then bill the nonoperators for their proportionate share. However, a provision for requiring advances from the nonoperators prior to cost incurrence or payment is generally included in the operating agreement.

THE ACCOUNTING PROCEDURE

The accounting procedure, which is normally attached to the JOA, is of extreme importance to the accountant because it details the procedures to be followed in charging costs to the joint operations. The Council of Petroleum Accountants Societies (COPAS) plays a critical role in the development and interpretation of accounting procedures. In addition, COPAS actively disseminates information and guidance relating to industry best practices for a wide variety of joint interest accounting topics.[1-6] Perhaps the most widely used COPAS accounting procedures are the procedures issued in 1984 and 1986.

From time to time, as industry operating methods, practices, and technology evolve, COPAS issues new accounting procedures and guidelines. The new procedures also seek to clarify issues prone to disputes. COPAS issued its latest procedure in 2005. The COPAS accounting procedures are as follows:

COPAS MFI-1	1962 Accounting Procedure
COPAS MFI-2	1968 Accounting Procedure
COPAS MFI-4	1975 Accounting Procedure
COPAS MFI-5	1976 Offshore Accounting Procedure
COPAS MFI-17	1984 Accounting Procedure
COPAS MFI-19	1986 Offshore Accounting Procedure
COPAS MFI-30	1995 Accounting Procedure
COPAS MFI-51	2005 Accounting Procedure

When a new accounting procedure is published, it does not automatically replace an existing accounting procedure in an existing JOA, because the JOA and related accounting procedure are a contract. In order for a new accounting procedure to become effective, the parties to the JOA must agree to replace the existing accounting procedure with a new or different accounting procedure. Accordingly, any given JOA may include the *2005 COPAS Accounting Procedure* or some prior generation accounting procedure, such as the *1984 COPAS Accounting Procedure*. The accountant is legally bound to follow the accounting procedure called for by the contract. It is imperative for the accountant to consult the JOA and become familiar with the terms of the specific accounting procedure that relates to any given joint property.

In practice, all of the above accounting procedures are relevant today, since they are a part of various JOAs that have been executed over time. A copy of the *2005 COPAS Accounting Procedure* [COPAS *Model Form (MF)-6*] is shown in Figure 12–2. This procedure also appears in *COPAS Model Form Interpretation (MFI)-51, 2005 COPAS Accounting Procedure*. MFI-51 includes the accounting procedure as well as a thorough interpretation of the various aspects of the procedure. The other accounting procedures are also available from COPAS.

Fig. 12–2. Accounting procedure
Source: Copyright 1984–2005 by the Council of Petroleum Accountants Societies, Inc. (COPAS). From COPAS MF-6. 2005 COPAS Accounting Procedure. 2005. COPAS documents are available for purchase at www.COPAS.org.

EXHIBIT " "
ACCOUNTING PROCEDURE
JOINT OPERATIONS

Attached to and made part of _____
_____
_____

I. GENERAL PROVISIONS – ACCOUNTING PROCEDURE

IF THE PARTIES FAIL TO SELECT EITHER ONE OF COMPETING "ALTERNATIVE" PROVISIONS, OR SELECT ALL THE COMPETING "ALTERNATIVE" PROVISIONS, ALTERNATIVE 1 IN EACH SUCH INSTANCE SHALL BE DEEMED TO HAVE BEEN ADOPTED BY THE PARTIES AS A RESULT OF ANY SUCH OMISSION OR DUPLICATE NOTATION.

IN THE EVENT THAT ANY "OPTIONAL" PROVISION OF THIS ACCOUNTING PROCEDURE IS NOT ADOPTED BY THE PARTIES TO THE AGREEMENT BY A TYPED, PRINTED OR HANDWRITTEN INDICATION, SUCH PROVISION SHALL NOT FORM A PART OF THIS ACCOUNTING PROCEDURE, AND NO INFERENCE SHALL BE MADE CONCERNING THE INTENT OF THE PARTIES IN SUCH EVENT.

1. DEFINITIONS

All terms used in this Accounting Procedure shall have the following meaning, unless otherwise expressly defined in the Agreement:

"**Affiliate**" means for a person, another person that controls, is controlled by, or is under common control with that person. In this definition, (a) control means the ownership by one person, directly or indirectly, of more than fifty percent (50%) of the voting securities of a corporation or, for other persons, the equivalent ownership interest (such as partnership interests), and (b) "person" means an individual, corporation, partnership, trust, estate, unincorporated organization, association, or other legal entity.

"**Agreement**" means the operating agreement, farmout agreement, or other contract between the Parties to which this Accounting Procedure is attached.

"**Controllable Material**" means Material that, at the time of acquisition or disposition by the Joint Account, as applicable, is so classified in the Material Classification Manual most recently recommended by the Council of Petroleum Accountants Societies (COPAS).

"**Equalized Freight**" means the procedure of charging transportation cost to the Joint Account based upon the distance from the nearest Railway Receiving Point to the property.

"**Excluded Amount**" means a specified excluded trucking amount most recently recommended by COPAS.

Copyright © 2005 by COPAS, Inc.

"**Field Office**" means a structure, or portion of a structure, whether a temporary or permanent installation, the primary function of which is to directly serve daily operation and maintenance activities of the Joint Property and which serves as a staging area for directly chargeable field personnel.

"**First Level Supervision**" means those employees whose primary function in Joint Operations is the direct oversight of the Operator's field employees and/or contract labor directly employed On-site in a field operating capacity. First Level Supervision functions may include, but are not limited to:

- Responsibility for field employees and contract labor engaged in activities that can include field operations, maintenance, construction, well remedial work, equipment movement and drilling
- Responsibility for day-to-day direct oversight of rig operations
- Responsibility for day-to-day direct oversight of construction operations
- Coordination of job priorities and approval of work procedures
- Responsibility for optimal resource utilization (equipment, Materials, personnel)
- Responsibility for meeting production and field operating expense targets
- Representation of the Parties in local matters involving community, vendors, regulatory agents and landowners, as an incidental part of the supervisor's operating responsibilities
- Responsibility for all emergency responses with field staff
- Responsibility for implementing safety and environmental practices
- Responsibility for field adherence to company policy
- Responsibility for employment decisions and performance appraisals for field personnel
- Oversight of sub-groups for field functions such as electrical, safety, environmental, telecommunications, which may have group or team leaders.

"**Joint Account**" means the account showing the charges paid and credits received in the conduct of the Joint Operations that are to be shared by the Parties, but does not include proceeds attributable to hydrocarbons and by-products produced under the Agreement.

"**Joint Operations**" means all operations necessary or proper for the exploration, appraisal, development, production, protection, maintenance, repair, abandonment, and restoration of the Joint Property.

"**Joint Property**" means the real and personal property subject to the Agreement.

"**Laws**" means any laws, rules, regulations, decrees, and orders of the United States of America or any state thereof and all other governmental bodies, agencies, and other authorities having jurisdiction over or affecting the provisions contained in or the transactions contemplated by the Agreement or the Parties and their operations, whether such laws now exist or are hereafter amended, enacted, promulgated or issued.

"**Material**" means personal property, equipment, supplies, or consumables acquired or held for use by the Joint Property.

"**Non-Operators**" means the Parties to the Agreement other than the Operator.

"**Offshore Facilities**" means platforms, surface and subsea development and production systems, and other support systems such as oil and gas handling facilities, living quarters, offices, shops, cranes, electrical supply equipment and systems, fuel and water storage and piping, heliport, marine docking installations, communication facilities, navigation aids, and other similar facilities necessary in the conduct of offshore operations, all of which are located offshore.

"**Off-site**" means any location that is not considered On-site as defined in this Accounting Procedure.

"**On-site**" means on the Joint Property when in direct conduct of Joint Operations. The term "On-site" shall also include that portion of Offshore Facilities, Shore Base Facilities, fabrication yards, and staging areas from which Joint Operations are conducted, or other facilities that directly control equipment on the Joint Property, regardless of whether such facilities are owned by the Joint Account.

"**Operator**" means the Party designated pursuant to the Agreement to conduct the Joint Operations.

"**Parties**" means legal entities signatory to the Agreement or their successors and assigns. Parties shall be referred to individually as "Party."

"**Participating Interest**" means the percentage of the costs and risks of conducting an operation under the Agreement that a Party agrees, or is otherwise obligated, to pay and bear.

"**Participating Party**" means a Party that approves a proposed operation or otherwise agrees, or becomes liable, to pay and bear a share of the costs and risks of conducting an operation under the Agreement.

"**Personal Expenses**" means reimbursed costs for travel and temporary living expenses.

"**Railway Receiving Point**" means the railhead nearest the Joint Property for which freight rates are published, even though an actual railhead may not exist.

"**Shore Base Facilities**" means onshore support facilities that during Joint Operations provide such services to the Joint Property as a receiving and transshipment point for Materials; debarkation point for drilling and production personnel and services; communication, scheduling and dispatching center; and other associated functions serving the Joint Property.

"**Supply Store**" means a recognized source or common stock point for a given Material item.

"**Technical Services**" means services providing specific engineering, geoscience, or other professional skills, such as those performed by engineers, geologists, geophysicists, and technicians, required to handle specific operating conditions and problems for the benefit of Joint Operations; provided, however, Technical Services shall not include those functions specifically identified as overhead under the second paragraph of the introduction of Section III (*Overhead*). Technical Services may be provided by the Operator, Operator's Affiliate, Non-Operator, Non-Operator Affiliates, and/or third parties.

2. **STATEMENTS AND BILLINGS**

The Operator shall bill Non-Operators on or before the last day of the month for their proportionate share of the Joint Account for the preceding month. Such bills shall be accompanied by statements that identify the AFE (authority for expenditure), lease or facility, and all charges and credits summarized by appropriate categories of investment and expense. Controllable Material shall be separately identified and fully described in detail, or at the Operator's option, Controllable Material may be summarized by major Material classifications. Intangible drilling costs, audit adjustments, and unusual charges and credits shall be separately and clearly identified.

The Operator may make available to Non-Operators any statements and bills required under Section I.2 and/or Section I.3.A (*Advances and Payments by the Parties*) via email, electronic data interchange, internet websites or other equivalent electronic media in lieu of paper copies. The Operator shall provide the Non-Operators instructions and any necessary information to access and receive the statements and bills within the timeframes specified herein. A statement or billing shall be deemed as delivered twenty-four (24) hours (exclusive of weekends and holidays) after the Operator notifies the Non-Operator that the statement or billing is available on the website and/or sent via email or electronic data interchange transmission. Each Non-Operator individually shall elect to receive statements and billings electronically, if available from the Operator, or request paper copies. Such election may be changed upon thirty (30) days prior written notice to the Operator.

3. **ADVANCES AND PAYMENTS BY THE PARTIES**

   A. Unless otherwise provided for in the Agreement, the Operator may require the Non-Operators to advance their share of the estimated cash outlay for the succeeding month's operations within fifteen (15) days after receipt of the advance request or by the first day of the month for which the advance is required, whichever is later. The Operator shall adjust each monthly billing to reflect advances received from the Non-Operators for such month. If a refund is due, the Operator shall apply the amount to be refunded to the subsequent month's billing or advance, unless the Non-Operator sends the Operator a written request for a cash refund. The Operator shall remit the refund to the Non-Operator within fifteen (15) days of receipt of such written request.

   B. Except as provided below, each Party shall pay its proportionate share of all bills in full within fifteen (15) days of receipt date. If payment is not made within such time, the unpaid balance shall bear interest compounded monthly at the prime rate published by the *Wall Street Journal* on the first day of each month the payment is delinquent, plus three percent (3%), per annum, or the maximum contract rate permitted by the applicable usury Laws governing the Joint Property, whichever is the lesser, plus attorney's fees, court costs, and other costs in connection with the collection of unpaid amounts. If the *Wall Street Journal* ceases to be published or discontinues publishing a prime rate, the unpaid balance shall bear interest compounded monthly at the prime rate published by the Federal Reserve plus three percent (3%) per annum. Interest shall begin accruing on the first day of the month in which the payment was due. Payment shall not be reduced or delayed as a result of inquiries or anticipated credits unless the Operator has agreed. Notwithstanding the foregoing, the Non-Operator may reduce payment, provided it furnishes documentation and explanation to the Operator at the time payment is made, to the extent such reduction is caused by:

   (1) being billed at an incorrect working interest or Participating Interest that is higher than such Non-Operator's actual working interest or Participating Interest, as applicable; or
   (2) being billed for a project or AFE requiring approval of the Parties under the Agreement that the Non-Operator has not approved or is not otherwise obligated to pay under the Agreement; or
   (3) being billed for a property in which the Non-Operator no longer owns a working interest, provided the Non-Operator has furnished the Operator a copy of the recorded assignment or letter in-lieu. Notwithstanding the foregoing, the Non-Operator shall remain responsible for paying bills attributable to the interest it sold or transferred for any bills rendered during the thirty (30) day period following the Operator's receipt of such written notice; or
   (4) charges outside the adjustment period, as provided in Section I.4 (*Adjustments*).

4. ADJUSTMENTS

A. Payment of any such bills shall not prejudice the right of any Party to protest or question the correctness thereof; however, all bills and statements, including payout statements, rendered during any calendar year shall conclusively be presumed to be true and correct, with respect only to expenditures, after twenty-four (24) months following the end of any such calendar year, unless within said period a Party takes specific detailed written exception thereto making a claim for adjustment. The Operator shall provide a response to all written exceptions, whether or not contained in an audit report, within the time periods prescribed in Section I.5 (*Expenditure Audits*).

B. All adjustments initiated by the Operator, except those described in items (1) through (4) of this Section I.4.B, are limited to the twenty-four (24) month period following the end of the calendar year in which the original charge appeared or should have appeared on the Operator's Joint Account statement or payout statement. Adjustments that may be made beyond the twenty-four (24) month period are limited to adjustments resulting from the following:

(1) a physical inventory of Controllable Material as provided for in Section V (*Inventories of Controllable Material*), or
(2) an offsetting entry (whether in whole or in part) that is the direct result of a specific joint interest audit exception granted by the Operator relating to another property, or
(3) a government/regulatory audit, or
(4) a working interest ownership or Participating Interest adjustment.

5. EXPENDITURE AUDITS

A. A Non-Operator, upon written notice to the Operator and all other Non-Operators, shall have the right to audit the Operator's accounts and records relating to the Joint Account within the twenty-four (24) month period following the end of such calendar year in which such bill was rendered; however, conducting an audit shall not extend the time for the taking of written exception to and the adjustment of accounts as provided for in Section I.4 (*Adjustments*). Any Party that is subject to payout accounting under the Agreement shall have the right to audit the accounts and records of the Party responsible for preparing the payout statements, or of the Party furnishing information to the Party responsible for preparing payout statements. Audits of payout accounts may include the volumes of hydrocarbons produced and saved and proceeds received for such hydrocarbons as they pertain to payout accounting required under the Agreement. Unless otherwise provided in the Agreement, audits of a payout account shall be conducted within the twenty-four (24) month period following the end of the calendar year in which the payout statement was rendered.

Where there are two or more Non-Operators, the Non-Operators shall make every reasonable effort to conduct a joint audit in a manner that will result in a minimum of inconvenience to the Operator. The Operator shall bear no portion of the Non-Operators' audit cost incurred under this paragraph unless agreed to by the Operator. The audits shall not be conducted more than once each year without prior approval of the Operator, except upon the resignation or removal of the Operator, and shall be made at the expense of those Non-Operators approving such audit.

The Non-Operator leading the audit (hereinafter "lead audit company") shall issue the audit report within ninety (90) days after completion of the audit testing and analysis; however, the ninety (90) day time period shall not extend the twenty-four (24) month requirement for taking specific detailed written exception as required in Section I.4.A (*Adjustments*) above. All claims shall be supported with sufficient documentation.

A timely filed written exception or audit report containing written exceptions (hereinafter "written exceptions") shall, with respect to the claims made therein, preclude the Operator from asserting a statute of limitations defense against such claims, and the Operator hereby waives its right to assert any statute of limitations defense against such claims for so long as any Non-Operator continues to comply with the deadlines for resolving exceptions provided in this Accounting Procedure. If the Non-Operators fail to comply with the additional deadlines in Section I.5.B or I.5.C, the Operator's waiver of its rights to assert a statute of limitations defense against the claims brought by the Non-Operators shall lapse, and such claims shall then be subject to the applicable statute of limitations; provided that such waiver shall not lapse in the event that the Operator has failed to comply with the deadlines in Section I.5.B or I.5.C.

B. The Operator shall provide a written response to all exceptions in an audit report within one hundred eighty (180) days after Operator receives such report. Denied exceptions should be accompanied by a substantive response. If the Operator fails to provide substantive response to an exception within this one hundred eighty (180) day period, the Operator will owe interest on that exception or portion thereof, if ultimately granted, from the date it received the audit report. Interest shall be calculated using the rate set forth in Section I.3.B (*Advances and Payments by the Parties*).

C. The lead audit company shall reply to the Operator's response to an audit report within ninety (90) days of receipt, and the Operator shall reply to the lead audit company's follow-up response within ninety (90) days of receipt; provided, however, each Non-Operator shall have the right to represent itself if it disagrees with the lead audit company's position or believes the lead audit company is not adequately fulfilling its duties. Unless otherwise provided for in Section I.5.E, if the Operator fails to provide substantive response to an exception within this ninety (90) day period, the Operator will owe interest on that exception or portion thereof, if ultimately granted, from the date it received the audit report. Interest shall be calculated using the rate set forth in Section I.3.B (*Advances and Payments by the Parties*).

D. If any Party fails to meet the deadlines in Sections I.5.B or I.5.C or if any audit issues are outstanding fifteen (15) months after Operator receives the audit report, the Operator or any Non-Operator participating in the audit has the right to call a resolution meeting, as set forth in this Section I.5.D or it may invoke the dispute resolution procedures included in the Agreement, if applicable. The meeting will require one month's written notice to the Operator and all Non-Operators participating in the audit. The meeting shall be held at the Operator's office or mutually agreed location, and shall be attended by representatives of the Parties with authority to resolve such outstanding issues. Any Party who fails to attend the resolution meeting shall be bound by any resolution reached at the meeting. The lead audit company will make good faith efforts to coordinate the response and positions of the Non-Operator participants throughout the resolution process; however, each Non-Operator shall have the right to represent itself. Attendees will make good faith efforts to resolve outstanding issues, and each Party will be required to present substantive information supporting its position. A resolution meeting may be held as often as agreed to by the Parties. Issues unresolved at one meeting may be discussed at subsequent meetings until each such issue is resolved.

If the Agreement contains no dispute resolution procedures and the audit issues cannot be resolved by negotiation, the dispute shall be submitted to mediation. In such event, promptly following one Party's written request for mediation, the Parties to the dispute shall choose a mutually acceptable mediator and share the costs of mediation services equally. The Parties shall each have present at the mediation at least one individual who has the authority to settle the dispute. The Parties shall make reasonable efforts to ensure that the mediation commences within sixty (60) days of the date of the mediation request. Notwithstanding the above, any Party may file a lawsuit or complaint (1) if the Parties are unable after reasonable efforts, to commence mediation within sixty (60) days of the date of the mediation request, (2) for statute of limitations reasons, or (3) to seek a preliminary injunction or other provisional judicial relief, if in its sole judgment an injunction or other provisional relief is necessary to avoid irreparable damage or to preserve the status quo. Despite such action, the Parties shall continue to try to resolve the dispute by mediation.

E. ☐ *(Optional Provision – Forfeiture Penalties)*
If the Non-Operators fail to meet the deadline in Section I.5.C, any unresolved exceptions that were not addressed by the Non-Operators within one (1) year following receipt of the last substantive response of the Operator shall be deemed to have been withdrawn by the Non-Operators. If the Operator fails to meet the deadlines in Section I.5.B or I.5.C, any unresolved exceptions that were not addressed by the Operator within one (1) year following receipt of the audit report or receipt of the last substantive response of the Non-Operators, whichever is later, shall be deemed to have been granted by the Operator and adjustments shall be made, without interest, to the Joint Account.

6. **APPROVAL BY PARTIES**

    A. **General Matters**

    Where an approval or other agreement of the Parties or Non-Operators is expressly required under other Sections of this Accounting Procedure and if the Agreement to which this Accounting Procedure is attached contains no contrary provisions in regard thereto, the Operator shall notify all Non-Operators of the Operator's proposal and the agreement or approval of a majority in interest of the Non-Operators shall be controlling on all Non-Operators.

    This Section I.6.A applies to specific situations of limited duration where a Party proposes to change the accounting for charges from that prescribed in this Accounting Procedure. This provision does not apply to amendments to this Accounting Procedure, which are covered by Section I.6.B.

    B. **Amendments**

    If the Agreement to which this Accounting Procedure is attached contains no contrary provisions in regard thereto, this Accounting Procedure can be amended by an affirmative vote of _____ (____) or more Parties, one of which is the Operator, having a combined working interest of at least _____ percent (____%), which approval shall be binding on all Parties, provided, however, approval of at least one (1) Non-Operator shall be required.

    C. **Affiliates**
    For the purpose of administering the voting procedures of Sections I.6.A and I.6.B, if Parties to this Agreement are Affiliates of each other, then such Affiliates shall be combined and treated as a single Party having the combined working interest or Participating Interest of such Affiliates.

    For the purposes of administering the voting procedures in Section I.6.A, if a Non-Operator is an Affiliate of the Operator, votes under Section I.6.A shall require the majority in interest of the Non-Operator(s) after excluding the interest of the Operator's Affiliate.

II. DIRECT CHARGES

The Operator shall charge the Joint Account with the following items:

1. RENTALS AND ROYALTIES

Lease rentals and royalties paid by the Operator, on behalf of all Parties, for the Joint Operations.

2. LABOR

A. Salaries and wages, including incentive compensation programs as set forth in COPAS MFI-37 ("Chargeability of Incentive Compensation Programs"), for:

   (1) Operator's field employees directly employed On-site in the conduct of Joint Operations,

   (2) Operator's employees directly employed on Shore Base Facilities, Offshore Facilities, or other facilities serving the Joint Property if such costs are not charged under Section II.6 (*Equipment and Facilities Furnished by Operator*) or are not a function covered under Section III (*Overhead*),

   (3) Operator's employees providing First Level Supervision,

   (4) Operator's employees providing On-site Technical Services for the Joint Property if such charges are excluded from the overhead rates in Section III (*Overhead*),

   (5) Operator's employees providing Off-site Technical Services for the Joint Property if such charges are excluded from the overhead rates in Section III (*Overhead*).

   Charges for the Operator's employees identified in Section II.2.A may be made based on the employee's actual salaries and wages, or in lieu thereof, a day rate representing the Operator's average salaries and wages of the employee's specific job category.

   Charges for personnel chargeable under this Section II.2.A who are foreign nationals shall not exceed comparable compensation paid to an equivalent U.S. employee pursuant to this Section II.2, unless otherwise approved by the Parties pursuant to Section I.6.A (*General Matters*).

B. Operator's cost of holiday, vacation, sickness, and disability benefits, and other customary allowances paid to employees whose salaries and wages are chargeable to the Joint Account under Section II.2.A, excluding severance payments or other termination allowances. Such costs under this Section II.2.B may be charged on a "when and as-paid basis" or by "percentage assessment" on the amount of salaries and wages chargeable to the Joint Account under Section II.2.A. If percentage assessment is used, the rate shall be based on the Operator's cost experience.

C. Expenditures or contributions made pursuant to assessments imposed by governmental authority that are applicable to costs chargeable to the Joint Account under Sections II.2.A and B.

D. Personal Expenses of personnel whose salaries and wages are chargeable to the Joint Account under Section II.2.A when the expenses are incurred in connection with directly chargeable activities.

E. Reasonable relocation costs incurred in transferring to the Joint Property personnel whose salaries and wages are chargeable to the Joint Account under Section II.2.A. Notwithstanding the foregoing, relocation costs that result from reorganization or merger of a Party, or that are for the primary benefit of the Operator, shall not be chargeable to the Joint Account. Extraordinary relocation costs, such as those incurred as a result of transfers from remote locations, such as Alaska or overseas, shall not be charged to the Joint Account unless approved by the Parties pursuant to Section I.6.A (*General Matters*).

F. Training costs as specified in COPAS MFI-35 ("Charging of Training Costs to the Joint Account") for personnel whose salaries and wages are chargeable under Section II.2.A. This training charge shall include the wages, salaries, training course cost, and Personal Expenses incurred during the training session. The training cost shall be charged or allocated to the property or properties directly benefiting from the training. The cost of the training course shall not exceed prevailing commercial rates, where such rates are available.

G. Operator's current cost of established plans for employee benefits, as described in COPAS MFI-27 ("Employee Benefits Chargeable to Joint Operations and Subject to Percentage Limitation"), applicable to the Operator's labor costs chargeable to the Joint Account under Sections II.2.A and B based on the Operator's actual cost not to exceed the employee benefits limitation percentage most recently recommended by COPAS.

H. Award payments to employees, in accordance with COPAS MFI-49 ("Awards to Employees and Contractors") for personnel whose salaries and wages are chargeable under Section II.2.A.

3. **MATERIAL**

Material purchased or furnished by the Operator for use on the Joint Property in the conduct of Joint Operations as provided under Section IV (*Material Purchases, Transfers, and Dispositions*). Only such Material shall be purchased for or transferred to the Joint Property as may be required for immediate use or is reasonably practical and consistent with efficient and economical operations. The accumulation of surplus stocks shall be avoided.

4. **TRANSPORTATION**

A. Transportation of the Operator's, Operator's Affiliate's, or contractor's personnel necessary for Joint Operations.

B. Transportation of Material between the Joint Property and another property, or from the Operator's warehouse or other storage point to the Joint Property, shall be charged to the receiving property using one of the methods listed below. Transportation of Material from the Joint Property to the Operator's warehouse or other storage point shall be paid for by the Joint Property using one of the methods listed below:

(1) If the actual trucking charge is less than or equal to the Excluded Amount the Operator may charge actual trucking cost or a theoretical charge from the Railway Receiving Point to the Joint Property. The basis for the theoretical charge is the per hundred weight charge plus fuel surcharges from the Railway Receiving Point to the Joint Property. The Operator shall consistently apply the selected alternative.

(2) If the actual trucking charge is greater than the Excluded Amount, the Operator shall charge Equalized Freight. Accessorial charges such as loading and unloading costs, split pick-up

costs, detention, call out charges, and permit fees shall be charged directly to the Joint Property and shall not be included when calculating the Equalized Freight.

5. SERVICES

The cost of contract services, equipment, and utilities used in the conduct of Joint Operations, except for contract services, equipment, and utilities covered by Section III (*Overhead*), or Section II.7 (*Affiliates*), or excluded under Section II.9 (*Legal Expense*). Awards paid to contractors shall be chargeable pursuant to COPAS MFI- 49 ("Awards to Employees and Contractors").

The costs of third party Technical Services are chargeable to the extent excluded from the overhead rates under Section III (*Overhead*).

6. EQUIPMENT AND FACILITIES FURNISHED BY OPERATOR

In the absence of a separately negotiated agreement, equipment and facilities furnished by the Operator will be charged as follows:

A. Operator shall charge the Joint Account for use of Operator-owned equipment and facilities, including but not limited to production facilities, Shore Base Facilities, Offshore Facilities, and Field Offices, at rates commensurate with the costs of ownership and operation. The cost of Field Offices shall be chargeable to the extent the Field Offices provide direct service to personnel who are chargeable pursuant to Section II.2.A (*Labor*). Such rates may include labor, maintenance, repairs, other operating expense, insurance, taxes, depreciation using straight line depreciation method, and interest on gross investment less accumulated depreciation not to exceed _____ percent (__%) per annum; provided, however, depreciation shall not be charged when the equipment and facilities investment have been fully depreciated. The rate may include an element of the estimated cost for abandonment, reclamation, and dismantlement. Such rates shall not exceed the average commercial rates currently prevailing in the immediate area of the Joint Property.

B. In lieu of charges in Section II.6.A above, the Operator may elect to use average commercial rates prevailing in the immediate area of the Joint Property, less twenty percent (20%). If equipment and facilities are charged under this Section II.6.B, the Operator shall adequately document and support commercial rates and shall periodically review and update the rate and the supporting documentation. For automotive equipment, the Operator may elect to use rates published by the Petroleum Motor Transport Association (PMTA) or such other organization recognized by COPAS as the official source of rates.

7. AFFILIATES

A. Charges for an Affiliate's goods and/or services used in operations requiring an AFE or other authorization from the Non-Operators may be made without the approval of the Parties provided (i) the Affiliate is identified and the Affiliate goods and services are specifically detailed in the approved AFE or other authorization, and (ii) the total costs for such Affiliate's goods and services billed to such individual project do not exceed $_____. If the total costs for an Affiliate's goods and services charged to such individual project are not specifically detailed in the approved AFE or authorization or exceed such amount, charges for such Affiliate shall require approval of the Parties, pursuant to Section I.6.A (*General Matters*).

B. For an Affiliate's goods and /or services used in operations not requiring an AFE or other authorization from the Non-Operators, charges for such Affiliate's goods and services shall require approval of the Parties, pursuant to Section I.6.A (*General Matters*), if the charges exceed $_____ in a given calendar year.

C. The cost of the Affiliate's goods or services shall not exceed average commercial rates prevailing in the area of the Joint Property, unless the Operator obtains the Non-Operators' approval of such rates. The Operator shall adequately document and support commercial rates and shall periodically review and update the rate and the supporting documentation; provided, however, documentation of commercial rates shall not be required if the Operator obtains Non-Operator approval of its Affiliate's rates or charges prior to billing Non-Operators for such Affiliate's goods and services. Notwithstanding the foregoing, direct charges for Affiliate-owned communication facilities or systems shall be made pursuant to Section II.12 (*Communications*).

If the Parties fail to designate an amount in Sections II.7.A or II.7.B, in each instance the amount deemed adopted by the Parties as a result of such omission shall be the amount established as the Operator's expenditure limitation in the Agreement. If the Agreement does not contain an Operator's expenditure limitation, the amount deemed adopted by the Parties as a result of such omission shall be zero dollars ($ 0.00).

8. **DAMAGES AND LOSSES TO JOINT PROPERTY**

All costs or expenses necessary for the repair or replacement of Joint Property resulting from damages or losses incurred, except to the extent such damages or losses result from a Party's or Parties' gross negligence or willful misconduct, in which case such Party or Parties shall be solely liable.

The Operator shall furnish the Non-Operator written notice of damages or losses incurred as soon as practicable after a report has been received by the Operator.

9. **LEGAL EXPENSE**

Recording fees and costs of handling, settling, or otherwise discharging litigation, claims, and liens incurred in or resulting from operations under the Agreement, or necessary to protect or recover the Joint Property, to the extent permitted under the Agreement. Costs of the Operator's or Affiliate's legal staff or outside attorneys, including fees and expenses, are not chargeable unless approved by the Parties pursuant to Section I.6.A (*General Matters*) or otherwise provided for in the Agreement.

Notwithstanding the foregoing paragraph, costs for procuring abstracts, fees paid to outside attorneys for title examinations (including preliminary, supplemental, shut-in royalty opinions, division order title opinions), and curative work shall be chargeable to the extent permitted as a direct charge in the Agreement.

10. **TAXES AND PERMITS**

All taxes and permitting fees of every kind and nature, assessed or levied upon or in connection with the Joint Property, or the production therefrom, and which have been paid by the Operator for the benefit of the Parties, including penalties and interest, except to the extent the penalties and interest result from the Operator's gross negligence or willful misconduct.

If ad valorem taxes paid by the Operator are based in whole or in part upon separate valuations of each Party's working interest, then notwithstanding any contrary provisions, the charges to the Parties will be made in accordance with the tax value generated by each Party's working interest.

Costs of tax consultants or advisors, the Operator's employees, or Operator's Affiliate employees in matters regarding ad valorem or other tax matters, are not permitted as direct charges unless approved by the Parties pursuant to Section I.6.A (*General Matters*).

Charges to the Joint Account resulting from sales/use tax audits, including extrapolated amounts and penalties and interest, are permitted, provided the Non-Operator shall be allowed to review the invoices and other underlying source documents which served as the basis for tax charges and to determine that the correct amount of taxes were charged to the Joint Account. If the Non-Operator is not permitted to review such documentation, the sales/use tax amount shall not be directly charged unless the Operator can conclusively document the amount owed by the Joint Account.

11. INSURANCE

Net premiums paid for insurance required to be carried for Joint Operations for the protection of the Parties. If Joint Operations are conducted at locations where the Operator acts as self-insurer in regard to its worker's compensation and employer's liability insurance obligation, the Operator shall charge the Joint Account manual rates for the risk assumed in its self-insurance program as regulated by the jurisdiction governing the Joint Property. In the case of offshore operations in federal waters, the manual rates of the adjacent state shall be used for personnel performing work On-site, and such rates shall be adjusted for offshore operations by the U.S. Longshoreman and Harbor Workers (USL&H) or Jones Act surcharge, as appropriate.

12. COMMUNICATIONS

Costs of acquiring, leasing, installing, operating, repairing, and maintaining communication facilities or systems, including satellite, radio and microwave facilities, between the Joint Property and the Operator's office(s) directly responsible for field operations in accordance with the provisions of COPAS MFI-44 ("Field Computer and Communication Systems"). If the communications facilities or systems serving the Joint Property are Operator-owned, charges to the Joint Account shall be made as provided in Section II.6 (*Equipment and Facilities Furnished by Operator*). If the communication facilities or systems serving the Joint Property are owned by the Operator's Affiliate, charges to the Joint Account shall not exceed average commercial rates prevailing in the area of the Joint Property. The Operator shall adequately document and support commercial rates and shall periodically review and update the rate and the supporting documentation.

13. ECOLOGICAL, ENVIRONMENTAL, AND SAFETY

Costs incurred for Technical Services and drafting to comply with ecological, environmental or safety Laws or standards recommended by Occupational Safety and Health Administration (OSHA) or other regulatory authorities. All other labor and functions incurred for ecological, environmental and safety matters, including management, administration, and permitting, shall be covered by Sections II.2 (Labor), II.5 (Services), or Section III (Overhead), as applicable.

Costs to provide or have available pollution containment and removal equipment plus actual costs of control and cleanup and resulting responsibilities of oil and other spills as well as discharges from permitted outfalls as required by applicable Laws, or other pollution containment and removal equipment deemed appropriate by the Operator for prudent operations, are directly chargeable.

14. ABANDONMENT AND RECLAMATION

Costs incurred for abandonment and reclamation of the Joint Property, including costs required by lease agreements or by Laws.

15. OTHER EXPENDITURES

Any other expenditure not covered or dealt with in the foregoing provisions of this Section II (*Direct Charges*), or in Section III (*Overhead*) and which is of direct benefit to the Joint Property and is incurred by the Operator in the necessary and proper conduct of the Joint Operations. Charges made under this Section II.15 shall require approval of the Parties, pursuant to Section I.6.A (*General Matters*).

III. OVERHEAD

As compensation for costs not specifically identified as chargeable to the Joint Account pursuant to Section II (*Direct Charges*), the Operator shall charge the Joint Account in accordance with this Section III.

Functions included in the overhead rates regardless of whether performed by the Operator, Operator's Affiliates or third parties and regardless of location, shall include, but not be limited to, costs and expenses of:

- warehousing, other than for warehouses that are jointly owned under this Agreement
- design and drafting (except when allowed as a direct charge under Sections II.13, III.1.A(ii), and III.2, Option B)
- inventory costs not chargeable under Section V (*Inventories of Controllable Material*)
- procurement
- administration
- accounting and auditing
- gas dispatching and gas chart integration
- human resources
- management
- supervision not directly charged under Section II.2 (*Labor*)
- legal services not directly chargeable under Section II.9 (*Legal Expense*)
- taxation, other than those costs identified as directly chargeable under Section II.10 (*Taxes and Permits*)
- preparation and monitoring of permits and certifications; preparing regulatory reports; appearances before or meetings with governmental agencies or other authorities having jurisdiction over the Joint Property, other than On-site inspections; reviewing, interpreting, or submitting comments on or lobbying with respect to Laws or proposed Laws.

Overhead charges shall include the salaries or wages plus applicable payroll burdens, benefits, and Personal Expenses of personnel performing overhead functions, as well as office and other related expenses of overhead functions.

1. **OVERHEAD—DRILLING AND PRODUCING OPERATIONS**

   As compensation for costs incurred but not chargeable under Section II (*Direct Charges*) and not covered by other provisions of this Section III, the Operator shall charge on either:

   ☐ **(Alternative 1)** Fixed Rate Basis, Section III.1.B.

   ☐ **(Alternative 2)** Percentage Basis, Section III.1.C.

   A. **Technical Services**

   (i) Except as otherwise provided in Section II.13 (*Ecological Environmental, and Safety*) and Section III.2 (*Overhead – Major Construction and Catastrophe*), or by approval of the Parties pursuant to Section I.6.A (*General Matters*), the salaries, wages, related payroll burdens and benefits, and Personal Expenses for **On-site** Technical Services, including third party Technical Services:

   ☐ **(Alternative 1 – Direct)** shall be charged <u>direct</u> to the Joint Account.

   ☐ **(Alternative 2 – Overhead)** shall be covered by the <u>overhead</u> rates.

   (ii) Except as otherwise provided in Section II.13 (*Ecological, Environmental, and Safety*) and Section III.2 (*Overhead – Major Construction and Catastrophe*), or by approval of the Parties pursuant to Section I.6.A (*General Matters*), the salaries, wages, related payroll burdens and benefits, and Personal Expenses for **Off-site** Technical Services, including third party Technical Services:

   ☐ **(Alternative 1 – All Overhead)** shall be covered by the <u>overhead</u> rates.

   ☐ **(Alternative 2 – All Direct)** shall be charged <u>direct</u> to the Joint Account.

   ☐ **(Alternative 3 – Drilling Direct)** shall be charged <u>direct</u> to the Joint Account, <u>**only**</u> to the extent such Technical Services are directly attributable to drilling, redrilling, deepening, or sidetracking operations, through completion, temporary abandonment, or abandonment if a dry hole. Off-site Technical Services for all other operations, including workover, recompletion, abandonment of producing wells, and the construction or expansion of fixed assets not covered by Section III.2 (*Overhead - Major Construction and Catastrophe*) shall be covered by the overhead rates.

   Notwithstanding anything to the contrary in this Section III, Technical Services provided by Operator's Affiliates are subject to limitations set forth in Section II.7 (*Affiliates*). Charges for Technical personnel performing non-technical work shall not be governed by this Section III.1.A, but instead governed by other provisions of this Accounting Procedure relating to the type of work being performed.

   B. **Overhead—Fixed Rate Basis**

   (1) The Operator shall charge the Joint Account at the following rates per well per month:

   Drilling Well Rate per month $_____ (prorated for less than a full month)

   Producing Well Rate per month $_____

(2) Application of Overhead—Drilling Well Rate shall be as follows:

    (a) Charges for onshore drilling wells shall begin on the spud date and terminate on the date the drilling and/or completion equipment used on the well is released, whichever occurs later. Charges for offshore and inland waters drilling wells shall begin on the date the drilling or completion equipment arrives on location and terminate on the date the drilling or completion equipment moves off location, or is released, whichever occurs first. No charge shall be made during suspension of drilling and/or completion operations for fifteen (15) or more consecutive calendar days.

    (b) Charges for any well undergoing any type of workover, recompletion, and/or abandonment for a period of five (5) or more consecutive work days shall be made at the Drilling Well Rate. Such charges shall be applied for the period from date operations, with rig or other units used in operations, commence through date of rig or other unit release, except that no charges shall be made during suspension of operations for fifteen (15) or more consecutive calendar days.

(3) Application of Overhead — Producing Well Rate shall be as follows:

    (a) An active well that is produced, injected into for recovery or disposal, or used to obtain water supply to support operations for any portion of the month shall be considered as a one-well charge for the entire month.

    (b) Each active completion in a multi-completed well shall be considered as a one-well charge provided each completion is considered a separate well by the governing regulatory authority.

    (c) A one-well charge shall be made for the month in which plugging and abandonment operations are completed on any well, unless the Drilling Well Rate applies, as provided in Sections III.1.B.(2)(a) or (b). This one-well charge shall be made whether or not the well has produced.

    (d) An active gas well shut in because of overproduction or failure of a purchaser, processor, or transporter to take production shall be considered as a one-well charge provided the gas well is directly connected to a permanent sales outlet.

    (e) Any well not meeting the criteria set forth in Sections III.1.B.(3) (a), (b), (c), or (d) shall not qualify for a producing overhead charge.

(4) The well rates shall be adjusted on the first day of April each year following the effective date of the Agreement; provided, however, if this Accounting Procedure is attached to or otherwise governing the payout accounting under a farmout agreement, the rates shall be adjusted on the first day of April each year following the effective date of such farmout agreement. The adjustment shall be computed by applying the adjustment factor most recently published by COPAS. The adjusted rates shall be the initial or amended rates agreed to by the Parties increased or decreased by the adjustment factor described herein, for each year from the effective date of such rates, in accordance with COPAS MFI-47 ("Adjustment of Overhead Rates").

C. Overhead—Percentage Basis

(1) Operator shall charge the Joint Account at the following rates:

  (a) Development Rate _____ Percent (%) of the cost of development of the Joint Property, exclusive of costs provided under Section II.9 (*Legal Expense*) and all Material salvage credits.

  (b) Operating Rate _____ Percent (%) of the cost of operating the Joint Property, exclusive of costs provided under Sections II.1 (*Rentals and Royalties*) and II.9 (*Legal Expense*); all Material salvage credits; the value of substances purchased for enhanced recovery; all property and ad valorem taxes, and any other taxes and assessments that are levied, assessed, and paid upon the mineral interest in and to the Joint Property.

(2) Application of Overhead—Percentage Basis shall be as follows:

  (a) The Development Rate shall be applied to all costs in connection with:

  [i] drilling, redrilling, sidetracking, or deepening of a well
  [ii] a well undergoing plugback or workover operations for a period of five (5) or more consecutive work-days
  [iii] preliminary expenditures necessary in preparation for drilling
  [iv] expenditures incurred in abandoning when the well is not completed as a producer
  [v] construction or installation of fixed assets, the expansion of fixed assets and any other project clearly discernible as a fixed asset, other than Major Construction or Catastrophe as defined in Section III.2 (*Overhead-Major Construction and Catastrophe*).

  (b) The Operating Rate shall be applied to all other costs in connection with Joint Operations, except those subject to Section III.2 (*Overhead-Major Construction and Catastrophe*).

2. OVERHEAD—MAJOR CONSTRUCTION AND CATASTROPHE

To compensate the Operator for overhead costs incurred in connection with a Major Construction project or Catastrophe, the Operator shall either negotiate a rate prior to the beginning of the project, or shall charge the Joint Account for overhead based on the following rates for any Major Construction project in excess of the Operator's expenditure limit under the Agreement, or for any Catastrophe regardless of the amount. If the Agreement to which this Accounting Procedure is attached does not contain an expenditure limit, Major Construction Overhead shall be assessed for any single Major Construction project costing in excess of $100,000 gross.

Major Construction shall mean the construction and installation of fixed assets, the expansion of fixed assets, and any other project clearly discernible as a fixed asset required for the development and operation of the Joint Property, or in the dismantlement, abandonment, removal, and restoration of platforms, production equipment, and other operating facilities.

Catastrophe is defined as a sudden calamitous event bringing damage, loss, or destruction to property or the environment, such as an oil spill, blowout, explosion, fire, storm, hurricane, or other disaster. The overhead rate shall be applied to those costs necessary to restore the Joint Property to the equivalent condition that existed prior to the event.

A. If the Operator absorbs the engineering, design and drafting costs related to the project:

   (1) ____% of total costs if such costs are less than $100,000; plus

   (2) ____% of total costs in excess of $100,000 but less than $1,000,000; plus

   (3) ____% of total costs in excess of $1,000,000.

B. If the Operator charges engineering, design and drafting costs related to the project directly to the Joint Account:

   (1) ____% of total costs if such costs are less than $100,000; plus

   (2) ____% of total costs in excess of $100,000 but less than $1,000,000; plus

   (3) ____% of total costs in excess of $1,000,000.

Total cost shall mean the gross cost of any one project. For the purpose of this paragraph, the component parts of a single Major Construction project shall not be treated separately, and the cost of drilling and workover wells and purchasing and installing pumping units and downhole artificial lift equipment shall be excluded. For Catastrophes, the rates shall be applied to all costs associated with each single occurrence or event.

On each project, the Operator shall advise the Non-Operator(s) in advance which of the above options shall apply.

For the purposes of calculating Catastrophe Overhead, the cost of drilling relief wells, substitute wells, or conducting other well operations directly resulting from the catastrophic event shall be included. Expenditures to which these rates apply shall not be reduced by salvage or insurance recoveries. Expenditures that qualify for Major Construction or Catastrophe Overhead shall not qualify for overhead under any other overhead provisions.

In the event of any conflict between the provisions of this Section III.2 and the provisions of Sections II.2 (*Labor*), II.5 (*Services*), or II.7 (*Affiliates*), the provisions of this Section III.2 shall govern.

3. **AMENDMENT OF OVERHEAD RATES**

   The overhead rates provided for in this Section III may be amended from time to time if, in practice, the rates are found to be insufficient or excessive, in accordance with the provisions of Section I.6.B (*Amendments*).

IV. MATERIAL PURCHASES, TRANSFERS, AND DISPOSITIONS

The Operator is responsible for Joint Account Material and shall make proper and timely charges and credits for direct purchases, transfers, and dispositions. The Operator shall provide all Material for use in the conduct of Joint Operations; however, Material may be supplied by the Non-Operators, at the Operator's option. Material furnished by any Party shall be furnished without any express or implied warranties as to quality, fitness for use, or any other matter.

1. DIRECT PURCHASES

Direct purchases shall be charged to the Joint Account at the price paid by the Operator after deduction of all discounts received. The Operator shall make good faith efforts to take discounts offered by suppliers, but shall not be liable for failure to take discounts except to the extent such failure was the result of the Operator's gross negligence or willful misconduct. A direct purchase shall be deemed to occur when an agreement is made between an Operator and a third party for the acquisition of Material for a specific well site or location. Material provided by the Operator under "vendor stocking programs," where the initial use is for a Joint Property and title of the Material does not pass from the manufacturer, distributor, or agent until usage, is considered a direct purchase. If Material is found to be defective or is returned to the manufacturer, distributor, or agent for any other reason, credit shall be passed to the Joint Account within sixty (60) days after the Operator has received adjustment from the manufacturer, distributor, or agent.

2. TRANSFERS

A transfer is determined to occur when the Operator (i) furnishes Material from a storage facility or from another operated property, (ii) has assumed liability for the storage costs and changes in value, and (iii) has previously secured and held title to the transferred Material. Similarly, the removal of Material from the Joint Property to a storage facility or to another operated property is also considered a transfer; provided, however, Material that is moved from the Joint Property to a storage location for safe-keeping pending disposition may remain charged to the Joint Account and is not considered a transfer. Material shall be disposed of in accordance with Section IV.3 (*Disposition of Surplus*) and the Agreement to which this Accounting Procedure is attached.

A. PRICING

The value of Material transferred to/from the Joint Property should generally reflect the market value on the date of physical transfer. Regardless of the pricing method used, the Operator shall make available to the Non-Operators sufficient documentation to verify the Material valuation. When higher than specification grade or size tubulars are used in the conduct of Joint Operations, the Operator shall charge the Joint Account at the equivalent price for well design specification tubulars, unless such higher specification grade or sized tubulars are approved by the Parties pursuant to Section I.6.A (*General Matters*). Transfers of new Material will be priced using one of the following pricing methods; provided, however, the Operator shall use consistent pricing methods, and not alternate between methods for the purpose of choosing the method most favorable to the Operator for a specific transfer:

(1) Using published prices in effect on date of movement as adjusted by the appropriate COPAS Historical Price Multiplier (HPM) or prices provided by the COPAS Computerized Equipment Pricing System (CEPS).

(a) For oil country tubulars and line pipe, the published price shall be based upon eastern mill carload base prices (Houston, Texas, for special end) adjusted as of date of movement, plus transportation cost as defined in Section IV.2.B (*Freight*).

(b) For other Material, the published price shall be the published list price in effect at date of movement, as listed by a Supply Store nearest the Joint Property where like Material is normally available, or point of manufacture plus transportation costs as defined in Section IV.2.B (*Freight*).

(2) Based on a price quotation from a vendor that reflects a current realistic acquisition cost.

(3) Based on the amount paid by the Operator for like Material in the vicinity of the Joint Property within the previous twelve (12) months from the date of physical transfer.

(4) As agreed to by the Participating Parties for Material being transferred to the Joint Property, and by the Parties owning the Material for Material being transferred from the Joint Property.

B. **FREIGHT**

Transportation costs shall be added to the Material transfer price using the method prescribed by the COPAS Computerized Equipment Pricing System (CEPS). If not using CEPS, transportation costs shall be calculated as follows:

(1) Transportation costs for oil country tubulars and line pipe shall be calculated using the distance from eastern mill to the Railway Receiving Point based on the carload weight basis as recommended by the COPAS MFI-38 ("Material Pricing Manual") and other COPAS MFIs in effect at the time of the transfer.

(2) Transportation costs for special mill items shall be calculated from that mill's shipping point to the Railway Receiving Point. For transportation costs from other than eastern mills, the 30,000-pound interstate truck rate shall be used. Transportation costs for macaroni tubing shall be calculated based on the interstate truck rate per weight of tubing transferred to the Railway Receiving Point.

(3) Transportation costs for special end tubular goods shall be calculated using the interstate truck rate from Houston, Texas, to the Railway Receiving Point.

(4) Transportation costs for Material other than that described in Sections IV.2.B.(1) through (3), shall be calculated from the Supply Store or point of manufacture, whichever is appropriate, to the Railway Receiving Point.

Regardless of whether using CEPS or manually calculating transportation costs, transportation costs from the Railway Receiving Point to the Joint Property are in addition to the foregoing, and may be charged to the Joint Account based on actual costs incurred. All transportation costs are subject to Equalized Freight as provided in Section II.4 (*Transportation*) of this Accounting Procedure.

C. **TAXES**

Sales and use taxes shall be added to the Material transfer price using either the method contained in the COPAS Computerized Equipment Pricing System (CEPS) or the applicable tax rate in effect for the Joint Property at the time and place of transfer. In either case, the Joint Account shall be charged or credited at the rate that would have governed had the Material been a direct purchase.

D. **CONDITION**

(1) Condition "A" – New and unused Material in sound and serviceable condition shall be charged at one hundred percent (100%) of the price as determined in Sections IV.2.A (*Pricing*), IV.2.B (*Freight*), and IV.2.C (*Taxes*). Material transferred from the Joint Property that was not placed in service shall be credited as charged without gain or loss; provided,

however, any unused Material that was charged to the Joint Account through a direct purchase will be credited to the Joint Account at the original cost paid less restocking fees charged by the vendor. New and unused Material transferred from the Joint Property may be credited at a price other than the price originally charged to the Joint Account provided such price is approved by the Parties owning such Material, pursuant to Section I.6.A (*General Matters*). All refurbishing costs required or necessary to return the Material to original condition or to correct handling, transportation, or other damages will be borne by the divesting property. The Joint Account is responsible for Material preparation, handling, and transportation costs for new and unused Material charged to the Joint Property either through a direct purchase or transfer. Any preparation costs incurred, including any internal or external coating and wrapping, will be credited on new Material provided these services were not repeated for such Material for the receiving property.

(2) Condition "B" – Used Material in sound and serviceable condition and suitable for reuse without reconditioning shall be priced by multiplying the price determined in Sections IV.2.A (*Pricing*), IV.2.B (*Freight*), and IV.2.C (*Taxes*) by seventy-five percent (75%).

Except as provided in Section IV.2.D(3), all reconditioning costs required to return the Material to Condition "B" or to correct handling, transportation or other damages will be borne by the divesting property.

If the Material was originally charged to the Joint Account as used Material and placed in service for the Joint Property, the Material will be credited at the price determined in Sections IV.2.A (*Pricing*), IV.2.B (*Freight*), and IV.2.C (*Taxes*) multiplied by sixty-five percent (65%).

Unless otherwise agreed to by the Parties that paid for such Material, used Material transferred from the Joint Property that was not placed in service on the property shall be credited as charged without gain or loss.

(3) Condition "C" – Material that is not in sound and serviceable condition and not suitable for its original function until after reconditioning shall be priced by multiplying the price determined in Sections IV.2.A (*Pricing*), IV.2.B (*Freight*), and IV.2.C (*Taxes*) by fifty percent (50%).

The cost of reconditioning may be charged to the receiving property to the extent Condition "C" value, plus cost of reconditioning, does not exceed Condition "B" value.

(4) Condition "D" – Material that (i) is no longer suitable for its original purpose but useable for some other purpose, (ii) is obsolete, or (iii) does not meet original specifications but still has value and can be used in other applications as a substitute for items with different specifications, is considered Condition "D" Material. Casing, tubing, or drill pipe used as line pipe shall be priced as Grade A and B seamless line pipe of comparable size and weight. Used casing, tubing, or drill pipe utilized as line pipe shall be priced at used line pipe prices. Casing, tubing, or drill pipe used as higher pressure service lines than standard line pipe, e.g., power oil lines, shall be priced under normal pricing procedures for casing, tubing, or drill pipe. Upset tubular goods shall be priced on a non-upset basis. For other items, the price used should result in the Joint Account being charged or credited with the value of the service rendered or use of the Material, or as agreed to by the Parties pursuant to Section I.6.A (*General Matters*).

(5) Condition "E" – Junk shall be priced at prevailing scrap value prices.

E. **OTHER PRICING PROVISIONS**

(1) Preparation Costs

Subject to Section II (*Direct Charges*) and Section III (*Overhead*) of this Accounting Procedure, costs incurred by the Operator in making Material serviceable including inspection, third party surveillance services, and other similar services will be charged to the Joint Account at prices which reflect the Operator's actual costs of the services. Documentation must be provided to the Non-Operators upon request to support the cost of service. New coating and/or wrapping shall be considered a component of the Materials and priced in accordance with Sections IV.1 (*Direct Purchases*) or IV.2.A (*Pricing*), as applicable. No charges or credits shall be made for used coating or wrapping. Charges and credits for inspections shall be made in accordance with COPAS MFI-38 ("Material Pricing Manual").

(2) Loading and Unloading Costs

Loading and unloading costs related to the movement of the Material to the Joint Property shall be charged in accordance with the methods specified in COPAS MFI-38 ("Material Pricing Manual").

3. **DISPOSITION OF SURPLUS**

Surplus Material is that Material, whether new or used, that is no longer required for Joint Operations. The Operator may purchase, but shall be under no obligation to purchase, the interest of the Non-Operators in surplus Material.

Dispositions for the purpose of this procedure are considered to be the relinquishment of title of the Material from the Joint Property to either a third party, a Non-Operator, or to the Operator. To avoid the accumulation of surplus Material, the Operator should make good faith efforts to dispose of surplus within twelve (12) months through buy/sale agreements, trade, sale to a third party, division in kind, or other dispositions as agreed to by the Parties.

Disposal of surplus Materials shall be made in accordance with the terms of the Agreement to which this Accounting Procedure is attached. If the Agreement contains no provisions governing disposal of surplus Material, the following terms shall apply:

- The Operator may, through a sale to an unrelated third party or entity, dispose of surplus Material having a gross sale value that is less than or equal to the Operator's expenditure limit as set forth in the Agreement to which this Accounting Procedure is attached without the prior approval of the Parties owning such Material.

- If the gross sale value exceeds the Agreement expenditure limit, the disposal must be agreed to by the Parties owning such Material.

- Operator may purchase surplus Condition "A" or "B" Material without approval of the Parties owning such Material, based on the pricing methods set forth in Section IV.2 (*Transfers*).

- Operator may purchase Condition "C" Material without prior approval of the Parties owning such Material if the value of the Materials, based on the pricing methods set forth in Section IV.2 (*Transfers*), is less than or equal to the Operator's expenditure limitation set forth in the Agreement. The Operator shall provide documentation supporting the classification of the Material as Condition C.

- Operator may dispose of Condition "D" or "E" Material under procedures normally utilized by Operator without prior approval of the Parties owning such Material.

4. SPECIAL PRICING PROVISIONS

A. PREMIUM PRICING

Whenever Material is available only at inflated prices due to national emergencies, strikes, government imposed foreign trade restrictions, or other unusual causes over which the Operator has no control, for direct purchase the Operator may charge the Joint Account for the required Material at the Operator's actual cost incurred in providing such Material, making it suitable for use, and moving it to the Joint Property. Material transferred or disposed of during premium pricing situations shall be valued in accordance with Section IV.2 (*Transfers*) or Section IV.3 (*Disposition of Surplus*), as applicable.

B. SHOP-MADE ITEMS

Items fabricated by the Operator's employees, or by contract laborers under the direction of the Operator, shall be priced using the value of the Material used to construct the item plus the cost of labor to fabricate the item. If the Material is from the Operator's scrap or junk account, the Material shall be priced at either twenty-five percent (25%) of the current price as determined in Section IV.2.A (*Pricing*) or scrap value, whichever is higher. In no event shall the amount charged exceed the value of the item commensurate with its use.

C. MILL REJECTS

Mill rejects purchased as "limited service" casing or tubing shall be priced at eighty percent (80%) of K-55/J-55 price as determined in Section IV.2 (*Transfers*). Line pipe converted to casing or tubing with casing or tubing couplings attached shall be priced as K-55/J-55 casing or tubing at the nearest size and weight.

V. INVENTORIES OF CONTROLLABLE MATERIAL

The Operator shall maintain records of Controllable Material charged to the Joint Account, with sufficient detail to perform physical inventories.

Adjustments to the Joint Account by the Operator resulting from a physical inventory of Controllable Material shall be made within twelve (12) months following the taking of the inventory or receipt of Non-Operator inventory report. Charges and credits for overages or shortages will be valued for the Joint Account in accordance with Section IV.2 (*Transfers*) and shall be based on the Condition "B" prices in effect on the date of physical inventory unless the inventorying Parties can provide sufficient evidence another Material condition applies.

1. **DIRECTED INVENTORIES**

   Physical inventories shall be performed by the Operator upon written request of a majority in working interests of the Non-Operators (hereinafter, "directed inventory"); provided, however, the Operator shall not be required to perform directed inventories more frequently than once every five (5) years. Directed inventories shall be commenced within one hundred eighty (180) days after the Operator receives written notice that a majority in interest of the Non-Operators has requested the inventory. All Parties shall be governed by the results of any directed inventory.

   Expenses of directed inventories will be borne by the Joint Account; provided, however, costs associated with any post-report follow-up work in settling the inventory will be absorbed by the Party incurring such costs. The Operator is expected to exercise judgment in keeping expenses within reasonable limits. Any anticipated disproportionate or extraordinary costs should be discussed and agreed upon prior to commencement of the inventory. Expenses of directed inventories may include the following:

   A. A per diem rate for each inventory person, representative of actual salaries, wages, and payroll burdens and benefits of the personnel performing the inventory or a rate agreed to by the Parties pursuant to Section I.6.A (*General Matters*). The per diem rate shall also be applied to a reasonable number of days for pre-inventory work and report preparation.

   B. Actual transportation costs and Personal Expenses for the inventory team.

   C. Reasonable charges for report preparation and distribution to the Non-Operators.

2. **NON-DIRECTED INVENTORIES**

   A. **OPERATOR INVENTORIES**

   Physical inventories that are not requested by the Non-Operators may be performed by the Operator, at the Operator's discretion. The expenses of conducting such Operator-initiated inventories shall not be charged to the Joint Account.

   B. **NON-OPERATOR INVENTORIES**

   Subject to the terms of the Agreement to which this Accounting Procedure is attached, the Non-Operators may conduct a physical inventory at reasonable times at their sole cost and risk after giving the Operator at least ninety (90) days prior written notice. The Non-Operator inventory report shall be furnished to the Operator in writing within ninety (90) days of completing the inventory fieldwork.

   C. **SPECIAL INVENTORIES**

   The expense of conducting inventories other than those described in Sections V.1 (*Directed Inventories*), V.2.A (*Operator Inventories*), or V.2.B (*Non-Operator Inventories*), shall be charged to the Party requesting such inventory; provided, however, inventories required due to a change of Operator shall be charged to the Joint Account in the same manner as described in Section V.1 (*Directed Inventories*).

The accounting procedure is an integral part of any JOA. The accounting procedure specifically addresses issues related to the maintenance of the joint account (defined later). In particular, it addresses the determination of appropriate charges and credits applicable to the joint operation.

The main sections normally included in the accounting procedure are discussed below. For a comprehensive explanation of the terms of any given accounting procedure, one should consult the relevant COPAS Model Form Interpretation (MFI) for that particular accounting procedure.

General provisions

The general provisions section deals with a variety of topics. Some selected terms are discussed below:

1. **Definitions.** This section defines terms used in the contract that are frequently subject to question or interpretation. Examples include first-level supervision, technical employee, and controllable material.

2. **Statements and billings.** The operator is to provide a monthly statement to all of the nonoperators. The statement should include a listing of all costs and expenditures incurred during the preceding month, the amount of advances received from the parties, each party's share of the costs and expenses, and the respective cash balances or deficits. The costs and expenditures incurred during the month should be identified as follows:

   a. Relating to a specific authorization for expenditure (AFE)
   b. Relating to a particular lease or facility
   c. Summarized as related to investment, e.g., wells-in-progress or expense

   Items of controllable material (e.g., generators, compressors, etc.) must be reported separately along with intangible drilling costs, audit adjustments, and any unusual charges or credits. (Controllable material is defined in each JOA and typically refers to items of tangible equipment with a relatively high cost, a life longer than one year, and with a salvage value. Often agreements refer to *COPAS MFI-28*, "Material Classification Manual," for a listing of items of controllable material.)

3. **Advances and payments by the parties.** The operator is given the right to require the nonoperators to prepay or advance the next month's estimated cash outlays—referred to as a **cash call**. Cash calls are common in international oil and gas operations and in situations where there are a number of wells being drilled and cash outlays are large. In the event that the cash call for any given month exceeds the actual amount of cash required for that month, each party's share of the balance is typically carried forward to reduce that party's cash call for the next month; however, the nonoperators may request a refund. If cash calls are not used, the operator will utilize a billing process whereby a billing is sent to the nonoperators itemizing costs for the month. The nonoperators must remit their proportionate share to the operator in a timely manner—typically within 10 days.

4. **Adjustments.** Payment of a billing or cash advance does not indicate that the nonoperator agrees with the correctness of a billing or statement. The nonoperator has 24 months from the end of the current year to raise exceptions to any charge. After that period, the statements are deemed to be true and correct.

5. **Expenditure audits.** All nonoperators have the right to audit the joint account and other records of the operator pertaining to the joint operation. The nonoperators have 24 months from the end of the year in which the disputed charge occurred to raise an exception or make a claim to the operator.

6. **Approvals by parties.** This section relates to voting provisions related to miscellaneous matters that require special approval by the parties.

7. **Allocations.** In the course of a joint operation, it frequently becomes necessary to allocate joint or common costs between the joint operation and other operations. For example, the operator may have solely-owned equipment that is used on jointly owned properties. The costs associated with the equipment may be allocated to all of the properties that it serves. Most contracts state such allocations are to be made on an equitable basis in accordance with accounting standards. Unfortunately, there are no formalized accounting standards relating to joint or common cost allocations in oil and gas operations. Therefore, it is necessary for the operator to be careful to use methods that are rational and equitable. The operator should be prepared to explain and justify such cost allocation methods if called upon to do so by the nonoperators.

Direct charges

All the working interest owners in a jointly owned property have an obligation to pay their proportionate share of the costs in return for a share of production. The term **joint account** refers to costs that have been identified with or allocated to a particular jointly owned property and, therefore, are the responsibility of that particular group of working interest owners. The term *joint account* does not necessarily refer to a separate account. Two specific types of costs are recognized and separated for the purpose of making charges to the joint account: direct and indirect.

Direct costs are costs that are specifically identified with the joint operation. **Indirect costs** are not individually identified with the joint operation per se. Rather, an operator is allowed to recover its indirect costs by charging the joint account some agreed-upon amount, e.g., a percentage of direct costs or a fixed amount per well drilled. A working knowledge of these two types of costs is necessary in order to understand the specific charges that can be made to the joint account.

Expenditures made for material and services on the joint property for the direct benefit of the joint property are the primary source of direct charges. The following are examples of general activities typically charged directly to the joint account: exploratory drilling; development drilling; installation of production equipment; operation, maintenance, and repair of wells and equipment; and rentals.

Costs incurred at a general or administrative level are the primary sources of indirect or overhead charges. These costs benefit the joint property but in an indirect manner. Examples of general activities are: home office administration, data processing, office services, human resources, and legal support.

The following is a more detailed discussion of some of the costs that are normally treated as direct costs (versus indirect costs) in various COPAS accounting procedures.

1. **Rentals and royalties.** Any type of lease rental or royalty paid by the operator on behalf of the nonoperators is a direct charge.

2. **Labor.** The salaries, wages, and related costs of employees of the operator engaged in the joint operations, whether temporarily or permanently assigned, are direct charges. These employees include the operator's field employees directly employed on-site doing work related to the joint property, as well as employees providing first-level supervision, technical services, and other activities that directly benefit the joint property. Since organizational charts and job assignments differ from company to company, it is often difficult to determine for certain whether any given employee is direct versus indirect. If it is determined that the employee is a direct charge, then that employee's salary must not also be included in the determination of the overhead rate for that property.

    a. First level of supervision. A commonly encountered issue is what levels of the management organization are general administrative overhead and what levels are directly related to the property served. As management gets further and further removed from the physical operations, the answers become less and less clear. Most operating agreements provide for the salaries and expenses of first-level supervision in the field to be a direct charge to the joint account. The operations in the field include drilling wells, repairing wells, recompletion of wells, producing wells, constructing and operating facilities, etc. One frequently encountered problem within the industry is identifying the employees who qualify as first-level supervision for these purposes. The 1975 *COPAS Accounting Procedure* and 1984 *COPAS Accounting Procedure* defined first-level supervision as those employees whose primary function in joint operations is the direct supervision of other employees and/or contract labor directly employed on the joint property in a field operating capacity. Generally, first-level supervisors do not have engineering or administrative staff but rely upon the staff associated with the administrative or functional/technical office for these services. The 2005 *COPAS Accounting Procedure* includes a comprehensive definition of first level supervision as follows:

    > *Those employees whose primary function in Joint Operations is the direct oversight of the Operator's field employees and/or contract labor directly employed On-site in a field operating capacity. First Level Supervision functions may include, but are not limited to:*
    >
    > - *Responsibility for field employees and contract labor engaged in activities that can include field operations, maintenance, construction, well remedial work, equipment movement and drilling*
    >
    > - *Responsibility for day-to-day direct oversight of rig operations*
    >
    > - *Responsibility for day-to-day direct oversight of construction operations*
    >
    > - *Coordination of job priorities and approval of work procedures*
    >
    > - *Responsibility for optimal resource utilization (equipment, materials, personnel)*
    >
    > - *Responsibility for meeting production and field operating expense targets*

- *Representation of the Parties in local matters involving community, vendors, regulatory agents, and landowners, as an incidental part of the supervisor's operating responsibilities*

- *Responsibility for all emergency responses with field staff*

- *Responsibility for implementing safety and environmental practices*

- *Responsibility for field adherence to company policy*

- *Responsibility for employment decisions and performance appraisals for field personnel*

- *Oversight of sub-groups for field functions such as electrical, safety, environmental, telecommunications, which may have group or team leaders.*

b. Technical labor. Charging the cost of the operator's technical employees to the joint account has been a long-standing issue in domestic joint venture accounting. Technical employees include employees having special and specific engineering, geological, or other professional skills, and whose primary function in joint operations is the handling of specific operating conditions and problems. The technical employees usually have no supervisory authority except that required to resolve the particular problem to which they are assigned. Charging of the salary and wages of technical employees is expressly provided for in each COPAS accounting procedure beginning in 1962, so long as the technical labor is not included as a component of the overhead rate being charged by the operator.

c. Employee benefits. Employee benefits are generally considered to be part of the employer's total labor cost. Most operating agreements allow the operator to charge the joint account with the current cost of established employee benefit plans as long as they are made available to all employees on a regular basis. This charge is usually expressed as a percentage of the total labor chargeable to the joint account. Another method of charging employee benefits to the joint account is on a *when and as paid* basis. However, the percentage assessment is typically allowed. COPAS issued *MFI-27*, "Employee Benefits Chargeable to Joint Operations and Subject to Percentage Limitation," to help clarify which costs are to be included in the percentage calculation and which are not. Costs that can be included in the percentage calculation include: bonuses (e.g., Christmas bonuses), medical and dental insurance, business travel insurance, pensions, profit-sharing plans, life insurance, tuition assistance, long-term disability insurance, and vision care plans. Among the costs not included are personal leaves, car pool subsidies, company car use, employee stock ownership plans, layoff benefits, and parking.

3. **Materials and supplies.** The costs of materials and supplies—net of any discounts—purchased or furnished by the operator for the joint operations are direct charges. The costs include, but are not limited to, export brokers' fees, transportation charges, loading and unloading fees, export and import duties, license fees, and in-transit losses not covered by insurance.

4. **Transportation.** The cost of transportation related to personnel who are directly serving the joint property is a direct charge, as is the cost of transportation of equipment and material.

5. **Services.** The cost of services is generally broken down between those services provided by a third party and those provided by an affiliate of the operator. The cost of services performed by third parties for the benefit of the joint operation are direct charges, provided the transactions that resulted in the charges are derived pursuant to an arm's length transaction. The cost of professional administrative, scientific, or technical personnel services provided by an affiliate of the operator in lieu of services provided by the operator's own personnel are direct charges, if provided for the direct benefit of the joint operations. The rates charged must be equal to the actual cost of the services, must exclude any element of profit, and should not be higher than charges of third parties for comparable services performed under comparable conditions.

6. **Exclusively owned equipment and facilities of the operator.** The operator may use equipment and facilities it exclusively owns on a joint property. In that case, the operator is allowed to charge the joint account rental rates based on the actual cost incurred by the operator, including factors relating to the cost of ownership. However, the rates charged may not exceed the average prevailing commercial rates of nonaffiliated third parties for like equipment and facilities used in the same area.

7. **Affiliates.** Special attention must be paid to those situations in which the operator acquires goods and services from an entity with whom the operator has any type of residual relationship. Any time the operator acquires goods and services from an affiliate, there is a concern that there should be no residual benefit that the operator derives due to its relationship. There should be full disclosure of the relationship, and in some cases, special approvals may be necessary in order for the related costs to be treated entirely as direct charges.

8. **Damages and losses.** Costs that can be associated with losses by casualty or theft are directly chargeable to the joint account. Any settlement received from an insurance carrier should be credited to the parties participating in any joint property insurance coverage.

9. **Legal expense.** Generally all legal expenses, except attorney fees, incurred for the benefit of the joint property are direct charges. Attorney fees must typically be approved by all of the parties in order to be treated as direct charges.

10. **Taxes, licenses, permits, etc.** Generally all costs incurred by the operator in relation to the acquisition, maintenance, renewal, or relinquishment of licenses, permits, or surface rights acquired by the joint operation are direct charges.

11. **Insurance.** The operator is typically required to carry insurance for the protection of the joint property. The cost of such insurance is a direct charge. If the operator opts to self-insure, then a special provision may apply in determining the amount to be charged to the joint account.

12. **Communications.** The costs of acquiring, leasing, installing, operating, repairing, or otherwise utilizing communication systems are direct charges if the equipment is necessary for the joint operations. Such equipment may include satellite, radio, and microwave facilities.

13. **Ecological and environmental costs.** Ecological and environmental costs incurred in relation to a particular operation are generally considered to be a cost of operating jointly owned property and are directly chargeable to the joint account.

14. **Abandonment and reclamation.** In many jurisdictions, various legal and regulatory requirements exist relating to the abandonment and reclamation of the property. Lease provisions may also exist. All costs required by such provisions are direct charges.

15. **Offices, camps, and miscellaneous facilities.** The costs of maintaining any offices, suboffices, camps, warehouses, housing, shore-based facilities—or other facilities of the operator and/or affiliates of the operator that are directly serving the joint operations—are directly chargeable to the joint account. If any such facilities serve other operations in addition to the particular joint operation in question, or any business other than the petroleum operations, the net costs are to be allocated to the operations served on an equitable and consistent basis. In recent years there has been considerable debate regarding identification of which offices are direct and which are indirect. The *2005 COPAS Accounting Procedure* includes a specific definition of field offices indicating that in order to be a direct charge, a field office must directly serve daily operations and maintenance activities of the joint property and serve as a staging area for directly chargeable field personnel.

16. **Other.** Inevitably, costs will occur that are not included in the above list. Many agreements include rather vague language as to what would fall into the other category. The *2005 COPAS Accounting Procedure* seeks to add clarity, indicating that in order to be treated as a direct charge, any other cost must meet the following criteria:

    a. The cost must not be addressed in the section on general provisions or direct charges.

    b. The cost must be expended for the direct benefit of the joint operations.

    c. The cost must be necessary for the proper conduct of the joint operations and be approved by the parties.

Overhead

Various methods have been utilized in the industry in the past to recover the indirect costs or **overhead** associated with an oil or gas operation. The charging of overhead is the means by which the operator recovers its cost for such items as clerical, administrative, engineering, accounting, home office expenses, and other costs not allowed as direct charges. The accounting procedure typically provides for three types of overhead:

- Overhead incurred in drilling and production operations
- Overhead incurred in construction operations
- Catastrophe overhead

The methods of computing overhead are either combined fixed rate or percentage basis.

The **combined fixed rate** basis is the most commonly used method of computing overhead in domestic production and drilling operations. The rate is referred to as "combined" because it is meant to cover an operator's expenses at all levels (e.g., home office, regional, district, etc.) and "fixed" because it does not vary in proportion to actual expenses. To determine the amount of overhead, the overhead rate for production operations is multiplied by the number of wells that are producing during all or any part of the month. The drilling overhead rate is multiplied times the number of wells being drilled, prorated for the number of days during the month that the wells are actually being drilled.

For example, assume the producing well rate is $100/month, and the drilling well rate is $300/month. If there are 50 wells on the lease that produced at least 1 day during the month, then the total producing well overhead for the month would be $100 × 50 wells ($5,000). If there are 2 wells being drilled on the lease, with 1 well having drilling in progress all month and the other having drilling operations for only 10 days, then the drilling well overhead is $400, i.e., $300 + 10/30 × $300. The rates are determined through negotiation and are typically adjusted annually to reflect changes in prices and costs.

The less common alternate is to provide for production and drilling overhead to be computed as a percentage of direct costs. If the percentage basis is used, there is typically one rate for drilling and development overhead and another rate for production overhead, e.g., 5% of all drilling and development costs plus 2% of all production expenditures.

The other types of overhead are construction overhead and catastrophe overhead. **Construction overhead** is provided to compensate the operator for indirect costs incurred while major construction projects are underway (e.g., expansion of a gathering system or construction of a central tank battery). **Catastrophe overhead** compensates the operator for indirect costs incurred in the event of an unexpected occurrence such as a hurricane, oil spill, explosion, fire, etc. Construction overhead and catastrophe overhead are typically calculated on a percentage sliding scale basis. For example:

5% of total costs up to and including $100,000, plus
3% of costs in excess of $100,000 but less than $1,000,000, plus
2% of costs equal to or in excess of $1,000,000

In international operations, overhead rates are frequently based on a sliding percentage of appropriate direct charges. Generally one set of rates applies to exploration operations, while a different set of rates apply to development operations and to production operations. The rates applicable to exploration operations are normally the highest, while the rates applicable to production operations are the lowest. These rates are applied to the cumulative annual expenditures. The following is an example of international exploration operations overhead rates:

Annual Expenditures for Exploration Operations

Direct Charges Incurred and Charged to Joint Operations	Percentage Rate of Direct Charges Charged to Joint Operations as Overhead
$0 to $5,000,000	5%
$5,000,001 to $15,000,000	3%
$15,000,001 to $25,000,000	2%
over $25,000,000	1%

Pricing of joint account material purchases, transfers, and dispositions

The operator is responsible for the provision and disposition of material to be used in the joint operations, and should make proper and timely charges and credits for all material movements affecting the joint property. In addition, the operator is obligated to make timely disposition of any idle and/or surplus material, with such disposal being made either through sale of the material to the operator or nonoperators, division in kind, or sales to outsiders. The operator may purchase, but is under no obligation to purchase, the interest of nonoperators in surplus material.

1. **Purchases.** Material that is purchased from a third party should be charged to the joint account at the price paid by the operator after the deduction of all discounts received. In some cases, by purchasing in bulk, the operator may be able to qualify for quantity discounts. The operator should pass a pro rata share of such discounts to the joint account.

2. **Material transfer pricing.** Material either (a) owned by one of the parties that is moved *to* the joint property or (b) that is transferred *from* the joint property or disposed of by the operator should be priced on the following basis exclusive of cash discounts (unless otherwise agreed to by the parties):

    - **New material (Condition A).** New material should be priced at the current new price in effect at the date of movement, as listed by a reliable supply store near the joint property or near the manufacturer. If applicable, transportation costs to the receiving point nearest the joint property are also charged to the lease.

    - **Good used material (Condition B).** Condition B material is material in sound and serviceable condition that is suitable for reuse without reconditioning. Material moved *to* the joint property is priced at 75% of the current new price. Material used on and moved *from* the joint property should be charged at 75% of the current new price. If the material was originally charged to the joint account as good used material (Condition B), the material should be transferred at 65% of the current new price.

    - **Used material (Condition C).** Condition C material is not in sound and serviceable condition, and is not suitable for its original function until after reconditioning. Condition C material transferred *to* a joint property should be priced at 50% of the current new price. The cost of reconditioning should be charged to the receiving property, provided Condition C value, plus the cost of reconditioning, does not exceed Condition B value.

    - **Condition D.** Material, excluding junk, no longer suitable for its original purpose, but usable for some other purpose should be priced on a basis commensurate with its use. In most cases, the operator may dispose of Condition D material without prior approval of nonoperators.

    - **Condition E.** Junk should be priced at prevailing junk prices.

    Transportation costs from the receiving point nearest the joint property (for Condition A material) or from the sending property to the receiving property (for all other conditions) should be charged to the joint account of the receiving property. The cost of the equipment itself and the transportation costs may be recorded in different subaccounts to separately identify the costs for cost management purposes.

3. **Disposition of material.** The disposition of surplus material can occur by any of the three following methods:
    a. *Material purchased by operator or nonoperator.* The operator or nonoperators may purchase material from the joint account on the basis of condition value.
    b. *Division in kind.* The operator and nonoperators may divide surplus material from the joint property in proportion to each party's interest.
    c. *Sales to outsiders.* Sales are accomplished on the basis of competitive bidding. Sales to outsiders normally occur after it has been determined that neither the operator nor nonoperators have a need for the material.

Inventories

The accounting procedure requires the operator to maintain detailed records of controllable material and to conduct regular physical inventories. The operator is responsible for maintaining an accurate record of controllable material. The listing of controllable material should be compared with a physical examination of existing assets at reasonable intervals, and appropriate action taken when discrepancies are identified.

Expense of conducting periodic inventories. Any expenses incurred by the operator in conducting periodic inventories should be charged to the joint account. If nonoperators elect to have a representative present, they do so at their own cost and expense.

Special inventories. Special inventories are generally required whenever there is a change in operator. The expenses related to conducting special inventories resulting from a change of operator are normally charged to the joint account. Special inventories also may be requested due to a sale or change of interest in the joint property. The expenses related to conducting other special inventories are generally borne by the parties requesting such an inventory. In either event, all parties must agree to the results of the inventory.

JOINT INTEREST ACCOUNTING

Most joint ventures are accounted for using the proportionate consolidation method. Under the proportionate consolidation method, each owner accounts for its pro rata portion of the assets, liabilities, revenues, and expenses of the venture. The discussion and examples following illustrate this process.

Booking charges to the joint account: accumulation of joint costs in operator's regular accounts

The most common method of booking joint costs is to initially record all costs in the operator's regular accounts (e.g., lease operating expense, wells and equipment, wells-in-progress, etc.). The costs are associated with specific properties via a system of property identification numbers. At the end of each month, the operator identifies all costs that have been charged to the jointly owned properties it operates. The operator then recognizes a receivable for the nonoperators' share of those costs and credits or *cuts back* its regular accounts for the nonoperators' portion of the costs. The operator's own portion of the costs incurred during the month is thus left in its regular accounts. This process is illustrated in the following example.

EXAMPLE

Accumulation of Joint Costs in Regular Accounts

Tyler Company owns 60%, South Company owns 10%, and North Company owns 30% of the joint working interest in Lease A. Tyler Company is the operator. Tyler Company incurs the following costs during October 2012 in connection with Lease A.

Salaries and wages, field employees	$5,000
Contract service, reacidizing	2,500
Purchase and installation of compressor unit	900
Property taxes paid	500
Equipment from operator's inventory installed on lease	600
Allowed overhead charge (two wells at $1,200 per well)	2,400

Entries during month by operator (Tyler Company)

Lease operating expense—joint lease	5,000	
Wages payable		5,000
Lease operating expense—joint lease	2,500	
A/P		2,500
Wells and equipment—joint lease	900	
A/P		900
Lease operating expense—joint lease	500	
Cash		500
Wells and equipment—joint lease	600	
Materials and supplies		600

Entries at end of month

Tyler Company

Lease operating expense—joint lease	2,400	
Overhead expense—control account*.		2,400
A/R—South Company (10% × $11,900)	1,190	
A/R—North Company (30% × $11,900)	3,570	
Lease operating expense—joint lease (40% × $10,400)		4,160
Wells and equipment—joint lease (40% × $1,500).		600

South Company

Lease operating expense (10% × $10,400)	1,040	
Wells and equipment (10% × $1,500)	150	
A/P—Tyler Company		1,190

North Company

Lease operating expense (30% × $10,400)	3,120	
Wells and equipment (30% × $1,500)	450	
A/P—Tyler Company		3,570

* The actual overhead costs were charged to the overhead expense—control account when incurred, with the allowed overhead charges billed to the lease at the end of the month.

Booking charges to the joint account: distribution of joint costs as incurred

Another method that might be used by the operator is to record the distribution of joint costs as incurred. Using this approach, the operator charges its regular (non–joint interest) accounts and recognizes a receivable from the nonoperators for their portion of the costs as each transaction occurs. The operator, however, actually bills the nonoperators for their portion of the costs only once a month.

EXAMPLE

Distribution of Joint Costs as Incurred

Assume the same ownership and operator as in the previous example. Tyler Company incurs minor workover costs of $10,000.

Entry

Lease operating expense (60% × $10,000)	6,000	
Accounts receivable—South (10% × $10,000)	1,000	
Accounts receivable—North (30% × $10,000)	3,000	
Cash		10,000

Nonconsent operations

A situation that occurs frequently and requires considerable accounting effort is a nonconsent operation. Nonconsent operations arise when one or more of the working interest owners do not consent to the drilling, deepening, reworking, or abandonment of a well. Another term frequently used to refer to this arrangement is **sole risk**. The JOA provides the procedures the operator must follow when one or more of the parties decide to go nonconsent. Since nonconsent operations occur frequently in relation to drilling a well, this discussion focuses on drilling situations. Nonconsent operations related to deepening, reworking, or abandonment are similar to drilling operations.

First, the party wishing to drill—usually the operator—must give written notice to all of the working interest owners of the proposed drilling operation. The parties have a period of time, typically 30 days, to reply. If one or more of the parties elects not to participate, the consenting parties are renotified and given the election to pay only their proportionate share of the costs, or in addition to their proportionate share, to pay all or part of the nonconsenting party's share. The party electing not to participate is referred to as a **carried working interest** or **carried party**. The working interest owners who agree to pay the carried party's share of the costs are referred to as the **carrying parties**. If none of the working interest owners agree to participate in drilling the well, the operator can either drill the well and carry all of the other owners himself or not drill the well.

When a working interest owner goes nonconsent, it does not relinquish its interest in the well. Rather its interest reverts temporarily to the carrying party or parties. The carrying parties are allowed to sell and keep the revenue from the carried party's share of oil and gas produced from the well until they have recovered the costs that they paid on behalf of the carried party plus a penalty. The penalty is provided in the JOA. A common penalty is 200% of costs, resulting in a total recovery of 300% of costs, i.e., a recovery of costs (100%) plus the penalty (200%). In practice, a 200% penalty is typically stated as a 300% penalty; thus, in practice, the penalty is actually stated in terms of total cost recovery. When the carrying parties have recovered the cost they paid on behalf of the carried party plus the penalty, the parties are said to have reached **payout**. From that point forward, all of the working interest owners participate in costs and revenues at their percentages previous to the nonconsent.

EXAMPLE

Carried Working Interest

Tyler Company, South Company, and North Company each own 33.33% of the joint working interest in Lease A. The royalty interest is 1/8. Tyler, the operator, notified North Company and South Company of its plans to drill Well No. 2 at an anticipated cost of $100,000. South Company elects to go nonconsent. Tyler and North agree to carry their proportionate share of South Company's costs. If the well is successful, they will be allowed to recover from South Company's share of production, the costs they carried plus a penalty of 300%. Upon payout, South Company will resume participation at 33.33%.

Tyler Company and North Company determine their proportionate share of South Company's costs and revenues in the following manner:

	Interests of Consenting Parties	Proportion of South's Costs and Revenues
Tyler Company	33.3333%	33.3333/66.6666 = 50%
North Company	33.3333%	33.3333/66.6666 = 50%
Total	66.6666%	

Assume that the well is drilled at a cost of $100,000. Tyler Company and North Company each pay $50,000, of which $33,333 is their own share and $16,667 ($100,000 × 33.3333% × 50%) is their portion of South Company's share.

On July 25, 2010, Well No. 2 is completed. Gross production and operating costs for the first three months of production are as follows: (Severance taxes have been ignored in this problem.)

Month	Gross Sales	Sales Price	Total Operating Expenses
August	1,250 bbl	$80/bbl	$10,000
September	2,250 bbl	$80/bbl	15,000
October	1,500 bbl	$80/bbl	25,000

The payout calculation made by Tyler Company (50% WI) is as follows:

	Aug.	Sept.	Oct.
Sales volume....................	1,250	2,250	1,500
Price/bbl......................	$ 80	$ 80	$ 80
Gross sales ($)	100,000	180,000	120,000
Net of royalty...................	7/8	7/8	7/8
Net sales ($)	87,500	157,500	105,000
Operating expense ($)	(10,000)	(15,000)	(25,000)
Net revenue ($)..................	$ 77,500	$142,500	$ 80,000

Net Revenue to Tyler

Total to Tyler (50%)..............	$ 38,750	$ 71,250	$ 40,000
Tyler's portion (33.3333%).........	25,833	47,500	26,667
South's portion (16.6667%).........	$ 12,917	$ 23,750	$ 13,333

Payout

South Company's share of well.............	$ 33,333
Tyler's proportion of South's cost	50%
Amount paid by Tyler Company.............	16,667
Penalty................................	× 300%
Recoverable by Tyler Company.............	$ 50,000
Amount to be recovered...................	$ 50,000
Recovered in August.....................	(12,917)
Balance to be recovered	$ 37,083
Recovered in September...................	(23,750)
Balance to be recovered	$ 13,333
Recovered in October	(13,333)
Balance to be recovered	$ 0

In November, South Company would be treated as a 33.33% working interest owner, paying its proportionate share of operating costs and receiving its proportionate share of revenues.

Entries for Tyler Company for July and August

July

Wells and equipment (50% × $100,000)........	50,000	
A/R—North Company (50% × $100,000)	50,000	
Accounts payable		100,000

August

Lease operating expense (50% × $10,000)	5,000	
A/R—North Company (50% × $10,000)	5,000	
Cash		10,000
Cash	100,000	
Royalty payable ($100,000 × 1/8)		12,500
A/P—North Company (50% × $100,000 × 7/8) ..		43,750
Oil revenue (50% × $100,000 × 7/8)		43,750

Note: Tyler Company does not set up a receivable from South Company for South's carried interest. Also, Tyler Company recognizes its share of revenue retained from South Company's interest as revenue—not as a reduction in its capitalized costs.

The operator provides the carried party with a payout statement on a monthly basis so both parties know when payout is near. Ideally, the operator would be able to determine the precise point during the month when payout occurs, and immediately begin to distribute operating costs and revenue to the carried party. Realistically, it is very difficult, if not impossible, to track payout that precisely. Accordingly, the carried party is usually notified at the end of the month, and some type of adjustment is made to equalize the costs and revenues. For example, in the problem above, if payout had been reached sometime during September or October rather than at the end of October, South Company could receive a cash payment from Tyler and North. Alternately, Tyler and North could pay a portion of South's operating expenses during November.

Accounting for materials

One of the most difficult and challenging problems facing a joint interest operator is the pricing of material transferred *to* and *from* the property being operated. Material to be used on a joint property can be purchased directly for the specific property, moved to or from the operator's warehouse, or transferred from another property. Material purchased directly for a property is charged to that property at the cost of the material, less all discounts received. Any transportation charges are also charged to the property. The following example illustrates the purchase of new material for a joint property.

EXAMPLE

Purchase of New Material

Tyler Company purchased casing to be used in the workover of a well on a joint property. The net price of the casing is $100,000. Loading, hauling, and unloading costs from the supplier's warehouse to the wellsite were $8,000. Tyler is the operator on the lease and has a 60% WI.

Entry

A/R—nonoperators (40% × $108,000)	43,200	
Lease operating expense (casing) (60% × $100,000)	60,000	
Lease operating expense (freight) (60% × $8,000)	4,800	
Cash		108,000

The preceding example gives the *net result* at the end of the month after distribution of joint costs.

Transfer from warehouse—Condition A. When material is transferred from the operator's warehouse or a joint property to another joint property, it must be priced according to the accounting procedure. The price used to record the transfer should approximate the current market price of the material or equipment. Several methodologies may be used to obtain market prices used to record transfers. These methodologies are listed in the *1995 COPAS Accounting Procedure* and include a COPAS database called the *Computerized Equipment Pricing System (CEPS)*, which contains a generic price calculated for each piece of material or equipment. The database is updated annually by a historical price multiplier. Other methods that can be used to approximate the current market price are vendor quotes, historical purchase prices, or manually applying the historical price multiplier to a published price. In addition, transfer prices may be mutually agreed upon by the parties.

Condition A is new material, and the joint property would normally be charged with the current market price of the material plus transportation charges from the warehouse to the property.

The following example illustrates the transfer of a pump from the operator's warehouse to a joint property owned 60% by Tyler, the operator. Note that the asset being transferred is inventory, not wells and equipment.

EXAMPLE

Transfer from Warehouse—Condition A

Assume that Tyler Company has a submersible pump in its warehouse with a cost of $9,000, including shipping and handling to the warehouse. The pump is new and is to be installed in a newly drilled well. The current market value is $9,000. The pump is transferred from Tyler's warehouse to a jointly owned property in which Tyler has a 60% working interest. Transportation costs from the warehouse to the property total $200.

Entry to record transfer of pump

A/R—nonoperators ($9,000 × 40%)	3,600	
Wells-in-progress—L&WE (pump) ($9,000 × 60%)	5,400	
Warehouse inventory		9,000

Entry to record transportation costs

A/R—nonoperators ($200 × 40%)	80	
Wells-in-progress—IDC (transportation) ($200 × 60%)	120	
A/P—freight company		200

Transfer from warehouse to joint property—Condition B. When used material is transferred to or from a jointly owned property and is not scrap, the material must be repriced. Industry practice requires that the repricing be based on the current market value of the material and not on its historical cost. By using the current market value, the material being moved to a new location, i.e., another lease or a warehouse, is repriced to a value that reflects the current price that piece of equipment would cost if purchased from a third party. The repricing involves multiplying the current market price by a factor that is determined based on the condition that the used material is in when it is transferred.

Condition B material is material in sound and serviceable condition that may be reused for its original purpose without reconditioning. This material should be charged from the sending property or warehouse *to* the receiving property at 75% of the current market price. The following example assumes the same facts as the previous example, except that the material is in Condition B rather than Condition A, and the current market value is $10,000 rather than $9,000. Since the material is being transferred from a warehouse, the difference between the operator's share of the original cost of the material and the condition value of the material is recorded as a miscellaneous revenue or expense.

EXAMPLE

Transfer from Warehouse to Joint Property—Condition B

Assume the pump in the previous example that was carried on the books of Tyler's warehouse at $9,000 has a current market price of $10,000. Also assume that the pump is used, is in sound condition, and is being installed in a newly drilled well. Ignore transportation costs.

The Condition B value of the pump would be $10,000 x 75% = $7,500.

Entry

A/R—nonoperators (40% × $7,500)	3,000	
Wells-in-progress—L&WE ($7,500 × 60%)	4,500	
Other revenue/expense ($9,000 – $7,500)	1,500	
Warehouse inventory		9,000

Transfer from one property to another property—Condition B. When material is transferred *from* one property to another property, the percentage used in determining the credit to the sending property depends on the original condition of the material when it was charged to that property. If Condition B material was originally Condition A material when it was first charged to the property, the material will be removed at 75% of the current market price. However, if Condition B material was originally Condition B material when it was first charged to the property, the material will be removed at 65% of the current market price. The reduction represents an assumed reduction in service value.

When material is transferred between jointly owned properties with different working interests, the nonoperators of the sending property are credited with their share of the condition value of the equipment. The operator removes the equipment from its accounts at the equipment's original historical cost. Neither the successful efforts nor full cost rules allow gain or loss recognition in this case. Thus the operator's share of the difference between the original cost of the equipment and the condition value is debited or credited to the accumulated DD&A account for the lease.

EXAMPLE

Transfer from Property to Another Property—Condition B

A pump originally costing $32,000 is transferred from Alpha Lease, a joint property on which Tyler Company serves as operator and has a 60% interest, to Beta Lease, another joint property operated by Tyler in which Tyler has a 90% interest. At the time of transfer, the current market price of the pump is $40,000. The pump is Condition B and was Condition A material when first charged to the property. Ignore transportation charges. Assume that the pump is being installed in a newly drilled well.

The condition value of the equipment is $40,000 × 75% = $30,000.

Entry

A/R—Beta Lease nonoperators ($30,000 × 10%) . . .	3,000	
Wells-in-progress—L&WE—Beta Lease ($30,000 × 90%)	27,000	
Accumulated DD&A—Alpha Lease [($32,000 – $30,000) × 60%]	1,200	
A/R*—Alpha Lease nonoperators ($30,000 × 40%)		12,000
Wells and equipment—Alpha Lease ($32,000 × 60%)		19,200

* When material is transferred off a lease, the customary accounting treatment is to credit (reduce) the account receivable from the nonoperators.

Transfer from property to another property—Condition C. As discussed earlier, Condition C material is not in sound and serviceable condition and is not suitable for its original use without reconditioning. Condition C material transferred *to* a property should be charged to the property at 50% of the current market price. The cost of any reconditioning is paid by the receiving property. Condition C material transferred off a property may be handled in two different ways, depending upon which property pays the reconditioning costs. Condition C material can be transferred at 50% of the current market price, with the receiving property paying for the reconditioning costs. Alternately, the material can be transferred at 75% of the current new price, with the sending property paying all of the reconditioning costs. (If the sending property reconditions the material, the material would then be in Condition B and would thus be transferred at 75% of the current market price.) Condition C material that is reconditioned should never have a total cost in excess of Condition B value.

EXAMPLE

Transfer from Joint Property to Another Joint Property—Condition C

Tyler Company is the operator on both Lease A and Lease B, and has a 40% working interest in Lease A and an 80% working interest in Lease B. A piece of equipment originally costing $30,000 is transferred from Lease A to Lease B. The equipment is being installed in a producing well but it is not a replacement or a repair. The current market price of the equipment is $36,000, and the equipment is transferred at Condition C. The working interest owners of Lease B will pay all costs of reconditioning.

The condition value of the equipment is $36,000 × 50% = $18,000.

Entry

A/R—nonoperators—Lease B (20% × $18,000)	3,600	
Wells and equipment—Lease B (80% × $18,000)	14,400	
Accumulated DD&A—Lease A [40% × ($30,000 – $18,000)]	4,800	
A/R—nonoperators—Lease A (60% × $18,000)		10,800
Wells and equipment—Lease A (40% × $30,000)		12,000

Assume that the cost of reconditioning the equipment is $5,000. This entire amount would be charged to Lease B.

Entry

A/R—nonoperators—Lease B (20% × $5,000)	1,000	
Wells and equipment—L&WE—Lease B (80% × $5,000)	4,000	
A/P—Vendor		5,000

In the preceding example, Tyler's accumulated DD&A account is debited for $4,800, which is Tyler's share of the difference between the original cost of the equipment and the condition value of the equipment: 40% × [30,000 – (50% × 36,000)].

It is possible for Condition C equipment to be used in some manner other than its original purpose. For example, production tubing that is Condition C could not be reused as production tubing without reconditioning. It could, however, be used as a flow line, since a flow line would not have to withstand the pressure that tubing used in a well would. In that case, the equipment would be transferred as Condition C, and there would be no reconditioning cost incurred by the receiving property.

In some cases, material transferred between properties may be priced in excess of the appropriate condition value of the current prices for the same new material. Premium pricing occurs when the operator's cost, either through reconditioning or actual price paid, exceeds the condition value of current prices for new material. The nonoperators must always be notified when premium pricing occurs.

Offshore Operations

Offshore operations are commonly joint working interest ventures because of the large dollar amounts and large amount of risk involved. Domestically, these operations are conducted in either federal- or state-owned waters. Bidding on leases from either the state or the federal government is normally required to obtain a lease. Bidding on offshore federal leases is done by sealed bids, with generally a separate bid for each tract.

Some different types of costs are incurred for offshore operations in comparison to onshore operations. These include the costs of using mobile rigs, fixed platforms, helicopter costs, and special safety equipment designed for offshore use. Many offshore costs require allocation between differing tracts, e.g., fuel costs, boat costs, helicopter costs, etc. COPAS accounting procedures have addressed these problems.

Normal offshore joint operations are accounted for by using one of the COPAS offshore accounting procedures. Such procedures include the *1976 COPAS Accounting Procedure Offshore Joint Operations*, *1986 COPAS Offshore Accounting Procedure Joint Operations*, *1995 COPAS Accounting Procedure*, or *2005 COPAS Accounting Procedure*.

Joint Interest Audits

Nonoperators in joint venture operations have the right to audit the accounts and records of the operator within 24 months following the close of a year. Companies involved in joint venture operations normally have full-time joint interest auditors on staff to audit the charges made by operating companies in joint venture operations. Audits are generally not called more than once each calendar year. The audit is normally initiated by the nonoperating partner with the largest interest in the venture. This nonoperator usually provides the lead auditor in conducting the audit. The other nonoperators may furnish auditors or share the expenses of the audit. The nonoperators must give written notice to the operator that an audit is requested, after which a date to begin the audit will be confirmed.

The joint interest auditor's role in an audit of costs is limited to examining the charges made to the joint accounts and the support for those charges. The procedures used in conducting joint interest audits will vary widely. However, the following suggested procedures are discussed in *COPAS Accounting Guideline (AG)-19*, "Expenditure Audits in the Petroleum Industry: Protocol & Procedure Guidelines."

1. **Operating agreement and AFE.** Review and brief the agreement for compliance of activities; review the AFE and compare it to actual expenditures.

2. **Minutes of operator's meetings.** Review and note significant and unusual actions that often give rise to errors.

3. **Company labor.** Obtain labor rates, contracts, and other labor policies. Make a test of payroll time sheets, giving special attention to labor rates, reasonableness of allocations to properties, and charges for holidays and vacation. On a selective basis, compare names, occupations, rates, hours, and distribution reflected on payroll time sheets to the daily time reports.

4. **Materials and services purchased externally.** Check that all invoices are properly authorized, all discounts taken, prices are reasonable, and supporting data exist for unusual charges.

5. **Material and supplies transferred to or from the operation.** Check all charges, paying particular attention to the condition of the material transferred and the prices assigned.

6. **Warehousing.** Determine whether periodic inventories are being taken, check all inventory adjustments, and make a periodic physical check of selected items of warehouse stock, tracing these items to the warehouse records.

7. **Overhead.** Review overhead charges and make sure they are in accordance with the joint operating agreement.

8. **Services and facilities.** Check the fairness of rates being charged for services and facilities.

9. **Capital and maintenance jobs.** Check that all work has been properly authorized, bids are being sought for jobs, costs are reasonable, and all expenditures in excess of the amount allowed under the joint operating agreement have been properly approved.

10. **Taxes and insurance.** Make sure taxes are assessed against the nonoperators fairly and that insurance is being carried in accordance with the joint operating agreement.

11. **Allocation of income and expenditures.** Insure that all charges and credits to the nonoperators are based on the ownership percentages specified in the joint operating agreement.

12. **Capital assets.** Insure that records are being properly maintained, that assets charged to the joint accounts are actually being used on the property or properties, and that these tangible assets are necessary on the projects.

Following the audit, the joint interest auditors prepare a report for the operator. There is no standard format for this report, but it will include information such as the property description, audit period, names of the auditors, and scope of the audit. Also included will be a list of audit exceptions and supporting materials for charges to which the auditors seek adjustment. The operator will then review the exceptions and make adjustments where appropriate. The operator typically has 180 days following receipt of the audit report to reply regarding adjustments.

The economic benefits from performing an audit often exceed the cost of the audit, especially in ventures where many drilling and development costs are being incurred. Consequently, joint interest audits by nonoperators are common, creating a separate field of expertise within oil and gas accounting.

While joint interest audits have traditionally concentrated on costs, audits of revenue are becoming more common. The need for revenue audits has increased in part because of the deregulation of natural gas pricing. Gas can now be sold through brokers or marketers, on the spot market, to transmission companies, or directly to end users. Consequently, the complexity involved in tracking numerous gas sales can cause errors to be made.

As with audits of costs, nonoperators have the right to audit the accounts and records of the operator relating to revenues within 24 months following the close of a year. Revenue audits concentrate on production information and sales agreements, and require a thorough understanding of the measurement and flow of the product. Revenue auditing is a relatively new field but will continue to become more important because of the complexities of today's markets.

PROBLEMS

1. When two or more parties own a joint working interest, who will usually manage the property?

2. When do parties normally enter into an operating agreement?

3. What are the duties of the operator? Nonoperator?

4. What is a carried working interest? Payout?

5. What is the difference between a direct cost and overhead?

6. What is combined fixed rate overhead?

7. Placid Oil Corporation operates the Reida Lease. The accounting procedure attached to the JOA allows Placid to recoup its overhead by the use of a combined fixed rate—well basis of $1,000/producing well and $10,000/drilling well.

    REQUIRED:

    a. How much total overhead would Placid bill the joint account if the Reida Lease had four wells that produced every day the previous month?

    b. What if three wells produced every day, and only one produced for 5 days?

    c. What if the only operation on the lease the previous month was the drilling of a well? Drilling operations commenced on the first day of the month. Operations were suspended for 4 days on the 20th, commenced again on the 24th, and continued through the end of the month. A month is considered to be 30 days.

8. Hostetler Energy owns 70%, Challenger Company owns 20%, and Hill Oil Company owns 10% of the working interest property 1004. Assume Hostetler Energy is the operator and incurs the following costs during the month of September 2013, in connection with the property:

Salaries and wages, field employees	$ 6,000
Salaries and wages, first-level field supervisors	2,000
Operator's cost of holiday, vacation, sickness, and disability benefits, 8% of above	640
Social Security tax, 7.5% of above	600
Employee benefits, group life insurance.	1,920
Material installed on property from Hostetler's inventory	300
Transportation of material and employees 1,200 miles @ $0.25/mile	300
Contract service, reacidizing (workover)	2,500
Purchase and installation of compressor unit	1,500
Repair of Christmas tree	500
Property taxes paid	500
Insurance premium paid.	800
Overhead, two wells @ $700/well	1,400

REQUIRED: Give the entries to record and distribute the costs, assuming regular accounts are used.

9. Bulldog Oil Corporation, the operator of Lease A, purchased casing with a list price of $60,000 for a joint interest property in which it has a 40% WI. The casing is to be used in a workover. The vendor gives a discount of 10% off list price and also has credit terms of 2/10, n/30 (i.e., a 2% discount can be taken if paid in 10 days, otherwise the full amount of the invoice less any returns or allowances is due in 30 days). Loading, hauling, and unloading costs amounted to $4,000. Prepare the entry to record the purchase.

10. Longhorn Oil Corporation transferred an item of equipment from its wholly owned warehouse to a jointly owned lease in which it has a 70% WI. The item of equipment is in Condition B, and the current market price for the equipment is $50,000. The item of equipment was carried on Longhorn's books at $40,000. Give the entry to record the transfer, ignoring transportation charges.

11. Core Petroleum owns 60%, Dwight Corporation owns 30%, and Webb Company owns 10% of the working interest property number 2008. Core Petroleum is the operator and bills Dwight and Webb monthly for their portion of costs incurred. During May 2016, Core incurred costs as follows:

Salaries and wages, field employees	$ 8,000
Salaries and wages, first-level field supervisors	4,000
Social security taxes, 7.5% of above	900
New separator, purchased and installed	20,000
Repairs to Christmas tree	1,000
Property taxes paid	2,000

Other Costs:

Employee benefits—21% of salaries and wages

Transportation of material and employees, 1,500 miles at $0.25/mile

Overhead, 22% of all costs listed above

REQUIRED: Give the entries to record and distribute the costs, assuming regular accounts are used.

12. List and briefly discuss the main points of the COPAS accounting procedures.

13. Grover Petroleum owns a piece of equipment, originally costing $60,000, that is currently being used on Lease A. Grover Petroleum owns a 40% working interest in Lease A and serves as the operator of the lease. The company plans to use the equipment on a lease wholly owned by Grover Petroleum. The equipment is transferred to the company's warehouse.

    REQUIRED: Prepare the entry to record the transfer under each of the following independent situations:

    a. The equipment is in Condition B and originally was Condition A when transferred to the property. The current market price is $80,000.

    b. The equipment is in Condition C, and Grover Petroleum will pay for the reconditioning. The current market price is $80,000.

14. Brown Oil Company is the operator of Lease A and Lease B and has a 60% WI in each lease. A piece of equipment, which originally cost $30,000, is transferred from Lease A to Lease B. The current market price of the equipment is $40,000. The equipment is in Condition C. The working interest owners of Lease B will pay for reconditioning.

    REQUIRED: Prepare an entry to record the transfer.

15. The Raupe Lease has the following working interest owners: Reed Corporation 50%, League Energy 25%, and Sunshine Oil Company 25%. There is a 1/8 royalty on the lease. On April 1, 2011, Reed Corporation, the operator, receives notice that League Energy is going nonconsent on the drilling of the Gusher No. 2. Reed Corporation and

Sunshine Oil Company agree to carry League's share proportionately. The nonconsent penalty is 300%. On August 1, the Gusher No. 2, which was drilled and completed at a cost of $750,000, goes on production. The production and operating information for the next few months is as follows:

Month	Production	Operating Costs	Sales Price/bbl
August	8,000 bbl	$ 75,000	$60
September	12,000 bbl	120,000	57
October	15,000 bbl	150,000	54
November	20,814 bbl	225,000	60

REQUIRED: Assuming severance tax is ignored:

a. Determine Reed Corporation's and Sunshine Oil Company's proportionate shares of drilling and equipping costs.

b. Prepare a table determining when League Energy will reach payout.

c. Prepare the journal entry that Reed Corporation will make during August to book its share of production revenue.

16. Session Gas Company owns a 33.3% working interest in a lease in West Texas. Roger Williams, a local farmer, owns a 1/8 royalty interest in the lease. Session is the operator, and its partners, Rocky Energy and Asteroid Petroleum, each own 33.3% of the working interest. Session analyzed the prospects for the lease and proposed drilling a gas well. Asteroid agreed, but Rocky decided to go nonconsent. Session and Asteroid both agreed to proportionately carry Rocky's working interest. The joint operating agreement stipulates that a 150% drilling and completion cost penalty will be assessed on any partner choosing not to participate in drilling the well.

On July 1, 2014, the Gusher No. 2 was drilled and completed at a total cost of $300,000. The following information is available concerning production and sales. Assume each company contracts to sell its gas for $6.00/Mcf.

Month	Sales Volume	Total Operating Expenses	Sold by Session	Sold by Asteroid
August	60,000 Mcf	$45,000	55,000 Mcf	5,000 Mcf
September	62,000 Mcf	45,000	50,000 Mcf	12,000 Mcf
October	68,000 Mcf	45,000	58,000 Mcf	10,000 Mcf
November	60,000 Mcf	45,000	40,000 Mcf	20,000 Mcf

REQUIRED: Ignoring severance tax:

a. Determine when Rocky will reach payout if payout is calculated based on the quantity actually sold. Hint: Session and Asteroid would have to compute payout separately.

b. Determine when Rocky will reach payout if payout is calculated using the amount to which each partner is entitled.

17. The Hill Lease is operated by Harper Corporation. The foreman of the Hill Lease must make a decision regarding the replacement of a pumping unit on the lease. The foreman has identified three possible alternatives:

   Alternative 1

   Purchase a new pumping unit. The foreman has located a vendor willing to sell the unit needed for a delivered cost of $65,000 FOB destination (i.e., the manufacturer pays freight to the destination point).

   Alternative 2

   The Odessa Lease, which is also operated by Harper Corporation, has a surplus pumping unit that could be used. If the Odessa Lease equipment is used, the Odessa Lease would be charged with the cost of repairing the unit that would then be transferred to the Hill Lease at Condition B.

   Alternative 3

   The Midland Lease, which is also operated by Harper Company, has a surplus pumping unit that could be used. If the Midland Lease unit is used, the equipment would be transferred unrepaired at Condition C.

   Other information:

	Harpers's Working Interest	Unit Repair Costs	Trucking Costs
Hill Lease	100%	n/a	$ 0
Odessa Lease	60%	$10,000	2,000
Midland Lease	30%	5,000	3,000

   The current market price for the unit is $70,000.

   REQUIRED: Determine what the foreman on the Hill Lease should do.

REFERENCES

1. Council of Petroleum Accountants Societies, Inc. 2005. *Model Form Interpretation (MFI)-51, 2005 COPAS Accounting Procedure.* Denver, CO: COPAS.

2. Council of Petroleum Accountants Societies, Inc. 1996. *Model Form Interpretation (MFI)-28,* "Material Classification Manual." Denver, CO: COPAS.

3. Council of Petroleum Accountants Societies, Inc. 1976. *Model Form Interpretation (MFI)-5,* "Accounting Procedure Offshore Joint Operations." Denver, CO: COPAS.

4. Council of Petroleum Accountants Societies, Inc. 1986. *Model Form Interpretation (MFI)-19,* "Offshore Accounting Procedure Joint Operations." Denver, CO: COPAS.

5. Council of Petroleum Accountants Societies, Inc. 1996. *Model Form Interpretation (MFI)-30,* "Accounting Procedure Interpretation." Denver, CO: COPAS.

6. Council of Petroleum Accountants Societies, Inc. 1998. *Accounting Guideline (AG)-19,* "Expenditure Audits in the Petroleum Industry: Protocol, & Procedure Guidelines." Denver, CO: COPAS.

13

CONVEYANCES

In the oil and gas industry, companies often sell, trade, or exchange their interests in oil and gas properties to other parties. Collectively these transactions are referred to as **conveyances**. Mineral interests are generally conveyed by three basic methods: by leasing, sales or exchanges, or sharing arrangements. The basic oil and gas mineral lease whereby an oil and gas company contracts with the owner of the mineral rights in a property—creating a working interest and a royalty interest—was discussed in chapter 1. Accounting for leasing activities was discussed in chapter 4. This chapter discusses the accounting that is required when working interests are subsequently conveyed by sales, exchanges, and sharing arrangements under both the successful efforts and full cost methods of accounting.

MINERAL INTERESTS

The mineral interest in a property may be divided into operating (working) interests and nonoperating (nonworking) interests. The previous chapters dealt primarily with simple situations that included only a basic working interest and a royalty interest created through leasing. A working interest can also be purchased, sold, carried, pooled, unitized, etc. In addition, when a working interest is conveyed, a nonoperating interest may be created. These nonoperating interests, which are in addition to royalty interests, can be classified as overriding royalty interests, production payment interests, or net profits interests. These interests are normally created or conveyed through sales, exchanges, or sharing arrangements. The interests are typically created to spread risks, share costs, obtain financing, secure sufficient acreage to meet state spacing requirements, or to perform secondary or tertiary recovery operations.

The next section contains a discussion and definition of different types of interests including examples that illustrate how revenues and costs would be distributed among the various interest owners.

Types of interests

Basic working interest (WI). One working interest owner and one lease make up the basic working interest. The working interest is responsible for paying all of the costs related to exploration, development, and production of the leased property.

Joint working interest. Two or more parties each own an undivided fraction of the working interest in a single lease. A joint working interest may result from one of the three methods mentioned above, i.e., (1) leasing, (2) sales or exchanges, or (3) sharing arrangements. The joint interest format is very popular because it provides a means of sharing the high risk and high capital investment associated with oil and gas ventures.

One party—usually the party with the largest percentage of the working interest—has the responsibility of managing the property, including developing and operating the property. This party is known as the operator. In a joint interest situation, the working interest owners enter into a joint operating agreement that specifies the rights and obligations of each party. Accounting for joint interest operations was discussed in chapter 12.

EXAMPLE

Joint Working Interest

Tyler Oil Company and Duster Oil Company signed a lease agreement with Frank Brown for a 3,000 acre lease in Texas. Mr. Brown received a 1/8 royalty interest. Tyler Oil Company has a 60% WI, and Duster Oil Company has a 40% WI. Subsequently, the two companies signed a joint operating agreement designating Tyler Oil Company as the operator of the lease. Tyler pays 60% and Duster pays 40% of the cost to explore, develop, and operate the property. Frank Brown will receive 1/8 of the gross revenue from the property. Tyler Oil Company and Duster Oil Company will split the remaining 7/8 of the revenue from the property (60% and 40%, respectively).

Basic royalty interest (RI). The basic type of interest retained by a mineral rights owner or fee interest owner when leasing the mineral rights to another party. The royalty interest owner receives a fraction of the gross revenue from the property but is not required to pay any of the costs of exploration, development, or production.

Overriding royalty interest (ORI). Frequently, working interest owners, through conveyance, create an ORI—a nonoperating interest created out of the working interest. An ORI is similar to a basic royalty interest in that the owner is not responsible for the cost of exploring, developing, or producing the property. The fundamental difference between a royalty interest and an ORI is that the royalty interest is created from the original mineral rights, and the ORI is created from the working interest. The ORI's share of revenue is a stated percentage of the share of revenue belonging to the working interest from which it was created. An ORI is created by either being retained or carved out. A retained ORI is created when the working interest owner sells or conveys its working interest in a property, and in the same transaction, retains an ORI. A carved-out ORI is created when the working interest owner keeps the working interest but creates an ORI that is conveyed to another party.

EXAMPLE

Overriding Royalty Interest

Tyler Oil Company signed a lease agreement with Sally Clark covering 640 acres in Texas. Sally Clark received a 1/8 royalty interest and Tyler Oil Company received a 100% WI. A few months later, Tyler Oil Company decides that at the present time it is not financially able to develop the property.

a. **Retained ORI.** Tyler Oil Company sells its working interest to Simms Oil Company for $50,000 and retains a 1/7 overriding royalty. As a result of this transaction, Simms Oil Company will pay 100% of the cost to explore, develop, and produce the property. Sally Clark will receive 1/8 of the gross revenue from the property, Tyler Oil Company will receive 7/8 × 1/7 of the revenue from the property, and Simms Oil Company will receive 7/8 × 6/7 of the revenue. Since Tyler Oil Company sold its working interest and retained an ORI, the ORI is referred to as a **retained ORI**.

b. **Carved-out ORI.** Assume instead that Tyler Oil Company decides to go ahead and develop the property. Since Tyler Oil Company is short of capital, it approaches a local investor who agrees to pay Tyler $10,000, and in exchange, Tyler Oil Company conveys a 1/16 ORI in the lease to the investor. The result of this transaction is that Tyler Oil Company will pay 100% of the cost to explore, develop, and produce the property. Sally Clark will receive 1/8 of the gross revenue from the property, the investor will receive 7/8 × 1/16, and Tyler Oil Company will receive 7/8 × 15/16. Since Tyler Oil Company kept its working interest and created an ORI, the ORI is referred to as a **carved-out ORI**.

Production payment interest (PPI). A production payment interest is a nonoperating interest created out of the working interest. Production payment interests are typically expressed in terms of a certain amount of money, a certain period of time, or a certain quantity of oil or gas. In other words, a production payment is limited to a specified amount of money, time, or quantity of oil or gas, after which the production payment interest ceases to exist. Therefore, unlike the other nonoperating interests, a production payment interest may terminate before the reservoir is depleted. The owner of a production payment interest is not responsible for any of the cost of exploring, developing, or producing a property. If a production payment is payable with money, the payment is stated as a percentage of the working interest's share of revenue, since the production payment interest was created from the working interest. If a production payment is payable in product (i.e., oil, gas, etc.), payment is typically stated as a percentage of the working interest's share of current production. Like ORIs, production payments are created by being carved out or by being retained.

EXAMPLE

Production Payment Interest

Tyler Oil Company signed a lease agreement with Sampson Irving covering 640 acres in Texas. Sampson Irving received a 1/8 royalty interest, and Tyler Oil Company received a 100% WI. A few months later, Tyler Oil Company decides that at the present time it is not financially able to develop the property.

a. **Retained production payment.** Tyler Oil Company decides to sell its working interest to Smith Company for a consideration of $500,000 and a payment of 100,000 barrels of oil to be paid out of the first 20% of Smith Company's working interest share of production. During the first year of production, the lease produces 40,000 barrels (gross), of which 5,000 barrels (40,000 barrels × 1/8), or the revenue from 5,000 barrels, belongs to Sampson Irving. Tyler Company will receive 7,000 barrels (40,000 barrels × 7/8 × 0.20), leaving the production payment balance owed to Tyler Oil Company of 93,000 barrels (100,000 barrels – 7,000 barrels). Smith Company's share of production is what is left after paying the royalty and production payment: 40,000 barrels × 7/8 × 80% = 28,000 barrels. Smith Company pays 100% of the costs to explore, develop, and produce the lease. Since Tyler Oil Company sold its working interest and retained a production payment interest, the production payment is referred to as a **retained production payment**.

b. **Carved-out production payments.** Assume instead that Tyler Oil Company decides to go ahead and develop the property. Since Tyler Oil Company is short of capital, it approaches a local investor. On 1/1/2011, the investor loans Tyler Oil Company $100,000. In exchange, Tyler Oil Company agrees to repay the investor the principle of $100,000, plus interest at a rate of 10%, out of the first 20% of Tyler Oil's share of production. The first month's production is sold for $200,000. Sampson Irving will receive $25,000 ($200,000 × 1/8), and Tyler Oil Company will pay the investor (the production payment owner) $35,000 ($200,000 × 7/8 × 0.20). The remaining balance payable to the production payment owner is $65,833 ($100,000 plus interest of $833 less $35,000 payment: $100,000 + ($100,000 × 10% × 1/12) – $35,000. Tyler Oil Company will pay 100% of the cost to explore, develop, and produce the property. Since Tyler Oil Company kept its working interest and created a production payment interest, the production payment is referred to as a **carved-out production payment interest**.

Net profits interest. A net profits interest is a nonoperating interest normally created out of the working interest by either carve out or retention, but more commonly by retention. A net profits interest may also be created when the mineral rights owner leases his interest. This type of interest is common offshore. A net profits interest is similar to an ORI. However, rather than being paid a percentage of the working interest owner's share of production, the net profits interest owner receives a stated share of the working interest owner's share of the net profits from the property. The holder of this type of interest is not responsible for paying his share of losses; however, such losses may be recovered by the working interest owner out of future net profit payments. The calculation of net profits, i.e., the allowed deductions from gross revenue to compute net profit, should be clearly indicated in the contract.

EXAMPLE

Net Profits Interest

Tyler Company obtains an offshore lease, specifically a net profit share lease, from the U.S. government. The lease contract specifies that the government will receive a lump sum bonus and a 10% share of the net profits produced from the lease. The contract provides for a detailed list of allowable charges from revenue in arriving at net profit. Tyler Oil Company owns the working interest and, accordingly, must pay 100% of the cost to explore, develop, and produce the property.

CONVEYANCES: GENERAL RULES

In SEC *Reg. S-X 4-10*, the SEC adopted all of the conveyance provisions in *SFAS No. 19*. Thus, the conveyance provisions in *SFAS No. 19* apply to all publicly traded companies, including those using the successful efforts method, and with some minor modification, those using the full cost method. These provisions are discussed in the following sections. The specific provisions for accounting for conveyances under full cost accounting are discussed later in the chapter.

The main accounting question to be answered in recording a conveyance is whether a gain or loss should be recognized. The conveyance rules contained in *SFAS No. 19* can generally be summarized as follows:

1. Generally, no gain or loss should be recognized in transactions that involve the pooling of assets in a joint undertaking involving a particular property or group of properties (*SFAS No. 19*, par. 44b).
2. Generally, no gain should be recognized, but a loss may be recognized in transactions where:
    a. Part of an interest in an oil and gas asset is sold and substantial uncertainty exists about the recovery of costs applicable to the retained interest (*SFAS No. 19*, par. 45a).
    b. Part of an interest is sold and the seller has a substantial obligation for future performance without proportional reimbursement (*SFAS No. 19*, par. 45b).
3. For transactions other than those listed above, gain or loss recognition is required so long as it is appropriate under GAAP. Examples include the sale of an entire interest in a proved property, the sale of the entire interest in an unproved property, nonmonetary exchanges that have commercial substance, and certain production payments.
4. Some transactions are in effect borrowings repayable in cash or cash equivalents. These transactions should be accounted for as borrowings regardless of their legal form. Typically, the most common examples of these transactions include some type of production payment.

The above types of transactions and the paragraphs in *SFAS No. 19* that specify the appropriate accounting treatment for these transactions are discussed below. When attempting to determine the appropriate accounting treatment for a transaction, it is necessary to carefully study the characteristics of the transaction rather than the particular name that a conveyance may go by. Paragraphs 47a through 47m of *SFAS No. 19* should then be examined to determine in which paragraph(s) that particular transaction is actually addressed.

The principles for accounting for conveyances generally apply to companies using either the successful efforts or the full cost method. However, the examples provided in the discussion below illustrate the accounting entries that would be made to record various types of conveyances by companies using the successful efforts method. The specific provisions that apply to full cost companies are discussed later in this chapter.

CONVEYANCES: EXCHANGES AND POOLINGS

In the oil and gas industry there are numerous types of nonmonetary transactions where companies trade assets used in oil and gas producing operations for other assets used in oil and gas producing operations. These transactions range from the trade of a single piece of equipment to the conveyance of the entire working interest in a property. These sharing arrangements are reciprocal transfers, since the assets are not bought or sold. Examples include farm-ins/farm-outs, free wells, carried working interests, poolings, and unitizations. The following discussion describes these different types of sharing arrangements and explains the accounting treatment prescribed by *SFAS No. 19*.

SFAS No. 153, *Exchanges of Nonmonetary Assets*, requires the use of fair value valuation for nonmonetary exchanges when the transaction has "commercial substance." A reciprocal transfer of a nonmonetary asset is considered an exchange only if the transferor has no substantial continuing involvement in the transferred asset such that the usual risks and rewards of ownership of the asset are transferred. Most oil and gas sharing arrangements would not be considered nonmonetary exchanges, since in most such arrangements, the parties retain some type of interest in the asset(s). If the arrangement were considered a nonmonetary exchange, fair value valuation would be required only if the transaction has "commercial substance." According to *SFAS No. 153*, par. 21:

> *A nonmonetary exchange has commercial substance if the entity's future cash flows are expected to significantly change as a result of the exchange. The entity's future cash flows are expected to significantly change if either of the following criteria is met:*
>
> *a. The configuration (risk, timing, and amount) of the future cash flows of the asset(s) received differs significantly from the configuration of the future cash flows of the asset(s) transferred.*
>
> *b. The entity-specific value of the asset(s) received differs from the entity specific value of the asset(s) transferred, and the difference is significant in relation to the fair values of the assets exchanged.*
>
> *A qualitative assessment will, in some cases, be conclusive in determining that the estimated cash flows of the entity are expected to significantly change as a result of the exchange.*

If an exchange transaction is of commercial substance, the transaction must normally be measured based on the fair market values at the date of the exchange of the assets, and gain or loss recognized.

EXAMPLE

Nonmonetary Exchange

Tyler Oil Company has a 100% WI in a proved property in West Texas that it has been operating for a number of years. Due to changes in Tyler's strategic focus, Tyler entered into a trade agreement with Farmer Company. According to this agreement, Tyler will assign its working interest in the West Texas property to Farmer, and in exchange, Farmer will assign to Tyler its 100% WI in a proved property it owns in Oklahoma.

Tyler has the following information relating to its West Texas lease:

Proved property cost	$100,000
Less: Accumulated DD&A—proved property	(45,000)
Net carrying value	$ 55,000
Wells and equipment	$750,000
Less: Accumulated DD&A—wells and equipment	(250,000)
Net carrying value	$500,000
Total net carrying value	$555,000

Tyler estimates the fair market value (FMV) of the Oklahoma lease (that it is receiving in the exchange) to be as follows:

FMV of proved property	$ 800,000
FMV of wells and equipment	200,000
Total fair market value	$1,000,000

If Tyler determines that the trade is a nonmonetary exchange that is of commercial substance, then in accordance with *SFAS No. 153*, the exchange must be recorded at fair market value. The entry to record the exchange is as follows:

Proved property (Oklahoma lease)	800,000	
Wells and equipment (Oklahoma lease)	200,000	
Accumulated DD&A—proved property	45,000	
Accumulated DD&A—wells and equipment	250,000	
Proved property (West Texas lease)		100,000
Wells and equipment (West Texas lease)		750,000
Gain on nonmonetary exchange		445,000

If, on the other hand, Tyler determines that the exchange is *not* of commercial substance, the exchange should be recorded by assigning the net carrying value of the property being given up to the net carrying value of the property being received. The $555,000 total net carrying value of the West Texas property becomes the net carrying value of the Oklahoma property. The equipment being received may be recorded at its FMV, with the balance assigned to the proved property, and no gain is recognized. The entry to record the exchange is as follows:

Proved property (Oklahoma lease)	355,000	
Wells and equipment (Oklahoma lease).	200,000	
Accumulated DD&A—proved property	45,000	
Accumulated DD&A—wells and equipment.	250,000	
Proved property (West Texas lease).		100,000
Wells and equipment (West Texas lease)		750,000

Farm-ins/farm-outs

In practice, the term **farm-out** is used to refer to a wide variety of transactions. Rather than being guided by the fact that any given transaction is referred to as a "farm-out," it is important to examine the terms of transaction in order to verify that the terms actually fit the description of a farm-out transaction provided in *SFAS No. 19*. The terms **farm-in** and **farm-out** typically refer to an arrangement in which the owner of a working interest (the farmor) assigns all or part of the working interest to another party (the farmee) in return for the exploration and development of the property. These arrangements may be structured in many different ways. For example, the farmor may convey its working interest to the farmee and retain an ORI. The farmee in return agrees to pay all the costs to explore, develop, and produce the property. This arrangement is a farm-out from the perspective of the farmor, who conveys the working interest and retains an ORI. It is a farm-in from the perspective of the farmee, who earns the working interest by drilling a well, etc.

Farm-in/farm-out agreements are nonmonetary transactions for which there is no gain or loss recognition. Par. 47b of *SFAS No. 19* describes the accounting for a farm-in/farm-out where the farmor retains an ORI. According to *SFAS No. 19*, par. 47b:

> *An assignment of the operating interest in an unproved property with retention of a nonoperating interest in return for drilling, development, and operation by the assignee is a pooling of assets in a joint undertaking for which the assignor shall not recognize gain or loss. The assignor's cost of the original interest shall become the cost of the interest retained. The assignee shall account for all costs incurred as specified by paragraphs 15–41 and shall allocate none of those costs to the mineral interest acquired. If oil or gas is discovered, each party shall report its share of reserves and production (paragraphs 50–56).*

Paragraphs 15–41, referenced in the above paragraph, refer to the accounting for acquisition, exploration, development, and production costs required under successful efforts accounting. These requirements were discussed and illustrated in earlier chapters. The accounting specified in paragraphs 50–56, referenced in the above paragraph, is illustrated in the following examples.

In the example below, Tyler Oil Company owns 100% of the working interest in an unproved lease. Tyler trades its working interest in the unproved property for a nonworking interest in the property, and in doing so, avoids the risk and cost associated with drilling and developing the property. Tyler Oil Company (the farmor) initially has the cost of the unproved property on its books. When the working interest is conveyed to Frisco Oil Company (the farmee), an entry is made on the books of Tyler Oil Company moving the balance from an unproved property with a working interest to an unproved property with an ORI-type account. If Frisco Oil Company is successful in finding oil and gas reserves, Tyler Oil Company will transfer the balance in the account to a proved ORI account. In contrast, Frisco Oil Company will have no balance in its unproved property account. Instead, Frisco Oil Company will debit wells-in-progress as a well is drilled. Later the costs in that account would be transferred to wells and equipment or dry-hole expense, depending on the outcome of the drilling.

EXAMPLE

Farm-in/Farm-out

Tyler Oil Company signed a lease agreement with Rita Jones covering 400 acres in Oklahoma. Ms. Jones received a bonus of $20,000 and a 1/8 royalty interest, and Tyler Oil Company received 100% of the working interest.

A few months later, Tyler Oil Company decides that at the present time it is not financially able to develop the property. Tyler Oil Company enters into an agreement with Frisco Oil Company wherein Tyler Oil Company assigns its working interest to Frisco Oil Company and retains a 1/7 ORI. Frisco agrees to drill a well on the property within the next six months and to pay all of the costs of drilling the well, as well as other costs to develop the property. As a result of this agreement, Tyler Oil Company owns a 1/7 ORI interest in the property, but Frisco Oil Company must pay all of the costs and bear all of the risks of drilling. Frisco drills a successful exploratory well at a cost of $100,000.

Applicable Paragraph: Paragraph No. 47b

Entries—Tyler Oil Company

On the date of the conveyance:

Unproved ORI.	20,000	
Unproved property		20,000

When and if the property becomes proved:

Proved ORI	20,000	
Unproved ORI.		20,000

Entries—Frisco Oil Company

On the date of the conveyance:

No entry

As the well is drilled:

Wells-in-progress.	100,000	
Cash		100,000

When and if the well is successful:

Wells and equipment	100,000	
Wells-in-progress.		100,000

Farm-ins/farm-outs with a reversionary working interest

A common variation of the basic farm-in/farm-out agreement provides for a share of the working interest to revert back to the farmor at some point in time. For example, a working interest owner may convey its working interest to another party and retain an ORI. The other party in return agrees to pay all of the cost to explore, drill, and develop the property. When the farmee has generated enough income from the property to recover the costs it expended on exploring, drilling, and developing the property, the farmor's ORI reverts back to a working interest. The two parties then share the working interest in the property according to previously agreed-upon percentages. This type of conveyance is referred to as a **farm-in/farm-out with a reversionary working interest**.

EXAMPLE

Farm-in/Farm-out with Reversionary Working Interest

Tyler Oil Company signed a lease agreement with Rita Jones covering 400 acres in Oklahoma. Ms. Jones received a bonus of $20,000 and a 1/8 royalty interest, and Tyler Oil Company received a 100% WI.

On 1/1/2010, Tyler Oil Company enters into an agreement with Frisco Oil Company wherein Tyler Oil Company assigns its working interest and retains a 1/7 ORI. Frisco agrees to drill a well on the property within the next six months and to pay all of the costs of drilling the well. If the well is successful, Frisco Oil Company will pay all the operating costs and retain the net profit (after payment of the royalty, ORI, and operating expenses) until it has recovered the cost of drilling and completing the well. At that point, Tyler Oil Company's ORI will revert to a 50% WI.

During May 2010, Frisco Oil Company drills Well No. 1 at a cost of $1,000,000. The well is successful. Estimated gross proved reserves total 400,000 barrels, and gross proved developed reserves are 200,000 barrels. During 2011 and 2012, 25,000 gross barrels per year are produced and sold for $60/bbl. Operating costs are $25/bbl. Severance tax is ignored.

Applicable Paragraph: Paragraph No. 47b

Computation of Barrels for Payout

$$7/8 \times 6/7 \times \$60\,Y) - \$25\,Y = \$1,000,000$$
$$Y = \underline{50,000 \text{ bbl}}$$

where Y is the number of gross barrels produced before payout

Number of WI barrels produced before payout (assuming a price of $60/bbl and cost of $25/bbl):

$$50,000 \times 7/8 \times 6/7 = \underline{37,500 \text{ bbl}}$$

Computation of Proved Reserves

Total WI share of proved reserves remaining after payout*:
$$7/8(400,000 - 50,000) = \underline{306,250 \text{ bbl}}$$

Tyler Company's portion of proved reserves:

Reserves to payout as ORI: 50,000 × 7/8 × 1/7	=	6,250 bbl
Remaining after payout as WI: 306,250 × 50%	=	153,125 bbl
Tyler's total proved reserves	=	159,375 bbl

* Remember that the ORI reverts to a working interest after payout. Also note that Tyler Company is the carried party and so has only property costs. The property costs will be amortized over the total proved reserves. Frisco is the carrying party and so has no property costs.

Computation of Proved Developed Reserves

Total WI share of proved developed reserves remaining after payout:
7/8(200,000 – 50,000) = <u>131,250 bbl</u>

Frisco Oil Company's portion of proved developed reserves:

Reserves to payout:	50,000 × 7/8 × 6/7	=	37,500 bbl
Remaining after payout:	131,250 × 50%	=	65,625 bbl
Frisco's total proved developed reserves		=	103,125 bbl

Note: Frisco Oil Company is the carrying party, and so it has only well costs and no property costs. Additionally, any working interest barrels before payout belong to Frisco, as the carrying party.

Computation of Payout

	2011	2012
Sales volume	25,000 bbl	25,000 bbl
Price/bbl	$ 60	$ 60
Gross sales ($)	$1,500,000	$1,500,000
Net of royalty	7/8	7/8
	1,312,500	1,312,500
Net of ORI	6/7	6/7
Net revenue to WI	1,125,000	1,125,000
Operating expense*	(625,000)	(625,000)
Net profit	$ 500,000	$ 500,000

*$25/bbl × 25,000 bbl

Payout	
Amount to be recovered	$1,000,000
Recovered in Year 1	(500,000)
Balance to be recovered	$ 500,000
Recovered in Year 2	(500,000)
Balance to be recovered	$ 0

Tyler Oil Company

Entry—1/1/2010
Unproved ORI.	20,000	
Unproved property		20,000
(to transfer the working interest to a nonworking interest)		

Entry—5/2010
Proved ORI	20,000	
Unproved ORI.		20,000
(to reclassify the property)		

Entries—2011

Cash (25,000 × 7/8 × 1/7 × $60)	187,500	
Production revenue		187,500
DD&A expense	392	
Accumulated DD&A—proved ORI		392

$$\frac{\$20,000}{159,375 \text{ bbl}} \times 3,125 \text{ bbl*} = \underline{\$392}$$

*Tyler's share of production = 25,000 × 7/8 × 1/7 = 3,125 bbl

Entries—2012

Cash (25,000 × 7/8 × 1/7 × $60)	187,500	
Production revenue		187,500
DD&A expense	392	
Accumulated DD&A—Proved ORI		392

$$\frac{\$20,000 - \$392}{159,375 \text{ bbl} - 3,125 \text{ bbl}} \times 3,125 \text{ bbl} = \underline{\$392}$$

Proved property	19,216	
Accumulated DD&A—proved ORI	784	
Proved ORI		20,000
(to transfer the nonworking interest to a working interest)		

Frisco Oil Company

On the date of the conveyance:
No entry

Entry—5/2010 (As the well is drilled)
Wells-in-progress.	1,000,000	
Cash		1,000,000

Entry—5/2010
Wells and equipment	1,000,000	
Wells-in-progress.		1,000,000

Entries—2011
Cash (25,000 × $60).	1,500,000	
Royalty payable (25,000 × 1/8 × $60)		187,500
ORI payable (25,000 × 7/8 × 1/7 × $60)		187,500
Production revenue (25,000 × 7/8 × 6/7 × $60). . . .		1,125,000

(to record revenue)

Operating expense (25,000 × $25)	625,000	
Cash		625,000

(to record operating expense)

DD&A expense	181,818	
Accumulated DD&A—wells		181,818

$$\frac{\$1,000,000}{103,125 \text{ bbl}} \times 18,750 \text{ bbl}^* = \underline{\$181,818}$$

*Frisco's share of production = 25,000 × 7/8 × 6/7 = 18,750 bbl

Entries—2012

Cash (25,000 × $60)	1,500,000	
Royalty payable (25,000 × 1/8 × $60)		187,500
ORI payable (25,000 × 7/8 × 1/7 × $60)		187,500
Production revenue (25,000 × 7/8 × 6/7 × $60). . . .		1,125,000
(to record revenue)		
Operating expense (25,000 × $25)	625,000	
Cash		625,000
(to record operating expense)		
DD&A expense	181,818	
Accumulated DD&A—wells		181,818

$$\frac{\$1,000,000 - \$181,818}{103,125 \text{ bbl} - 18,750 \text{ bbl}} \times 18,750 \text{ bbl} = \underline{\$181,818}$$

12/31/2012:

Frisco Oil Company would adjust its property database and records to indicate that 50% of the working interest reverted back to Tyler Oil Company, and the property is now a jointly owned property.

Free wells

Sharing arrangements that result in free wells can take different forms. For example, the owner of a working interest transfers a portion of its working interest to a second party in exchange for the second party agreeing to drill (and possibly equip) a well free of cost to the assignor. The result is that the assignor ultimately has an interest in a well for which it has incurred no costs. These agreements may include more than one well, or the free well may need to be a producing well.

Accounting for a free well arrangement is similar to the accounting for a farm-in/farm-out described previously. Each party accounts for the costs that they actually expend, and no gain or loss recognition is allowed. According to *SFAS No. 19*, par. 47c:

> An assignment of a part of an operating interest in an unproved property in exchange for a "free well" with provision for joint ownership and operation is a pooling of assets in a joint undertaking by the parties. The assignor shall record no cost for the obligatory well; the assignee shall record no cost for the mineral interest acquired. All drilling, development, and operating costs

incurred by either party shall be accounted for as provided in paragraphs 15–41 of this Statement. If the conveyance agreement requires the assignee to incur geological or geophysical expenditures instead of, or in addition to, a drilling obligation, those costs shall likewise be accounted for by the assignee as provided in paragraphs 15–41 of this Statement. If reserves are discovered, each party shall report its share of reserves and production (paragraphs 50–56).

EXAMPLE

Free Well

In 2010, Tyler Oil Company entered into a lease agreement with Lenny Franks, paying Mr. Franks a $50,000 bonus and a 1/8 royalty interest. In 2012, Tyler Oil Company assigned 25% of its working interest to Daring Company in return for Daring Company agreeing to drill and equip a well on that lease. Daring Company drilled the well in 2013 at a cost of $600,000 for IDC and equipment. The well was successful. Gross proved reserves were initially estimated to be 400,000 barrels, and gross proved developed reserves were estimated to be 300,000 barrels. Gross production during 2013 was 10,000 barrels, production costs were $25/bbl, and the selling price was $60/bbl. Assume Daring Company is the operator and that the purchaser of the oil assumes the responsibility of distributing revenue and 5% severance taxes.

Applicable Paragraph: Paragraph No. 47c

Tyler Oil Company

Entries—2012

No entry for conveyance

Entries—2013

Proved property	50,000	
Unproved property		50,000
(to reclassify property)		
A/R (10,000 × $60 × 7/8 × 0.95 × 0.75)	374,062	
Severance tax expense		
(10,000 × $60 × 7/8 × 0.05 × 0.75)	19,688	
Oil sales (10,000 × $60 × 7/8 × 0.75)		393,750
(to record revenue)		

Operating expense (10,000 × $25 × 0.75)	187,500	
Payable to Daring Company		187,500
(to record lifting costs)		
DD&A expense	1,250	
Accumulated DD&A		1,250
(to record DD&A)		

$$\text{Acquisition costs only: } \$50,000 \times \frac{10,000 \text{ bbl} \times 0.75 \times 7/8}{400,000 \text{ bbl} \times 0.75 \times 7/8} = \underline{\$1,250}$$

Daring Company (operator)

Entries—2012

No entry for conveyance

Entries—2013

Wells and equipment	600,000	
Cash		600,000
A/R (10,000 × $60 × 7/8 × 0.95 × 0.25)	124,688	
Severance tax expense (10,000 × $60 × 7/8 × 0.05 × 0.25)	6,562	
Oil sales (10,000 × $60 × 7/8 × 0.25)		131,250
Operating expense (10,000 × $25 × 0.25).	62,500	
Joint interest receivable	187,500	
Cash (10,000 × $25)		250,000
DD&A expense	20,000	
Accumulated DD&A		20,000
(to record DD&A)		

$$\text{Drilling costs only: } \$600,000 \times \frac{10,000 \text{ bbl} \times 0.25 \times 7/8}{300,000 \text{ bbl} \times 0.25 \times 7/8} = \underline{\$20,000}$$

Carried interests or sole risk

The most common occurrences of **carried interests** arise in joint interest operations when one or more of the working interest owners elect not to participate in the drilling of a well. This situation is also commonly referred to as **sole risk**. Any party electing not to participate is referred to as a **carried interest** or **carried party**. The working interest owners who agree to pay the carried party's share of the costs are referred to as the **carrying parties**. When a working interest is carried, the carried working interest owner does not permanently relinquish its interest in the property. Rather, its interest is temporarily transferred to the carrying parties and will revert back when the carrying parties reach payout. In some cases, a working interest owner might not consent to the drilling of a well in an effort to avoid the risk of drilling. To minimize the likelihood of this occurring, most joint operating agreements include a penalty provision. Through this provision, the carrying parties are allowed to recover the carried party's share of drilling and completion costs, plus a penalty (often 100% to 200% of the carried party's share of costs). This type of carried working interest is discussed in detail in chapter 12.

Another situation where a carried interest may arise is a farm-in/farm-out, where the farmor assigns a portion of his working interest in exchange for the farmee agreeing to drill and equip a well. Since the farmor has a working interest but does not participate in the drilling of the well, his working interest is said to be carried by the farmee. In these arrangements the farmee recoups the farmor's share of drilling costs (which the farmee paid) from the farmor's share of net profits. In other words, payout is computed. In a carried interest arising from this type of arrangement, there is no penalty assessed on the farmor's share of costs.

These two types of arrangements are accounted for in the same manner. Moreover, the accounting is identical to a farm-in/farm-out with a reversionary interest, except that there is no ORI created. According to *SFAS No. 19*, par. 47d:

> *A part of an operating interest in an unproved property may be assigned to effect an arrangement called a "carried interest" whereby the assignee (the carrying party) agrees to defray all costs of drilling, developing, and operating the property and is entitled to all of the revenue from production from the property, excluding any third party interest, until all of the assignee's costs have been recovered, after which the assignor will share in both costs and production. Such an arrangement represents a pooling of assets in a joint undertaking by the assignor and assignee. The carried party shall make no accounting for any costs and revenue until after recoupment (payout) of the carried costs by the carrying party. Subsequent to payout the carried party shall account for its share of revenue, operating expenses, and (if the agreement provides for subsequent sharing of costs rather than a carried interest) subsequent development costs. During the payout period the carrying party shall record all costs, including those carried, as provided in paragraphs 15–41 and shall record all revenue from the property including that applicable to the recovery of costs carried. The carried party shall report as oil or gas reserves only its share of proved reserves estimated to remain after payout, and unit-of-production amortization of the carried*

party's property cost shall not commence prior to payout. Prior to payout the carrying party's reserve estimates and production data shall include the quantities applicable to recoupment of the carried costs (paragraphs 50–56).

In accounting for a carried interest, no gain or loss is to be recognized. The carrying party pays all of the cost to drill and equip a well, the full amount of which is capitalized to the wells-in-progress account. The carried party pays and records none of the drilling-related costs. If the well is dry, the carrying party clears the wells-in-progress account to dry-hole expense. If the well is successful, the costs are cleared to the carrying party's wells and equipment account. As production occurs, the carrying party keeps all of the working interest's proceeds from selling the production and pays all of the operating expenses. As was illustrated in chapter 12, the carrying party is entitled to keep the carried party's share of net profit until the carrying party has recouped the carried party's share of drilling and completion costs, which the carrying party paid, plus any applicable penalty. At the point that payout is reached, the carried party resumes its role as a regular working interest partner in the operation.

Another important and difficult feature of this type of arrangement is the computation of DD&A before payout. Before payout, the carried party does not recognize DD&A because it did not bear any of the cost of drilling or completing the well and because it is not receiving any revenue from production. In the carrying party's computation of DD&A, the estimate of proved developed reserves to be used is the carrying party's normal working interest percentage of reserves *plus* the amount of the carried party's share of reserves that the carrying party estimates it will be entitled to in order to recover the carried party's share of costs plus any penalty, i.e., until payout.

EXAMPLE

Carried Interest Created by a Farm-Out/Farm-In

Lowfund Company owns 100% of the working interest in a lease that it acquired from Joseph Walker. Lowfund paid a $100,000 bonus to Mr. Walker, who agreed to a 1/5 royalty interest. Tyler Oil Company agrees to drill a well on Lowfund Company's unproved lease in return for 50% WI in the lease and the right to recover all of its costs. In Year 1, Tyler Oil Company incurs $575,000 in IDC and equipment costs in drilling the well. The well is successful and finds gross proved reserves of 900,000 barrels, and gross proved developed reserves of 400,000 barrels. During each of the first five years, 12,500 gross barrels of oil per year are produced and sold for $60/bbl from the lease and lifting costs totaled $25/bbl. Ignore severance taxes and assume reserve estimates do not change. Assume Tyler distributes revenue.

Applicable Paragraph: Paragraph No. 47d

Computation of Barrels for Payout

$(4/5 \times \$60Y) - \$25Y = \$575{,}000$

$Y = \underline{25{,}000 \text{ barrels}}$

Where Y = number of barrels produced before payout (assuming a constant price of $60/bbl and cost of $25/bbl)

Number of WI barrels produced before payout = $(25{,}000 \times 4/5) = \underline{20{,}000 \text{ barrels}}$

Computation of Proved Reserves

Total working interest share of proved reserves remaining after payout:

$4/5(900{,}000 - 25{,}000) = \underline{700{,}000 \text{ barrels}}$

Lowfund Company's portion of proved reserves remaining after payout:

$700{,}000 \times 50\% = \underline{350{,}000 \text{ barrels}}$

Lowfund Company's total proved reserves = $\underline{350{,}000 \text{ barrels}}$

Note: Lowfund Company is the carried party and so has only property costs and no well costs. Additionally, any working interest barrels before payout belong to the carrying party.

Computation of Proved Developed Reserves

Total working interest share of proved developed reserves remaining after payout:

$4/5(400{,}000 - 25{,}000) = \underline{300{,}000 \text{ barrels}}$

Tyler Oil Company's portion of proved developed reserves remaining after payout:

$(300{,}000 \times 50\%) = \underline{150{,}000 \text{ barrels}}$

Tyler Oil Company's total proved developed reserves:

$150{,}000 + (25{,}000 \times 4/5) = \underline{170{,}000 \text{ barrels}}$

Note: Tyler Company is the carrying party and so has only well costs and no property costs. Additionally, any working interest barrels before payout belong to Tyler, as the carrying party.

Computation of Payout	Year 1	Year 2
Sales volume, bbl	12,500	12,500
Price/bbl	$ 60	$ 60
Gross sales ($)	$750,000	$750,000
Net of royalty	4/5	4/5
Net sales	$600,000	$600,000
Operating expense	312,500	312,500
Net profit	$287,500	$287,500

Net Profit to Tyler		
Total to Tyler (100%)	$287,500	$287,500
Tyler's portion (50%)	143,750	143,750
Lowfund's portion	$143,750	$143,750

	Payout:
Cost of well	$575,000
Lowfund Company's proportion	50%
Lowfund's share of cost	$287,500
Amount to be recovered	$287,500
Recovered in Year 1	(143,750)
Balance to be recovered	$143,750
Recovered in Year 2	(143,750)
Balance to be recovered	$ 0

Entries

No entries for Lowfund Company until after payout (except to reclassify property as proved).

Tyler Oil Company, YEAR 1

Wells and equipment	575,000	
Cash		575,000
A/R—purchaser (12,500 × $60)	750,000	
Royalty payable (12,500 × 1/5 × $60)		150,000
Sales revenue (12,500 × 4/5 × $60)		600,000

Operating expense (12,500 × $25)	312,500	
Cash		312,500
DD&A expense	33,824	
Accumulated DD&A		33,824

Drilling costs only: $\dfrac{\$575,000}{170,000 \text{ bbl}} \times (12,500 \text{ bbl} \times 4/5)^* = \underline{\$33,824}$

*Working interest share of production = 12,500 × 4/5 = 10,000 bbl

Note: Tyler Company has no property costs and therefore calculates DD&A only on its drilling costs.

Tyler Oil Company, YEAR 2

A/R—purchaser (12,500 × $60)	$750,000	
Royalty payable (12,500 × 1/5 × $60)		150,000
Sales revenue (12,500 × 4/5 × $60)		600,000
Operating expense (12,500 × $25)	312,500	
Cash		312,500
DD&A expense	33,824	
Accumulated DD&A		33,824

Drilling costs only: $\dfrac{\$575,000 - \$33,824}{170,000 \text{ bbl} - 10,000 \text{ bbl}} \times (12,500 \text{ bbl} \times 4/5) = \underline{\$33,824}$

Note: Tyler's proved developed reserves in the denominator have been reduced by its portion of production in Year 1.

Tyler Oil Company, YEAR 3—After Payout

A/R—purchaser (12,500 × $60)	750,000	
Royalty payable (12,500 × 1/5 × $60)		150,000
A/P—Lowfund (12,500 × 4/5 × $60 × 50%)		300,000
Sales revenue (12,500 × 4/5 × $60 × 50%)		300,000
Operating expense (12,500 × $25 × 50%)	156,250	
A/R—Lowfund (12,500 × $25 × 50%)	156,250	
Cash		312,500

DD&A expense	16,912	
Accumulated DD&A		16,912

Drilling costs only: $\dfrac{\$575{,}000 - 2(\$33{,}824)}{170{,}000 \text{ bbl} - 2(10{,}000 \text{ bbl})} \times 5{,}000 \text{ bbl}^* = \underline{\$16{,}912}$

*Tyler's share of production = 12,500 bbl × 4/5 × 50% = 5,000 bbl

Note: Once payout is reached, Lowfund Company receives 50% of the working interest owner's share of revenue and pays 50% of the costs incurred.

Lowfund Company, YEAR 3

A/R—Tyler Oil Co. (12,500 × 4/5 × $60 × 50%)	300,000	
Sales revenue		300,000
Operating expense (12,500 × $25 × 50%)	156,250	
A/P—Tyler Oil Co		156,250
DD&A expense	1,429	
Accumulated DD&A		1,429

Leasehold costs only: $\dfrac{\$100{,}000}{350{,}000 \text{ bbl}} \times 5{,}000 \text{ bbl} = \underline{\$1{,}429}$

Note: Lowfund Company has no well costs and therefore calculates DD&A only on its leasehold costs.

Joint venture operations

In chapter 12, accounting for joint venture operations was discussed in detail. It was noted that in a joint venture operation, each company typically accounts for its proportionate share of costs and revenue. Par. 47e of *SFAS No. 19* describes a situation that is a joint venture situation, specifically where working interest owners swap their working interests in different properties. This type of transaction is a pooling of interest in a joint undertaking to find and produce oil and gas; thus, no gain or loss is to be recognized by any of the parties. According to *SFAS No. 19*, par. 47e:

> *A part of an operating interest owned may be exchanged for a part of an operating interest owned by another party. The purpose of such an arrangement, commonly called a joint venture in the oil and gas industry, often is to avoid duplication of facilities, diversify risks, and achieve operating efficiencies. No gain or loss shall be recognized by either party at the time*

of the transaction. In some joint ventures, which may or may not involve an exchange of interests, the parties may share different elements of costs in different proportions. In such an arrangement a party may acquire an interest in a property or in wells and related equipment that is disproportionate to the share of costs borne by it. As in the case of a carried interest or a free well, each party shall account for its own cost under the provisions of this Statement. No gain shall be recognized for the acquisition of an interest in joint assets, the cost of which may have been paid in whole or in part by another party.

The following example illustrates a conveyance where a portion of the working interest in one lease is traded for a portion of the working interest in another lease. Each party merely transfers a portion of the investment in the property that the party is conveying to the property that the party is receiving.

EXAMPLE

Exchange of Working Interests

Tyler Oil Company acquired 100% of the working interest in Lease A for $80,000. Leveland Company acquired 100% of the working interest in Lease B for $200,000. Tyler and Leveland enter into an agreement whereby Tyler assigns 50% of its working interest in Lease A to Leveland, and in return, Leveland conveys to Tyler Oil Company 30% of its working interest in Lease B.

Applicable Paragraph: Paragraph No. 47e

Entries to record conveyance

Tyler Oil Company:

Oil and gas properties—Lease B (50% × $80,000). . . .	40,000	
Oil and gas properties—Lease A		40,000

Leveland Company:

Oil and gas properties—Lease A (30% × $200,000) . . .	60,000	
Oil and gas properties—Lease B		60,000

Poolings and unitizations

In a **pooling or unitization**, the working interests as well as the nonworking interests in two or more properties are typically combined. Each interest owner owns the same type of interest (but a smaller percentage) in the total combined property as they held previously in the separate property. The terms *pooling* and *unitization* are often used interchangeably. However, the most common usage of the term *pooling* is the combining of unproved properties, whereas the term *unitization* is commonly used to refer to a larger combination involving an entire producing field or reservoir for purposes of enhanced oil and gas recovery.

In a *pooling*, two or more properties are combined to form a single operating unit. After the pooling, if the working interests are held by two or more parties, a joint operating agreement is entered into, and one of the parties is designated as operator. Pooling may be voluntary or nonvoluntary and usually brings together small tracts to obtain sufficient acreage so that a well may be drilled under a state's specific well spacing rules. As a rule, both working interests and nonworking interests are combined, and each party receives an interest stated as a percentage of ownership in the combined acreage of the same type as contributed. For example, a royalty interest owner receives a royalty interest in the combined acreage in exchange for the royalty interest in a particular property.

EXAMPLE

Pooling

Tyler Oil Company enters into an agreement to lease 80 acres from Rita Jones. Jones receives a $15,000 bonus and a 1/8 royalty interest (RI). Tyler owns a 100% WI. Duster Oil Company enters into an agreement to lease 80 acres from Carolyn Pinkie. Pinkie receives a $5,000 bonus and a 1/8 royalty interest. Duster owns a 100% WI. The two leases are contiguous. In order to develop and operate the properties efficiently, Tyler and Duster agree to pool their leases and enter into a joint operating agreement. Ownership interests are recalculated based on acreage contributed.

Applicable Paragraph: Paragraph No. 47e

The pooled property consists of 160 acres. The parties' interests in the pooled property are as follows:

Tyler Oil Company	50% WI
Duster Oil Company	50% WI
Rita Jones	1/16 RI
Carolyn Pinkie	1/16 RI

This transaction would result in no accounting entries for either party. The only action required would be to change the database or property records to reflect the change in property description and the change in interest.

Unitizations

A *unitization* is similar to a pooling, although unitization usually refers to the combination of leases that have been at least partially developed. In a unitization, the parties enter into a unitization agreement that defines the areas to be unitized and specifies the rights and obligations of each party. One party, known as the unit operator, has the responsibility of operating the unit. The purpose of unitizations is more economical and efficient development and operation. In particular, a unitization may be necessary to conduct secondary or tertiary recovery operations.

Unitizations (or poolings) may be voluntary or mandatory according to state law. In some states, unitizations may be forced if some percentage, such as 65%, of the involved parties agrees to the unit. In other states, one party alone may instigate a unitization. Typically, both working interests and nonworking interests are combined, with each party receiving an interest stated as a percentage of ownership in the unit of the same type as contributed.

Two significant issues must be resolved in a unitization. One problem is the determination of oil and gas reserves underlying each property contributed by each party. The relative amount of the reserves contributed by each party is typically used to determine the parties' **participation factors**, i.e., percentage of working interests in the unitized property. These participation factors are very important, since the amount of costs paid by each party and revenues received by each party are based on the participation factors.

The other problem in a unitization is the determination of the fair market value to be assigned to each lease's wells, equipment, and facilities that are contributed to the unit. These contributions must be equalized, because the participation factors do not account for the fact that the properties may be in varying stages of development. For example, if a working interest owner has a participation factor (percentage ownership interest) of 40%, then that party should have contributed to the unit 40% of the agreed-upon value of wells and equipment. If the party contributed less than 40%, then the party must pay cash; if the party contributed more than 40%, then the party will receive cash.

To equalize investment, i.e., to equalize contributions, the participation factors are multiplied by the total agreed-upon fair market value of the IDC and equipment contributed by all parties. This calculation is used to arrive at the assigned value of each participant's interest in the unit-wide IDC and equipment. The participant's assigned value is matched against the fair market value of the property transferred by the participant to determine the amount each participant will pay or receive in cash. Cash received is treated as a recovery of cost, and cash paid is treated as additional investment. No gain or loss is recognized. According to *SFAS No. 19*, par. 47f:

> In a unitization all the operating and nonoperating participants pool their assets in a producing area (normally a field) to form a single unit and in return receive an undivided interest (of the same type as previously held) in that unit. Unitizations generally are undertaken to obtain operating efficiencies and to enhance recovery of reserves, often through improved recovery operations. Participation in the unit is generally proportionate to the oil and gas reserves contributed by each. Because the properties may be in different stages of development at the time of unitization, some participants

may pay cash and others may receive cash to equalize contributions of wells and related equipment and facilities with the ownership interests in reserves. In those circumstances, cash paid by a participant shall be recorded as an additional investment in wells and related equipment and facilities, and cash received by a participant shall be recorded as a recovery of cost. The cost of the assets contributed plus or minus cash paid or received is the cost of the participant's undivided interest in the assets of the unit. Each participant shall include its interest in reporting reserve estimates and production data (paragraphs 50–56).

EXAMPLE

Unitization

Three companies own adjacent leases that share a common reservoir. The companies have the following costs on their books as of May 1, 2012:

Company	IDC	L&WE	Leasehold	Total
A	$ 50,000	$30,000	$10,000	$ 90,000
B	150,000	50,000	20,000	220,000
C	none	none	30,000	30,000
	$200,000	$80,000	$60,000	$340,000

On May 1, the three companies enter into a unitization agreement. After an engineering study and negotiations between the companies, the following participation factors and fair market values and are agreed upon as follows:

Company	IDC	L&WE	Leasehold*	Total	Participation Factors
A	$ 60,000	$ 40,000	Ignore	$100,000	40%
B	100,000	60,000	Ignore	160,000	25%
C	0	0	Ignore	0	35%
	$160,000	$100,000		$260,000	100%

*Most agreements do not provide for an equalization of leasehold investments.

Applicable Paragraph: Paragraph No. 47f

Equalization of Investment

a	b	c	d	e	f
		Agreed Value	Assigned Value: Factor x Total		
	Participation	Contributed	Agreed Value	Receive	Pay
Company	Factor (given)	(given)	(col. b x $260,000)	(c – d)	(d – c)
A	40%	$100,000	$104,000		$ 4,000
B	25%	160,000	65,000	$95,000	
C	35%	0	91,000		91,000
	100%	$260,000	$260,000	$95,000	$95,000

Entries to record equalization of investments:

COMPANY A

Wells and equipment	4,000	
Cash		4,000

COMPANY B

Cash	95,000	
Wells and equipment		95,000

COMPANY C

Wells and equipment	91,000	
Cash		91,000

Entries to record additional development or production costs would be similar to joint interest examples.

CONVEYANCES: SALES

A sale of an oil and gas property can involve either a proved property or an unproved property and can be of the entire interest or only of part of an interest. Further, the interest sold can be a working interest or a nonworking interest. How the sale is accounted for depends upon whether a proved or unproved property is involved and whether an entire or partial interest is sold. Figure 13–1 summarizes the accounting treatment of sales of oil and gas properties.

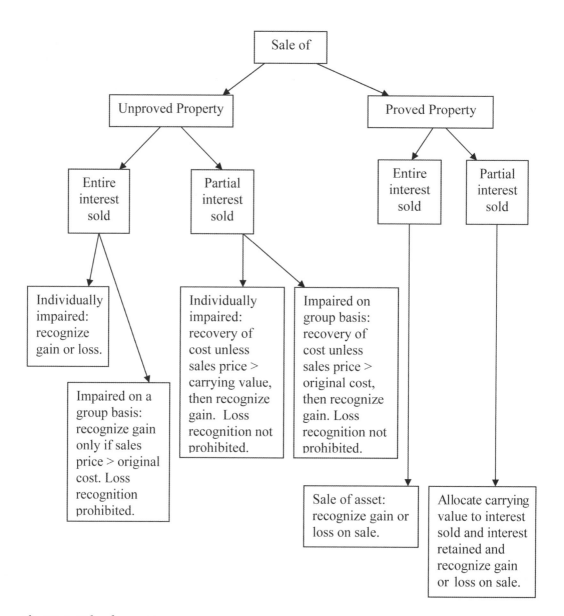

Fig. 13–1. Sale of property

Unproved property sales

In order to understand the accounting treatment for sales of unproved properties, it is necessary to divide the topic between sales of the entire interest in a property and sales involving only a partial interest in the unproved property.

Sales of entire interest in unproved property. The accounting treatment of a sale of the entire interest in an unproved property depends on whether impairment of the property has been assessed individually or on a group basis. If the entire interest in an unproved property that has been assessed on an individual basis is sold, a gain or loss should be recognized. If, however, the entire interest in an unproved property that has been assessed on a group basis is sold, a gain should be recognized only if the selling price exceeds the original cost. Recognizing a loss is prohibited in the case of a property assessed on a group basis, because an individual carrying value is not known. To avoid recognizing a loss, the allowance for impairment account is debited. According to *SFAS No. 19*, par. 47g:

> *If the entire interest in an unproved property is sold for cash or cash equivalent, recognition of gain or loss depends on whether, in applying paragraph 28 of this Statement, impairment had been assessed for that property individually or by amortizing that property as part of a group. If impairment was assessed individually, gain or loss shall be recognized. For a property amortized by providing a valuation allowance on a group basis, neither a gain nor loss shall be recognized when an unproved property is sold unless the sales price exceeds the original cost of the property, in which case gain shall be recognized in the amount of such excess.*

EXAMPLE

Sale of Entire Unproved Property

Tyler Oil Company owns 100% of the working interest in an undeveloped lease in Texas for which it paid $80,000. In each case below, Tyler sells its entire interest in the property.

Applicable Paragraph: Paragraph No. 47g

Individually Impaired

The property is individually significant, and impairment of $30,000 had been recorded in a prior year. The following entry would be made if the property were sold for $60,000:

Entry

Cash	60,000	
Impairment allowance.	30,000	
Unproved property		80,000
Gain on sale of unproved property		10,000

If instead the property were sold for $40,000, the following entry would be made:

Entry

Cash	40,000	
Impairment allowance	30,000	
Loss on sale of unproved property	10,000	
Unproved property		80,000

Property Impaired as a Group

Now assume that the property was individually insignificant and had been amortized as part of a group of unproved properties. The following entry would be made if the property were sold for $60,000:

Entry

Cash	60,000	
Impairment allowance*	20,000	
Unproved property		80,000

*The impairment allowance account is debited to avoid recognizing a loss.

If instead, the property were sold for $90,000, the following entry would be made:

Entry

Cash	90,000	
Unproved property		80,000
Gain on sale of unproved property		10,000

Sales of a partial interest in unproved property. Sale of only part of an interest in an unproved property receives special accounting treatment, because substantial uncertainty exists as to the future recoverability of the cost of the interest retained. In this type of conveyance, the selling price should be treated as a recovery of cost with no gain recognized. (Recognizing a loss is not prohibited.) However, if the property has been assessed on an individual basis, and the selling price of the partial interest exceeds the net carrying value of the entire property, a gain should be recognized. If the property has been assessed on a group basis, a gain should be recognized if the selling price of the partial interest exceeds the original cost of the entire interest.

According to *SFAS No. 19*, par. 47h:

> *If a part of the interest in an unproved property is sold, even though for cash or cash equivalent, substantial uncertainty usually exists as to recovery of the cost applicable to the interest retained. Consequently, the amount received shall be treated as a recovery of cost. However, if the sales price exceeds the carrying amount of a property whose impairment has been assessed individually in accordance with paragraph 28 of this Statement, or exceeds the original cost of a property amortized by providing a valuation allowance on a group basis, gain shall be recognized in the amount of such excess.*

EXAMPLE

Sale of Portion of Unproved Property—Individual Impairment

Tyler Oil Company has an unproved lease for which it paid $100,000. The property was individually significant, and individual impairment of $20,000 had been assessed. Tyler Oil Company conveys a 25% WI in return for $10,000 cash.

Applicable Paragraph: Paragraph No. 47h

Entry

Cash	10,000	
Unproved property		10,000

No gain or loss is recognized. Since the property is unproved, there is substantial uncertainty regarding the costs applicable to the retained interest. However, since the property is individually significant, recognition of additional impairment may be called for since the 25% working interest was sold for only $10,000.

Next, assume that the 25% interest is sold for $110,000.

Entry to record the sale

Cash	110,000	
Allowance for impairment	20,000	
Unproved property		99,990*
Gain on sale of unproved property		30,010

* In order to maintain adequate control for the property in the accounts, most companies leave some nominal amount (e.g., $10) in the unproved property account.

In this case, a gain is recognized because the selling price of the partial interest that was sold exceeds the net book value of the entire interest.

EXAMPLE

Sale of Portion of Unproved Property—Group Impairment

Tyler Oil Company has an unproved lease for which it paid $100,000. The property is **not** considered to be individually significant. Tyler Oil Company conveys a 25% WI in return for $40,000 cash.

Applicable Paragraph: Paragraph No. 47h

Entry

Cash	40,000	
Unproved property		40,000

No gain or loss is recognized. Since the property is unproved, there is substantial uncertainty regarding the costs applicable to the retained interest.

Next, assume that the 25% WI is sold for $110,000.

Entry to record the sale

Cash	110,000	
Unproved property		99,990
Gain on sale of unproved property		10,010

In this case, a gain is recognized because the selling price of the partial interest that was sold exceeds the original cost of the entire interest.

Par. 47h also applies to partial sales of an unproved property where the entire working interest is sold, and a nonworking interest, such as an ORI, is retained. The recognition of gain or loss again depends upon whether the interest in the working interest was individually impaired or amortized as part of a group.

EXAMPLE

Sale of Working Interest in Unproved Property with Retention of ORI

Tyler Oil Company owns 100% of the working interest in an unproved lease that it acquired for $100,000. The property is individually significant, and individual impairment of $20,000 had been assessed. Tyler Oil Company conveys the working interest to another party and retains a 1/7 ORI in return for $30,000 cash.

Applicable Paragraph: Paragraph No. 47h

Entry

Cash	30,000	
Allowance for impairment	20,000	
Unproved ORI	50,000	
Unproved property		100,000

No gain or loss is recognized. Since the property is unproved, there is substantial uncertainty regarding the costs applicable to the ORI that was retained. The entire allowance balance is written off, and the difference between the net book value and the cash received is attributed to the unproved ORI that was retained.

Now, assume that the working interest in the unproved property costing $100,000 is not considered to be individually significant. Tyler Oil Company sells the working interest and retains a 1/7 ORI for a cash payment of $110,000.

Applicable Paragraph: Paragraph No. 47h

Entry to record the sale

Cash	110,000	
Unproved ORI	10*	
Unproved property		100,000
Gain on sale of unproved property		10,010

*In order to maintain adequate control for the property, $10 is placed in the unproved ORI account.

In this case, a gain is recognized, because the selling price of the working interest that was sold exceeds the original cost of the entire interest.

The rules regarding sales of partial and entire interests in unproved properties apply to partial and entire sales of nonworking interests as well as working interests. The accounting treatment for the sale of a partial or entire nonworking interest in an unproved property is the same as the accounting when a working interest is involved.

EXAMPLE

Sale of Overriding Royalty Interest in Unproved Property

Tyler Oil Company has an ORI in an unproved lease for which it paid $60,000. The ORI has not been impaired. In each case below, Tyler sells the entire ORI for cash.

Applicable Paragraph: Paragraph No. 47g

If Tyler Oil Company sells the entire ORI for a cash consideration of $50,000, a loss of $10,000 would be recognized.

Entry

Cash	50,000	
Loss on sale of ORI	10,000	
Investment in ORI		60,000

However, if Tyler Oil Company sells the entire ORI for a cash consideration of $80,000, a gain of $20,000 would be recognized.

Entry

Cash	80,000	
Investment in ORI		60,000
Gain on sale of ORI		20,000

EXAMPLE

Sale of Portion of ORI in Unproved Property

Tyler Oil Company has an ORI in an unproved property for which it paid $80,000. The ORI has not been impaired. In each case below, Tyler sells a portion of the ORI for cash.

Applicable Paragraph: Paragraph No. 47h

If Tyler Oil Company sells 60% of the ORI for a cash consideration of $65,000, no gain or loss would be recognized.

Entry
Cash	65,000	
Investment in ORI		65,000

If Tyler Oil Company sells 60% of the ORI for a cash consideration of $105,000, a gain would be recognized, and a nominal amount would be left in the ORI account for control purposes.

Entry
Cash	105,000	
Investment in ORI		79,990
Gain on sale of ORI		25,010

Proved property sales

Sales and purchases of an entire interest in a proved property. If an entire proved property that has been accounted for as a separate amortization base is sold, a gain or loss should be recognized. This type of sale is treated the same as the sale of any depreciable asset. According to *SFAS No. 19*, par. 47i:

> The sale of an entire interest in a proved property that constitutes a separate amortization base is not one of the types of conveyances described in paragraphs 44 or 45. The difference between the amount of sales proceeds and the unamortized cost shall be recognized as a gain or loss.

The following example illustrates the accounting when the entire interest in a proved property is sold.

EXAMPLE

Sale and Purchase of Entire Interest in a Proved Property

Tyler Oil Company owns 100% of the working interest in a proved property with the following costs:

Leasehold	$ 60,000
IDC	200,000
Equipment	75,000
Total accumulated DD&A	50,000

Tyler Oil Company sells this property to Hannah Corporation for $300,000. The property constitutes a separate amortization base. The fair market value of the equipment is estimated to be $55,000.

Applicable Paragraph: Paragraph No. 47i

Entry—Tyler

Cash	300,000	
Accumulated DD&A—total	50,000	
Proved property		60,000
Wells and equipment—IDC		200,000
Wells and equipment—L&WE		75,000
Gain on sale of property		15,000

Hannah Corporation would allocate $55,000 of the purchase price to the equipment it purchased. Since it is not possible to separate the purchased IDC from the value of the proved leasehold interest, the amount not assigned to equipment is recorded as the cost of the proved property.

Entry—Hannah

Wells and equipment—L&WE	55,000	
Proved property	245,000	
Cash		300,000

The same rules apply to the sale of an entire nonworking interest in a proved property.

EXAMPLE

Sale of Entire Overriding Royalty Interest in Proved Property

Tyler Oil Company owns a 1/7 ORI in a proved property with *unrecovered* costs of $40,000. The interest constitutes a separate amortization base. Tyler Oil Company sells the entire ORI for $30,000.

Applicable Paragraph: Paragraph No. 47i

Entry to record the sale

Cash	30,000	
Loss on sale of ORI	10,000	
Investment in ORI, net		40,000

Sales and purchases of a partial interest in a proved property. Paragraph 47j describes the accounting when a partial interest in a proved property is sold. (Par. 47k describes the accounting for a more specific sale of a partial interest in a proved property where the working interest is sold and a nonworking interest retained. This situation is discussed later.) When a partial interest in a proved property is sold, the unamortized cost of the property should be apportioned between the interest sold and the interest retained based on the relative fair market value of the two interests. Any difference between the selling price and the cost apportioned to the interest sold should be recognized as a gain or loss. If the interest sold is an undivided interest, then the fair market value of the interest sold and the fair market value of the interest retained will be exactly the same proportion as the proportion of the working interest sold and retained. (See the following example.) The same apportionment and gain or loss recognition rules apply regardless of whether a working interest or a nonworking interest is retained. According to *SFAS No. 19*, par. 47j:

> *The sale of a part of a proved property, or of an entire proved property constituting a part of an amortization base, shall be accounted for as the sale of an asset, and a gain or loss shall be recognized, since it is not one of the conveyances described in paragraphs 44 or 45. The unamortized cost of the property or group of properties a part of which was sold shall be apportioned to the interest sold and the interest retained on the basis of the fair values of those interests. However, the sale may be accounted for as a normal retirement under the provisions of paragraph 41 with no gain or loss recognized if doing so does not significantly affect the unit-of-production amortization rate.*

EXAMPLE

Sale and Purchase of Partial Interest in a Proved Property

Tyler Oil Company owns a 100% WI in a proved property with the following costs:

Leasehold	$ 60,000
IDC	200,000
Equipment	75,000
Total accumulated DD&A	50,000

Tyler Oil Company conveys 75% of the working interest in this property to Fisher Energy for $300,000. The property constitutes a separate amortization base.

Applicable Paragraph: Paragraph No. 47j

In this case, the costs would be allocated between the interest sold and the interest retained on the basis of relative fair market values. Since the interest is an undivided working interest, 75% of the costs will be allocated to the interest sold, and 25% will be allocated to the interest retained.

Entry—Tyler

Cash	300,000	
Accumulated DD&A—total (75% × $50,000)	37,500	
Proved property (75% × $60,000)		45,000
Wells and equipment—IDC (75% × $200,000)		150,000
Wells and equipment—L&WE (75% × $75,000)		56,250
Gain on sale of property		86,250

Fisher Energy must record the purchase of the working interest in the wells and equipment. Fisher Energy can reasonably estimate the fair market value of the equipment but has no basis for allocating the remainder of the costs between the IDC versus leasehold. Therefore, industry practice generally has the remainder of the costs being allocated entirely to the leasehold and amortized over total proved reserves. Assume in this case that the fair market value of the equipment is $100,000.

Purchase price allocated to equipment:	$100,000 × 75% = $75,000
Purchase price allocated to proved property:	$300,000 – $75,000 = $225,000

Entry—Fisher

Wells and equipment—L&WE	75,000	
Proved property	225,000	
Cash		300,000

In the less common situation in which a divided interest is sold, the relative fair market values (FMV) of the leasehold, equipment, and IDC must be determined. These values are then used to allocate the costs and determine any gain or loss.

EXAMPLE

Sale of Divided Interest in a Proved Property

Tyler Oil Company owns 100% of the working interest in a 640 acre proved property with the following net costs:

Leasehold	$ 55,000
IDC	190,000
Equipment	77,000

Tyler Oil Company sells 100% of the working interest, including the wells and equipment in the western 320 acres of the property for $150,000. An appraisal is performed with the following results:

	FMV of Portion Sold	FMV of Portion Retained	Total FMV
Leasehold	$ 30,000	$ 90,000	$120,000
IDC	80,000	100,000	180,000
Equipment	40,000	30,000	70,000
Total	$150,000	$220,000	

Applicable Paragraph: Paragraph No. 47j

Allocation of costs to interest sold:

FMV_S = FMV of interest sold FMV_R = FMV of interest retained

$$\frac{FMV_S}{FMV_S + FMV_R} \times CV \text{ of property} = \text{Cost assigned to interest sold}$$

Leasehold: $30,000/$120,000 × $55,000 = $13,750
IDC: $80,000/$180,000 × $190,000 = $84,444
Equipment: $40,000/$70,000 × $77,000 = $44,000

Entry

Cash	150,000	
Proved property		13,750
Wells and equipment—IDC		84,444
Wells and equipment—L&WE		44,000
Gain on sale of property		7,806

Sales of a working interest in a proved property with retention of a nonworking interest. When the entire working interest in a proved property is sold and a nonworking interest is retained, a gain or loss is determined in the same manner as a partial sale of a proved property. The costs are allocated between interest sold and interest retained based on relative fair market values of the interests. According to *SFAS No. 19*, par. 47k:

> *The sale of the operating interest in a proved property for cash with retention of a nonoperating interest is not one of the types of conveyances described in paragraphs 44 or 45. Accordingly, it shall be accounted for as the sale of an asset, and any gain or loss shall be recognized. The seller shall allocate the cost of the proved property to the operating interest sold and the nonoperating interest retained on the basis of the fair values of those interests.*

EXAMPLE

Sale of Working Interest in a Proved Property with Retention of a Nonworking Interest

Tyler Oil Company sold its 100% WI in a proved property for $600,000 and retained an ORI. Tyler's net cost basis in the property was $555,000. The fair market value of the entire original working interest was $900,000.

Tyler has the following net investment in its accounts:

Leasehold	$155,000
IDC	312,000
Equipment	88,000
Total	$555,000

Applicable Paragraph: Paragraph No. 47k

Since the fair market value of the entire property is $900,000, and the working interest was sold for $600,000, the costs would be allocated as follows:

Allocation of costs to interest sold:
$600,000/$900,000 × $555,000 = $370,000

Allocation of costs to interest retained:
$300,000/$900,000 × $555,000 = $185,000

Entry

Cash	600,000	
Investment in ORI	185,000	
Proved property		155,000
Wells and equipment—IDC		312,000
Wells and equipment—L&WE		88,000
Gain on sale of property ($600,000 – $370,000)		230,000

The following example also illustrates the sale of a partial interest in a proved property. However in this example, a nonworking interest is sold and a portion of the working interest is retained. In this situation, unlike the previous example, the accounts associated with the original working interest should remain on the books, but reduced by the amount of the nonworking interest sold.

EXAMPLE

Sale of Overriding Royalty Interest in Proved Property

Tyler Oil Company owns a 100% WI in the Alfalfa lease that has a 1/8 royalty interest. The property is proved and has the following capitalized costs:

Leasehold	$100,000
IDC	500,000
Equipment	125,000
Accumulated DD&A (on leasehold, IDC, and equipment)	(225,000)
Total book value	$500,000

Tyler Company carved out an overriding royalty interest for a consideration of $300,000. The fair market value of the working interest after the carve-out was $1,200,000.

Applicable Paragraph: Paragraph No. 47j

Allocation

FMV_S = FMV of interest sold FMV_R = FMV of interest retained

$$\frac{FMV_S}{FMV_S + FMV_R} \times CV \text{ of property} = \text{Cost assigned to interest sold}$$

$$\frac{\$300,000}{\$300,000 + \$1,200,000} \times \$500,000 = 0.2 \times \$500,000 = \underline{\$100,000}$$

$$\frac{FMV_R}{FMV_R + FMV_S} \times CV \text{ of property} = \text{Cost assigned to interest retained}$$

$$\frac{\$1,200,000}{\$300,000 + \$1,200,000} \times \$500,000 = 0.8 \times \$500,000 = \underline{\$400,000}$$

Entry

Cash	300,000	
Accumulated DD&A (0.2 × $225,000)	45,000	
Proved property (0.2 × $100,000)		20,000
Wells and equipment—IDC (0.2 × $500,000)		100,000
Wells and equipment—L&WE (0.2 × $125,000)		25,000
Gain ($300,000 – $100,000)		200,000

Conveyances: Production Payments

Production payment interests may be created by being retained or carved out of the working interest and may be satisfied by monetary payment or by delivery of a specified quantity of oil or gas. Accounting for a production payment depends upon whether the production payment is retained or carved out, how it will be paid, and whether the production payment is an economic interest. Generally only those production payment interests that are payable by delivery of a specified quantity of oil or gas are considered economic interests and are subject to DD&A. Production payment interests that are payable out of cash typically are not economic interests and are not subject to DD&A. Figure 13–2 overviews the general approach to accounting for both carved out and retained production payments payable in either money or volumes of oil or gas.

Retained production payments

A retained production payment may be payable out of oil or gas or in money from a specified share of production from a proved property. The accounting treatment depends on whether the payment is stated in terms of money or in terms of a volume of production. It also depends on whether the repayment of the production payment is reasonably assured or not reasonably assured. Each of these scenarios is discussed below.

SFAS No. 19, par. 47l specifies the following relating to a retained production payment payable in money:

> *The sale of a proved property subject to a retained production payment that is expressed as a fixed sum of money payable only from a specified share of production from that property, with the purchaser of the property obligated to incur the future costs of operating the property, shall be accounted for as follows:*
>
> i. If satisfaction of the retained production payment is reasonably assured. *The seller of the property, who retained the production payment, shall record the transaction as a sale, with recognition of any resulting gain or loss. The retained production payment shall be recorded as a receivable, with interest accounted for in accordance with the provisions of APB Opinion No. 21, "Interest on Receivables and Payables." The purchaser shall record as the cost of the assets acquired the cash consideration paid plus the present value (determined in accordance with APB Opinion No. 21) of the retained production payment, which shall be recorded as a payable. The oil and gas reserve estimates and production data, including those applicable to liquidation of the retained production payment, shall be reported by the purchaser of the property (paragraphs 50–56).*
>
> ii. If satisfaction of the retained production payment is not reasonably assured. *The transaction is in substance a sale with retention of an overriding royalty that shall be accounted for in accordance with paragraph 47(k).*

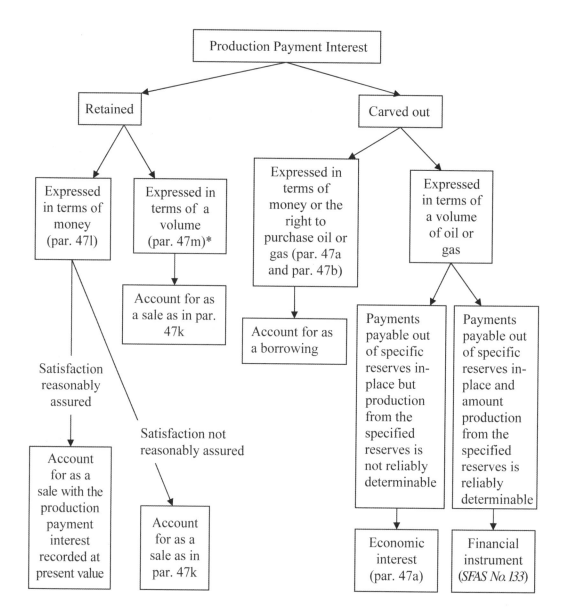

*Note: A PPI stated in terms of a quantity of oil and gas is an economic interest.

Fig. 13–2. Production payment interests

Retained production payments payable in money—reasonably assured. As seen above, when a retained production payment is payable out of a fixed sum of money, the accounting treatment depends upon whether the payment is reasonably assured or not. If the payment is reasonably assured, the seller records a receivable for the production payment in the amount of the present value of the production payment and recognizes a gain or loss on the sale of the working interest. The buyer records a payable equal to the present value of the production payment and capitalizes as the cost of the asset the present value of the production payment, plus any cash paid. Both parties compute the present value using their own cost of capital as the interest rate.

To compute the present value of the production payment, the future cash flows to the production payment owner must be estimated. To do this, a projection of future production over time must first be made. The production payment owner's share of revenue from that production is then discounted to arrive at the present value of the production payment. As the production payment is paid, the owner of the retained production payment records the payment received as interest revenue and a reduction to the receivable. The working interest owner records the payment made as interest expense and a reduction to the liability. From the seller's standpoint, its production payment interest is not a mineral interest in the property and therefore is not subject to DD&A.

EXAMPLE

Retained Production Payment Payable in Money—Reasonably Assured

Tyler Oil Company has a 100% WI in a fully developed lease subject to a 1/8 royalty interest. As of January 1, 2011, Tyler's net capitalized costs related to the property are $500,000, and total gross proved developed reserves are 250,000 barrels.

On January 1, 2011, Tyler Oil Company sold its entire working interest to Roberts Oil Company for $1,000,000 and a retained production payment of $360,000. The production payment is payable to Tyler Oil Company out of 25% of the proceeds of the working interest owner's share of production. It is estimated that the production payment will be paid off in three years. Assume that the present value of the production payment of $360,000 would be $300,000 using Tyler Company's cost of capital (10%) and $325,000 using Roberts Company's cost of capital (5%). During 2011, gross production totaled 11,200 barrels, operating costs totaled $148,980, and the selling price was $60/bbl. Ignore severance taxes.

Applicable Paragraph: Paragraph No. 47i

Entries

Tyler Oil Company

Cash	1,000,000	
Production payment receivable (at PV using 10%). . . .	300,000	
Net capitalized costs of property		500,000
Gain on sale ($1,300,000 – $500,000)		800,000
(to record conveyance)		

Roberts Company

Proved property, etc. ($1,000,000 + $325,000)	1,325,000	
Production payment payable (at PV using 5%). . . .		325,000
Cash		1,000,000
(to record conveyance)		
A/R—purchaser (11,200 × $60).	672,000	
Royalty payable (11,200 × 1/8 × $60)		84,000
Oil revenue (11,200 × 7/8 × $60)		588,000
(to record revenue from the sale of oil)		
Operating expense.	148,980	
Cash		148,980
(to record production costs)		
DD&A expense	59,360	
Accumulated DD&A		59,360

Allocation

Roberts' share of proved developed reserves:
250,000 bbl × 7/8 = 218,750 bbl

Roberts' share of production:
11,200 bbl × 7/8 = 9,800 bbl

$$\text{DD\&A expense} = \frac{\$1,325,000}{218,750 \text{ bbl}} \times 9,800 \text{ bbl} = \underline{\$59,360}$$

Production payment payable ($147,000 – $16,250) . . .	130,750	
Interest expense ($325,000 × 5%).	16,250	
Cash (25% × 9,800 bbl × $60).		147,000
(to record partial payment of production payment payable)		

Tyler Oil Company

Cash (25% × 9,800 bbl × $60).	147,000	
Interest revenue ($300,000 × 10%)		30,000
Production payment receivable ($147,000 – $30,000)		117,000

Retained production payments payable in money—*not* reasonably assured. If the retained production payment is not reasonably assured, the conveyance is, in substance, a sale of a proved property with a retained ORI. The production payment's life in this instance is likely to be coextensive with the life of the lease and, therefore, has the same characteristics as an ORI. Consequently, this type of production payment should be treated as an ORI rather than as a receivable or a payable. Further, even though the production payment is payable in cash, the production payment is considered to be an economic interest subject to DD&A. To the extent that the production payment's fair market value can be determined, the unamortized cost of the working interest should be allocated between the interest sold and the interest retained. A gain or loss should be recognized on the interest sold.

EXAMPLE

Retained Production Payment Payable in Money—*Not* Reasonably Assured

Tyler Oil Company owns a 100% WI in a proved property with net capitalized costs of $100,000. Tyler Oil Company sells the working interest for $425,000 cash and a retained production payment of $200,000. The production payment is payable out of the first 70% of the proceeds of the working interest owner's share of production. Satisfaction of the retained production payment is not reasonably assured, but its fair market value is estimated to be $75,000.

Applicable Paragraph: Paragraph No. 47iii

Allocation

Capitalized costs allocated to interest retained:

$$\frac{\$75,000}{\$425,000 + \$75,000} \times \$100,000 = \underline{\$15,000}$$

Capitalized costs allocated to interest sold:

$$\frac{\$425,000}{\$425,000 + \$75,000} \times \$100,000 = \underline{\$85,000}$$

Entry

Cash	425,000	
Production payment interest	15,000	
Net capitalized costs.		100,000
Gain ($425,000 – $85,000).		340,000

SFAS No. 19 does not specifically specify the treatment of a sale of an *unproved* property subject to a retained production payment expressed as a fixed sum of money. However, if the property is unproved, the production payment cannot be reasonably assured. It appears that the appropriate accounting treatment would be to treat the production payment as an ORI, with no gain or loss recognized on the sale unless the selling price exceeds the carrying value of the original interest if assessed individually, or exceeds the original cost if assessed on a group basis.

Retained production payments payable in production. A retained production payment payable by delivery of a specified quantity of petroleum products should be accounted for as a sale. The unamortized cost of the property should be allocated between the interest sold and the interest retained, based on relative fair market values. A gain or loss is recognized on the interest sold. This type of production payment is an economic interest and is therefore subject to DD&A. According to *SFAS No. 19*, par. 47m:

> *The sale of a proved property subject to a retained production payment that is expressed as a right to a specified quantity of oil or gas out of a specified share of future production shall be accounted for in accordance with paragraph 47(k).*

EXAMPLE

Retained Production Payment Payable in Product

Tyler Oil Company owns a working interest in a proved property with the following capitalized costs:

Leasehold	$160,000
IDC	400,000
Equipment	140,000
Total accumulated DD&A	(100,000)
Total book value	$600,000

Tyler Oil Company sells the working interest to Stephens Company for $800,000 and retains a production payment of 20,000 barrels of oil payable out of 40% of the working interest owner's share of production. The fair market value of the entire original working interest was $1,000,000.

Applicable Paragraph: Paragraph No. 47m

Allocation

Cost allocated to interest sold:

$$\frac{\$800,000}{\$800,000 + \$200,000^*} \times \$600,000 = \underline{\$480,000}$$

*$1,000,000 − $800,000

Cost allocated to interest retained:

$$\frac{\$200,000}{\$800,000 + \$200,000} \times \$600,000 = \underline{\$120,000}$$

Entry—Tyler Oil Company

Cash	800,000	
Production payment interest	120,000	
Accumulated DD&A	100,000	
Proved property		160,000
Wells and equipment—IDC		400,000
Wells and equipment—L&WE		140,000
Gain ($800,000 − $480,000)		320,000

Carved-out production payments

In a carved-out production payment, the working interest owner keeps its working interest in the property and carves out or creates a production payment in exchange for cash. The buyer of the production payment interest agrees to receive either money or a stated amount of product in the future. The accounting treatment for carved-out production payments depends on whether the payment is stated in money or in product.

Carved-out production payments payable in money. Carved-out production payments payable in terms of money are generally designed as a means of financing. They may take many forms, including a production loan or an exploration advance.

A production loan may be negotiated between an oil and gas company and a lending institution. The lending institution loans the company money, and the company agrees to repay the loan from the proceeds from the production from a particular property. However, if the production from the property is insufficient to liquidate the loan, then the company must repay the loan with money from other sources. This type of transaction should be accounted for as a payable from the perspective of the oil and gas company and a receivable from the perspective of the lending institution. There is no gain or loss recognition and no revenue effect recognized on the books of the oil and gas company.

Occasionally a pipeline company or utility company may advance funds to an oil or gas company in exchange for the right to purchase all or part of the oil or gas produced from a property. In this case, the receipt of cash is recorded as a payable on the books of the oil and gas company. As the product is produced and delivered to the production payment owner, revenue is recognized and the liability is liquidated. According to *SFAS No. 19*, par. 43a:

> *Enterprises seeking supplies of oil or gas sometimes make cash advances to operators to finance exploration in return for the right to purchase oil or gas discovered. Funds advanced for exploration that are repayable by offset against purchases of oil or gas discovered, or in cash if insufficient oil or gas is produced by a specified date, shall be accounted for as a receivable by the lender and as a payable by the operator.*

The next example illustrates money advanced to the operator in return for the right to purchase oil or gas. This situation, although sometimes referred to as a carved-out production payment is, in fact, a borrowing.

EXAMPLE

Production Advance

Tyler Oil Company receives $100,000 from Myers Company in return for the right to purchase oil and gas in the future.

Applicable Paragraph: Paragraph No. 47a

Entries

Tyler Oil Company:

Cash	100,000	
Production payment payable		100,000

Myers Company:

Production payment receivable	100,000	
Cash		100,000

Another variation of a carved-out production payment payable in money occurs when an oil and gas company carves out and sells a production payment interest in a property in order to obtain financing from a lender. Unlike the first two cases, here the financing is repayable *exclusively* from the property from which it was carved out. In other words, the borrowing is without recourse to the company's other assets. The production payment is to be repaid if, when, and as production occurs. Clearly, the intent of the oil and gas company is to obtain financing—not to sell production. In addition, the lender is clearly making a loan rather than purchasing an interest in oil and gas reserves. Accordingly, the transaction is to be accounted for as a loan from both the perspective of the oil and gas company and the lender. In all of these cases, the oil and gas company includes the reserves necessary to liquidate the production loan or exploration advance among their reserves for purposes of disclosure and for computing amortization. In other words, the production payment interest is not accounted for as a mineral interest. According to *SFAS No. 19*, par. 43b:

> *Funds advanced to an operator that are repayable in cash out of the proceeds from a specified share of future production of a producing property, until the amount advanced plus interest at a specified or determinable rate is paid in full, shall be accounted for as a borrowing. The advance is a payable for the recipient of the cash and a receivable for the party making the advance. Such transactions, as well as those described in paragraph 47(a) below, are commonly referred to as production payments. The two types differ in substance, however, as explained in paragraph 47(a).*

EXAMPLE

Carved-Out Production Payment Payable in Money

Tyler Oil Company owns 100% of the working interest in a fully developed lease that has a 1/8 royalty interest. The lease has the following capitalized costs and reserve data as of January 1, 2011:

Unrecovered costs	$400,000
Proved developed reserves, gross	800,000 bbl

On January 1, 2011, Tyler Oil Company carves out a $900,000 production payment to ABC Bank. The production payment is payable to ABC Bank out of 20% of the proceeds of Tyler's share of production, with interest of 10% on the unpaid balance. During 2011, gross production totaled 12,400 barrels of oil, production costs were $25/bbl, and the selling price was $80/bbl. Ignore production taxes.

Applicable Paragraph: Paragraph No. 43b

Entries—Tyler Oil Company

Cash	900,000	
Production payment payable		900,000
(to record production payment)		
Cash (12,400 × $80)	992,000	
Royalty payable (12,400 × 1/8 × $80)		124,000
Oil revenue (12,400 × 7/8 × $80)		868,000
(to record revenue)		
Operating expense ($25 × 12,400)	310,000	
Cash		310,000
(to record production costs)		
Production payment payable ($173,600 – $90,000) . . .	83,600	
Interest expense (10% × $900,000)	90,000	
Cash ($868,000 × 0.20)		173,600
(to record partial payment of production payment)		
DD&A expense	6,200	
Accumulated DD&A		6,200
(to record DD&A)		

Allocation:

Tyler's share of proved developed reserves: 800,000 bbl × 7/8 = <u>700,000 bbl</u>

Tyler's share of production: 12,400 bbl × 7/8 = <u>10,850 bbl</u>

$$\text{DD\&A expense} = \frac{\$400,000}{700,000 \text{ bbl}} \times 10,850 \text{ bbl} = \underline{\$6,200}$$

Entries—ABC Bank

Production payment receivable	900,000	
Cash		900,000
(to record production payment)		
Cash	173,600	
Production payment receivable		83,600
Interest revenue		90,000
(to record partial collection of production payment)		

These provisions from *SFAS No. 19* are consistent with FASB Emerging Issues Task Force (EITF) *Issue No. 88-18*, "Sales of Future Revenues."[1] EITF *Issue No. 88-18* addresses situations involving advances for future production and concludes that financing arrangements that are, in essence, financing arrangements should be accounted for as debt. According to EITF *Issue No. 88-18*, each transaction is to be carefully analyzed:[2]

> *Classification as debt or deferred income depends on the specific facts and circumstances of the transaction. The Task Force also reached a consensus that the presence of any one of the following factors independently creates a rebuttable presumption that classification of the proceeds as debt is appropriate:*
>
> 1. *The transaction does not purport to be a sale (that is, the form of the transaction is debt).*
>
> 2. *The enterprise has significant continuing involvement in the generation of the cash flows due to the investor (for example, active involvement in the generation of the operating revenues of a product line, subsidiary, or business segment).*
>
> 3. *The transaction is cancellable by either the enterprise or the investor through payment of a lump sum or other transfer of assets by the enterprise.*
>
> 4. *The investor's rate of return is implicitly or explicitly limited by the terms of the transaction.*
>
> 5. *Variations in the enterprise's revenue or income underlying the transaction have only trifling impact on the investor's rate of return.*
>
> 6. *The investor has any recourse to the enterprise relating to the payments due the investor.*

Carved-out production payments payable in product or volumetric production payment (VPP). A carved-out production payment payable by a specified quantity of oil or gas out of future production is, in essence, a sale of reserves in place. If the production from the specified reserves in place is inadequate, the producer has no obligation to make up the shortfall. The seller has a substantial obligation for future performance, i.e., to produce the oil or gas and transfer a portion of it to the buyer of the production payment. As a result, no gain or loss should be recognized at the time of the conveyance. In this situation, the seller records the funds received not as income, but as unearned revenue. As production occurs and the production payment is made, the seller recognizes a portion of the unearned revenue as earned revenue. The buyer or owner of the production payment capitalizes the cost of the production payment at the amount paid and amortizes the capitalized cost as reserves are produced and delivered to the buyer of the production payment. According to *SFAS No. 19*, par. 47a:

> *Some production payments differ from those described in paragraph 43(b) in that the seller's obligation is not expressed in monetary terms but as an obligation to deliver, free and clear of all expenses associated with operation of the property, a specified quantity of oil or gas to the purchaser out of a specified share of future production. Such a transaction is a sale of a mineral interest for which gain shall not be recognized because the seller has a substantial obligation for future performance. The seller shall account for the funds received as unearned revenue to be recognized as the oil or gas is delivered. The purchaser of such a production payment has acquired an interest in a mineral property that shall be recorded at cost and amortized by the unit-of-production method as delivery takes place. The related reserve estimates and production data shall be reported as those of the purchaser of the production payment and not of the seller (paragraphs 50–56).*

Par. 47a has been the source of debate since *SFAS No. 19* was issued. The FASB clearly indicates that the transaction is to be recorded by the seller as unearned revenue. Revenue is to be recognized as the oil or gas is produced and the production payment delivered to the buyer. Related reserves are to be reported by the buyer of the production payment interest. Applied literally, this treatment results in a mismatching of expenses and revenues for the seller. As the production payment is delivered and the unearned revenue is recognized as revenue, there is no matching of DD&A expense against the production payment revenue. This results from the fact that the capitalized cost of the property from which the production payment is carved out is not changed when the production payment is carved out. But as production occurs, the capitalized cost is amortized over the proved developed reserves *minus* the production payment reserves. Thus, there is no matching of DD&A expense related to the reserves that are carved out and then subsequently recognized as revenue as production occurs.

One possible solution, although not explicitly allowed for in *SFAS No. 19*, is to divide the capitalized costs between the working interest retained and the production payment carved out. (Note that the use of this approach does not change total capitalized costs, it simply separates the costs into different accounts.) Then, as the production payment is paid and the unearned revenue is recognized as revenue, the capitalized costs allocated to the carved-out production payment are amortized and matched with revenue. The other capitalized costs for the property are amortized over the non–production payment reserves. Remember, *SFAS No. 19* states that the reserves related to the production payment must be reported by the buyer of the payment and not by the seller.

The owner or buyer of the carved-out production payment capitalizes the cost when the interest is acquired. The production payment is an economic interest subject to DD&A, and the company amortizes the production payment interest as the barrels are produced and delivered. This process is illustrated below.

EXAMPLE

Carved-Out Production Payment Payable in Product

Tyler Oil Company owns 100% of the working interest in a fully developed lease with a 1/8 royalty interest. The lease has the following capitalized costs and reserve data as of January 1, 2014.

Unrecovered costs	$4,000,000
Proved developed reserves, gross	800,000 bbl

On January 1, 2014, Tyler Oil Company carves out a $1,800,000 production payment (PPI) to Alpha Company. The production payment is payable to Alpha Company by delivery of 30,000 barrels out of the first 25% of Tyler's share of production. Tyler has no obligation to make up any shortfall if quantities of production are not sufficient to satisfy the production payment interest. During 2014, gross production totaled 16,000 barrels of oil, production costs totaled $250,000, and the average selling price was $60/bbl. Ignore production taxes.

Applicable Paragraph: Paragraph No. 47a

Allocations

If the fair market value of the 30,000 barrels carved-out production payment is $1,800,000 ($60/bbl), then the fair market value of the retained interest in the proved developed reserves can be approximated at $40,200,000 [(800,000 × 7/8 – 30,000) × $60].

The capitalized costs could be allocated to the interest carved out and the interest retained as follows:

$$\frac{\$1,800,000}{\$40,200,000 + \$1,800,000} \times \$4,000,000 = \underline{\$171,429} \text{ to carved-out production payment}$$

$$\frac{\$40,200,000}{\$40,200,000 + \$1,800,000} \times \$4,000,000 = \underline{\$3,828,571} \text{ to retained working interest}$$

Entries—Tyler Oil Company

Cash	1,800,000	
Unearned revenue		1,800,000
(to record production payment)		

Carved-out production payment	171,429	
Capitalized costs.		171,429
(to record allocation of capitalized costs to carved-out production payment)		

Cash (16,000 × 1/8 × $60 + 16,000 × 7/8 × 0.75 × $60)*	750,000	
Royalty payable (16,000 × 1/8 × $60)		120,000
Oil revenue (16,000 × 7/8 × 0.75 × $60)		630,000
(to record sale of production)		

* 3,500 barrels (i.e., 16,000 barrels × 7/8 × 0.25) were delivered to Alpha and therefore were not sold by Tyler.

Unearned revenue (16,000 × 7/8 × 0.25 × $60).	210,000	
Oil revenue		210,000
(to record delivery of the PPI's share of production at $60/bbl)		

Operating expense	250,000	
Cash		250,000
(to record production costs)		

DD&A expense	60,000	
Accumulated DD&A		60,000
(to record DD&A on capitalized costs retained)		

$$\frac{\$4{,}000{,}000 - \$171{,}429}{(800{,}000 \times 7/8) - 30{,}000 \text{ bbl}} \times (16{,}000 \times 7/8 \times 0.75) = \underline{\$60{,}000}$$

DD&A expense	20,000	
Carved-out production payment		20,000
(to record DD&A on costs allocated to carved-out production payment)		

$$\frac{\$171{,}429}{30{,}000 \text{ bbl}} \times (16{,}000 \times 7/8 \times 0.25) = \underline{\$20{,}000}$$

Note the 30,000 barrels of proved developed reserves related to the production payment would be reported by Alpha Company and *not* by Tyler Oil Company. If Tyler Oil Company does not transfer $171,429 to a carved-out production payment account, then that amount would remain with the other capitalized costs and would be amortized over Tyler's share of proved developed reserves minus the 30,000 barrels related to the production payment.

Entries—Alpha Company

Production payment interest	1,800,000	
Cash		1,800,000
(to record production payment)		
Cash	210,000	
Oil revenue (16,000 × 7/8 × 0.25 × $60)		210,000
(to record income for 3,500 barrels received and sold at $60)		
DD&A expense	210,000	
Production payment interest		210,000
(to record DD&A on production payment interest)		

$$\frac{\$1,800,000}{30,000 \text{ bbl}} \times 3{,}500 \text{ bbl} = \underline{\$210{,}000}$$

Another form of volumetric production payment occurs when the repayment is to be made in product but is not limited to specific production and/or reserves in place. This type of arrangement is not treated as a sale of a mineral interest, but rather a prepaid sales transaction. Here the proceeds are accounted for as deferred revenue, but the producer retains the reserves and recognizes sales revenues as production occurs.

The production payments described above are classic or traditional-type volumetric production payments. These arrangements became popular in the 1970s and 1980s when the buyers of volumetric production payments were interested in obtaining a guaranteed supply of oil or natural gas to be used in the course of normal business operations. For example, in an environment with high demand and relatively low supply, a refiner might "buy" a volumetric production payment from an E&P company so as to assure uninterrupted supply of crude oil into its refinery.

Today, these types of production payments are again popular. However, they are becoming increasingly complex and require special accounting analysis. Some volumetric production payments are purely financial-type transactions. Here the producer or "seller" receives an advance cash payment related to future production; however, the buyer or "counterparty" is not interested in receiving physical quantities of oil and natural gas. Rather, they are engaging in a purely financial arrangement. These transactions must be carefully analyzed by both the buyer and the seller, since all, or a portion, of the transaction may constitute an embedded derivative as defined by SFAS No. 133, "Accounting for Derivative Instruments and Hedging Activities."3

A derivative is a financial instrument that derives its value from the value of some other financial instrument or variable. For example, a stock option is a derivative because it derives its value from the value of the underlying stock. Companies often use derivatives to manage their risk, since they can use the cash flows or fair values from the instruments to offset the risks related to changes in cash flows or fair values of specific "at-risk" assets. In SFAS No. 133, the FASB concluded that derivative financial instruments represent assets and liabilities and, as such, are to be recognized on the balance sheet at fair value.

Paragraph 12 of SFAS No. 133 indicates that although a financial instrument may not be a derivative instrument, it may be necessary to carefully analyze the transaction in order to determine whether it involves an "embedded derivative." Some volumetric production payments are related to hybrid instruments composed of both a host debt instrument along with a commodity forward hedge. Recently, the Derivatives Implementation Group (DIG) issued DIG Implementation Issue No. B11. It indicates that carved-out volumetric production payments for which the quantity of the oil or natural gas to be delivered is reliably determinable are to be accounted for as derivatives.[4] These arrangements must be bifurcated or divided into two separate components, and the volumetric production payment accounted for in accordance with the provisions of SFAS No. 133.

According to DIG Implementation Issue No. B11, carved-out volumetric production payments payable out of specific reserves in place (but where, if the production from the specified reserves in place is inadequate, the producer has no obligation to make up the shortfall) involve situations where the quantity of production is not reliably determinable. Accordingly, these arrangements should continue to be accounted for in accordance with SFAS No. 19, par. 47a (as described above). Additionally, companies may rely on the provisions of SFAS No. 133 par. 10b wherein a volumetric production payment that constitutes a part of a normal purchase or normal sale is exempted from SFAS No. 133. As a result, it should be accounted for in accordance with SFAS No. 19.

CONVEYANCES—FULL COST

As indicated earlier, most conveyances are accounted for in the same way under full cost as under successful efforts. The differences are discussed in the following paragraphs.

The full cost accounting method provides for cost accumulation on a country-by-country basis. As discussed in an earlier chapter, acquisition, exploration, and development costs are capitalized in a full-cost cost pool (a country). Therefore, the individual costs theoretically lose their particular lease, field, or reservoir character. Sales and abandonments of individual properties should generally give rise only to adjustments in the cost pool. In general, no gains or losses should be recognized under full cost accounting for sales of oil and gas properties unless such adjustments to capitalized costs arising from these transactions would materially distort the amortization rate. In *Reg. S-X 4-10*, the SEC states a material distortion will generally not occur unless 25% or more of the reserves are sold. According to Reg. S-X 4-10 (i)(6)(i):

> *(i) Sales and abandonments of oil and gas properties. Sales of oil and gas properties, whether or not being amortized currently, shall be accounted for as adjustments of capitalized costs, with no gain or loss recognized, unless such adjustments would significantly alter the relationship between capitalized costs and proved reserves of oil and gas attributable to a cost center. For instance, a significant alteration would not ordinarily be expected to occur for sales involving less than 25% of the reserve quantities of a given cost center. If gain or loss is recognized on such a sale, total capitalized costs within the cost center shall be allocated between the reserves sold and reserves retained on the same basis used to compute amortization, unless there are substantial economic differences between the properties sold and those retained, in which case capitalized costs shall be allocated on the basis of the relative fair values of the properties. Abandonments of oil and gas properties shall be accounted for as adjustments of capitalized costs, that is, the cost of abandoned properties shall be charged to the full cost center and amortized (subject to the limitation on capitalized costs in paragraph (b) of this section).*

The following example illustrates the sale and abandonment of properties when the amortization rate is not materially distorted.

EXAMPLE

Sale and Abandonment—Unproved and Proved Properties

On December 31, 2012, Tyler Company had total capitalized costs of $6,000,000 in the U.S. cost pool and accumulated DD&A of $2,000,000. During January 2013, the following transactions occurred:

Unproved property abandoned:	
Original cost	$ 60,000
Proved property abandoned:	
Leasehold cost	30,000
Equipment	300,000
IDC	400,000
Salvage value of equipment	40,000
Proved property sold:	
Selling price	100,000
Leasehold cost	20,000
Equipment	150,000
IDC	200,000

Entry

Cash	100,000	
Surplus stock	40,000	
Accumulated DD&A (plug)	1,020,000	
Unproved properties.		60,000
Proved properties, leasehold.		50,000
Wells and equipment—L&WE		450,000
Wells and equipment—IDC.		600,000

Theoretically, under full cost accounting, a write-off of capitalized amounts is not necessary. However, most companies would remove the costs that apply to properties sold or abandoned as shown in the preceding example. If the individual costs are not removed, then the proceeds received from the sale of the proved property and the items salvaged during abandonment would be debited to the cash account and the surplus stock account and credited to either the accumulated DD&A account or the cost pool account.

Sometimes a sale is great enough to cause a significant alteration in the relationship between capitalized costs and proved reserves within a cost center. When this happens, a gain or loss should be recognized as shown in the following example.

EXAMPLE

Sale of Significant Reserves: DD&A Rate Distortion

Tyler Company had total capitalized costs in Romania of $25,000,000 and accumulated DD&A of $10,000,000. Reserves in Romania amounted to 600,000 barrels. Two hundred thousand (200,000) barrels of reserves were sold for $12,000,000. The selling price per barrel was $60/bbl ($12,000,000/200,000) versus an amortization rate of $25/bbl [($25,000,000 –$10,000,000)/600,000]. Further, 1/3—more than 25% of the reserves—has been sold. Consequently, a significant change has occurred between capitalized costs and proved reserves, and a gain or loss should be recognized. Since 1/3 of the reserves has been sold, 1/3 of the capitalized costs should be allocated to the reserves sold.

Entry

Cash	12,000,000	
Accumulated DD&A (1/3 × $10,000,000)	3,333,333	
Proved properties (1/3 × $25,000,000)		8,333,333
Gain on sale of proved properties		7,000,000

The gain was $35/bbl ($60 – $25) on the 200,000 barrels sold for a total gain of $7,000,000.

Another problem in full cost accounting for sales and abandonments concerns transactions involving unproved properties that were acquired for resale or promotion purposes. *Reg. S-X 4-10* (i)(6)(iii)(A) requires the following treatment:

> Except as provided in subparagraph (i)(6)(i) of this section, all consideration received from sales or transfers of properties in connection with partnerships, joint venture operations, or various other forms of drilling arrangements involving oil and gas exploration and development activities (e.g., carried interest, turnkey wells, management fees, etc.) shall be credited to be the full cost account, except to the extent of amounts that represent reimbursement of organization, offering, general and administrative expenses, etc., that are identifiable with the transaction, if such amounts are currently incurred and charged to expense.

Under this rule, sales, abandonments, and drilling arrangements should, in almost all circumstances, *not* give rise to recognition of a gain or loss. Even if the unproved property is acquired for the purpose of selling or transferring to a drilling fund, no gain or loss should generally be recognized. The only entry is an adjustment to the full cost pool. These rules also severely limit the recognition of income from management fees or service activities where an interest is held in the property. (For more information, see appendix B.)

In comparison of the treatment of conveyances under successful efforts versus full cost, any conveyance that gives rise to a gain or loss under the successful efforts method generally is treated as an adjustment to the cost pool under the full cost method. All other conveyances are usually handled the same under full cost as under successful efforts.

PROBLEMS

Assume all production and reserves are gross barrels or Mcfs.

1. What is the major purpose of unitization?

2. Distinguish between a carried interest situation, a free-well transaction, and a farm-in/farm-out transaction.

3. Identify the types of interests that are created in the following situations. If the interest is an overriding royalty or a production payment interest, also state whether it is a retained or carved-out interest.

    a. Zeke Company owns the working interest in a proved property with net capitalized costs of $100,000. Zeke sold the lease for $250,000 cash and a payment of $150,000, plus interest of 10% to be paid out of the first 60% of the oil produced.

    b. Wildcat Oil Company acquired an undeveloped lease for which it paid $30,000. Financially unable to develop the lease, Wildcat agreed to allow Friendly Company to earn a 30% working interest by paying 100% of the cost of drilling and completing a well.

    c. Young Oil Company owns a 100% WI in Lease A, which has a 1/8 royalty interest. On February 1, 2013, Young carved out a $400,000 payment, payable out of 60% of the net proceeds of the working interest's share of production, plus interest of 10% on the unpaid balance.

d. Four companies own adjacent leases that share a common reservoir. The companies decide to operate the properties as one in order to obtain improved operating efficiency. Following negotiations by engineers, geologists, and others, the companies agree upon participation factors and market values of contributed IDC and equipment.

e. Mabel Oil Company acquired an unproved property at a cost of $50,000. Mabel later sold the working interest and kept a nonworking interest. As a result, Mabel will receive 1/16th of the revenue of the working interest from which the interest was created.

f. Company ABC assigned a 40% WI in an unproved property to Company XYZ in return for Company XYZ bearing all costs of drilling, developing, and operating the property. Company XYZ is entitled to all of the revenue from production (net of royalty) until Company XYZ has recovered all of its costs, at which time the property becomes a joint working interest.

4. The following transactions occurred during 2018:

   a. Joyner Oil Company and Brown Oil Company jointly purchased a 2,000 acre lease in Oklahoma for $60,000. Joyner has a 60% WI, and Brown, a 40% WI. Joyner will be the operator of the lease.

   b. Rayburn Oil Company owns 100% WI in a lease (with a 1/6 royalty interest) with capitalized costs of $600,000. Rayburn assigns the working interest to Fugate Oil for $800,000 and keeps a 15% overriding royalty interest.

   c. Rayburn Oil Company owns 100% WI in a lease (with a 1/6 royalty interest) in Oklahoma and transfers a 1/16 overriding royalty interest to the controller of the company.

   d. Sells Oil Company owns the working interest in a lease in Nueces County, Texas and assigns 40% of the working interest to the Knight Oil Company in return for Knight drilling and equipping a well on the property. After the well is completed, Sells and Knight will share revenues and costs.

   e. Knight Oil Company assigns the working interest in Lease A to Sells Oil Company for $700,000, and in return, Knight will receive 20,000 barrels of oil from the first 30% of Sells' production.

   f. Sells Oil Company sells a production payment interest of 20,000 barrels of oil for $300,000, which is to be paid from the first 30% of production.

   g. Cantu Oil Company owns a working interest in Harris County, Texas. Cantu transfers 60% of the working interest to Stephens Oil Company in exchange for Stephens bearing all costs of drilling, completing, and operating the property until payout. Stephens Oil Company will receive 100% of the working interest's share of revenue until drilling and developing costs have been recovered, at which time the property will be operated as a joint property.

   REQUIRED: Identify the following types of interests created by the 2018 transactions. If the interest is an overriding royalty or a production payment interest, also state whether it is a retained or carved-out interest.

5. Hays, Bush, and King signed a lease agreement with Big Pink, the owner of the mineral rights. Big Pink received a 1/7 royalty interest. The companies' working interests are 50%, 30%, and 20%, respectively. The companies signed an operating agreement designating Hays as the operator of the lease. Assuming revenues of $42,000 and costs of $14,000 the first year of operations, determine how much each party will receive in revenue and pay in costs the first year of operations.

6. Yale Oil Company owns the working interest in a small lease in Louisiana that has a 1/6 royalty interest. The royalty interest owner is Mr. Bell. Yale also owns the working interest in numerous leases in Texas. Not having the facilities in Louisiana to develop the Louisiana lease, Yale assigns his entire working interest in the lease to Smith for a consideration of $300,000, and retains a 1/15 overriding royalty. Assuming revenues of $42,000 and costs of $12,000 the first year of operations, determine how much each economic interest owner will receive in revenue and pay in costs the first year of operations.

7. Mabel Oil Company owns 100% of the working interest in a lease that has a 1/7 royalty interest. The royalty interest owner is Mr. Kyle. Needing additional funds to develop the property, Mabel sold Pitt Company 60,000 barrels of oil for a consideration of $800,000. The oil is to be paid out of the first 20% of the working interest's share of production. During the first four years of production, 140,000 barrels are produced each year. How many barrels of oil does each interest receive in each of the first four years?

8. Wildcat Oil Company acquired an undeveloped lease for which it paid $30,000. The lease is burdened with a 1/8 royalty. Financially unable to develop the lease, Wildcat *sold* 60% of its working interest to two parties for $200,000 ($100,000 each), agreeing to use the money to drill and equip a well. When the well is completed, each of the three companies will share future development and operating costs. The well cost $200,000 and was successful. Estimated proved reserves were 125,000 barrels, and proved developed reserves were 87,500 barrels (12/31). Wildcat is the operator, and 2,500 barrels were produced and sold in the first year of operations. The selling price was $80/bbl, operating costs were $20/bbl, and the severance tax rate was 5%. The purchaser assumed the responsibility of paying severance taxes and the royalty interest owner.

   a. Determine how much revenue and operating costs each party should record for the first year of operations assuming successful efforts companies.

   b. Give all entries necessary for Wildcat Oil Company and the buyers.

   c. Give the entries assuming Wildcat Oil Company is a full cost company.

9. Lomax Company assigned 40% of the working interest in an unproved property with a 1/8 royalty interest to Mabel Company in return for Mabel Company bearing all costs of drilling, developing, and operating the property (until payout). Mabel Company is entitled to all of the revenue from production until Mabel Company has recovered all of its costs, at which time the property becomes a joint working interest. Acquisition costs of the property incurred by Lomax Company were $80,000. Mabel Company incurs $450,000 in IDC and equipment costs in drilling the well. The well is successful. Production begins early in the second year, and 9,000 barrels are produced during each of the first four years of production. The selling price was $80/bbl, and operating costs were $20/bbl. Determine how much revenue and operating costs each part should record for the first three years of operations.

10. Company Z (a successful efforts company) owns 100% WI in a lease in which Dudley Smith owns a 1/8 royalty interest. Company Z conveys to Company Q 30% of the working interest in exchange for a cash consideration of $10,000. The undeveloped lease has capitalized costs of $40,000 and an impairment allowance of $15,000.

    a. Give the entry for Company Z to record the conveyance.
    b. Give the entry if the interest had sold for $30,000.

11. Company Z owns a 100% WI in a proved property with net capitalized costs of $100,000. Company Z sold the lease for $250,000 cash and a production payment of $150,000, plus interest of 10% to be paid out of the proceeds from the first 60% of the oil produced. Satisfaction of the retained production payment is not reasonably assured. The fair market value of the production payment interest is estimated to be $50,000. Give the entry to record the conveyance assuming that Company Z uses the successful efforts method.

12. Philco Company owns a proved property with the following costs:

Leasehold	$ 80,000
IDC	500,000
Equipment	120,000
Total accumulated DD&A (separate amortization base)	250,000

    Philco Company sells 100% of the working interest in the property to Company Q for $600,000.

    a. Give the entry for Philco Company to record the sale assuming that Philco uses the successful efforts method.
    b. Give the entry for Philco Company to record the sale if the property had sold for $400,000.
    c. Give the entry for part a and part b, assuming Philco Company is a full cost company, and the reserves sold constitute 15% of Philco's share of reserves in the cost center.

13. Wildcat Oil Company leased undeveloped acreage from David Jones for $30,000, with Jones receiving a 1/8 royalty interest. Financially unable to develop the lease, Wildcat enters into a farm-in/farm-out agreement with Jayhawk Company. Jayhawk agrees to drill and complete a well in return for 60% of the working interest and the right to recover all of its costs. Jayhawk drills and completes the well for $100,000. Estimated proved reserves are 125,000 barrels, and proved developed reserves are 25,000 barrels. Jayhawk is the operator, and production totals 1,000 bbl/month for the first six months. Assume that the average selling price is $80/bbl, and lifting costs average $20/bbl. Ignore severance taxes and assume reserve estimates do not change. Jayhawk assumes the responsibility of paying the royalty interest.

    a. Calculate payout.
    b. Assuming that both companies are successful efforts companies, give all of the entries, including monthly DD&A expense for the first three months, that would be made by Wildcat Oil Company and by Jayhawk Company.

14. Bingo Oil Company owns 100% of the working interest in a fully developed lease on which there is a 1/8 royalty interest. The lease has the following capitalized costs and reserve data as of January 1, 2015:

Unrecovered costs	$600,000
Proved developed reserves.	200,000 bbl

On January 1, 2015, Bingo Oil Company carves out a $500,000 production payment to Capital Bank. The production payment is payable to Capital Bank out of 60% of the proceeds of the Bingo's share of production with interest of 10% on the unpaid balance. During 2015, production totaled 5,000 barrels of oil, production costs were $20/bbl, and the selling price was $80/bbl. Ignore production taxes and assume Bingo pays the royalty interest owner.

   a. Give all the entries made by Bingo Oil Company (a successful efforts company) relating to the above lease and to account for the carved-out production payment during 2015.

   b. Give all the entries made by Capital Bank to account for the production payment during 2015.

15. Zink Company owns 100% of the working interest in a fully developed lease on which there is a 1/8 royalty interest. The lease has the following capitalized costs and reserve data as of January 1, 2017:

Unrecovered costs	$500,000
Proved developed reserves.	125,000 bbl

On January 1, 2017, Zink Company carves out a $400,000 production payment to Delta Company. The production payment is payable to Delta Company by delivery of 6,250 barrels out of the first 50% of Zink's share of production. During 2017, production totaled 4,000 barrels of oil, production costs totaled $50,000, and the average selling price was $80/bbl. Ignore production taxes and assume Zink pays the royalty interest owner.

   a. Give all the entries made by Zink Company (a successful efforts company) relating to the above lease and to account for the carved-out production payment during 2017.

   b. Give all the entries made by Delta Company (a successful efforts company) to account for the production payment during 2017.

16. In 2016, Beta Company purchased the working interest of an unproved lease for $50,000. In 2017, Beta Company recognized impairment of $20,000 on this lease. In 2018, Beta Company sold the entire working interest to Company Q for:

   a. $25,000
   b. $55,000

   Assume instead that the property was assessed on a group basis and sold for:

   c. $25,000
   d. $55,000

Give the entry to record the sale in each of the above situations, assuming that Beta Company is a successful efforts company.

17. Mair Company sold 50% of the working interest in a proved lease to Company Q for $450,000. Mair Company's net cost basis in this proved property was $200,000. Mair Company uses the successful efforts method. Give the entry for Mair Company to record the sale.

18. Four companies own adjacent leases that share a common reservoir. The companies each have a 100% working interest in their respective leases in which they have the following investment:

Company	IDC	L&WE	Leasehold	Total
A	$200,000	$ 80,000	$ 30,000	$ 310,000
B	100,000	60,000	20,000	180,000
C	300,000	150,000	40,000	490,000
D	0	0	50,000	50,000
	$600,000	$290,000	$140,000	$1,030,000

The companies decide to unitize in order to obtain improved operating efficiency. Following negotiations by engineers, geologists, and others, the following participation factors and market values were agreed upon:

Company	IDC	L&WE	Participation Total	Factor
A	$230,000	$100,000	$330,000	10%
B	140,000	50,000	190,000	40%
C	260,000	180,000	440,000	20%
D	0	0	0	30%
	$630,000	$330,000	$960,000	100%

a. Determine equalization of investment and prepare entries for all the parties, assuming all of the companies are successful efforts companies.

b. Give the entries, assuming Company D is a full cost company.

19. Zepher Company acquired 100% of the working interest in an unproved property at a cost of $50,000. Zepher later sold the working interest, retaining an overriding royalty interest (ORI). Give the entry to record the conveyance of the working interest, assuming Zepher is a successful efforts company and received $30,000.

20. Brown Company, a successful efforts company, has a 1/8 royalty interest in an unproved property. Assuming Brown Company's net capitalized cost in the property is $100,000, give the entry to record the sale of the entire royalty interest for:

   a. $80,000
   b. $110,000

21. Greene Oil Company, a successful efforts company, owns 100% of the working interest in a 320 acre proved property with the following net, unamortized costs:

Leasehold	$ 60,000
IDC	200,000
Equipment	88,000

Greene Oil Company sells 100% of the working interest, including the wells and equipment, on the western 160 acres of the property for $200,000. An appraisal is performed with the following results:

	FMV of Portion Sold	FMV of Portion Retained	Total FMV
Leasehold	$ 30,000	$ 80,000	$110,000
IDC	130,000	90,000	220,000
Equipment	40,000	10,000	50,000
Total	$200,000	$180,000	

Give the entry to record the sale.

22. Wolfforth Company sold its 100% WI in a proved property for $600,000 and retained an ORI. Wolfforth's net cost basis in the property was $500,000. The fair market value of the entire original working interest was $700,000.

Wolfforth had the following net investment in its accounts:

Leasehold.	$100,000
IDC.	300,000
Equipment	100,000
Total	$500,000

Prepare Wolfforth Company's entry to record the sale, assuming Wolfforth is a successful efforts company.

23. Flower Company owns a 100% WI in an unproved property for which it paid $80,000. The property is burdened with a 1/8 royalty. Flower Company agrees to farm out the working interest to Barrel Company and retain a 1/7 ORI in return for Barrel Company agreeing to drill, develop, and operate the property. During 2015, Barrel Company incurs costs of $200,000 in drilling a well and $100,000 to equip the well. The total proved reserves are estimated to be 75,000 barrels, and proved developed reserves are 25,000 barrels. Production during 2015 totaled 5,000 barrels, which was sold for $80/bbl. Ignore severance tax and assume Barrel pays the royalty interest and ORI owners.

  a. Give the entries that would be made by Flower Company, assuming it uses the successful efforts method.

  b. Give the entries that would be made by Barrel Company, assuming it uses the successful efforts method.

24. French Company signed a lease agreement with Rita Mack covering 900 acres in Oklahoma. Ms. Mack received a bonus of $50,000 and a 1/5 royalty interest, and French Company received 100% of the working interest.

    On 1/1/2018, French Company enters into an agreement with Donald Oil Company wherein French Company assigns its working interest and retains a 1/4 ORI. Donald agrees to pay all of the cost to drill a well on the property. If the well is successful, Donald Oil Company will pay all of the operating costs and retain the net profit (after payment of the royalty, ORI, and operating expenses) until it has recovered the cost of drilling and completing the well. At that point, French Company's ORI will revert to a 45% WI. Donald pays the royalty interest and ORI owners.

    During November 2018, Donald Oil Company drills Well No. 1 at a cost of $210,000. The well is successful. Estimated proved reserves total 100,000 barrels, and proved developed reserves are 50,000 barrels. During 2019 and 2020, 3,750 bbl/year are produced and sold for $80/bbl. Operating costs are $20/bbl. Ignore severance tax and assume reserve estimates do not change.

    Compute payout. Prepare all of the entries that would be made by both French Company and Donald Oil Company during 2018, 2019, and 2020, assuming both companies are successful efforts companies.

25. Fielder Oil Company, a successful efforts company, has an unproved lease for which it paid $150,000. The property was individually significant, and individual impairment of $50,000 had been assessed. Make the journal entries to record Fielder Oil Company's conveyance of 50% of the working interest in return for:

    a. $10,000 cash

    b. $160,000 cash

26. Bammel Oil Company, a successful efforts company, has an unproved lease for which it paid $50,000. The property is *not* considered to be individually significant. Make the journal entries to record Bammel Oil Company's conveyance of 20% of the working interest in return for:

    a. $10,000 cash

    b. $70,000 cash

27. Gamble Company, a full cost company, has an unproved lease for which it paid $100,000. Give the entry to record the sale of the property, assuming Gamble Company sold the property for:

    a. $80,000

    b. $130,000

    Give the entry to record the sale, assuming Gamble Company sold only 25% of the working interest in the property and received:

    c. $60,000

    d. $110,000

28. Tiger Oil Company, a successful efforts company, owns an ORI in an unproved property that cost $20,000. The ORI has not been impaired. Assume Tiger sold its entire interest in the ORI for the following amounts:

   a. $15,000
   b. $25,000

   Prepare journal entries to record the sales under the different assumptions.

29. Higgins Company receives $200,000 from Garza Company in payment for the right to purchase natural gas in the future.

   Prepare journal entries for each company, assuming that they both use the successful efforts method.

30. Hein Oil Company, a successful efforts company, owns 100% of the working interest in a proved property that has the following capitalized costs:

Proved Property	$240,000
IDC	600,000
Equipment	400,000
Total Accumulated DD&A	200,000

   A 1/8 royalty on the property is owned by Sammy Jones. Hein sells the working interest for $1,200,000 and retains a production payment interest of $300,000. The production payment interest is payable from the proceeds from 30% of the working interest share of production. The fair market value of the entire original working interest is $1,500,000.

   Prepare the journal entry by Hein to record the above transaction, assuming that payment of the production payment interest is not reasonably assured.

31. Carpenter Oil Company owns a 100% WI in a proved property in Wise County, Texas. Carpenter sells the working interest to Knight Oil Company for $400,000, plus a retained production payment interest of $300,000, payable in cash from 60% of the working interest's portion of the revenue. The capitalized cost of the working interest owner's proved property is $800,000, and accumulated DD&A is $200,000. The present value of the production payment interest is $250,000 and is reasonably assured of payout.

   a. Prepare the entry to record the sale of the proved property, assuming that Carpenter uses the successful efforts method of accounting.
   b. Prepare the entry for Carpenter, assuming that the full cost method of accounting is being used.

32. Sells Oil Company, a full cost company, has total capitalized costs in Venezuela of $20,000,000 and accumulated DD&A of $4,000,000. Proved reserves in Venezuela are 2,000,000 barrels. Reserves of 500,000 barrels are sold to Oyona Oil Company for $8,000,000.

   Prepare the entry for Sells Oil Company.

References

1. Financial Accounting Standards Board. 1988. Emerging Issues Task Force (EITF) *Issue No. 88-18.* "Sales of Future Revenues." Stamford, CT: FASB.

2. Ibid. p. 2.

3. Financial Accounting Standards Board. 1998. *Statement of Financial Accounting Standards No. 133.* "Accounting for Derivative Instruments and Hedging Activities." Stamford, CT: FASB.

4. Financial Accounting Standards Board. 2006. *Statement 133 Implementation Issue No. B11.* "Embedded Derivatives: Volumetric Production Payments." Stamford, CT: FASB.

14

OIL AND GAS DISCLOSURES[1]

Statement of Financial Accounting Standards (SFAS) No. 69, "Disclosures about Oil and Gas Producing Activities," requires both successful efforts and full cost companies to prepare a comprehensive set of disclosures dealing with historical-based and future-based information. Companies required to present this information are publicly traded companies with significant oil and gas producing activities that meet one or more criteria. These criteria are specified in *SFAS No. 69*, par. 8, amended by *SFAS No. 131*, "Disclosures about Segments of an Enterprise and Related Information:"[2]

> *For purposes of this Statement, an enterprise is regarded as having significant oil and gas producing activities if it satisfies one or more of the following tests. The tests shall be applied separately for each year for which a complete set of annual financial statements is presented.*
>
> a. *Revenues from oil and gas producing activities, as defined in paragraph 25 (including both sales to unaffiliated customers and sales or transfers to the enterprise's other operations), are 10 percent or more of the combined revenues (sales to unaffiliated customers and sales or transfers to the enterprise's other operations) of all of the enterprise's industry segments.*

b. Results of operations for oil and gas producing activities, excluding the effect of income taxes, are 10 percent or more of the greater of:

   i. The combined operating profit of all industry segments that did not incur an operating loss

   ii. The combined operating loss of all industry segments that did incur an operating loss.

c. The identifiable assets of oil and gas producing activities (tangible and intangible enterprise assets that are used by oil and gas producing activities, including an allocated portion of assets used jointly with other operations) are 10 percent or more of the assets of the enterprise, excluding assets used exclusively for general corporate purposes.

Required Disclosures

SFAS No. 69 requires publicly traded companies with significant oil and gas producing activities to disclose supplementary information in their annual financial statements related to the following items:

Historical-based

1. Proved reserve quantity information
2. Capitalized costs relating to oil and gas producing activities
3. Costs incurred for property acquisition, exploration, and development activities
4. Results of operations for oil and gas producing activities

Future value–based

5. A standardized measure of discounted future net cash flows relating to proved oil and gas reserve quantities
6. Changes in the standardized measure of discounted future net cash flows relating to proved oil and gas reserve quantities

Each of these disclosures must be presented in the aggregate. Disclosures 1, 3, 4, and 5 must also be presented for each geographical area where the company has significant operations. This chapter includes a discussion of the requirements related to each of these disclosures, as well as an illustrative example utilizing data related to Tyler Oil Company. The detailed requirements for each disclosure are discussed below, followed by an example of that specific disclosure utilizing the following information.

Illustrative Example

Tyler Oil Company, a successful efforts company, began operations early in 20XA with the acquisition of three unproved leases: A, B, and C. Each of the leases is burdened with a 1/5 royalty interest. During 20XA, Tyler incurred G&G costs and began drilling an exploratory well on each of the three leases. The well on Lease A was successfully completed, but no oil was produced and sold during 20XA. The well on Lease B was in progress at year-end, and the well on Lease C was a dry hole. No equipment was salvaged.

During 20XB, Tyler successfully completed the well on Lease B and produced and sold oil from both Lease A and Lease B. All production was sold in the year produced. All reserves on Leases A and B were fully developed at the end of 20XB. Tyler plans to further explore Lease C and paid a delay rental on Lease C. The facts representing Tyler Company's first two years of operations, which occurred solely in the United States, follow. All reserve quantities and production data given apply only to Tyler's interest.

Remember that the working interest's revenue is based on the working interest owner's share of total production, in this case gross barrels × 4/5. However, since the working interest pays 100% of the costs, production costs are based on total gross barrels. Thus, if production costs are given as a dollar amount per barrel, total production costs would normally be gross barrels times the production costs. In this chapter, in order to simplify and make more understandable some of the calculations, production costs per working interest barrel are given. Production costs per working interest barrel are calculated by dividing total production costs by the working interest owner's share of the barrels produced.

Also recall that dismantlement, restoration, and abandonment costs, commonly referred to as **decommissioning costs**, must be estimated and recorded as part of the related asset for both successful efforts and full cost companies. In order to facilitate tax calculations, future decommissioning costs are given separately from the related asset cost. Please note that due to the overwhelming number of calculations involved in preparing *SFAS No. 69* disclosures, several simplifying assumptions were made in this example; for example, the omission of decommissioning costs for Lease B and the use of unrealistically short production lives for both Lease A and Lease B.

EXAMPLE

12/31/20XA

	Item	Lease A	Lease B	Lease C
a.	Acquisition costs	$ 30,000	$ 20,000	$ 40,000
b.	G&G costs	$ 50,000	$ 40,000	$ 20,000
c.	Drilling costs:			
	IDC	$200,000	$100,000	$120,000
	Tangible	90,000		10,000
d.	Drilling results	Proved reserves	Incomplete	Dry
e.	Estimated production of estimated proved reserves, bbl	20XB– 7,500 20XC– 10,000 20XD– 12,500		
	Total proved reserves, bbl	30,000		
f.	Reserve estimate, bbl 12/31/XA:			
	Proved, at 12/31 (decreases by estimated production)	20XA– 30,000 20XB– 22,500 20XC– 12,500		
	Proved developed at 12/31/XA (decreases by estimated production and increases as a result of estimated development)	20XA– 13,750 20XB– 15,000 20XC– 12,500		
g.	Estimated tangible development costs	20XB– $45,000 20XC– 25,000		
h.	Estimated decommissioning costs	20XD– $15,000		
i.	Current market price of oil	$88/bbl		
j.	Estimated future production costs based on year-end costs	20XB–$150,000 20XC– 200,000 20XD– 250,000		
k.	Estimated current production costs/bbl	$20/bbl		

	Item	12/31/20XB		
		Lease A	Lease B	Lease C
a.	Drilling costs			
	Tangible		$50,000	
b.	Drilling results		Proved reserves	
c.	Actual and estimated production, bbl (where A = actual; R = revised; E = estimated)	20XB– 8,000 A 20XC– 10,750 R 20XD– 12,000 R	20XB– 8,750 A 20XC– 12,500 E	
d.	Actual and estimated tangible development costs (where A = actual; R = revised; E = estimated)	20XB– $46,000 A 20XC– $0 R	20XB– $50,000 A	
e.	Estimated decommissioning costs	20XD– $15,000		
f.	Market price of oil sold	$100/bbl	$100/bbl	
g.	Current market price of oil at end of year	$104/bbl	$104/bbl	
h.	Production cost of oil sold	$192,000	$210,000	
i.	Production cost per barrel of oil sold	$24/bbl	$24/bbl	
j.	Estimated future production costs at year-end	20XC–$258,000 20XD–$288,000	20XC–$300,000	
k.	Current production cost per barrel at year end	$24/bbl	$24/bbl	
l.	Delay rental			$5,000

Assume a tax rate of 40% and that Tyler does not qualify for percentage depletion because it is an integrated producer. For purposes of the required capitalization and amortization of 30% of IDC, assume nine months of amortization in 20XA. Because of the relatively short lives of the properties, also assume Tyler elects for tax purposes to use the unit-of-production method for calculating depreciation. Tyler uses proved reserves for calculating tax depletion and proved developed reserves for tax depreciation. In the case solution, the alternative minimum tax is ignored. Deferred income taxes are also ignored. Instead, it is assumed that income tax expense is based on pretax financial accounting income multiplied by the tax rate. Further, it is assumed that carryforwards are permitted when it is likely that the carryforward benefit will be realized.

Each mandated disclosure is illustrated below. Disclosures for 20XA and 20XB are presented together for each disclosure. The solution is presented assuming that Tyler is using both of the following:

1. The successful efforts method of accounting, with a property as the cost center
2. The full cost method of accounting, with all possible costs included in DD&A

Proved Reserve Quantity Information

Perhaps the most widely examined *SFAS No. 69* disclosure is the reserve report. A rudimentary understanding of how reserves are estimated and the various categories of reserves is crucial to understanding the financial statements of oil and gas companies. In addition to the provision found in *SFAS No. 69*, the SEC has imposed other requirements on registrant companies. These can be found in *Reg. S-X 4-10* and in various SEC documents, some of which are referenced in the discussion below.

There are two broad categories of reserve estimation methodologies used by engineers and geologists, with both methodologies involving a great deal of uncertainty. These two categories are deterministic versus probabilistic methodologies. A reserve estimation methodology is referred to as **deterministic** if a single best estimate of reserves is made based on known geological, engineering, and economic data. The methodology is referred to as **probabilistic** if known geological, engineering, and economic data are used to generate a range of estimates and their associated probabilities.

The Society of Petroleum Engineers (SPE) and the World Petroleum Congress (WPC) have developed definitions of reserves estimated using these two methodologies. These definitions have been studied by and, to varying degrees, adopted by various accounting boards around the world. Reserves estimated using deterministic methodologies include proved reserves and the two subcategories of proved reserves: proved developed reserves and proved undeveloped reserves. Reserves estimated using probabilistic methodologies include proven and probable reserves and possible reserves. The quantities of reserves estimated using each of these methodologies differ due to the differing assumptions and approaches used in making the estimates. More information regarding these engineering methodologies is available on the Web site of the Society of Petroleum Engineers (www.spe.org) and the World Petroleum Congress (www.world-petoleum.org).

The SPE, WPC, American Association of Petroleum Geologists (AAPG), and Society of Petroleum Evaluation Engineers (SPEE) identified the need to establish a common framework for developing reserve definitions. In March 2007, this group issued the *Petroleum*

Management Resources System and continues to work with the SEC and the International Accounting Standards Board (IASB) in formulating and refining various reserve definitions and reserve reporting guidance. Information on this project is available at www.spe.org.

In November 2000, the SEC addressed a number of issues regarding the reserves being reported by companies in their *SFAS No. 69* reserve report. In *2000 Current Issues and Rulemaking Projects*, par. (16)(i), the SEC communicated their position regarding the use of deterministic versus probabilistic methods in estimating reported reserves. The SEC indicated a preference for the deterministic approach:[3]

> *Probabilistic methods of reserve estimating have become more useful due to improved computing and more important because of its acceptance by professional organizations such as the SPE. The SEC staff feels that it would be premature to issue any confidence criteria at this time. The SPE has specified a 90% confidence level for the determination of proved reserves by probabilistic methods. Yet, many instances of past and current practice in deterministic methodology utilize a median or best estimate for proved reserves. Since the likelihood of a subsequent increase or positive revision to proved reserve estimates should be much greater than the likelihood of a decrease, we see an inconsistency that should be resolved. If probabilistic methods are used, the limiting criteria in the SEC definitions, such as Lowest Known Horizon (LKH), are still in effect and shall be honored. Probabilistic aggregation of proved reserves can result in larger reserve estimates (due to the decrease in uncertainty of recovery) than simple addition would yield. We require a straight forward reconciliation of this for financial reporting purposes.*

Reserve definitions

The only reserves that may be reported under *SFAS No. 69* and *Reg. S-X 4-10* are proved reserves, with proved reserves being further classified as being developed or undeveloped. These are defined as follows in SEC *Reg. S-X 4-10*:

> **Proved reserves.** *Proved oil and gas reserves are the estimated quantities of crude oil, natural gas, and natural gas liquids which geological and engineering data demonstrate with reasonable certainty to be recoverable in future years from known reservoirs under existing economic and operating conditions, i.e., prices and costs as of the date the estimate is made. Prices include consideration of changes in existing prices provided only by contractual arrangements, but not on escalations based upon future conditions.*

> **Proved developed reserves.** *Proved developed oil and gas reserves are reserves that can be expected to be recovered through existing wells with existing equipment and operating methods. Additional oil and gas expected to be obtained through the application of fluid injection or other improved recovery techniques for supplementing the natural forces and mechanisms of primary*

recovery should be included as "proved developed reserves" only after testing by a pilot project or after the operation of an installed program has confirmed through production response that increased recovery will be achieved.

Proved undeveloped reserves. *Proved undeveloped oil and gas reserves are reserves that are expected to be recovered from new wells on undrilled acreage, or from existing wells where a relatively major expenditure is required for recompletion. Reserves on undrilled acreage shall be limited to those drilling units offsetting productive units that are reasonably certain of production when drilled. Proved reserves for other undrilled units can be claimed only where it can be demonstrated with certainty that there is continuity of production from the existing productive formation. Under no circumstances should estimates for proved undeveloped reserves be attributable to any acreage for which an application of fluid injection or other improved recovery technique is contemplated, unless such techniques have been proved effective by actual tests in the area and in the same reservoir.*

Use of end of year prices

In recent years, the prices of oil and gas have been subject to a substantial amount of volatility and seasonality. The fluctuation in prices has led to questions regarding the price to be used by engineers in estimating proved reserves. Note that the definition of proved reserves indicates that the reserve estimates are to be made using "prices and costs as of the date the estimate is made." In estimating proved reserves, the SPE and the WPC practices allow some latitude in the determination of the specific price to be used. They indicate that in some cases, average prices for the period may be more appropriate than year-end prices. Despite considerable criticism, the SEC has adopted the position that the price to be used in estimating reserves is the year-end price. Apparently the SEC's position is based on the fact that year-end prices are required by *SFAS No. 69* in calculating the standardized measure of the future cash flows from production of proved oil and gas reserves. Thus, the SEC has concluded that the same prices should also be used in the estimation of the proved reserve quantities. Figures 7–2 and 7–3 in chapter 7 show the differences between year-end prices versus average annual prices for crude oil and for natural gas for the 15-year period from 1990 through 2004. In many of those years, the differences are quite large. These differences could have a substantial impact on the quantities of proved reserves as well as the standardized measure of the future cash flows from production of proved oil and gas reserves that companies report. In late 2007, the SEC launched a project aimed at reexamining its reserve reporting guidance, including its mandated definition of proved reserves and its use of end-of-year prices.

Reserve quantity disclosure

The reserve quantity disclosure provides information regarding quantities of companies' estimated proved and proved developed oil and gas reserves. Specifically, the purpose of the disclosure is to explain changes in quantities of proved reserves from one year to the next. According to *SFAS No. 69*, par. 10–11:

Net quantities of an enterprise's interests in proved reserves and proved developed reserves of (a) crude oil (including condensate and natural gas liquids) and (b) natural gas shall be disclosed as of the beginning and the end of the year. "Net" quantities of reserves include those relating to the enterprise's operating and nonoperating interests in properties as defined in paragraph 11(a) of Statement 19. Quantities of reserves relating to royalty interests owned shall be included in "net" quantities if the necessary information is available to the enterprise; if reserves relating to royalty interests owned are not included because the information is unavailable, that fact and the enterprise's share of oil and gas produced for those royalty interests shall be disclosed for the year. "Net" quantities shall not include reserves relating to interests of others in properties owned by the enterprise.

Changes in the net quantities of an enterprise's proved reserves of oil and of gas during the year shall be disclosed. Changes resulting from each of the following shall be shown separately with appropriate explanation of significant changes:

a. *Revisions of previous estimates. Revisions represent changes in previous estimates of proved reserves, either upward or downward, resulting from new information (except for an increase in proved acreage) normally obtained from development drilling and production history or resulting from a change in economic factors.*

b. *Improved recovery. Changes in reserve estimates resulting from application of improved recovery techniques shall be shown separately, if significant. If not significant, such changes shall be included in revisions of previous estimates.*

c. *Purchases of minerals in place.*

d. *Extensions and discoveries. Additions to proved reserves that result from (1) extension of the proved acreage of previously discovered (old) reservoirs through additional drilling in periods subsequent to discovery and (2) discovery of new fields with proved reserves or of new reservoirs of proved reserves in old fields.*

e. *Production.*

f. *Sales of minerals in place.*

Since this disclosure presents only quantities of reserves and not dollar values, the disclosure should be the same for a company using the successful efforts method or the full cost method. Thus, this disclosure should aid in the comparison of full cost versus successful efforts companies, while also providing information related specifically to companies' upstream business.

The reserve report for Tyler Oil Company appears below.

Reserve Quantity Information
for the Years Ended December 31, 20XB and 20XA

	20XB	20XA
Proved reserves:		
Beginning of year	30,000	0
Revisions of previous estimates	750[a]	0
Improved recovery	0	0
Extensions and discoveries	21,250	30,000
Production	(16,750)	0
Purchases of minerals in place	0	0
Sales of minerals in place	(0)	(0)
End of year	35,250	30,000
Proved developed reserves:		
Beginning of year	13,750	0
End of year	35,250	13,750
Quantities applicable to long-term supply agreements in which the company acts as producer or operator:		
Proved reserves—		
End of year	X	X
Received during the year	X	X
Equity in proved reserves of equity investees	X	X

Notes:

20XB

a. Estimate at 12/31/XB of previously discovered reserves in place 12/31/XB, Lease A. . . . . 22,750 bbl

   Plus: 20XB production of previously discovered reserves 8,000

   Estimate of previously discovered reserves in place on 12/31/XA, made 12/31/XB 30,750

   Less: Estimate of reserves, made 12/31/XA of reserves in place 12/31/XA . . . . (30,000)

   Revisions of previous estimates, increase . . . . 750 bbl

Note that since the estimate made in 20XA involved Lease A only, calculation of the revision to that previous estimate can only involve Lease A.

If a company has both oil reserves and gas reserves and the quantities of each mineral are significant, the reserve quantity information should be reported separately. The reserve information should not be reported based on equivalent units.

The reserves included should be those related to properties in which the company has both working interests and nonworking interests, if the information is available. Reserve quantities for both working interests and nonworking interests owned should typically be included in the disclosure *net*. For example, assume a company has the following interests:

- A 70% working interest in a property with a 1/8 royalty interest and 10,000 total gross barrels of oil reserves
- A 1/5 royalty interest in a property with 15,000 total gross barrels of oil reserves

The net reserves would be computed as follows:

Property 1:	$70\% \times 7/8 \times 10,000$	=	6,125 bbl
Property 2:	$1/5 \times 15,000$	=	3,000 bbl
	Net reserves	=	9,125 bbl

Reserves that are owned by or entitled to another company should never be included. The reserves to be included in the report are limited to proved reserves in which the enterprise has an ownership interest, or if the underlying contract is a PSC, an entitlement interest. (See chapter 15.)

As with reserves, production figures reported were previously required to always reflect production quantities net of royalties. However, in May 2001, the SEC *Industry Guide for the Extractive Industries* indicated in paragraph (3)(B) that in special situations (e.g., production from foreign operations), production figures inclusive of royalties may be reported. This may be done if doing so, in that particular situation, enhances the information. If reported production quantities include royalties, companies are to make it clear that they have deviated from the normal treatment.

CAPITALIZED COSTS RELATING TO OIL AND GAS PRODUCING ACTIVITIES

SFAS No. 69 requires companies to disclose aggregated company-wide capitalized costs relating to its oil and gas producing activities and the company-wide related accumulated depreciation, depletion, amortization, and valuation allowances as of the end of the year. According to *SFAS No. 69*, par. 19:

> *If significant, capitalized costs of unproved properties shall be separately disclosed. Capitalized costs of support equipment and facilities may be disclosed separately or included, as appropriate, with capitalized costs of proved and unproved properties.*

This disclosure includes all oil and gas assets capitalized at year-end. Unproved property costs, if significant, must be separately disclosed. Costs of support equipment and facilities may be disclosed either separately or included with the capitalized costs of proved and unproved properties.

Since the capitalized costs of upstream activities are set out in this disclosure, this disclosure enables the comparison of companies with both upstream and downstream operations to those of companies with solely upstream operations. This disclosure will be different depending on whether the company is a full cost company or a successful efforts company.

This disclosure for Tyler Oil Company appears below, assuming first that Tyler is a successful efforts company and then assuming Tyler uses the full cost method.

ILLUSTRATIVE EXAMPLE—SUCCESSFUL EFFORTS

Capitalized Costs Relating to Oil and Gas Producing Activities at December 31, 20XB and 20XA

	20XB	20XA
Capitalized costs:		
Unproved oil and gas properties	$ 40,000	$160,000[a]
Proved oil and gas properties	601,000[d]	335,000[b]
Total capitalized costs	641,000	495,000
Less: Accumulated DD&A	(189,710)[e]	(0)[c]
Net capitalized costs	$451,290	$495,000
Enterprise's share of equity method investees' net capitalized costs	X	X

Notes:

20XA

	Lease B	Lease C	Total
a. Acquisition costs	$ 20,000	$40,000	$ 60,000
Wells-in-progress	100,000		100,000
Total	$120,000	$40,000	$160,000

	Lease A
b. Acquisition costs	$ 30,000
Wells and equipment	290,000
Decommissioning	15,000
Total	$335,000

c. No production during 20XA.

20XB		Lease A	Lease B	Total
d.	Acquisition costs	$ 30,000	$ 20,000	$ 50,000
	Wells and equipment	290,000	150,000	440,000
	Development costs	46,000	50,000	96,000
	Decommissioning	15,000	0	15,000
	Total	$381,000	$220,000	$601,000

e. All reserves on both Lease A and Lease B are fully developed at 12/31/XB; all production was from these two leases.

Lease A: Costs to be amortized at 12/31/XB include acquisition costs, costs of drilling the well, development costs, and future decommissioning costs. Remember costs of decommissioning (which includes dismantlement, restoration, and abandonment) must be included in DD&A computations by both successful efforts and full cost companies. Proved reserves (and proved developed reserves, since the lease is fully developed) at 12/31/XB are 10,750 plus 12,000 barrels (see letter c of problem statement for 12/31/20XB).

Lease B: Costs to be amortized include acquisition costs, cost of drilling the well, and development costs. Proved reserves (and proved developed reserves, since the lease is fully developed) at 12/31/XB are 12,500.

Lease A: $\dfrac{\$30,000 + \$290,000 + \$46,000 + \$15,000}{22,750 \text{ bbl} + 8,000 \text{ bbl}} \times 8,000 \text{ bbl} = \$ 99,122$

Lease B: $\dfrac{\$20,000 + \$150,000 + \$50,000}{12,500 \text{ bbl} + 8,750 \text{ bbl}} \times 8,750 \text{ bbl} = \underline{90,588}$

$\underline{\underline{\$189,710}}$

The following disclosure presents the capitalized costs, assuming Tyler is a full cost company, and assuming Tyler amortizes all possible costs.

ILLUSTRATIVE EXAMPLE—FULL COST

Capitalized Costs Relating to Oil and Gas Producing Activities at December 31, 20XB and 20XA

	20XB	20XA
Capitalized costs:		
Unproved oil and gas properties	$195,000[d]	$350,000[a]
Proved oil and gas properties	691,000[e]	385,000[b]
Total capitalized costs	886,000	735,000
Less: Accumulated DD&A	(285,394)[f]	(0)[c]
Net capitalized costs	$600,606	$735,000
Enterprise's share of equity method investees' net capitalized costs	X	X

Notes:

20XA

		Lease B	Lease C	Total
a.	Acquisition costs	$ 20,000	$ 40,000	$ 60,000
	G&G costs	40,000	20,000	60,000
	Drilling costs	100,000	130,000	230,000
	Total	$160,000	$190,000	$350,000

		Lease A
b.	Acquisition costs	$ 30,000
	G&G costs	50,000
	Wells and equipment	290,000
	Decommissioning	15,000
	Total	$385,000

c. No production during 20XA.

20XB

		Lease C
d.	Acquisition costs	$ 40,000
	G&G costs	20,000
	Drilling costs	130,000
	Delay rental	5,000
	Total	$195,000

		Lease A	Lease B	Total
e.	Acquisition costs	$ 30,000	$ 20,000	$ 50,000
	G&G costs	50,000	40,000	90,000
	Wells and equipment	290,000	150,000	440,000
	Development costs	46,000	50,000	96,000
	Decommissioning	15,000	0	15,000
	Total	$431,000	$260,000	$691,000

f. Reserves are fully developed on 12/31/XB (all capitalized costs relating to all three leases, including decommissioning costs, are amortized).

$$\frac{\$886,000}{35,250 \text{ bbl} + 16,750 \text{ bbl}} \times 16,750 \text{ bbl} = \underline{\$285,394}$$

COSTS INCURRED FOR PROPERTY ACQUISITION, EXPLORATION, AND DEVELOPMENT ACTIVITIES

The previous disclosure reported only the *capitalized* acquisition, exploration, and development costs. These costs are a cumulative total of costs capitalized from day one of operations less accumulated DD&A and write-downs. In contrast, this next disclosure reports all of the acquisition, exploration, and development costs incurred during the *current year*, regardless of whether the costs were capitalized or expensed. As a result, this disclosure should be equivalent whether the company uses the successful efforts method or the full cost method. The disclosure thus aids in the comparison of firms regardless of the method of accounting being used. According to *SFAS No. 69*, par. 21:

> *Each of the following types of costs for the year shall be disclosed (whether those costs are capitalized or charged to expense at the time they are incurred under the provisions of paragraphs 15–22 of Statement 19):*
>
> a. Property acquisition costs
>
> b. Exploration costs
>
> c. Development costs.

If significant costs have been incurred to acquire proved properties with proved reserves, those costs must be disclosed separately from the costs of acquiring unproved properties.

The disclosure of "Costs Incurred for Property Acquisition, Exploration, and Development Activities" of Tyler Oil Company appears below. Disclosure of production costs is not reported in this disclosure; instead, it is reported in the disclosure of results of operations.

ILLUSTRATIVE EXAMPLE

Costs Incurred in Property Acquisition, Exploration, and Development Activities
for the Years Ended December 31, 20XB and 20XA

	20XB	20XA
Costs incurred:		
Acquisition of proved properties	$ 0	$ 0
Acquisition of unproved properties	0	90,000[a]
Exploration	55,000[c]	630,000[b]
Development	96,000[d]	0
Total costs incurred	$151,000	$720,000
Enterprise's share of equity method investee's costs of acquisition, exploration, and development...	X	X

Notes:

20XA

	Lease A	Lease B	Lease C	Total
a. Acquisition costs	$ 30,000	$ 20,000	$ 40,000	$ 90,000
b. G&G costs	$ 50,000	$ 40,000	$ 20,000	$110,000
Exploratory drilling	290,000	100,000	130,000	520,000
Total	$340,000	$140,000	$150,000	$630,000

20XB

	Lease B	Lease C	Total
c. Exploratory drilling	$ 50,000		$ 50,000
Delay rental		$ 5,000	5,000
Total	$ 50,000	$ 5,000	$ 55,000

	Lease A	Lease B	Total
d. Development costs	$ 46,000	$ 50,000	$ 96,000

Note that this disclosure reports costs incurred only during the year and classifies the costs according to how they were accounted for when incurred. For example in 20XA, although Lease A is proved by year-end, at the time of acquisition, Lease A was unproved. Therefore, the cost of acquisition is reported as an acquisition of unproved property cost. In 20XB, although Tyler now has both unproved and proved properties on its books, no costs of acquisition are reported, because the properties were acquired in the previous year, not in 20XB.

RESULTS OF OPERATIONS FOR OIL AND GAS PRODUCING ACTIVITIES

The disclosure of the results of operations for oil and gas producing activities is an income statement–type report that includes only the revenues and costs associated with upstream oil and gas exploration and production activities. As such, this disclosure should aid in the comparison of companies with only upstream activities to companies with both upstream and downstream activities. The report will differ depending on whether the company uses the full cost or the successful efforts method. According to *SFAS No. 69*, par. 24–28:

> *The results of operations for oil and gas producing activities shall be disclosed for the year. That information shall be disclosed in the aggregate and for each geographic area for which reserve quantities are disclosed (paragraph 12). The following information relating to those activities shall be presented:*
>
> *a. Revenues*
>
> *b. Production (lifting) costs*
>
> *c. Exploration expenses*
>
> *d. Depreciation, depletion, and amortization, and valuation provisions*
>
> *e. Income tax expenses*
>
> *f. Results of operations for oil and gas producing activities (excluding corporate overhead and interest costs)*
>
> *Revenues shall include sales to unaffiliated enterprises and sales or transfers to the enterprise's other operations (for example, refineries or chemical plants). Sales to unaffiliated enterprises and sales or transfers to the enterprise's other operations shall be disclosed separately. Revenues shall include sales to unaffiliated enterprises attributable to net working interests, royalty interests, oil payment interests, and net profits interests of the reporting enterprise. Sales or transfers to the enterprise's other operations shall be based on market prices determined at the point of delivery from the producing unit. Those market prices shall represent prices equivalent*

to those that could be obtained in an arm's-length transaction. Production or severance taxes shall not be deducted in determining gross revenues, but rather shall be included as part of production costs. Royalty payments and net profits disbursements shall be excluded from gross revenues.

Income taxes shall be computed using the statutory tax rate for the period, applied to revenues less production (lifting) costs, exploration expenses, depreciation, depletion, and amortization, and valuation provisions. Calculation of income tax expenses shall reflect permanent differences and tax credits and allowances relating to the oil and gas producing activities that are reflected in the enterprise's consolidated income tax expense for the period.

Results of operations for oil and gas producing activities are defined as revenues less production (lifting) costs, exploration expenses, depreciation, depletion, and amortization, valuation provisions, and income tax expenses. General corporate overhead and interest costs shall not be deducted in computing the results of operations for an enterprise's oil and gas producing activities. However, some expenses incurred at an enterprise's central administrative office may not be general corporate expenses, but rather may be operating expenses of oil and gas producing activities, and therefore should be reported as such. The nature of an expense rather than the location of its incurrence shall determine whether it is an operating expense. Only those expenses identified by their nature as operating expenses shall be allocated as operating expenses in computing the results of operations for oil and gas producing activities.

The amounts disclosed in conformity with paragraphs 24–27 shall include an enterprise's interests in proved oil and gas reserves (paragraph 10) and in oil and gas subject to purchase under long-term supply, purchase, or similar agreements and contracts in which the enterprise participates in the operation of the properties on which the oil or gas is located or otherwise serves as the producer of those reserves (paragraph 13).

The illustrative disclosure of results of operations for oil and gas producing activities appears below. Since some costs that are expensed under successful efforts are capitalized under full cost accounting, this disclosure would be different if Tyler were a successful effort company versus a full cost company. Disclosures assuming both accounting methods are presented below—first for successful efforts, then for full cost.

ILLUSTRATIVE EXAMPLE—SUCCESSFUL EFFORTS

Results of Operations from Oil and Gas Producing Activities for the Years Ended December 31, 20XB and 20XA

	20XB	20XA
Revenues from oil and gas producing activities		
Sales to unaffiliated parties	$1,675,000[c]	$ 0
Transfers to affiliated entities	0	0
Revenues	1,675,000	(0)
Production (lifting) costs	(402,000)[d]	(0)
Exploration expenses	(5,000)[e]	(240,000)[a]
Depreciation, depletion, amortization, and valuation provisions	(189,710)[f]	(0)
Pretax income from producing activities	1,078,290	(240,000)
Income tax expenses/estimated loss carryforward benefit	(431,316)[g]	96,000[b]
Results of oil and gas producing activities (excluding corporate overhead and interest costs)	$ 646,974	$(144,000)
Enterprise's share of equity method investees' results of operations	X	X

Notes:

20XA		Lease A	Lease B	Lease C	Total
a.	G&G costs	$ 50,000	$ 40,000	$ 20,000	$110,000
	Dry exploratory hole			130,000	130,000
	Total	$ 50,000	$ 40,000	$150,000	$240,000

b.	Pretax loss	$(240,000)
	Tax rate	40%
	Estimated loss carryforward benefit	$ 96,000*

* The estimated loss carryforward benefit is recognized in the current period because realization is more likely than not. It is assumed that adequate reserves by Tyler Company provide evidence that this benefit will be realized. With respect to deferred taxes (which are being ignored in this chapter), the $240,000 is a timing difference that generates a future tax benefit. The $240,000 may be considered analogous to a future deductible amount. A deferred tax asset of $96,000 would be recognized, and since a loss situation exists, income tax expense would be credited by the same amount.

20XB

c. $1,675,000 = 8,000 \text{ bbl} \times \$100 + 8,750 \text{ bbl} \times \$100$

d. $192,000 + $210,000 (production costs include severance taxes)

e. Delay rental = $5,000

f. See the note in part e of the illustrative example for capitalized costs using the successful efforts method. Note that any impairment expense would be included here.

g.
Pretax income..............................	$1,078,290
Tax rate...................................	40%
Income tax expense	$ 431,316

The following disclosure presents the results of operations for Tyler Company, assuming the accounting method being used is full cost. Note that Tyler does not report any exploration expenses, because companies using full cost accounting generally capitalize all exploration costs incurred.

ILLUSTRATIVE EXAMPLE—FULL COST

Results of Operations from Oil and Gas Producing Activities for the Years Ended December 31, 20XB and 20XA

	20XB	20XA
Revenues from oil and gas producing activities		
Sales to unaffiliated parties.................	$1,675,000[b]	$ 0
Transfers to affiliated entities	0	0
Revenues...........................	1,675,000	(0)
Production (lifting) costs.......................	(402,000)[c]	(0)
Exploration expenses	(0)	
Depreciation, depletion, amortization, and valuation provisions	(285,394)[d]	(0)
Pretax income from producing activities...........	987,606	(0)
Income tax expenses/estimated loss carryforward benefit.........................	(395,042)[e]	0[a]
Results of oil and gas producing activities (excluding corporate overhead and interest costs)...........	$ 592,564	$ 0
Enterprise's share of equity method investees' results of operations	X	X

Notes:

20XA

a. Pretax income..................... $ 0
    Tax rate........................... 40%
    Income tax expense .... $ 0

20XB

b. $1,675,000 = 8,000 bbl × $100 + 8,750 bbl × $100

c. $192,000 + $210,000 (production costs include severance taxes)

d. See the note in part f of the illustrative example for capitalized costs using the full cost method. Note that any full cost ceiling write-downs would be included here.

e. Pretax income..................... $ 987,606
    Tax rate........................... 40%
    Income tax expense .... $ 395,042

The future value-based disclosures mandated by *SFAS No. 69* are much more complicated and require significantly more effort than the historical-based disclosures. The future value-based disclosures consist of two disclosures. The first is the standardized measure of discounted future net cash flows relating to proved oil and gas reserve quantities (SMOG). The second disclosure enumerates the changes that occur in the standardized measure from one year to the next.

(**Note to the instructor:** If, because of time constraints, part of this chapter is deleted, the authors suggest that the computation of future income tax be deleted. This deletion results in significantly fewer computations but still retains the basic concepts underlying the future-based disclosures. Two homework problems are included in which the computation of future income tax is not required.)

STANDARDIZED MEASURE OF DISCOUNTED FUTURE NET CASH FLOWS RELATING TO PROVED OIL AND GAS RESERVE QUANTITIES

SMOG is a unique disclosure introduced by *SFAS No. 69*. The disclosure resulted from the SEC's concern that the value of an oil and gas company's most significant asset—its reserves—is not included in its historical cost financial statements. This disclosure is intended to represent the present value of future net cash flows from the development, production, and sale of the reserves in the ground. According to *SFAS No. 69*, par. 30:

A standardized measure of discounted future net cash flows relating to an enterprise's interests in (a) proved oil and gas reserves (paragraph 10) and (b) oil and gas subject to purchase under long-term supply, purchase, or similar agreements and contracts in which the enterprise participates in the operation of the properties on which the oil or gas is located or otherwise serves as the producer of those reserves (paragraph 13) shall be disclosed as of the end of the year. The standardized measure of discounted future net cash flows relating to those two types of interests in reserves may be combined for reporting purposes. The following information shall be disclosed in the aggregate and for each geographic area for which reserve quantities are disclosed in accordance with paragraph 12:

a. *Future cash inflows. These shall be computed by applying year-end prices of oil and gas relating to the enterprise's proved reserves to the year-end quantities of those reserves. Future price changes shall be considered only to the extent provided by contractual arrangements in existence at year-end.*

b. *Future development and production costs. These costs shall be computed by estimating the expenditures to be incurred in developing and producing the proved oil and gas reserves at the end of the year, based on year-end costs and assuming continuation of existing economic conditions. If estimated development expenditures are significant, they shall be presented separately from estimated production costs.*

c. *Future income tax expenses. These expenses shall be computed by applying the appropriate year-end statutory tax rates, with consideration of future tax rates already legislated, to the future pretax net cash flows relating to the enterprise's proved oil and gas reserves, less the tax basis of the properties involved. The future income tax expenses shall give effect to permanent differences and tax credits and allowances relating to the enterprise's proved oil and gas reserves.*

d. *Future net cash flows. These amounts are the result of subtracting future development and production costs and future income tax expenses from future cash inflows.*

e. *Discount. This amount shall be derived from using a discount rate of 10 percent a year to reflect the timing of the future net cash flows relating to proved oil and gas reserves.*

f. *Standardized measure of discounted future net cash flows. This amount is the future net cash flows less the computed discount.*

It should be noted that the standardized measure as required by *SFAS No. 69* does not represent the "true value" of a company's proved reserves. The assumptions underlying the estimate do not incorporate assumptions regarding future prices, costs, and technology, and all companies must use the same 10% discount rate. Thus, the figure is truly a standardized estimate intended to aid in comparison between firms. Also note that this disclosure deals with future *cash flows*. Therefore, although the term *income tax expense* is used, what is effectively meant is *future income tax payable*.

The standardized measure is calculated as follows:

Future cash inflows (year-end prices × estimated future production).	$XX,XXX
Future development and production costs (estimated costs to be incurred in developing and producing proved reserves, based on year-end costs).	(XX,XXX)
Future pretax net cash flows.	$XX,XXX
Future income tax (year-end statutory tax rate applied to future pretax net cash flows giving effect to tax deductions, tax credits, and allowances).	(XX,XXX)
Future net cash flows	X,XXX
Discount (future net cash flows discounted at 10%).	(X,XXX)
Standardized measure of discounted future net cash flows.	$ X,XXX

SFAS No. 69 requires all present value calculations to be at a rate of 10% a year. The following brief example, which calculates the nondiscounted and discounted value of estimated future revenue only, illustrates the basic approach used in determining the standardized measure.

EXAMPLE

Future Revenue at 12/31/XA

	20XB	20XC	20XD	Total
Future production, bbl.	7,500	10,000	12,500	30,000
× Year-end price/bbl.	$ 88	$ 88	$ 88	$ 88
Future revenue	660,000	880,000	1,100,000	2,640,000
× PV* factors at 10%	0.9091	0.8264	0.7513	
PV* (future revenue).	$600,006	$727,232	$ 826,430	$2,153,668

*present value

The illustrative example of the SMOG disclosure appears below. Although precision is precluded by the very nature of future-based information, exact computational approaches (along with some simplifying alternative approaches) are given for the future-based disclosures. Approaches that yield **exact** answers are given rather than simpler (and possibly more practical) approaches, because the authors feel an exact approach facilitates a better understanding of the required disclosures.

Supporting computations for amounts shown for 20XA are presented in Tables 14–1 and 14–2. The supporting computations for the 20XB amounts in the disclosure below are presented in Tables 14–3 and 14–4. If the computations for future income tax are being ignored, the present value of future net inflows before income tax would be calculated instead of the standardized measure.

ILLUSTRATIVE EXAMPLE

Standardized Measure of Discounted Future Net Cash Flows Relating to Proved Oil and Gas Reserves at December 31, 20XB and 20XA

	20XB	20XA
Future cash inflows.	$3,666,000	$2,640,000
Future costs:		
Production	(846,000)	(600,000)
Development	(0)	(70,000)
Decommissioning	(15,000)	(15,000)
Future net inflows before income tax.	2,805,000	1,955,000
*Future income taxes	(1,006,031)	(551,000)
*Future net cash flows.	1,798,969	1,404,000
*10% discount factor.	(213,801)	(248,797)
*Standardized measure of discounted net cash flows.	$1,585,168	$1,155,203
Enterprise's share of equity method investees' standardized measure of discounted future net cash flows relating to proved oil and gas reserves	X	X
*Alternate:		
PV (Future cash inflows before income tax)	$2,471,874	$1,591,360

Table 14–1 Schedule of Estimated Future Cash Flows 12/31/XA

		20XB	20XC	20XD	Total
a.	**Undiscounted Values:**				
	Future production, bbl (given)	7,500	10,000	12,500	30,000
	Future revenue (production x $88)	$660,000	$880,000	$1,100,000	$2,640,000
	Future production costs (given)	(150,000)	(200,000)	(250,000)	(600,000)
	Future development costs (given)	(45,000)	(25,000)	(0)	(70,000)
	Future decommissioning (given)	(0)	(0)	(15,000)	(15,000)
	Future cash inflows before income tax	465,000	655,000	835,000	1,955,000
	*Future income tax (Table 14–2)	(29,400)	(233,689)	(287,911)	(551,000)
	*Future net cash inflows	$435,600	$421,311	$547,089	$1,404,000
b.	**Present Values:** (PV factor × above values)				
	PV factors	0.9091	0.8264	0.7513	
	PV (future revenue) ..	$600,006	$727,232	$826,430	$2,153,668
	PV (future production costs)	(136,365)	(165,280)	(187,825)	(489,470)
	PV (future development costs)	(40,909)	(20,660)	(0)	(61,569)
	PV (future decommissioning) ..	(0)	(0)	(11,269)	(11,269)
	PV (future cash inflows before income tax) .	422,732	541,292	627,336	1,591,360
	*PV (future income tax)	(26,728)	(193,121)	(216,308)	(436,157)
	*PV (future net cash inflows)	$396,004	$348,171	$411,028	$1,155,203
	*Discount [future net cash inflows—PV(future net cash inflows)]				$248,797

*Do not calculate if ignoring future income tax calculations.

Throughout this example, the cash flows were assumed to occur at the end of each year, as indicated by the PV factors. For example, at the end of 20XA, the present value of the revenue expected in 20XB was calculated using a present value factor at a 10% interest rate and one year. It would be reasonable to assume that the cash flows occurred evenly throughout the year, and therefore, to use a midyear PV factor.

Table 14–2 Computation of Future Income Taxes 12/31/XA[++]

	20XB	20XC	20XD	Total
Future revenue, net of production costs (Table 14–1a)	$510,000	$680,000	$850,000	$2,040,000
Future IDC expensed	(0)	(0)	(0)	(0)
Amortization of IDC (Table 14–2b)	(18,000)	(18,000)	(18,000)	(54,000)
Depreciation (Table 14–2c)	(45,000)	(51,111)	(63,889)	(160,000)
Depletion (Table 14–2d)	(20,000)	(26,667)	(33,333)	(80,000)
Decommissioning costs	(0)	(0)	(15,000)	(15,000)
Loss carryforward (Table 14–2a)	(353,500)	(0)	(0)	(353,500)
Taxable income	73,500	584,222	719,778	1,377,500
Tax rate	40%	40%	40%	40%
Gross tax	29,400	233,689	287,911	551,000
Credits	(0)	(0)	(0)	(0)
Net tax	$ 29,400	$233,689	$287,911	$ 551,000

[++]Do not calculate if ignoring future income tax calculations.

a. **Actual Income Tax, 12/31/XA**

Revenue	$ 0
Production	(0)
IDC and dry hole	(340,000)*
Amortization of IDC	(13,500)[+]
Depreciation expense	(0)
Depletion expense	(0)
Taxable income (loss)	$(353,500)
Tax rate	40%
Net tax/carryforward	$(141,400)

*($300,000 × 70% + $120,000 + $10,000)

[+] ($300,000 × 30% × 9/60)

b. **Computation of Future Amortization of IDC, 12/31/XA**

$$\$300{,}000 \times 30\% \times 12/60 = \underline{\$18{,}000}$$

c. **Computation of Future Depreciation, 12/31/XA:** Computed using the unit-of-production method, proved developed reserves, and equipment costs incurred to date plus future estimated tangible costs.

Year	Rate*		Estimated Production		Depreciation
20XB	6.00000	×	7,500	=	$45,000
20XC	5.11111	×	10,000	=	51,111
20XD	5.11112	×	12,500	=	63,889

*Depreciation Rates:

20XB
$$\frac{\$90{,}000 + \$45{,}000}{15{,}000 \text{ bbl} + 7{,}500 \text{ bbl}} = \frac{\$135{,}000}{22{,}500 \text{ bbl}} = 6.00000$$

20XC
$$\frac{\$135{,}000 + \$25{,}000 - \$45{,}000}{12{,}500 \text{ bbl} + 10{,}000 \text{ bbl}} = \frac{\$115{,}000}{22{,}500 \text{ bbl}} = 5.11111$$

20XD
$$\frac{\$115{,}000 - \$51{,}111}{12{,}500 \text{ bbl}} = \frac{\$63{,}889}{12{,}500 \text{ bbl}} = 5.11112$$

d. **Computation of Future Depletion, 12/31/XA:** Based on proved reserves and assumed all future production is sold in year produced.

Year	Rate*		Estimated Production		Depreciation
20XB	2.66667	×	7,500	=	$20,000
20XC	2.66667	×	10,000	=	26,667
20XD	2.66667	×	12,500	=	33,333

*Tax basis and depletion rate:

Tax basis of Lease A: Acquisition costs . . . $30,000
  G&G costs 50,000
  $80,000

*Depletion rate = $\dfrac{\$80{,}000}{30{,}000 \text{ bbl}} = 2.66667$

Note that since costs and reserve estimates do not change, if a new rate were computed for 20XC and 20XD, it would be the same as the rate shown above, i.e., 2.66667.

Table 14–3 presents the schedule of total estimated future cash flows at 12/31/XB. The cash flows are broken down to the flows relating to reserves discovered in prior years (Lease A), reserves discovered in the current year (Lease B), and total reserves.

Table 14–3 Schedule of Estimated Future Cash Inflows 12/31/XB

	Relating to Reserves Proved in Prior Year		
	20XC	20XD	Total
a. Undiscounted Values:			
Future production (barrels) (given).................	10,750	12,000	22,750
Future revenue (production × $104).................	$1,118,000	$1,248,000	$2,366,000
Future production costs (given).................	(258,000)	(288,000)	(546,000)
Future development costs (given).................	(0)	(0)	(0)
Future decommissioning (given).................	(0)	(15,000)	(15,000)
Future cash inflows before income tax.................	860,000	945,000	1,805,000
*Future income tax (Table 14–4).............	—	—	—
*Future net cash inflows	—	—	—
b. Present Values: (PV factor × above values)			
PV factors	0.9091	0.8264	
PV (future revenue)	$1,016,374	$1,031,347	$2,047,721
PV (future production costs)	(234,548)	(238,003)	(472,551)
PV (future development costs)....	(0)	(0)	(0)
PV (future decommissioning costs)	(0)	(12,396)	(12,396)
PV (future cash inflows before income tax)	781,826	780,948	1,562,774
*PV (future income tax).........	—	—	—
*PV (future net cash inflows).....	—	—	—
*Discount [future net cash inflows – PV(future net cash inflows)]...................			

*Do not calculate if ignoring future income tax calculations.

Relating to Reserves Proved in Current Year	Relating to Reserves Proved in Prior and Current Years		
20XC	20XC	20XD	Total
12,500	23,250	12,000	35,250
$1,300,000	$2,418,000	$1,248,000	$3,666,000
(300,000)	(558,000)	(288,000)	(846,000)
(0)	(0)	(0)	(0)
(0)	(0)	(15,000)	(15,000)
1,000,000	1,860,000	945,000	2,805,000
—	(668,948)	(337,083)	(1,006,031)
—	$1,191,052	$ 607,917	$1,798,969
0.9091	0.9091	0.8264	
$1,181,830	$2,198,204	$1,031,347	$3,229,551
(272,730)	(507,278)	(238,003)	(745,281)
(0)	(0)	(0)	(0)
(0)	(0)	(12,396)	(12,396)
909,100	1,690,926	780,948	2,471,874
—	(608,141)	(278,565)	(886,706)
—	$1,082,785	$ 502,383	$1,585,168
			$ 213,801

---

*Do not calculate if ignoring future income tax calculations.

Table 14–4 Computation of Future Income Taxes 12/31/XB**

	20XC	20XD	Total
Future revenue, net of production costs (Table 14–3)	$1,860,000	$960,000	$2,820,000
Amortization of IDC	(18,000)	(18,000)	(36,000)
Depreciation (Table 14–4b)	(106,369)	(53,073)	(159,442)
Depletion (Table 14–4c)	(63,261)	(31,220)	(94,481)
Decommissioning	(0)	(15,000)	(15,000)
Taxable income	1,672,370	842,707	2,515,077
Tax rate	40%	40%	40%
Gross tax	668,948	337,083	1,006,031
Credits	(0)	(0)	(0)
Net tax	$ 668,948	$337,083	$1,006,031

a. **Actual Income Tax, 12/31/XB**

	Total
Revenue	$1,675,000
Production	(402,000)
IDC	(0)
Amortization of IDC	(18,000)
Depreciation of expense	(76,558)*
Depletion expense	(45,519)+
Delay rental	(5,000)
Loss carryforward	(353,500)
Taxable income	774,423
Tax rate	40%
Net tax	$ 309,769

** Note: Do not calculate if ignoring future income tax calculations.

Depreciation expense:

Lease A $\dfrac{\$90{,}000 + \$46{,}000}{22{,}750 \text{ bbl} + 8{,}000 \text{ bbl}} \times 8{,}000 \text{ bbl} = \$35{,}382$

Lease B $\dfrac{\$50{,}000 + \$50{,}000}{12{,}500 \text{ bbl} + 8{,}750 \text{ bbl}} \times 8{,}750 \text{ bbl} = \underline{41{,}176}$

Total $\underline{\$76{,}558}$

$^+$*Depletion expense:*

Lease A $\dfrac{\$80{,}000}{22{,}750 \text{ bbl} + 8{,}000 \text{ bbl}} \times 8{,}000 \text{ bbl} = \$20{,}813$

Lease B $\dfrac{\$60{,}000}{12{,}500 \text{ bbl} + 8{,}750 \text{ bbl}} \times 8{,}750 \text{ bbl} = \underline{\$24{,}706}$

Total $\underline{\$45{,}519}$

b. Computation of Future Depreciation, 12/31/XB

	Year	Rate*		Estimated Production		Depreciation
Lease A	20XC	4.42277	×	10,750	=	$47,545$^+$
	20XD	4.42275	×	12,000	=	53,073
Lease B	20XC	4.70592	×	12,500	=	58,824$^+$

* *Tax basis and depreciation rates:*

Undepreciated tax basis of Lease A and Lease B equipment, 12/31/20XC:

	Lease A	Lease B	
Tangible successful drilling	$ 90,000	$ 50,000	
Development costs	46,000	50,000	
Total	$136,000	$100,000	
Depreciation, 20XB actual.	(35,382)	(41,176)	
Net tax basis..................	$100,618	$ 58,824	$159,442

Depreciation rates:

Lease A 20XC: $\dfrac{\$100{,}618}{10{,}750 \text{ bbl} + 12{,}000 \text{ bbl}} = 4.42277$

20XD: $\dfrac{\$100{,}618 - \$47{,}545}{12{,}000 \text{ bbl}} = 4.42275$

Lease B 20XC: $\dfrac{\$58{,}824}{12{,}500 \text{ bbl}} = 4.70592$

⁺ Depreciation for 20XC = $47,545 (Lease A) + $58,824 (Lease B) = $106,369

c. Computation of Future Depletion, 12/31/XB

	Year	Rate*		Estimated Production		Depletion
Lease A	20XC	2.601626	×	10,750	=	$27,967⁺
	20XD	2.601626	×	12,000	=	31,220
Lease B	20XC	2.823520	×	12,500	=	35,294⁺

* *Tax basis and depletion rates:*

Undepleted tax basis of Leases A and B, 12/31/XC:

	Lease A	Lease B	
Acquisition costs	$ 30,000	$ 20,000	
G&G Costs.	50,000	40,000	
Depletion, 20XB actual	(20,813)	(24,706)	
Net tax basis.	$ 59,187	$ 35,294	$ 94,481

Depletion rates:

Lease A $\dfrac{\$59{,}187}{10{,}750 \text{ bbl} + 12{,}000 \text{ bbl}} = 2.601626$

Lease B $\dfrac{\$35{,}294}{12{,}500 \text{ bbl}} = 2.823520$

⁺ Depletion for 20XC = $27,967 (Lease A) + $35,294 (Lease B) = $63,261

CHANGES IN THE STANDARDIZED MEASURE OF DISCOUNTED FUTURE NET CASH FLOWS RELATING TO PROVED OIL AND GAS RESERVE QUANTITIES

In addition to computing the standardized measure figure, companies must also explain the changes in the standardized measure from one year to the next. For example, SMOG is determined for a year using year-end quantity estimates and year-end prices and costs. The standardized measure for the following year would be determined for that year using new year-end quantity estimates and new year-end prices and costs. Thus, any changes in quantity estimates and prices and costs cause SMOG to change. According to *SFAS No. 69*, par. 33, the following sources of change must be reported separately if individually significant:

a. Net change in sales and transfer prices and in production (lifting) costs related to future production

b. Changes in estimated future development costs

c. Sales and transfers of oil and gas produced during the period

d. Net change due to extensions, discoveries, and improved recovery

e. Net change due to purchases and sales of minerals in place

f. Net change due to revisions in quantity estimates

g. Previously estimated development costs incurred during the period

h. Accretion of discount

i. Other—unspecified

j. Net change in income taxes

In computing the amounts under each of the above categories, the effects of changes in prices and costs shall be computed before the effects of changes in quantities. As a result, changes in quantities shall be stated at year-end prices and costs. The change in computed income taxes shall reflect the effect of income taxes incurred during the period as well as the change in future income tax expenses. Therefore, all changes except income taxes shall be reported pretax.

Some of these changes may be easily visualized and understood by referring to the illustrative example. If the selling price of oil increased from $88/bbl at the end of 20XA to $104/bbl at the end of 20XB, there would be a change in the standardized measure due to a change in prices and costs of $16/bbl. If total estimated production at the end of 20XB were 22,750 barrels (estimated at 12/31/20XB) instead of 22,500 (estimated at 12/31/20XA), there would be a change due to revision of quantity. If the total barrels to be produced stayed the same but the timing of production was changed—6,250 barrels in 20XC and 16,250 barrels in 20XD—there would be a change due to the change in production timing. This last change is typically included in the "other" category.

An important source of change is accretion of discount. Accretion of discount is a change inherent in the concept of present value. If, for instance, the preceding example were redone at the end of 20XB, assuming no changes except that a year had passed, the cash flows would be one year closer. Therefore the 20XB values would be multiplied by 1, 20XC values by 0.9091, and 20XD values by 0.8264. Accretion of discount is calculated by multiplying the discount rate of 10% by the beginning-of-the-year standardized measure. A beginning standardized measure at 12/31/XA, plus or minus all changes to the standardized measure, should equal the ending standardized measure at 12/31/XB. The effect of adding accretion of discount to the beginning standardized measure, which is discounted to a present value as of 12/31/XA, is to restate the beginning measure to a present value as of 12/31/XB. The ending standardized measure is already at a present value as of 12/31/XB. Therefore, to avoid adding and subtracting *mixed dollars*, all other changes should be discounted to 12/31/XB, as diagramed below:

Beginning standardized measure, PV at 12/31/XA	$X,XXX
Accretion of discount	XXX
Beginning standardized measure, restated, PV at 12/31/XB.	X,XXX
Plus or minus all other changes, PV at 12/31/XB.	XXX
Ending standardized measure, PV at 12/31/XB	$X,XXX

Another source of change is designated as "other" and is the plug amount necessary to account for the total change in the standardized measure. "Other" consists of changes not individually identified and errors that are a result of simplifications in assumptions or computations. Simplifications may be made in practice because the increased precision obtained without simplifications may not be felt in many cases to warrant the additional computations and time that may be required.

The illustrative example for the changes in the SMOG disclosure is below. The disclosure is a reconciliation of the beginning SMOG balance to the ending SMOG balance. Supporting computations and explanations to the disclosure are given in a schedule format. In two cases, an alternate simplified approach is presented in addition to an exact approach, because of the widespread use in practice of the simplified approach and the significant savings in calculation time. An analysis or proof of individual changes or grouped changes to the standardized measure based on the *exact* solution is also provided to show that all changes to an item have been accounted for.

ILLUSTRATIVE EXAMPLE

Changes in Standardized Measure of Discounted Future Net Cash Flows from Proved Reserve Quantities for the Years Ended December 31, 20XB and 20XA

	20XB Exact Solution	20XA Exact Solution
Standardized measure, beginning of year*	$1,155,203	$ 0
Sales and transfers, net of production costs (Schedule A)	(1,273,000)	(0)
Net change in sales and transfer prices, net of production costs (Schedule B)	293,052	0
Extensions, discoveries, and improved recovery, net of future production and development cost (Schedule C)	1,524,100	1,591,360
Changes in estimated future development cost (Schedule D)	21,728	0
Development costs incurred during the period that reduced future development costs (Schedule E)	96,000	0
Revisions of quantity estimates (Schedule F)	59,490	0
Accretion of discount (Schedule G)	159,136	0
Net change in income taxes⁺ (Schedule H)	(450,549)	(436,157)
Purchase of reserves in place	0	0
Sale of reserves in place	(0)	(0)
Other	8	0
Standardized measure, end of period*	$1,585,168	$1,155,203

* If ignoring future income tax calculations, use the PV (future net inflows before income tax) instead of the standardized measure.

⁺ Do not calculate if ignoring future income tax calculations.

SFAS No. 69 specifies that the sources of change illustrated in the above disclosure, if individually significant, should be presented separately. The amounts shown in the previous disclosure are supported by schedules and discussions presented in the following sections. In reviewing these schedules, remember that accretion of discount, which is one source of change, is added for 20XB to the beginning SMOG balance at 1/1/XB. The beginning SMOG is thus restated to a PV at 12/31/XB. All changes to the SMOG are also discounted to 12/31/XB for 20XB to avoid adding and subtracting mixed dollars.

Analysis of reasons for changes in value of standardized measure 12/31/XB

In the discussion below and later in the chapter, net selling price per barrel is calculated. Net selling price is revenue minus production costs. Revenue accruing to the working interest owner is based on working interest barrels only, i.e., the working interest's share of production. Thus, in order to determine the net selling price per barrel, the production costs per barrel must also be based on working interest barrels.

Sales and transfers, net of production costs. The production and sale of reserves decrease the amount and value of reserves in the ground. If the reserves had been discovered in a prior year, the production and sale of the reserves would also decrease the standardized measure, since the SMOG is a valuation of reserves in the ground.

On 12/31/XA, it was estimated that 7,500 working interest barrels would be sold during 20XB. Those barrels were valued net at $68/bbl (selling price minus production costs per working interest barrel: $88 – $20) and discounted using a PV factor of 0.9091, resulting in a carrying value per barrel of $61.82 in the 12/31/XA standardized measure presented in the SMOG disclosure. The accretion of discount change increases the carrying value of $61.82 at 12/31/XA to $68 ($61.82 × 110%) per barrel. If the estimate made on 12/31/XA had been accurate, i.e., 7,500 barrels sold at a net price of $68/bbl, the sale of the reserves would have exactly canceled out the carrying value of those reserves in the 12/31/XA standardized measure, adjusted by accretion of discount. However, previously discovered reserves of 8,000 barrels were sold in 20XB at a net price of $76 ($100 – $24) per barrel. The actual barrels sold and the actual net price are used in Schedule A, "Sales and transfers." The differences between the previous estimates and the actual barrels sold and the actual selling price are included in other schedules. Specifically, the difference between estimated and actual barrels produced is a revision of quantity or production timing. The difference between estimated and actual net selling price is a change in prices and costs, as discussed later.

In addition, of the reserves discovered on Lease B in 20XB, 8,750 barrels were produced and sold during 20XB. Although these barrels are not included in either the beginning or ending SMOG, in practice they are included for informational purposes as an increase to proved reserves in Schedule C, "Extensions, discoveries, and improved recovery." Therefore, the barrels must also be included in Schedule A, "Sales and transfers." The effect of including the barrels in Schedule C is to increase the SMOG, while the effect of inclusion in Schedule A is to decrease the SMOG. Thus, the effects of including the barrels cancel out, as must be the case, since the barrels discovered and produced in the same year did not affect either the beginning or ending SMOG.

Schedule A-20XB: Sales and transfers, net of production costs

Sales ([8,000 + 8,750] bbl × $100)	=	$1,675,000
Production costs ($192,000 + $210,000)	=	(402,000)
		$1,273,000

Sales and transfers for 20XA would be zero, since there was no production or sales during 20XA.

Change in prices and costs. The following example illustrates the points made in the next two paragraphs. It is a simple example (not related to the current problem) of the calculation of the effects of a change in prices (revenue) and costs (production costs) and a revision of quantity.

EXAMPLE

	Barrels	Net Price	Net Revenue
Future production, 12/31/XA	3,000	$50	$150,000
Future production, 12/31/XB (assume no production during 20XB).	2,000	40	80,000

Analysis: Reconciliation of beginning balance to ending balance

Beginning future net revenue (3,000 × $50)	$150,000	
Change in prices (3,000 × $10)	(30,000)	
	120,000	($40/bbl)
Revision in quantity (1,000 × $40)	(40,000)	
Ending future net revenue (2,000 × $40)	$ 80,000	

In order to determine the effect of a change in one variable, all other variables must be held constant. Thus, to isolate that portion of total change in future net revenues due to a change in prices and costs, the quantity of production is held constant. Therefore, the net change in prices and costs per barrel should be applied to the beginning-of-the-year estimate of future production. If the end of the year production estimate is used, a change due to revision in quantity that is a separately identified change will be introduced into the changes in prices and costs calculation. This results in misstating the effect of a change in prices and costs, as well as partially double-counting the effect of a revision in quantity.

The order of calculation of changes in the standardized measure is significant for certain calculations. As seen in the preceding example, the change in prices and costs was calculated first. The effect of this change was to restate all beginning barrels of future production from the beginning net price of $50/bbl to the year-end net price of $40/bbl. Revision in quantity, calculated next, was thus stated at the year-end net price of $40/bbl. *SFAS No. 69* requires that the effect of changes in prices and costs be computed before the effect of revisions in quantities, so that revisions in quantities will be valued in year-end prices and costs.

In the illustrative example of Tyler Oil Company, the change in prices and costs for 20XA is zero, because Tyler is a new company, and there were no estimates prior to 20XA. The calculation of the effect of changes in prices and costs for 20XB is more complicated than the preceding illustration. Tyler obtained and sold production during the year. The net price of production sold during 20XB, $76/bbl (selling price minus production costs per working interest barrel: $100 – $24), differs from the year-end net price of $80/bbl ($104 – $24). To get a precise measure of the effect of changes in prices and costs, the estimated 20XB production must be valued based on $76/bbl, not $80/bbl. Remember that the effect of changes in prices and costs must be calculated before the effect of changes in quantities. Consequently, production estimates made as of 12/31/XA must be used instead of estimates made as of 12/31/XB.

In Version 1, in order to get an exact answer, the effects of the change are discounted on a year-by-year basis. In Version 2, the discount calculation is simplified by discounting the effects of the change by applying the average discount at year-end. The year-end average discount is used instead of the beginning-of-the-year average discount because all changes, as discussed earlier, should be discounted to year-end dollars.

Schedule B-20XB: Net change in sales and transfer prices, net of production costs

Version 1 (discounted year-by-year, actual net price is used)

	20XB	20XC	20XD	Total
Future production, barrels (estimated 12/31/XA) . . .	7,500	10,000	12,500	30,000
× Net change per barrel . . .	$ 8.00*	$ 12.00**	$ 12.00	
Net change	60,000	120,000	150,000	$330,000
PV factor.	1.00	0.9091	0.8264	
PV at 12/31/XB (net change).	$60,000	$109,092	$123,960	$293,052

* 12/31/XA net price per barrel of $68 ($88 – $20) vs. 20XB actual net price per barrel of $76 ($100 – $24).

**12/31/XA net price per barrel of $68 vs. 12/31/XB net price per barrel of $80 ($104 – $24).

Version 2 (Same as Version 1, except discount calculation is simplified)

a. Calculation of reserves to which changes apply:

   Estimate of reserves made 12/31/XA of reserves in place 12/31/XA . . . . 30,000 bbl

b. Calculation of net change in price and cost factors:

   For 20XB estimated production:

Change in price ($100 actual – $88 beginning)	$ 12.00/bbl
Less: Change in cost ($24 actual – $20 beginning)	4.00/bbl
Increase.	$ 8.00/bbl

   For all other reserves except 20XB estimated production:

Change in price ($104 ending – $88 beginning)	$ 16.00/bbl
Less: Change in cost ($24 ending – $20 beginning)	4.00/bbl
Increase.	$ 12.00/bbl

c. Undiscounted net change due to change in price and cost:

7,500 bbl × $8.	$ 60,000
22,500 bbl* × $12.	$270,000
	$330,000

   *(30,000 bbl – 7,500 bbl)

d. Discounted net change due to change in price and cost: (apply average discount to net change)

$$\frac{\text{PV (future revenues – future production costs)}\ (12/31/XB)(\text{table 14–3b})}{(\text{future revenues – future production costs})\ (12/31/XB)(\text{table 14–3a})} \times \text{net change}$$

$$= \frac{\$3{,}229{,}551 - \$745{,}281}{\$3{,}666{,}000 - \$846{,}000} \times \$330{,}000 = \underline{\$290{,}712}$$

A company may choose not to perform the extra calculations involved in valuing 20XB estimated production at the actual net price for 20XB versus the year-end net price for 20XB. In this case, production may be either (1) eliminated from the calculation, with the change calculated on reserves remaining after production or (2) valued at the year-end net price.

The choice that yields the smallest error depends upon the facts of the situation, in particular whether the actual net price is closer in value to the beginning or year-end net price. If a decision is made to compute the change only on reserves after production, either actual or estimated production may be deducted from the beginning estimate. Again, the choice that yields the smallest error depends upon the facts of the situation.

Changes from extensions, discoveries, and improved recovery. Extensions, discoveries, and improved recovery of proved reserves increase the amount and value of reserves in the ground and thus increase the standardized measure. The reserves are included in the SMOG net of production, development, and decommissioning costs. Calculation of the increase to the SMOG is relatively simple. However, the computation may be complicated slightly when part of the reserves discovered in a year are developed and/or produced in that same year.

In calculating the standardized measure at the end of the year, any development costs and production costs incurred and sales revenue earned during the current year are not included in the computation—those amounts are included in the historical cost data. If the development and production costs incurred and the sales revenue earned during the current year are related to reserves discovered during the same year, these costs and revenues would not have been estimated previously. These costs and revenues, therefore, do not affect either the beginning or the end of the year standardized measure.

Since there would be no change in the standardized measure associated with these items, the authors feel that it is more consistent and a truer reflection of the actual changes to ignore these costs and revenues in the calculation of changes to the standardized measure. However, these costs and revenues are, in practice, being included in extensions, discoveries, and improved recovery. Thus, the authors have included these costs and revenue to be consistent with actual practice. Specifically, if the current-year development costs, production costs, and revenues that are related to current discoveries are included in the calculations of changes to the SMOG, then the following must be done:

- The related costs and revenues must be included in valuing the change from extensions, discoveries, and improved recovery (Schedule C).
- The related development costs must be included as an increase to the change from development costs incurred during the period that reduced future development costs (Schedule E).
- The related sales revenue, net of production costs, must be included in the change from sales and transfers (Schedule A).

These items would be included in the above schedules as follows:

Schedule C: + Revenue – Production costs – Development costs

Schedule E: + Development costs

Schedule A: – Revenue + Production costs

The net effect of including these items must be zero, because these items do not affect either the beginning or ending SMOG. From the preceding, it can be seen that the revenue and all of the costs cancel out, and thus the net effect of including these items is zero.

Most of the actual calculations needed for Schedule C were done previously and can be found in Table 14–1 or Table 14–3. For 20XA, refer to Table 14–1; for 20XB, refer to that portion of Table 14–3 relating to reserves proved in the current year. The numbers from Table 14–3 relating to reserves proved in the current year, however, are only for the reserves still in the ground at 12/31/XB. In the illustrative example, production and development costs were incurred and sales revenue was earned during 20XB relating to reserves discovered during 20XB. Therefore, to include the reserves on Lease B that were discovered and produced in 20XB, the numbers taken from Table 14–3 must be adjusted for the 8,750 barrels discovered and produced in the current year, as shown below.

	From Table 14–3		20XB Production		
PV (future revenue)...	$1,181,830	+	$875,000	=	$2,056,830
PV (future production costs)............	(272,730)	+	(210,000)	=	(482,730)
PV (future development costs)............	(0)	+	(50,000)	=	(50,000)
PV (future decommissioning costs)............	(0)	+	(0)	=	(0)

Schedule C-20XB, 20XA: Extensions, discoveries, and improved recovery, net of future production and development costs

	20XB	20XA
PV (future revenue)	$2,056,830	$2,153,668
PV (future production costs).............	(482,730)	(489,470)
PV (future development costs)	(50,000)	(61,569)
PV (future decommissioning costs)........	(0)	(11,269)
	$1,524,100	$1,591,360

Note to the instructor: The present value amounts shown in Schedule C were calculated in Tables 14–1 and 14–3 as intermediate steps to determining the PV of the future net cash inflows. If an alternate Schedule C (for instructional purposes only) as follows is computed, the PV of the future net cash inflows can be determined more directly, resulting in substantial computational savings on the part of students.

PV (future revenue)	PV($1,300,000)	+	$875,000	=	PV($2,175,000)
PV (future production costs)	PV(300,000)	+	(210,000)	=	PV(510,000)
PV (future development costs)	(0)	+	(50,000)	=	PV(50,000)
PV (future decommissioning costs)	(0)	+	(0)	=	(0)
	PV($1,000,000)	+	$615,000	=	PV($1,615,000)

	20XB	20XA
PV (future revenue)	PV($2,175,000)	PV($2,640,000)
PV (future production costs)	PV(510,000)	PV(600,000)
PV (future development costs)	(50,000)	PV(70,000)
PV (future decommissioning costs)	(0)	PV(15,000)
	PV($1,615,000)	PV($1,955,000)

PV($1,955,000) = $1,591,360 (Table 14–1)

PV($1,615,000) = $909,100 (Table 14–3) + $615,000 = $1,524,100

Changes in estimated future development costs. Changes in future development costs (including decommissioning costs) result from three sources:

a. Revisions to previously estimated development costs, including revisions to those estimated for the current year
b. Future development costs associated with reserves discovered in the current year
c. Development costs incurred in the current year, which were previously estimated and which reduced future development costs

Revisions to previously estimated development costs (source a above) arise from a change in the estimated amount of development activities or in the estimated cost of development activities. Source b is not included in the calculation of changes in estimated future development costs, because it is included in changes from extensions, discoveries, and improved recovery (Schedule C). Source c also is not included in the calculation, because it is presented as a separate category in the analysis of changes in the standardized measure (Schedule E).

The following schedule analyzes changes to estimated future development costs.

Schedule D-20XB: Changes in estimated future development costs (including decommissioning costs), source a revisions only

a	b	c	d	e	f
Year	Estimated as of 12/31/XA	Estimated as of 12/31/XB	Change in Estimates	PV Factors	PV of Change in Estimate
20XB	$45,000	$46,000	$ 1,000	1.0000	$ 1,000
20XC	25,000	0	(25,000)	0.9091	(22,728)
20XD	15,000	15,000	0	0.8264	0
	$85,000	$61,000	$(24,000)		$(21,728)*

*Increase to the standardized measure.

The $21,728 is a decrease in estimated future development *costs* and so is an increase in standardized measure.

Development costs incurred during the period that reduce future development costs. Development costs that were estimated previously and were incurred during the current period are no longer future costs. Thus, they act to reduce estimated future development costs and increase the standardized measure. (Estimated future development costs are a deduction in the computation of the standardized measure.)

In the illustrative example at 12/31/XA, future undiscounted development costs of $45,000 were estimated as of 12/31/XA to be incurred during 20XB. The standardized measure at 12/31/XA was reduced by $40,909, the present value of the development costs. Accretion of discount increases this value to $45,000. If the estimate made on 12/31/XA had been correct, then the estimated future development costs for 20XA would have been offset exactly by the actual development costs incurred. The standardized measure would have increased by $45,000 from 20XA to 20XB with respect to this item.

In the illustrative example, even though Tyler's actual development costs incurred during 20XB were $46,000, the net effect on the standardized measure is still an increase of $45,000. This is because only $45,000 had been estimated previously and deducted in arriving at the standardized measure. Thus, no matter what actual costs had been incurred—assuming the development activity itself had been performed—the standardized measure is increased by only the estimated development costs, not the actual cost incurred. Schedule E, "Development costs incurred during the period that reduced future development costs," reports the actual development costs incurred. The difference between the estimated and actual development costs, $1,000 in the illustrative example, is included in the previous schedule, "Changes in estimated future development costs." In this case, the difference would be an increase to future development costs and thus a reduction to the standardized measure. Therefore, after including development costs incurred during the period ($46,000) and changes in estimated future development costs for 20XB ($1,000), the standardized measure has been increased by $46,000 and reduced by $1,000, for a net increase of $45,000. This is shown as follows:

	Increase (Decrease) in Estimated Future Costs	Increase (Decrease) in Standardized Measure
Actual 20XB cost in excess of estimated 20XB costs (Schedule D)	$ 1,000	$ (1,000)
Development costs incurred during 20XB (Schedule E)	(46,000)	46,000
	$(45,000)	$45,000

As discussed earlier, revenues and costs from reserves both discovered and produced in 20XB are being included in the changes in SMOG disclosure. Specifically, the related development costs are included in Schedule C, with a resulting decrease to the SMOG. To cancel that effect, the related development costs must also be included in Schedule E, even though these costs had not been estimated previously and thus do not reduce future development costs. Schedule E's amounts must be increased in the amount of $50,000 for the development costs not previously estimated, which were incurred on Lease B in 20XB and were related to the reserves proved in 20XB.

Schedule E-20XB: Development costs incurred during the period that reduced future development costs

Development costs incurred during the period relating to reserves
discovered in prior years, actual dollars ($46,000 + $50,000) . . . $96,000

When trying to understand and account for the change in the standardized measure due to a particular item, it may be helpful to reconcile the change in that particular item or group of items from the beginning of the year to the end of the year. This reconciliation can serve as a proof, when a precise approach is used, that all changes to that item or group of items have been accounted for, and accounted for correctly. Even when a precise approach is not used, the reconciliation can give an indication of whether all changes have been accounted for. Equally important, the reconciliation increases understanding of the standardized measure and changes in the standardized measure.

The following reconciliation analyzes the changes in development costs.

Analysis of Changes in Development Costs
(Including Decommissioning Costs)

PV of beginning estimate of future development costs including decommissioning costs ($61,569 + $11,269) (table 14–1)	$ 72,838
Accretion of discount (10% × 72,838).	7,284
PV of changes in estimated future development costs and decommissioning costs, Schedule D	(21,728)
Development costs incurred during the period, which reduced future development costs (Schedule E)	(46,000)*
Other (rounding)	2
PV of ending estimate of future development costs ncluding decommissioning costs ($0 + $12,396) (table 14–3). . . .	$ 12,396

* The $50,000 of development costs related to reserves both discovered and produced in 20XB was not previously estimated and is not included in either the beginning or ending estimate of future development costs. Thus, the $50,000 was omitted from the analysis.

Revision of quantity. The standardized measure and changes to the standardized measure are all reported as present values. Therefore, to obtain an exact answer, both the quantity and the timing of the revisions must be considered. In other words, to get an exact answer, the change in the standardized measure due to the change in the timing of production must be computed at the same time as the change due to the revisions of quantity. The alternative approach presented following the exact version does not compute the effect of a change in timing. The effect of a change in timing, then, would be included in the "other" category.

Schedule F-20XB: Revisions of quantity estimates and changes in timing of production

Version 1

	20XB	20XC	20XD	Total
Future production, bbl (estimated 12/31/XB)	8,000[a]	10,750	12,000	30,750
Future production, bbl (estimated 12/31/XA)	7,500	10,000	12,500	30,000
Revision, increase (decrease)	500	750	(500)	750
Price per barrel, net of production costs, 12/31/XB	$ 76[b]	$ 80[c]	$ 80	
Undiscounted change from quantity revisions	$38,000	$60,000	$(40,000)	$58,000
PV factor..................	1.000	0.9091	0.8264	
PV (change from quantity revision and change in timing)	$38,000	$54,546	$(33,056)	$59,490

[a] Actual barrels produced in 20XB
[b] Actual net price per barrel for 20XB production
[c] Net price per barrel at 12/31/20XB

Note that in the preceding schedule, part of the change in production timing/revision of quantity involved the difference between estimated and actual production for 20XB. In this case, 500 barrels more were actually produced and sold than estimated. Because the actual net selling price per working interest barrel of these barrels was not the same as the year-end net price, the 500 barrels were valued at the actual net selling price in order to get an exact answer. *SFAS No. 69*, however, requires that the effects of changes in prices and costs be computed before the effect of changes in quantities, so changes in quantities will be valued at year-end prices and costs. Therefore, the precise approach of Version 1, while yielding an exact answer, does not conform to the statement's requirements in this regard. (In the illustrative example, if the actual barrels produced and sold had been less than the barrels estimated, then the barrels that had been overestimated would still be on hand at the end of the year, and they would be valued at year-end prices.)

An alternate approach, presented as Version 2, calculates the change due to the revision of quantity estimates without regard to timing. The effects of the change are discounted by applying the average discount at year-end. The actual selling price is used to value the difference between actual and estimated production, and the year-end price is used to value the remainder of the revision.

Version 2 (Calculation of discount is simplified)

1. **Revision in quantity**

Proved reserves in place 12/31/XA, estimated on 12/31/XA		30,000 bbl
Reserves in place 12/31/XB, proved in prior years, estimated on 12/31/XB	22,750 bbl	
Production during 20XB	8,000	
Revised estimate of reserves in place 12/31/XA, estimated on 12/31/XB		30,750
Revision in estimate, increase (decrease)		750 bbl

2. **Undiscounted effect of revision**

500 bbl × ($100 – $24) (actual net selling price)	$ 38,000
250 bbl × ($104 – $24) (end of year price and cost)	20,000
Total undiscounted increase from revision	$ 58,000

3. **Present value of revision (apply average discount to revision)**

$$\frac{\text{PV (future revenues} - \text{future production costs) (12/31/XB)(table 14–3b)}}{\text{(future revenues} - \text{future production costs) (12/31/XB)(table 14–3a)}} \times \text{revision}$$

$$\frac{\$3{,}229{,}551 - \$745{,}281}{\$3{,}666{,}000 - \$846{,}000} \times \$58{,}000 = \underline{\$51{,}095}$$

Upon completing the effect of revisions in quantity, a reconciliation of beginning and ending estimated future revenues relating to reserves discovered in prior years may be prepared as follows. This reconciliation can be used to verify that every source of change has been explained.

Analysis of Changes in Price and Cost and Quantity Revisions and Sales and Transfers

	$ (PV)	bbl
PV of beginning future revenues, net of production costs estimated 12/31/XA ($2,153,668 − $489,470) (Table 14–1)............................	$1,664,198	30,000
Accretion of discount (10% × $1,664,198)	166,420	
Sales, net of production costs (Schedule A)	(608,000)*	(8,000)
PV of net change in prices and production costs, calculated without PV simplification (Schedule B, Version 1)....................	293,052	
Revision of quantity and timing, calculated without PV simplification (Schedule F, Version 1)	59,490	750
Other (rounding)	10	
PV of ending future revenues, net of production costs relating to reserves discovered in prior years ($2,047,721 − $472,551) (Table 14–3b)........	$1,575,170	22,750

* The sales net of production costs related to reserves both discovered and produced in 20XB were not previously estimated and are not included in either the beginning or ending estimate of future revenues net of production costs. Thus, that amount was omitted from the analysis.

Accretion of discount. As discussed earlier, accretion of discount is a change that is inherent in the concept of present value, and one that adjusts the beginning standardized measure from a present value as of December 31, 20XA, to a present value as of December 31, 20XB. Accretion of discount is calculated by applying the 10% discount rate to the value to be adjusted. *SFAS No. 69* requires all changes except income taxes to be reported on a pretax basis. Therefore, accretion of discount should be computed on a pretax basis. There is thus another portion of accretion of discount associated with taxes that would be included with the change in income taxes.

Schedule G-20XB: Accretion of discount (excluding accretion on income tax)

PV of beginning future net cash flows before income taxes (Table 14–1b).	$1,591,360
Discount rate	10%
Accretion of discount (increase).	$ 159,136

Net change in income taxes. The change in estimated future income tax for 20XB results from the following:

1. Income tax actually incurred during the current period (increase to the SMOG)
2. The difference between (a) the estimate of future income taxes made on December 31, 20XB, plus actual income taxes from 20XB, and (b) the estimate made on December 31, 20XA.

The estimate made on December 31, 20XB, plus actual income taxes from 20XB is equivalent to an estimate made of future taxes as of 12/31/XA based on 12/31/XB facts. Actual income taxes from 20XB are included in (1) above as an increase to the SMOG and in (2) above as a reduction to the SMOG. Thus, the actual income taxes from 20XB in (1) and (2) would cancel out and consequently may be ignored.

Schedule H-20XB: Net change in income taxes*, including accretion of discount

PV of future income taxes estimated 12/31/XB (Table 14–3b).	$886,706
Less: PV of future income taxes estimated 12/31/XA (Table 14–1b).	436,157
Net change in future income taxes, increase (decrease). . .	$450,549

*Do not calculate if ignoring future income tax calculations.

A reconciliation of the change in income tax is not provided, because a reconciliation is unnecessary, given the method by which the change in income tax is calculated.

The net change in income taxes for 20XA, the first year of operations for Tyler, is computed as shown below.

Schedule H-20XA: Net change in income taxes*

PV of future income taxes, estimated 12/31/XA (Table 14–1b).	$436,157
Less: PV of future income taxes, 1/1/XA	0
Net change in future income taxes, increase (decrease). . .	$436,157

*Do not calculate if ignoring future income tax calculations.

Conclusion

A primary purpose of supplemental disclosures is usefulness to users of the financial statements. Wright and Brock evaluated all of the academic research that has been published regarding the relevance and reliability of reserve quantity disclosures and reserve value disclosures.[4] The studies investigating relevance of reserve *quantity* disclosures indicate the disclosures are useful. Regarding reliability, studies indicate that the disclosures are not perceived as being accurate; however, there is no evidence that the figures are biased. Since the reserve quantities are not particularly accurate or reliable, it is not surprising that the studies examining the reliability of reserve *value* disclosures concluded that the disclosures are likewise not considered accurate or reliable.

Regarding relevance, the studies are mixed. Most studies failed to find evidence that the standardized measure disclosure provided information content to users of financial statements. Recent studies have attempted to break the standardized measure or the change in standardized measure into their components and test the reasons for the change. These studies found some evidence that the information was being used by investors, but the evidence was weak and not convincing.

In summary, there appears to be strong support for reserve quantity disclosures. Support is not as strong or consistent as to the usefulness of the standardized measure and changes to the standardized measure disclosures. Moreover, it is these latter disclosures that require significant preparation time. Summing up the overwhelming nature of these disclosures, one analyst commented in a survey, "I would rather have gasoline a few cents cheaper at the pump than to have these disclosures."

PROBLEMS

1. What companies are required to present the disclosures specified by *SFAS No. 69*?

2. Which disclosures are different for a successful efforts company compared to a full cost company?

3. Which disclosures must be presented for each geographic area?

4. Casing Oil, a successful efforts company, began operations on January 1, 20XA. Assume the following facts about Casing's first two years of operations. All reserve and production quantities apply only to Casing Oil's interest. Prepare the required disclosures under *SFAS No. 69*.

12/31/20XA

	Item	Lease R	Lease S
a.	Acquisition costs	$ 20,000	$ 25,000
b.	G&G costs	$ 40,000	$ 50,000
c.	Drilling costs:		
	IDC	$250,000	
	Tangible	50,000	$50,000
	Life of equipment	10 years	
d.	Drilling results:	Proved reserves	Dry
e.	Estimated production of estimated proved reserves, bbl	20XB– 10,000 20XC– 7,500 20XD– 9,000	
f.	Reserve estimate, bbl 12/31/XA: Proved, at 12/31/ (decreases by estimated production)	20XA– 26,500 20XB– 16,500 20XC– 9,000	
	Proved developed at 12/31/(decreases by estimated production and increases as a result of estimated development)	20XA– 15,000 20XB– 16,500 20XC– 9,000	
g.	Estimated tangible development costs	20XB–$40,000	
h.	Estimated decommissioning costs	20XD–$20,000	
i.	Current market price of oil	$46/bbl	
j.	Estimated future production costs based on year-end costs	20XB–$80,000 20XC–$60,000 20XD–720,000	
k.	Estimated current production costs /bbl	$8/bbl	

12/31/20XB

	Item	Lease R	Lease S
a.	Actual and estimated production, bbl A = actual; R = revised; E = estimated	20XB— 6,000 A 20XC— 10,000 R 20XD— 11,000 R	
b.	Actual and estimated tangible development costs A = actual; R = revised; E = estimated	20XB—$35,000 A	
c.	Estimated decommissioning costs	20XD—$20,000	
d.	Market price of oil sold	$48/bbl	
e.	Current market price of oil at end of year	$50/bbl	
f.	Production cost of oil sold	$72,000	
g.	Production cost per barrel of oil sold	$12/bbl	
h.	Estimated future production costs at year-end	20XC—$100,000 20XD— 110,000	
i.	Current production cost per barrel at year-end	$10/bbl	
j.	Delay rental		$2,000

Assume a tax rate of 40%, and that Casing Oil does not qualify for percentage depletion because it is an integrated producer. For purposes of the required capitalization and amortization of 30% of IDC, assume nine months of amortization in 20XA. Because of the short life of Lease R, also assume Casing elects to use the unit-of-production method for calculating depreciation. Use proved reserves for depletion and proved developed reserves for depreciation. Ignore the alternative minimum tax and deferred taxes. (What is the significance of no estimated future development costs on Lease R as of 12/31/XB?)

5. Buckley Oil Company, a successful efforts company, began operations January 1, 20XA. Assuming the following facts about Buckley's first two years of operations, prepare the required disclosures under *SFAS No. 69*. All reserve and production quantities apply only to Buckley Oil's interest. Ignore the computations for future income tax.

12/31/20XA

Item	Lease Q	Lease T
a. Acquisition costs	$ 50,000	$ 40,000
b. G&G costs	$ 30,000	$ 35,000
c. Drilling costs:		
IDC	$150,000	$100,000
Tangible	80,000	10,000
Life of equipment	10 years	
d. Drilling results:	Proved reserves	Incomplete
e. Estimated production of estimated proved reserves, bbl	20XB– 7,500 20XC–15,000 20XD–10,000	
f. Reserve estimate, bbl 12/31/XA Proved, at 12/31/(decreases by estimated production)	20XA– 32,500 20XB– 25,000 20XC– 10,000	
Proved developed at 12/31/XA (decreases by estimated production and increases as a result of estimated development)	20XA– 27,500 20XB– 25,000 20XC– 10,000	
g. Estimated tangible development costs (life, 10 years)	20XB–$25,000	
h. Estimated decommissioning costs	20XD–$15,000	
i. Current market price of oil	$ 52/bbl	
j. Estimated future production costs based on year-end costs	20XB–$ 75,000 20XC– 150,000 20XD– 100,000	
k. Estimated current production costs /bbl	$10/bbl	

12/31/20XB

	Item	Lease Q	Lease T
a.	Drilling costs: Tangible		$60,000
b.	Drilling results:		Proved reserves
c.	Actual and estimated production, bbl A = actual; R = revised; E = estimated	20XB– 4,000A 20XC– 12,500 R 20XD–11,000 R	20XC– 17,500 E
d.	Actual and estimated tangible development costs A = actual; R = revised; E = estimated	20XB–$18,000 A (life, 10 years)	20XB– $30,000 A
e.	Estimated decommissioning costs	20XD– $15,000	
f.	Market price of oil sold	$54/bbl	
g.	Current market price of oil at end of year	$58/bbl	$58/bbl
h.	Production cost of oil sold	$56,000	
i.	Production cost per barrel of oil sold	$14/bbl	
j.	Estimated future production costs at year-end	20XC–$175,000 20XD– 154,000	20XC–$245,000
k.	Current production cost per barrel at year-end	$14/bbl	$14/bbl

Assume a tax rate of 40%, and that Buckley does not qualify for percentage depletion because it is an integrated producer. Ignore deferred taxes and the alternative minimum tax. (What is the significance of no estimated future development costs on Lease Q and Lease T as of 12/31/XB?)

6. Wildcat Oil Company began operations on January 1, 20XA. The following facts relate to Wildcat's first two years of operations. All reserve and production quantities apply only to Wildcat Oil's interest.

12/31/20XA

	Item	Lease R	Lease S	Lease T
a.	Acquisition costs	$ 50,000	$ 40,000	$ 60,000
b.	G&G costs	$ 30,000	$ 35,000	$ 40,000
c.	Drilling costs: IDC Tangible Life of equipment	$160,000 80,000 10 years	$100,000	$135,000 $10,000
d.	Drilling results:	Proved reserves	Incomplete	Dry

	Item	Lease R	Lease S	Lease T
e.	Estimated production of estimated proved reserves, bbl	20XB– 12,500 20XC– 15,000 20XD– 10,000		
f.	Reserve estimate, bbl 12/31/XA Proved, at 12/31/ decreases by estimated production)	20XA– 37,500 20XB– 25,000 20XC– 10,000		
	Proved developed at 12/31/ (decreases by estimated production and increases as a result of estimated development)	20XA– 15,000 20XB– 20,000 20XC– 10,000		
g.	Estimated tangible development costs	20XB–$25,000 20XC– 30,000		
h.	Estimated decommissioning costs	20XD–$15,000		
i.	Current market price of oil	$56/bbl		
j.	Estimated future production costs based on year-end costs	20XB–$125,000 20XC– 150,000 20XD– 100,000		
k.	Estimated current production costs /bbl	$10/bbl		

12/31/20XB

	Item	Lease R	Lease S	Lease T
a.	Drilling costs: tangible			$60,000
b.	Drilling results:			Proved reserves
c.	Actual and estimated production, bbl A = actual; R = revised; E = estimated	20XB– 9,000 A 20XC– 12,500 R 20XD– 8,000 R	20XB–12,500 A 20XC–17,500 E	
d.	Actual and estimated tangible development costs A = actual; R = revised; E = estimated	20XB–$20,000 A 20XC– 0 R	20XB–$30,000 A	
e.	Estimated decommissioning costs	20XD–$15,000		

	Item	Lease R	Lease S	Lease T
f.	Market price of oil sold	$48/bbl	$48/bbl	
g.	Current market price of oil at end of year	52/bbl	52/bbl	
h.	Production cost of oil sold	$108,000	$150,000	
i.	Production cost per barrel of oil sold	$12/bbl	$12/bbl	
j.	Estimated future production costs at year-end	20XC– $175,000 20XD– 112,000	20XC– $245,000	
k.	Current production cost per barrel at year-end	$14/bbl	$14/bbl	
l.	Delay rental			$9,000

Assume a tax rate of 40% and that Wildcat does not qualify for percentage depletion because it is an integrated producer. For purposes of the required capitalization and amortization of 30% of IDC, assume nine months of amortization in 20XA. Because of the short lives of Lease R and Lease S, also assume Wildcat elects to use the unit-of-production method for calculating depreciation. Use proved reserves for depletion and proved developed reserves for depreciation. Ignore the alternative minimum tax and deferred taxes. (What is the significance of no estimated future development costs on Lease R and Lease S as of 12/31/XB?)

a. Prepare the required disclosures under *SFAS No. 69*, assuming Wildcat is a successful efforts company.

b. Assume instead that Wildcat is a full cost company that amortizes all possible costs. Prepare only those disclosures that would differ under full cost accounting compared to successful efforts.

7. Tiger Oil began operations on January 1, 20XA. Assume the following facts about Tiger's first two years of operations. All reserve and production quantities apply only to Tiger Oil's interest. Ignore the computations for future income tax.

12/31/20XA

	Item	Lease M	Lease N	Lease O
a.	Acquisition costs	$ 70,000	$ 50,000	$ 40,000
b.	G&G costs	$ 90,000	$ 65,000	$ 80,000
	Test-well contribution			15,000
c.	Drilling costs:			
	IDC	$300,000	$200,000	$180,000
	Tangible	130,000		20,000
	Life of equipment	10 years		
d.	Drilling results:	Proved reserves	Incomplete	Dry
e.	Estimated production of estimated proved reserves, bbl	20XB– 27,500 20XC– 20,000 20XD– 15,000		
f.	Reserve estimate, bbl 12/31/XA Proved, at 12/31/(decreases by estimated production)	20XA–62,500 20XB–35,000 20XC–15,000		
	Proved developed at 12/31/ (decreases by estimated production and increases as a result of estimated development)	20XA– 25,000 20XB– 35,000 20XC– 15,000		
g.	Estimated tangible development costs	20XB–$60,000 20XC– 45,000		
h.	Estimated decommissioning costs	20XD–$20,000		
i.	Current market price of oil	$40/bbl		
j.	Estimated future production costs based on year-end costs	20XB–$330,000 20XC– 240,000 20XD– 180,000		
k.	Estimated current production costs /bbl	$12/bbl		

12/31/20XB

a.	Drilling costs: tangible		$ 70,000	
b.	Drilling results:		Proved reserves	

	Item	Lease M	Lease N	Lease O
c.	Actual and estimated production, bbl A = actual; R = revised; E = estimated	20XB–14,000 A 20XC–23,500 R 20XD–17,500 R	20XB–20,000 A 20XC–30,000 E	
d.	Actual and estimated tangible development costs A = actual; R = revised; E = estimated	20XB–$63,000 A 20XC– $0 R	20XB–$75,000 A	
e.	Estimated decommissioning costs	20XD–$20,000		
f.	Market price of oil sold	$42/bbl	$42/bbl	
g.	Current market price of oil at end of year	$46/bbl	$46/bbl	
h.	Production cost of oil sold	$168,000	$240,000	
i.	Production cost per barrel of oil sold	$12/bbl	$12/bbl	
j.	Estimated future production costs at year-end	20XC–$282,000 20XD– 210,000	20XC–$360,000	
k.	Current production cost per barrel at year-end	$12/bbl	$12/bbl	
l.	Delay rental			$8,000

Assume a tax rate of 40% and that Tiger does not qualify for percentage depletion because it is an integrated producer. Ignore deferred taxes and the alternative minimum tax.

a. Prepare the required disclosures under *SFAS No. 69*, assuming Tiger is a successful efforts company.

b. Assume instead that Tiger is a full cost company that amortizes all possible costs. Prepare only those disclosures that would differ under full cost compared to successful efforts.

REFERENCES

1. The illustrative example in this chapter is largely based on: Gallun, Rebecca A., and Della Pearson. 1984. "A Comprehensive Look at FASB Statement 69." *Journal of Extractive Industries Accounting.* Spring: pp. 115–165.

2. Financial Accounting Standards Board. 1997. *Statement of Financial Accounting Standards No. 131.* "Disclosures about Segments of an Enterprise and Related Information." Stamford, CT: FASB.

3. Securities and Exchange Commission. 2000. *2000 Current Issues and Rulemaking Projects.* November 14. Washington, D.C.: Securities Exchange Commission, Division of Corporate Finance.

4. Wright, Charlotte, and H. Brock. 1999. "Relevance versus Reliability of Oil and Gas Reserve Quantity and Value Disclosures: The Results of Two Decades of Research." *Journal of Petroleum Accounting and Financial Management.* Fall/Winter. Vol. 18, No. 3: pp. 86–110.

15

ACCOUNTING FOR INTERNATIONAL PETROLEUM OPERATIONS

A company electing to engage in operations outside the United States will likely encounter many issues and difficulties not encountered in domestic operations. For example, a company must consider that the costs of operating in a foreign country are likely to be higher than undertaking similar operations domestically. In some locations, there may be risks associated with factors such as the stability of the government, stability of the currency, adequacy of the supply of materials and equipment, and availability of a qualified labor pool. In addition, local laws and customs may affect the way business is conducted in the country. For example, an oil and gas exploration and production company will encounter different laws and customs regarding the ownership of the minerals, as well as a wide variety of different tax laws and regulations. The local government or state-owned entity, rather than an individual, will, with limited exception, own the minerals. Consequently, an oil and gas company must typically enter into a contract with the local government for the right to explore for and produce oil and gas reserves. These contracts are complex and vary widely from country to country.

All of these issues impact oil and gas companies engaging in exploration and production activities outside the United States. This chapter provides an overview of some of the issues and difficulties encountered in international oil and gas operations. Of special interest are contracts between oil and gas companies and governments that dictate how costs, revenues, and reserves are to be shared.

Petroleum Fiscal Systems

Since mineral rights are typically owned by the government of the foreign country or "host" country, the government and the oil and gas company must come to an agreement as to what collective payments are to be received by the government in return for allowing the company to operate in its country. Collectively these payments are referred to as the **fiscal system** of the country. Examples of such payments include the following:

- Up-front bonuses paid to the host country
- Royalties paid to the host country
- Federal and provincial income taxes
- Various other taxes collected by the host country, including duties and special petroleum taxes
- Production sharing, wherein oil or gas is allocated between the parties for the purpose of recovery of capital or operating costs, or both
- Infrastructure development for the host country

The exact nature of payments that the government receives is determined by the legal system in the country. Countries that collect payment for oil and gas produced primarily in the form of royalties and taxes are referred to as having **concessionary systems**. The United States has a concessionary system, as does the United Kingdom, Norway, and a host of other countries. In the United States and in limited situations in a few other countries, it is possible for individuals to own mineral rights. Outside the United States, mineral rights typically are owned exclusively by the host government. Some countries allow foreign oil companies to own minerals in place, while others do not. In countries relying primarily on taxes and royalties, the contract between the government that owns the minerals and the oil and gas company wishing to explore, develop, and produce the minerals is referred to as a **concessionary contract or concessionary agreement**.

Countries where the government does not rely entirely on taxes and royalties are referred to as having **contractual systems**. In a contractual system, the oil and gas company must contract with the local government for the right to share in revenue from oil and gas production. The government typically retains title to the minerals throughout exploration, development, and production. The company either receives revenues from the sale of its share of production or is allowed to receive oil and gas in the form of repayment for cost recovery, a share of profits, and/or services rendered.

A wide variety of contracts are found in contractual systems, including **production sharing contracts (PSCs)** and **service contracts**, with PSCs being the most popular. Conceptually, in a PSC, the foreign oil and gas company, referred to as the **contractor**, is allowed to recover certain costs and receives a share of the profits. (The term *contractor* can be used to refer to a single company. If more than one foreign company owns a working interest in the property, the term is used collectively to refer to all of the foreign companies that share in the working interest.) The contractor typically receives payment in-kind (in the form of oil or gas). In a typical service contract, the contractor receives money representing a fee for conducting exploration, development, and production activities. In practice, contracts have numerous different terms and conditions that often make it difficult to classify them as being either

PSCs or service contracts. Typical terms found in PSCs and service contracts are discussed in detail later in the chapter.

Almost every contract has its own unique terms and characteristics. Often it is not only difficult to classify the contracts as being PSCs or service contracts, but also as being concessionary or contractual in nature. Each country negotiates the terms it believes are most beneficial to the citizens of the country. Consequently, contracts and agreements generated by any given country may be contractual in nature but have some aspects that resemble a concessionary contract and vice versa. It is important to remember that any discussion of these contracts reflects the most usual case. Contracts can be very different, since they are the product of intense negotiation and can include whatever clauses and conditions the negotiators agree to.

CONCESSIONARY SYSTEMS

As mentioned above, other than in a limited number of countries, an E&P company must contract with the host government for the right to explore, develop, and produce. In a concessionary system, the contractor conducts exploration, drilling, and possibly development and production activities at its sole risk and cost. In other words, the company pays all of the cost associated with exploration and drilling, without any prospect for reimbursement if oil or gas is not discovered. If oil or gas is discovered and produced, title to the oil or gas in a concessionary system will, at some point, pass to the contractor. The contractor will, in turn, pay a royalty to the government as production occurs. A contract under a concessionary system contains basic terms that are similar to a U.S. lease agreement.

Another form of revenue to the government is taxes. Generally, all countries have some form of income taxes. If the government decides to increase the petroleum-related payments received by the state, it may enact other types of taxes and levies assessed on petroleum exploration and production activities. These include, for example, production or severance taxes, petroleum revenue taxes, value added taxes, and resource rent taxes. In a concessionary system, the primary source of payments to the state is in the form of royalties and taxes. Consequently, the countries having concessionary systems are sometimes referred to as tax/royalty countries. The following example illustrates a simple concession agreement.

EXAMPLE

Concession Agreement

Tyler Company enters into a concession agreement with the Canadian government. Tyler pays the government, in U.S. dollars, a $5,000,000 signing bonus and agrees to pay the government royalties of 10% of gross production and a 5% severance tax. Tyler bears all of the costs associated with exploration, development, and production.

Tyler spends $5,000,000 on exploration and drilling costs, and in 2012 has gross revenue of $7,000,000 and production costs of $1,000,000. The income tax laws allow deduction of all production costs, with exploration and drilling costs deductible over a five-year period. The tax rate is 40%.

Gross revenue for 2012 would be shared by the parties as follows:

	Tyler Company	**Government**
Gross revenue....................	$7,000,000	
Royalty 10%.....................	(700,000)	$ 700,000
Production tax 5%................	(350,000)	350,000
Net revenue	$5,950,000	
Operating expenses	(1,000,000)	
1/5 of exploration and drilling costs ...	(1,000,000)	
Taxable income	$3,950,000	
Income taxes, 40%................	(1,580,000)	1,580,000
Net to Tyler/Government	$2,370,000	$2,630,000

CONCESSIONARY AGREEMENTS WITH GOVERNMENT PARTICIPATION

One variation of a concessionary agreement involves the host government participating in the oil and gas operations as a working interest owner. This type of arrangement is generally referred to as **government participation**. The government typically sets up a state-owned oil company to participate in the operations as a working interest owner. This arrangement may also be referred to as a joint venture arrangement. As with other joint operations, a joint operating agreement is typically executed between the parties. In this particular type of arrangement, the contractor may agree to pay 100% of the exploration-type expenditures and "carry" the state-owned company through the exploration phase. In other words, the contractor pays all of the costs related to exploration, exploratory drilling, and any other costs specified in the contract.

If commercial reserves are found, the government retains the right to participate or **back in** to the development and production operations as a working interest owner at an interest of up to 51%. This means the government can elect to become a joint venture partner with the contracting company after the results of initial exploration and drilling are known. If the state-owned oil company elects to participate, it then becomes liable for its proportionate share of all *future* drilling, development, and production costs. The agreement may allow the contractor to recover all or a portion of its up-front exploration-related expenditures. If this is the case, there are two methods of recovery. One is direct payment by the government to the contractor. The other, more frequently used method is to allow the contractor to recover some or all of its costs by the contractor keeping the state oil company's share of production until the contractor has recouped the allowed costs. Afterwards, the state-owned company shares in costs and production just like any other working interest owner. Under this arrangement, the government still receives a royalty on gross production, along with income taxes and other fiscal obligations required by the laws and regulations of the country.

CONTRACTUAL SYSTEMS

In some countries, the legal system as it pertains to mineral ownership is based on the principle that natural resources are owned by all citizens of the country, and the government should act in such a manner as to maximize the value of the resources flowing back to the people. As a result, the government owns and retains title to all minerals. The contractor comes into the country, spends money, and takes risks (i.e., exploration, drilling, development, and production). In return, if successful, the contractor is entitled to an amount of oil, gas, or money (tied to production) sufficient to enable it to recoup its costs and make a profit. In some countries, the contractor may be entitled to an ownership interest in the minerals at the point of sale. In most instances, the contractor may not ever receive an ownership interest in the minerals in place, but instead is entitled to a share of the revenue from the sale of the minerals. Some agreements require that all or part of any oil or gas produced within the country be sold locally.

Government involvement in operations

In a contractual system, the government plays an active role in exploration, development, and production, typically by acting through a state-owned oil company. The contractor (or one of the contracting companies) usually acts as the operator. Contracts frequently call for the formation of a joint management group to oversee operations and vote on all major operating decisions. Such a joint management group is typically made up of representatives from the contractor (or each of the companies that compose the contractor), the state-owned oil company, and the government (i.e., representatives from the Ministry of Petroleum or some other government agency). The contractor is typically required to submit an annual work program and budget to the joint management group for review and approval. In addition to the annual budget, the joint management group generally makes all major decisions regarding the management of the project. These decisions include approval of all major expenditures, evaluation of the results of exploration, planning and drilling of wells, and determination of the commerciality of drilling results.

The contractor is typically required to provide all technology and financing. In most contractual systems, any equipment or facilities brought into the country by the contractor become the property of the local government. (This does not apply to equipment and facilities that are owned by service companies, equipment brought into the country temporarily, or to leased equipment.) In some cases, title to the equipment and facilities passes to the government at the time the goods are brought into the country or upon installation. In other cases, title passes to the government when the costs of the equipment and facilities have been recovered by the contractor.

The contractor is allowed to bring some employees into the country; however, the majority of employees must be hired locally. The contractor must have a plan for training the local employees. This training obligation generally continues throughout the life of the contract and constitutes a significant cost of operations.

As mentioned earlier, the two types of agreements prevalent in contractual systems are production sharing and risk service contracts, with the production sharing contracts being, by far, the most common. These contracts are discussed in the following sections.

Production Sharing Contracts

Production sharing contracts (PSCs) first emerged in the 1960s when governments began to evaluate alternative strategies to maximize the value of their resources. Under a concessionary agreement (where the government is not a working interest owner), the government has little or no involvement in the management and decision making related to day-to-day drilling and operations. Governments were generally looking for ways to increase their total share of petroleum-related revenues and profits and to expand their role in the management of petroleum operations. These factors, as well as other issues, including legal constraints related to the ownership of minerals, largely led to the trend toward production sharing contracts.

Signature and production bonuses

A common feature of both concessionary agreements and PSCs is that the contractor agrees to pay the government an up-front bonus for signing the agreement. This bonus is often referred to as a **signing bonus** or **signature bonus**. Typically, signature bonuses are payable in a lump sum of money but sometimes may involve payment in the form of equipment.

In some instances, a lump sum of money is paid at the signing of the contract, and subsequent payments are made to the government when production reaches an agreed-upon level. These later payments are referred to as **production bonuses**. For example, a contractor may agree to pay the government a $3 million bonus at signing, $2 million when production reaches 2,000 bbl/day, and $1 million when production exceeds 4,000 bbl/day. In other words, rather than paying $6 million at signing, the bonuses are phased in as certain levels of production are achieved. This arrangement theoretically allows the contractor to use its capital in exploration and drilling, with production bonuses being paid as money is coming in from the production and sale of oil or gas. On the other hand, the government is taking on some degree of risk, since it receives a lower amount when the contract is signed and may not receive any additional bonuses if production does not occur or if it does not reach the specified level. However, that risk is balanced with the expectation that exploration, development, and production, overall, will increase and will result in higher royalties, taxes, etc.

Royalties

An interesting feature of production sharing contracts is the inclusion of a royalty provision. Payment of a royalty is a logical concept in a concessionary system where title to oil and gas passes to the contractor at some point. In a contractual system, title to the oil and gas reserves may never pass to the contractor; nonetheless, many PSCs contain royalty provisions. In practice, royalty provisions range from as low as zero to 15% or higher.

Since a royalty represents a payment to the government off the top of gross receipts, many argue that royalties can actually discourage capital investment in new drilling and development. In marginal situations, royalties may actually work against further development and production. For example, if the royalty is 10%, a maximum of 90% of gross revenue is available to cover capital costs and operating costs. This may discourage an oil and gas company from developing a marginal field or may result in the abandonment of a marginal field earlier than would otherwise be the case. To help offset this effect to some extent, some contracts contain **sliding scale royalties.** A sliding scale royalty provides for a lower royalty amount when production is lower and increases as production increases. Thus, in marginal situations where production is lower, the lower royalty may allow production that would otherwise not have been profitable. The following example illustrates a sliding scale royalty.

EXAMPLE

Sliding Scale Royalty

Average Daily Production	Royalty
Up to 7,000 bbl/day	5%
7,001 to 14,000 bbl/day	10%
Above 14,000 bbl/day	15%

In this example, when production is low, the royalty payment is low, and when production increases, the royalty increases. By using a sliding scale, when production is low, there is more cash available to the parties for additional exploration and development.

In some instances when a field reaches the point where proceeds only marginally cover costs, governments may abolish or set aside the royalty provision for that contract area, thus improving the economics for the contractor. In the government's view, it may be preferable to forego the royalty in order to induce the contractor to continue to produce. If the field is plugged and abandoned, the government receives no continuing economic benefit. On the other hand, if production continues, even absent the royalty, the government continues to receive other payments and benefits, such as taxes and employment of personnel.

Government participation

As in concessionary systems, some PSCs allow the government to participate in oil and gas projects through a state-owned oil company. In most cases, the contractor must pay all costs and bear all risks during the exploration phase of a project. The government does not reimburse the contractor for its share of the exploration costs. Instead, the contractor must look to production and cost recovery to recoup those costs. If there is no production, the contractor bears all of the costs without any provision for reimbursement.

If oil and gas is discovered, the government—through the state-owned oil company—can elect to participate at whatever level of working interest it chooses, up to a maximum of 51%. If the government backs into the project, their role is similar to any other working interest owner. In other words, the state oil company is responsible for its proportionate share of development and operating costs.

Cost recovery

Cost recovery is a feature common to most PSCs. The contract should specify which costs are recoverable, the order of recoverability, any limits on recoverability, and whether costs not recovered in one period can be carried forward into the next period. Since title to the oil or gas reserves may never pass to the contractor, cost recoverability is the mechanism whereby the contractor is able to recoup the costs that it has expended on a project. The oil (or gas) that goes to the parties to allow them to recover their costs is referred to as **cost oil** or **cost gas**. In most PSCs, the contractor must pay 100% of the cost incurred in the exploratory phase. This type of PSC is typical. If the field is declared a commercial field, then the government, through the state-owned oil company, can elect to participate as a working interest owner at any level it chooses, frequently up to a maximum of 51%. If sufficient production occurs, the contractor is eventually able to recover 100% of the exploration expenditures, its proportionate share of development expenditures, and its proportionate share of production expenditures.

Typically, there is a ceiling or maximum amount of production available for cost recovery. This maximum amount of production that can be used for cost recovery is referred to as a **cost oil cap**. For example, a contract may indicate that the maximum amount of production that can be used for cost recovery is 60% of annual gross production. This means that, regardless of the amount of oil and gas produced and the amount of recoverable costs that are unrecovered, only 60% of production can be used for cost recovery. In most contracts (though not all), recoverable costs that are not recovered in any given year can be carried forward to future years. Some contracts *amortize* or *depreciate* the amount of capital costs (i.e., exploration and development expenditures) recoverable in any year (e.g., only 1/5 of allowable capital costs are recoverable per year in any of the first five years of production). Other contracts simply employ an annual maximum to cap the amount of recoverable capital costs in any given year.

Occasionally contracts allow **interest cost recovery,** or recovery of interest incurred on capital expenditures. A PSC may allow interest cost recovery on "deemed" or imputed interest on costs incurred during the development phase but not during the exploration phase. One view is that the contractor may have its money invested for a lengthy period of time before the costs are fully recovered. Therefore, the contractor should be compensated in the form of an interest-type payment. The alternative view is that it is the contractor's responsibility to acquire sufficient funds to cover capital requirements, and thus interest cost recovery is not justified.

Government ownership of all equipment and facilities raises interesting questions regarding who is responsible for future abandonment and reclamation costs. Some PSCs include a funding provision. These provisions may require the contractor to put money into a sinking fund to be used in the future to pay for abandonment and reclamation. If these sinking fund payments are considered to be recoverable costs, as with other exploration or development costs, the government is actually paying for dismantlement and reclamation. In other words, assuming the contractor fully recovers all of its costs, the companies deposit money into a fund and ultimately get to cost recover those deposits from future production. Theoretically, if all costs are fully recovered, the contractor's net cost is ultimately zero, and there is money in the fund available to pay for dismantlement and reclamation as the costs are actually incurred.

Most contracts indicate the order in which costs recovery is to occur. The order of cost recovery is important, since it determines how quickly the contractor is able to recover certain costs. For example, assume a contractor paid 100% of exploration expenditures and shared development expenditures with the state-owned company. Logically, the contractor would prefer to be able to recover all exploration expenditures before development expenditures are recoverable. A common order of cost recovery is similar to the following:

- Current year operating costs
- Unrecovered exploration expenditures
- Unrecovered development expenditures
- Capitalized interest (if allowed)
- Any investment credit or capital uplift (defined later)
- Future abandonment cost fund

Profit oil

The gross revenue from production typically goes first to pay royalties. Next, certain production-related taxes may be taken out (e.g., production taxes and VAT), and then costs are recovered. The gross revenue remaining after deducting royalties, taxes, and cost oil is referred to as **profit oil** (or **profit gas**). The profit oil (or gas) is shared between the parties based on the terms and conditions set forth in the contract. For example, assume that a contract provides for a 5% production tax, a 10% royalty, and a cost oil cap of 50% of annual gross production. Initially the profit oil is equal to 35% [100% – (5% + 10% + 50%)]. Later, after exploration and development costs have been recovered, and if the recovery of operating costs does not exceed the cost oil cap (i.e., 50%), the percentage of profit oil increases. In other words, the "excess" cost oil becomes profit oil.

Per a PSC contract, the profit oil is shared by the contractor, the state-owned oil company, and the government. Typically, a percentage of the profit oil goes directly to the government with the companies (the contractor and the state-owned oil company) sharing the remainder in proportion to their working interests. The percentage of profit oil that goes to each of the parties is negotiated in each contract and often involves a complicated mathematical formula.

EXAMPLE

PSC Cost Recovery

Assume that Tyler Company, a U.S. company, is involved in petroleum operations in Trinidad. Tyler has a 49% WI, while the Local Oil Company has a 51% WI. Annual gross production is to be split in the following order:

1. Royalty is 15% of annual gross production and is to be paid in-kind.
2. VAT is equal to 5% of annual gross production and is to be paid in-kind.
3. Cost oil is limited to 60% of gross production, with costs to be recovered in the following order:
    a. Operating expenses
    b. Exploration costs (paid entirely by Tyler Company)
    c. Development costs (paid 49% by Tyler Company and 51% by Local Oil Company)
4. Any excess production remaining after cost recovery becomes profit oil:
    a. The government receives 12% of the profit oil.
    b. The remainder is split between Tyler and Local Oil Company based on their working interests.

For 2012 assume the following:

- Recoverable operating costs total $6,000,000.
- Exploration costs (unrecovered to date) total $60,000,000.
- Development costs (unrecovered to date) total $600,000,000.
- Any costs not recovered in the current year may be carried forward to be recovered in future years.
- The annual gross production for the year is 3,000,000 barrels (bbl) of oil.
- The agreed-upon price is $60/bbl.

Note that since payment to the parties is in-kind, it is necessary to convert the costs into barrels by dividing costs by a price per barrel. The contract will specify how the parties are to negotiate and agree upon the price per barrel to be used for this purpose.

The allocation of production to the parties would be as follows:

	Total Annual Gross Production (in bbl)	Govt. (in bbl)	Local Company 51% (in bbl)	Tyler Company 49% (in bbl)
Royalty and VAT:	600,000			
Royalty (15%)		450,000	450,000	
VAT (5%)		150,000	150,000	
Cost oil (limit: 60% × 3,000,000)	1,800,000			
Operating costs: $6,000,000/$60 bbl = 100,000 bbl		100,000	51,000	49,000
Exploration costs: $60,000,000/$60/bbl = 1,000,000 bbl		1,000,000		1,000,000
Development costs: 700,000 bbl*		700,000	357,000	343,000
Profit oil: 3,000,000 bbl × [1− (15% + 5% + 60%)/100%]	600,000			
To Government: 600,000 bbl × 12%		72,000	72,000	
Allocable profit oil: 600,000 bbl × 88% = 528,000 bbl:		528,000		
528,000 bbl × 51%			269,280	
528,000 bbl × 49%				258,720
Total	3,000,000	3,000,000	672,000	677,280	1,650,720

* In this example, the cost oil is 60% x 3,000,000 barrels or 1,800,000 barrels. At a rate of $60/bbl, it will take a total of $600,000,000/$60/bbl = 10,000,000 barrels to recover the total development costs. However, this year, after allocating 100,000 barrels for recovery of operating costs and 1,000,000 barrels for recovery of exploration costs, only 700,000 barrels were available for recovery of development expenditures. The unrecovered development expenditures are carried forward to be recovered in future years.

	Unrecovered costs
Operating costs	$ 0
Exploration costs	0
Development costs: $600,000,000 − (700,000 × $60)	$558,000,000

Also note that Tyler Oil Company will be liable for income taxes on all oil and gas operations determined in accordance with the local income tax laws.

Other terms and fiscal incentives

Governments sometimes provide incentives to companies in an effort to maximize the amount of money the companies will invest in exploration, drilling, and development. These incentives may appear in PSCs or result from other negotiations.

Capital uplifts. A capital uplift is an incentive offered by the government to encourage the contractor to maximize investment. A capital uplift, sometimes referred to as an *investment credit*, is an additional amount of cost recovery on capital expenditures over and above actual amounts spent. For example, if a company spends $1,000,000 in recoverable capital expenditures, and there is a 10% capital uplift in the contract, the company will be allowed to recover 110% of actual spending, or $1,100,000.

Ringfencing. Generally each contract area stands alone when computing cost recovery. That is, in determining cost recovery, only costs that are expended relative to a particular contract area are recoverable from production from that specific contract area. This restriction is referred to as *ringfencing*. In other words, there is an imaginary boundary around the contract area—neither costs nor production can be transferred outside the boundary. If production in that area is insufficient to allow for full recovery, costs cannot be transferred to another contract area where production is higher and recovered from that production. In these instances, the contract areas are said to be "ringfenced." An incentive that governments may provide is to *un-ringfence* or allow *cross-fence recovery*. This incentive is most effective when the government is seeking to increase exploration in a particular area by allowing a company to immediately recover certain exploration expenditures in the new, frontier area against production from a different, currently producing area.

Ringfencing and cross-fence allowance may also be used as tax-related incentives in computing certain petroleum taxes. In some petroleum tax regimes, gross revenue, deductions, and taxable income are determined on a project-by-project basis. If the projects are ringfenced, then deductible costs cannot be transferred outside the project area where they were incurred. If cross-fence (or cross-field) deduction is allowed, then deductible costs from one project area—perhaps where there is no taxable income—can be transferred to another project area and used to reduce taxable income there. This provision is one means by which governments may encourage new exploration. The cross-fence allowance permits a company exploring in a new area (where it has no taxable income, and thus an expenditure is of no current tax benefit) to receive immediate tax benefit by deducting the exploration costs against the taxable income generated in another project area.

Domestic market obligation. Some contracts specify that a certain percentage of the contractor's share of profit oil be sold to the local government, perhaps at a price that is less than the current market price. This requirement is referred to as the *domestic market obligation* and is often included in situations where the demand for crude oil in the country is greater than the government's share of production. The contractor's domestic market obligation reduces the shortfall, resulting in a lessening of the government's need to rely on imported oil or oil from more expensive sources.

Royalty holidays and tax holidays. *Royalty holidays* and *tax holidays* are incentives governments may use to encourage contractors to maximize investment early in the life of production. The government may specify a period of time (e.g., the first four years of production) during which the royalty provision is waived, resulting in the contractor paying no royalty on production during that period of time. This incentive provides the contractor with a break from the payment of royalties, or a *royalty holiday*. It also leaves the contractor with more money to reinvest in additional drilling and development. Therefore, the holiday benefits both the government and the contractor. Likewise, the government may specify a period of time during which the contractor is exempt from certain taxes. This is referred to as a *tax holiday*.

These are just a few of the most popular terms that may appear in PSCs. A discussion of all of the possible terms and variations would be too lengthy to include in this chapter. For additional discussion of PSCs, the reader is referred to *International Petroleum Accounting*, 1st ed., by Wright and Gallun.[1]

SERVICE CONTRACTS

The second type of agreement prevalent in a contractual system is a service agreement. Service agreements can be classified as being either **risk service contracts or nonrisk service contracts**. In nonrisk service agreements, the contractor provides services in the form of such activities as exploration, development, and production and is paid a fee by the government that covers all costs. In practice, nonrisk service agreements are rare. Risk service contracts are much more common. In a risk service contract, the contractor bears all of the costs and risks related to exploration, development, and production activities. In return, if production is achieved, the contractor is allowed to recover its costs as production is sold. In addition, the government pays the contractor a fee for its "services." The fee is typically based on production. The terms and features of risk service contracts are similar to those appearing in PSCs. Risk service agreements are popular in many countries in South America. As with production sharing contracts, a significant task facing the accountant is the proper determination of recoverable operating and capital costs.

EXAMPLE

Risk Service Agreement

Tyler Company enters into a risk service agreement with the Bolivian government. Tyler pays the government, in U.S. dollars, a $1,000,000 signing bonus and bears all of the costs associated with exploration, development, and production. The contract defines costs incurred in the exploration and development phase of each project area as being capital costs (CAPEX), and all costs incurred in the production phase as being operating costs (OPEX).

Each year in which production occurs, the government agrees to pay Tyler Company a fee comprised of the following:

- All OPEX incurred in the current year
- 1/10th of all unrecovered CAPEX
- $0.50/bbl on production from 0 to 4,000 bbl/day, $0.75/bbl on production from 4,001 to 10,000 bbl/day, and $1.00/bbl on production above 10,000 bbl/day

The maximum total fee that will be paid in any year is $1.35/bbl times the total number of barrels produced. Any unrecovered OPEX or CAPEX (unrecovered due to the maximum fee) may be carried forward to future years.

In 2013, production begins on Field No. 1. To date Tyler has spent $10,000,000 on CAPEX and during 2013, spends $2,000,000 in OPEX. Production during 2013 equals 4,000,000 barrels, or 4,000,000/365 = 10,959 bbl/day.

The fee that Tyler would receive for 2013 would be determined as follows:

OPEX	$2,000,000
CAPEX $10,000,000/10	1,000,000
4,000 × 365 days × $0.50	730,000
6,000 × 365 days × $0.75	1,642,500
959 × 365 days × $1	350,035
Total fee	$5,722,535

The total fee per barrel is computed as follows:

$$\frac{\$5,722,535}{4,000,000 \text{ bbl}} = \$1.4306 \text{ per bbl}$$

The $1.4306 fee per barrel is greater than the maximum of $1.35; therefore, the actual fee paid to Tyler Oil Company is:

$1.35 × 4,000,000 bbl = $5,400,000

The difference between the calculated fee of $5,722,535 and the maximum fee of $5,400,000 is $322,535. That amount is considered to be unrecovered CAPEX and is carried forward to the next year.

Tyler Oil Company will also be liable for income taxes on all oil and gas operations determined in accordance with the local income tax laws.

JOINT OPERATING AGREEMENTS

As with domestic operations, when two or more international parties are involved in a joint operation, they must execute some type of joint operating agreement. [*Model Form International Joint Operating Agreements* are published by the Association of International Petroleum Negotiators (AIPN).] In international operations, the function of the joint operating agreement depends on the parties and the detail included in the underlying concession agreement, PSC, or risk service contract.

The contract between the government and the contractor provides the basis for all subsequent contracts and agreements. Sometimes the PSC or concession agreement is thoroughly written and effectively also serves as the joint operating agreement. In other cases, a separate joint operating agreement is executed. Further, if the contractor is comprised of more than one company, the PSC, concession agreement, etc., may serve as the basic agreement between the governmental entity, including the state-owned oil company, and the contractor. A separate joint operating agreement is then executed between the parties acting as the contractor.

Whether the PSC, concession agreement, or risk service contract serves as the joint operating agreement or a separate joint operating agreement is executed, an accounting procedure must exist. Accounting procedures, as they appear in most domestic joint operating agreements, were discussed in detail in chapter 12. As detailed in chapter 12, the accounting procedure defines, among other things, how costs are to be shared between the parties, determination of overhead, material transfers and pricing, and inventories. An international joint operating agreement should address these same major categories. In addition, the accounting procedure included in a PSC or risk service contract should define which costs are recoverable and which costs are not recoverable.

Recoverable versus nonrecoverable costs

In joint interest accounting, one of the key tasks is to determine the proper amount of costs and revenues to be shared by each of the parties. In accounting for a PSC or risk service contract, evaluation of costs and allocation of revenues are likewise issues that involve considerable accounting attention. As a general rule of thumb, costs that are defined as direct costs in domestic joint operating agreements are likely to be recoverable under the PSC or risk service contract. In a domestic joint operating agreement, recoupment of indirect costs by the operator is specified via application of overhead rates. Overhead rates (often sliding scale rates) are also frequently used in PSCs and risk service contracts to determine the amounts of recoverable indirect costs.

Chapter 12 provides a thorough discussion of the types of costs that are typically classified as direct in domestic joint operating agreements. As just discussed, many of these costs would likewise appear in international accounting procedures as either direct costs or recoverable costs. There are a number of potentially sizeable costs encountered in international operations. A few of these are listed below:

- Transportation costs, in general, are likely to be much larger in international operations.

- Rig mobilization and demobilization costs are likely to be significant, since rigs may have to be moved long distances.

- Companies typically have a home office in their home country. In addition, they may have large administrative offices within the country of operations. The costs associated with the in-country offices are generally direct and recoverable. If there is more than one project or different projects with different owners, then the cost of the in-country office will have to be allocated to the projects on some reasonable basis.

- Costs associated with expatriate employees (i.e., employees working in a foreign country) who are relocated into the country include travel, moving, living quarters, education for dependent children, etc.

- Relocation costs associated with expatriate employees who move from one foreign assignment to another are frequently an issue that must be spelled out in the contract. For example, if an employee is relocated from an assignment in Poland to an assignment in China, and from there to an assignment in Thailand, how much of the total relocation costs would be chargeable to the operations in China?

- In most domestic joint operating agreements, charges relating to technical employees are chargeable to the joint operation only if the employee is physically on location at the joint property a minimum amount of time, typically eight hours. In international operations, technical employees working in offices located outside of the host country may nonetheless still be chargeable to the joint operation, so long as their work is directly benefiting the joint property.

FINANCIAL ACCOUNTING ISSUES

Financial accounting versus contract accounting

SFAS No. 19 and *Reg. S-X 4-10* were written from the perspective of concessionary contracts. However, over the years, accountants have recognized that, while the contract terms and conditions are quite different for PSCs and risk service contracts, the financial accounting should be fundamentally the same. Consequently, financial accounting for PSCs and risk service contracts is no different from financial accounting for concessionary contracts. Even though in a PSC or risk service contract the government owns the oil and gas reserves, the exploration, development, and production activities should be accounted for as oil and gas producing activities and not as contractor services. The production revenue that accrues to the contractor is to be accounted for as revenue and not as cost recovery.

In chapters 2 through 7 of this book, definitions that appear in *SFAS No. 19* and *Reg. S-X 4-10* were discussed in detail (e.g., exploratory wells and development wells). It should be noted that the definitions utilized in a contract are not necessarily the same as the definitions appearing in *SFAS No. 19* and *Reg. S-X 4-10*. For example, in a PSC, an exploratory well may be defined as any well drilled in the exploratory phase, and a development well as any well drilled in the development phase. Depending on the specific circumstances, it is entirely possible that a well drilled in the exploratory phase as defined in the PSC could be a development well per *SFAS No. 19* and Reg. S-X 4-10. Or, even more often, a well drilled in the development phase as defined in the PSC could be an exploratory well per the FASB and SEC definition. Obviously, the contract definitions apply when accounting for the contract, and the financial accounting definitions apply when involved in financial accounting.

Disclosures of proved reserves—*SFAS No. 69*

Estimation of proved reserves. Reserve estimation under a PSC or risk service contract is much more complex than reserve estimation under a concessionary contract. Basically, if estimating working interest reserves under a concessionary contract, gross recoverable reserves would first be estimated. Then, reserves attributable to royalty interests or other nonoperating interest would be subtracted. The remainder would be allocated to the working interest owners based on their relative working interests.

When estimating working interest reserves under a PSC or risk service contract, the net proved reserves remaining after deducting reserves related to royalties and other nonoperating interests would be allocated between the parties. The allocation is based on the amounts to which they are entitled as per the contract terms. To compute *entitlement reserves*, one would have to resolve issues such as the following:

- The amount of recoverable costs that have been or will be incurred
- The amount of costs spent in each phase (i.e., exploration, development, or production)
- The cost recovery terms in the contract
- The profit oil sharing terms in the contract
- The price assumptions to be used in converting the recoverable costs into quantities

A slight change in the prices that are used to convert recoverable costs to quantities of reserves can cause a significant shift in the total estimate of proved reserves to which the working interest owners are entitled.

Disclosure of reserves. As indicated earlier, outside the United States, in most instances the mineral rights are owned by the government. In the United States (and typically in countries having concessionary fiscal policies), ownership of any oil and gas that may be present in the ground may pass to the contractor when the lease or concession agreement is executed. In countries with contractual systems, the government typically retains ownership to all minerals. Thus, legal title to any oil and gas reserves that are discovered may never pass to the contractor. The latter situation is often a source of concern to a contractor using U.S. GAAP when attempting to determine which reserves should be disclosed in its financial statements. *SFAS No. 69* indicates that in order for a company to include reserves in its financial statement disclosures, the company should have an ownership interest in the reserves. *SFAS No. 69*, par. 10, specifically prohibits disclosure of reserves that are owned by another party. Over the years, paragraph 10 has been the source of much discussion. Under question is the determination of whether it is appropriate to report reserves located in countries where the government contract is a PSC, service agreement, or some other form of contract where the reserves are legally owned by the host government.

In March 2001, the SEC staff issued *Interpretations and Guidance*. Paragraph (II)(F)(3)(l) of that document clarified the SEC's position regarding these reserves. The SEC indicated that two possible methods exist for determining oil and gas reserves under PSCs: (1) the working interest method and (2) the economic interest method. Under the working interest method, the contractor's share of reserves is determined by multiplying the contractor's working interest by the estimate of total proved reserves, net of any royalty. Under the economic interest method, the share of reserves to which a company is entitled is determined by dividing its cost oil and profit oil by the year-end oil price. The lower the oil price, the higher the barrel entitlement, and vice versa. The working interest method has always been the accepted method; in this document, the SEC staff concluded the economic interest method is acceptable for PSCs, and use of the method is not in violation of paragraph 10 of *SFAS No. 69*. It is not clear whether paragraph (II)(F)(3)(l) applies to contracts other than PSCs. Thus, lack of consensus still exists as to what reserves, if any, to disclose under a risk service agreement.

Operations in some countries are subject to risks associated with instability in the government that could result in the government taking over operations or otherwise forcing an operator out of the country. Due to these concerns, *SFAS No. 69*, par. 12, requires companies to report their net proved reserves separately for each significant foreign location:

> *If some or all of its reserves are located in foreign countries, the disclosure of net quantities of reserves of oil and of gas and changes in them required by paragraph 10 and 11 shall be separately disclosed for (a) the enterprise's home country (if significant reserves are located there) and (b) each foreign geographic area in which significant reserves are located. Foreign geographic areas are individual countries or groups of countries as appropriate for meaningful disclosure in the circumstances.*

In practice, companies disclose their net share of proved reserves by major geographic area, but few provide information further segregating the reserves related to PSCs or risk service contracts. This practice may result, in part, from the fact that the disclosure of the "Standardized Measure of Discounted Future Net Cash Flow from the Production of Proved Oil and Gas Reserves" (specified by *SFAS No. 69*) does not require reporting reserves based on contract type.

INTERNATIONAL ACCOUNTING STANDARDS

International Financial Reporting Standards (IFRS) are issued by the International Accounting Standards Board (IASB). All European Union countries follow International Accounting Standards, as do many other countries around the world. However, currently there is no IFRS that comprehensively addresses accounting issues in the upstream oil and gas industry. The IASB is evaluating alternatives and has expressed the intention of issuing such a standard at some time in the future. Meanwhile, companies using International Accounting Standards may opt to use *SFAS No. 19* in accounting for their exploration and evaluation activities. It does not appear that the full cost method will be acceptable to the IASB. Therefore, many full cost companies reporting in countries that follow International Accounting Standards have chosen to switch to successful efforts.

Regardless of the country in which the parent company is located, most foreign governments require companies operating within their country to file financial statements prepared in accordance with local statutory accounting rules. For example, a French company operating in Venezuela might be required to account for its Venezuelan operations using local Venezuelan statutory accounting rules and, for purposes of parent company reporting, to also account for the Venezuelan operations using International Accounting Standards.

Historically the SEC required foreign companies whose securities are traded in the U.S. to reconcile their foreign-GAAP financial statements with U.S. GAAP. Commencing November 2007, the SEC dropped this requirement for foreign registrants using IASB standards.

For a more through discussion of International Accounting Standards, see *International Petroleum Accounting* by Wright and Gallun.[2]

PROBLEMS

1. Define the following:

    domestic market obligation

    ringfencing

    cost oil

    profit oil

    capital uplift

    production bonus

    sliding scale royalty

    tax holiday

    royalty holiday

    government participation

2. Distinguish between the following:

   concessionary system

   contractual system

3. Distinguish between the following:

   nonrisk service contract

   risk service contract

4. Describe the similarities and differences between PSCs and risk service contracts.

5. Discuss the various methods that governments utilize to generate revenues and other benefits from mineral resources.

6. Explain how estimation of a company's net proved reserves differs when operating under a concessionary contract versus a PSC.

7. Jones Oil Company operates under a PSC agreement in the South China Sea. Jones has 49% of the working interest, and Sinhai Oil Company (which is owned by the Chinese government) has 51% of the working interest. The agreement calls for annual gross production to be split in the following order:

   a. VAT equal to 7% of annual gross production

   b. Royalty of 13% of annual gross production

   c. Cost oil is limited to 62% of annual gross production, with costs to be recovered in the following order:

      1) Operating expenses

      2) Exploration expenditures (Jones Oil Company, 100%)

      3) Development costs (Jones Oil Company, 49%, and Sinhai Oil Company, 51%)

   d. Annual gross production remaining after cost recovery becomes profit oil and is split:

      1) The government receives 15% of profit oil.

      2) The remaining 85% is shared by Jones and Sinhai based on their working interests.

   During 2012:

   Recoverable operating costs equal $4,000,000.

   Unrecovered exploration costs equal $10,000,000.

   Unrecovered development costs equal $100,000,000.

   The annual gross production for the year is 2,000,000 barrels of oil.

   REQUIRED:

   a. Assuming the price to be used to convert costs into barrels is $100/bbl, allocate the production to the parties.

b. Assuming the price to be used to convert costs into barrels is $60/bbl, allocate the production to the parties.

8. Assume that Protex Company, a U.S. company, is involved in petroleum operations in Thailand. Protex Company has a 40% WI, while the Local Oil Company has a 60% WI. Annual gross production is to be split in the following order:

   a. Royalty is 15% of annual gross production and is to be paid in-kind
   b. VAT is equal to 5% of annual gross production and is to be paid in-kind
   c. Cost oil is limited to 50% of gross production, with costs to be recovered in the following order:
      1) Operating expenses paid 40% by Protex Company and 60% by Local Oil Company.
      2) Exploration costs (paid entirely by Protex Company).
      3) Development costs: after completion of exploration, Local Oil Company opted to participate at 60%. Therefore, development and operating costs were paid 40% by Protex Company and 60% by Local Oil Company.
   d. Any excess remaining after cost recovery become profit oil:
      1) Of the profit oil, 25% goes to the government.
      2) The remainder is split between Protex and Local Oil Company based on their working interests.

   For 2014, assume the following:
   - Recoverable operating costs total $2,600,000.
   - Exploration costs unrecovered to date total $260,000,000.
   - Development costs unrecovered to date total $1,300,000,000.
   - Any costs not recovered in the current year may be carried forward to be recovered in future years.
   - The annual gross production for the year is 10,000,000 barrels of oil.
   - The agreed upon price is $65/bbl.

   REQUIRED: Allocate the production between the parties.

9. Ibis Company enters into a concession agreement with the British government. Ibis pays the government a $10,000,000 (U.S.) signing bonus and agrees to pay the government royalties of 8% of gross production and 5% severance tax. Ibis bears all of the costs associated with exploration, development, and production.

   During 2013, Ibis spends $7,000,000 on exploration and drilling costs. Gross revenue was $5,000,000, and production costs were $2,000,000. The income tax laws allow deduction of all production costs, with exploration and drilling costs deductible over a seven-year period. The tax rate is 40%.

   REQUIRED: Show how the gross revenue for 2013 would be shared by the parties.

10. Fortune Company enters into a risk service agreement with the Chilean government. Fortune pays the government, in U.S. dollars, a $5,000,000 signing bonus and also agrees to pay all of the costs associated with exploration, development, and production. [The contract defines costs incurred in the exploration and development phase of each project area as being capital costs (CAPEX), and all costs incurred in the production phase as being operating costs (OPEX).]

Each year in which production occurs, the government agrees to pay Fortune a fee comprised of the following:

a. All OPEX incurred in the current year

b. 1/10th of all unrecovered CAPEX

c. $0.40/bbl on production from 0 to 4,000 bbl/day, $0.60/bbl on production from 4,001 to 10,000 bbl/day, and $0.90/bbl on production above 10,000 bbl/day.

The maximum total fee that will be paid in any year is $1.20/bbl times the total number of barrels produced. Any unrecovered OPEX or CAPEX (unrecovered due to the maximum fee) can be carried forward to future years.

Assume that in 2012, production commences on the Llama Field. At that time, Fortune had spent $10,000,000 on CAPEX, and during 2012, spends $3,000,000 in OPEX. Production equals 5,000,000 barrels or 5,000,000/365 = 13,699 bbl/day.

REQUIRED: Compute the fee that Fortune Company would receive for 2012.

REFERENCES

1. Wright, Charlotte J., and Rebecca A. Gallun. 2005. *International Petroleum Accounting.* 1st ed. Tulsa, OK: PennWell.

2. Ibid.

16

ANALYSIS OF OIL AND GAS COMPANIES' FINANCIAL STATEMENTS

The financial statements of oil and gas producing companies contain a great deal of information that is not contained in the financial statements of companies in other industries. In earlier chapters, the reporting requirements in the United States were discussed, specifically the use of the two alternative historical cost accounting methods that are supplemented by extensive disclosures. These reporting requirements resulted primarily from the SEC's conclusion that neither the successful efforts method nor the full cost method effectively communicates information that is useful in projecting the future cash flows of oil and gas producing companies. Instead, the extensive disclosures mandated by *SFAS No. 69* are specifically designed to provide additional information that is presumed to be useful to an investor or potential investor. In addition, the SEC requires additional disclosures as described in *SEC Industry Guide 2*, "Disclosure of Oil and Gas Operations" (discussed later). But, by allowing companies to use a historical cost method to account for exploration, development, and production activities, the SEC preserved the historical cost information for potential use by financial statement users in comparing oil and gas producing companies to companies in other industries.

In this chapter, various ratios and techniques employed by investment analysts, company personnel, and others in evaluating the financial performance of oil and gas producing companies are discussed. The chapter focuses on the ratios that are unique to oil and gas producing companies.

SOURCE OF DATA

The financial statements and 10Ks—especially the *SFAS No. 69* disclosures—are the primary source of the data necessary to compute most of the ratios unique to oil and gas companies. As discussed in chapter 14, *SFAS No. 69* disclosures must be presented in the annual reports of both successful efforts and full cost companies. Specifically, the following disclosures are required:

Historical cost–based:

1. Proved reserve quantity information
2. Capitalized costs relating to oil and gas producing activities
3. Costs incurred for property acquisition, exploration, and development activities
4. Results of operations for oil and gas producing activities

Future value–based:

5. A standardized measure of discounted future net cash flows relating to proved oil and gas reserve quantities
6. Changes in the standardized measure of discounted future net cash flows relating to proved oil and gas reserve quantities

In addition, the SEC requires certain additional information to be included in 10K filings. These requirements are detailed in the *SEC Industry Guide 2*, "Disclosure of Oil and Gas Operations."[1] Some of the key items to be disclosed in 10K are detailed following.

Production. Production is to be reported for each of the last three fiscal years, broken down into the same geographic areas required by *SFAS No. 69*. The information is to include the following:

1. The average sales price (including transfers) per unit of oil produced and of gas produced. The transfer price of oil and gas produced should be determined in accordance with *SFAS No. 69*.
2. The average production cost (lifting cost) per unit of production. The average production cost per unit of production should be computed using production costs disclosed pursuant to *SFAS No. 69*. Note: for this calculation, production is to be converted to BOEs using the same conversion process that was used to compute BOEs for purposes of DD&A.

Productive wells and acreage. As of a reasonably current date or as of the end of the most recent fiscal year, companies must report a variety of information regarding productive wells and acreage. This information includes the following:

1. Total gross and net productive wells, expressed separately for oil and gas. One or more completions in the same bore hole are counted as one well. Productive wells are defined as producing wells and wells capable of production.
2. The total gross and net developed acres (i.e., acres spaced or assignable to productive wells) by geographic areas.

Undeveloped acreage. As of a reasonably current date or as of the end of the most recent fiscal year, companies must report the following:

1. The amounts of undeveloped acreage, expressed in both gross and net acres by appropriate geographic area. Undeveloped acreage is defined as being lease acres on which wells have not been drilled or completed to a point that would permit the production of commercial quantities of oil and gas, regardless of whether or not such acreage contains proved reserves. Undeveloped acreage should not be confused with undrilled acreage "held by production" under the terms of a lease.
2. The minimum remaining terms of leases and concessions, if material.

Drilling activity. Drilling activity information is to be reported for each of the last three fiscal years by appropriate geographic areas. For purposes of this disclosure, a dry well (hole) is defined as "an exploratory or a development well found to be incapable of producing either oil or gas in sufficient quantities to justify completion as an oil or gas well." Productive wells are defined as exploratory or development wells that are not dry wells. Information regarding drilling must include the following:

1. The number of net productive and dry exploratory wells
2. The number of net productive and dry development wells

Present activities. Present activities include the number of wells in the process of being drilled (including wells temporarily suspended), waterfloods in process of installation, pressure maintenance operations, and any other related operations of material importance by appropriate geographic areas.

Delivery commitments. Some companies are obligated to provide a fixed and determinable quantity of oil or gas in the near future under existing contracts or agreements. These companies must include material information concerning the estimated availability of oil and gas from all principal sources.

COMPARING FINANCIAL REPORTS

Financial statements and other reports are evaluated for several different purposes. Oil and gas companies compete in capital markets for new investment dollars. Investors must determine which companies to invest in, and lenders must determine to which companies to lend money. To make these determinations, investors and lenders use various criteria, including the analysis of certain key ratios or performance measures. Investors and lenders then compare the ratios to identify the firms that have the greatest potential for successful performance in the future.

In addition, companies are continuously seeking to improve their own performance. One technique a company can use to evaluate its performance is to compare itself to other similar companies. This comparison is commonly referred to as **benchmarking**. The process involves identifying a pool of companies who are "peer" companies, i.e., companies that are similar to the company in size, operations, and various other factors. Performance measures or key ratios are computed for all of those companies and analyzed to determine how the company compares to its peer companies. In this way, comparative strengths and weaknesses can be identified. A plan can then be established and implemented to build on the strengths and correct the weaknesses. Benchmarking is a critical step in improving the quality of business processes.

Whether the financial statement analysis is performed by outside investors or for internal benchmarking, it is important to keep in mind that an analysis that involves trends over several years is typically much more informative than the calculation of ratios for a single year. Recall from chapter 2 that one of the unique characteristics of the oil and gas industry is the often long time span between when exploration expenditures are made and when the results are known, i.e., whether oil or gas has been found. The use of ratios computed over several years will not resolve the timing problem of expenditures made in one year and results known in another. However, it will somewhat mitigate the effects and is superior to single year analysis.

The ratios commonly used in the oil and gas industry for financial statement analysis and benchmarking are discussed in the remainder of this chapter. The ratios discussed are reserve ratios, reserve cost ratios, reserve value ratios, and financial ratios.

Reserve Ratios

Reserve replacement ratio

One key performance measure is the reserve replacement ratio. This ratio measures a company's success in replacing production and accordingly measures a company's ability to continue to operate in the future. The primary means by which reserves can be replaced are by discoveries, extensions, in place purchases, and revisions. Replacement of reserves through revisions is troublesome and is viewed by some as being an artificial source of growth.

A firm that is not replacing the reserves it produces will ultimately deplete its pool of available reserves and be forced to either purchase reserves in place or simply cease to do business. Since purchasing reserves in place is typically more expensive than acquisition of reserves through discoveries and extensions, it is desirable for a company to consistently add to its reserves at a rate equal to or higher than its rate of production. Accordingly, a company replacing the reserves it produces should have a reserve replacement ratio of at least 1.

The reserve replacement ratio is somewhat problematic, since it does not reflect the value of the reserve replacements. Another problem is determining which categories of reserves to include in the calculation. Specifically, reserve additions resulting from revisions in previous estimates and purchases of reserves in place are often treated in different ways. Consequently, there are multiple ways to calculate the reserve replacement ratio, three of which are described below. The most basic formula is as follows:

$$\text{Reserve replacement ratio} = \frac{\text{Extensions and discoveries} + \text{Improved recovery}}{\text{Production}}$$

Computed in this manner, the reserve replacement ratio considers only reserve additions resulting from current discoveries, extensions, and improved recovery. Some analysts argue that economic conditions may result in revisions in previously discovered reserves, and that those revisions should not be overlooked in examining a company's ability to replace its reserves. On the other hand, many of the conditions that result in revisions in estimates are beyond the control of a company's management. Thus, inclusion of those reserves does not accurately reflect management's role in reserve replacement. An alternative reserve replacement ratio that includes any current year revisions in reserve estimates in the numerator of the formula is as follows:

$$\text{Reserve replacement ratio} = \frac{\text{Extensions and discoveries} + \text{Improved recovery} + \text{Revisions in previous estimates}}{\text{Production}}$$

Another calculation includes reserves that are purchased in place (rather than acquired through discovery). Some analysts argue that, in certain situations, purchasing reserves is an alternative that may be in the best interest of the company. If reserves acquired by purchase are included in the numerator of the reserve replacement ratio, then any reserves sold in place (as opposed to through production) should be added to the denominator as follows:

$$\text{Reserve replacement ratio} = \frac{\text{Extensions and discoveries} + \text{Improved recovery} + \text{Revisions in previous estimates} + \text{Purchases of reserves in place}}{\text{Production} + \text{Sales of reserves in place}}$$

Presumably, a firm could not consistently replace its reserves by purchasing reserves in place. Therefore, the above formula might only be useful during those periods when in place purchases are material.

EXAMPLE

Reserve Replacement Ratio

Assume that the following information appears in a *SFAS No. 69* disclosure on Tyler Company's 2012 annual report. This basic data is used throughout the chapter to calculate various ratios.

Estimated Quantities of Net Proved Crude Oil and Natural Gas Liquids (Worldwide Totals) *in Thousands of Barrels*

Year ended Dec. 31	2010	2011	2012
Beginning of year proved reserves	2,553	2,654	2,729
Revisions of previous estimates	60	130	221
Improved recovery	210	132	115
Purchases of reserves in place	50	15	10
Sales of reserves in place	(23)	(32)	(37)
Extensions and discoveries	53	73	65
Production	(249)	(243)	(275)
End of year proved reserves	2,654	2,729	2,828

The reserve replacement ratios calculated using the three alternative methods for 2010, 2011, and 2012 are as follows:

2010:

$$\text{Reserve replacement ratio} = \frac{53 + 210}{249} = 1.056$$

$$\text{With revisions} = \frac{53 + 210 + 60}{249} = 1.297$$

$$\text{With revisions and in place sales} = \frac{53 + 210 + 60 + 50}{249 + 23} = 1.371$$

2011:

$$\text{Reserve replacement ratio} = \frac{73 + 132}{243} = 0.844$$

$$\text{With revisions} = \frac{73 + 132 + 130}{243} = 1.379$$

$$\text{With revisions and in place sales} = \frac{73 + 132 + 130 + 15}{243 + 32} = 1.273$$

2012:

$$\text{Reserve replacement ratio} = \frac{65 + 115}{275} = 0.655$$

$$\text{With revisions} = \frac{65 + 115 + 221}{275} = 1.458$$

$$\text{With revisions and in place sales} = \frac{65 + 115 + 221 + 10}{275 + 37} = 1.317$$

If a company has both oil and gas reserves, the reserve replacement ratio and the reserve life ratio (discussed below) typically are calculated separately for each mineral. The ratios may be calculated using equivalent units if called for by the particular analysis being done.

Reserve life ratio

The reserve life ratio is used to approximate or measure the number of years that production could continue at the current rate if no new reserves were added. The higher the reserve life ratio, the longer a firm could continue to generate enough cash flow to cover its financial obligations even if it curtailed exploration and discovery activities—assuming current production generates enough cash flow to cover financial obligations. The reserve life ratio is calculated as follows:

$$\text{Reserve life ratio} = \frac{\text{Total proved reserves at beginning of year}}{\text{Production}}$$

The ratio can be computed separately for oil reserves and for gas reserves. The average reserve life ratio for large companies is approximately eight to nine years for oil, and longer for gas.

EXAMPLE

Reserve Life Ratio

Using the basic data from the previous example for Tyler Company, the reserve life ratio for each year is as follows:

$$2010: \quad \frac{2{,}553}{249} = 10.25 \text{ years}$$

$$2011: \quad \frac{2{,}654}{243} = 10.92 \text{ years}$$

$$2012: \quad \frac{2{,}729}{275} = 9.92 \text{ years}$$

Net wells to gross wells ratio

Companies must disclose their total gross wells and net wells in their 10Ks filed with the SEC. Gross wells is the number of wells in which the company has any working interest, regardless of how large or small the interest may be. Net wells is the net interest in the wells, i.e., the total of all working interests in all wells. It is computed by multiplying each well in which a company has a working interest by the relevant working interest percentage and summing the result. For example, a company might have a 100% WI in 5 wells, a 75% working interest in 3 wells, and a 50% WI in 2 wells. It would then have 10 gross wells (i.e., 5 + 3 + 2) and 8.25 net wells (i.e., 5 × 1.0 + 3 × 0.75 + 2 × 0.50).

The net wells to gross wells ratio is calculated as follows:

$$\text{Net to gross wells} = \frac{\text{Net wells}}{\text{Gross wells}}$$

The ratio of net wells to gross wells is used as a gauge of future profitability. The rationale is if a company owns a large interest in each well, the company is likely to be the operator, benefit from being the operator, and as the operator, have a greater say in operations. In addition, the company is likely to be more profitable as a result of consolidated interests (i.e., having relatively larger interests in fewer properties rather than having to spread its resources over more properties with smaller interests). A high ratio indicates that a company owns relatively large working interests in wells, while a low ratio indicates that a company owns many small working interests.

EXAMPLE

Net Wells to Gross Wells Ratio

The table below shows Tyler's interests in various wells and the calculation of gross wells and net wells:

Gross Wells	Working Interest	Net Wells
15	100%	15.0
10	75%	7.5
15	60%	9.0
20	50%	10.0
5	40%	2.0
2	20%	0.4
6	15%	0.9
2	10%	0.2
75		45.0

Tyler Company has an interest in 75 wells, so its gross wells is 75. Tyler's net wells is 45, and its ratio of net wells to gross wells is as follows:

$$\text{New wells to gross wells} = \frac{45}{75} = 0.60$$

Average reserves per well ratio

The average reserves per well ratio is another measure used to evaluate a company's future profitability. The measure calculates the average reserves per well. A high average reserves per well ratio indicates that a given quantity of reserves can be produced with fewer wells and thus be produced more efficiently and more profitably. Consequently, the higher the average reserves per well ratio, the greater the company's future profit potential. The formula is given below:

$$\text{Average reserves per well ratio} = \frac{\text{Total proved reserves at beginning of year}}{\text{Net wells}}$$

EXAMPLE

Average Reserves per Well Ratio

Using the proved reserves presented earlier for Tyler Company and the net well calculation presented earlier, the average reserves per well for each of the three years are (assuming the net wells do not change over the years):

$$2010: \quad \frac{2{,}553{,}000}{45} = 56{,}733.333 \text{ bbl/well}$$

$$2011: \quad \frac{2{,}654{,}000}{45} = 58{,}977.778 \text{ bbl/well}$$

$$2012: \quad \frac{2{,}729{,}000}{45} = 60{,}644.444 \text{ bbl/well}$$

When a company has both oil and gas reserves, the average reserves per well ratio and the average daily production per well ratio (discussed below) are normally computed by converting the reserves to a common unit of measure based on energy content, specifically, barrels of energy (BOE). Most frequently, the British thermal unit conversion (Btu) (approximately 6 to 1) is used to compute BOE (chapters 6 and 7). For example, if Tyler Company also had proved gas reserves at the beginning of 2010 of 3,000,000 Mcf, the average reserves per well ratio for 2010, assuming the same net wells, would be as follows:

$$\text{Average reserves per well ratio} = \frac{2{,}553{,}000 + 3{,}000{,}000/6}{45} = 67{,}844.44 \text{ BOE/well}$$

Average daily production per well

The average daily production rate per well is yet another measure of a company's future profitability. The higher the average daily production per well, the more efficiently and profitably the reserves can be produced. The average daily production per well is computed as follows:

$$\text{Average daily production per well} = \frac{\text{Annual production}/365}{\text{Net wells}}$$

EXAMPLE

Average Daily Production per Well

Using the reserve and production information presented thus far for Tyler Company, the average daily production per well for each of the three years is as follows:

$$2010: \quad \frac{249{,}000/365}{45} \quad = \quad 15.1598 \text{ bbl/day/well}$$

$$2011: \quad \frac{243{,}000/365}{45} \quad = \quad 14.7945 \text{ bbl/day/well}$$

$$2012: \quad \frac{275{,}000/365}{45} \quad = \quad 16.7428 \text{ bbl/day/well}$$

RESERVE COST RATIOS

The reserve cost ratios are based on costs and on reserves. When a company has both oil and gas reserves, the reserve cost ratios are computed by converting the reserves to a common unit of measure based on energy content (BOE).

Finding costs ratios

Finding costs per BOE is one of the most common, but difficult to define, performance measures used in evaluating oil and gas operations. Finding costs per BOE is the most frequently cited ratio utilized in evaluating the efficiency of a company in adding new reserves. The basic ratio consists of the *finding costs of adding new reserves* divided by the *new reserves added*. The difficulty with calculating and using the finding costs per BOE ratio results from several factors.

First, there is no consensus regarding which costs should be included as finding costs. Second, companies use different methods of accounting for oil and gas exploration and development operations, i.e., full cost and successful efforts. Consequently, even if there were a specific definition of finding costs, the amounts would likely still differ, since the various accounting methods treat the costs differently in terms of expense and capitalization. Third, there is typically a timing difference between the period(s) when the finding costs were actually expended and when the new reserves are actually reported in the financial statements. This timing difference poses a difficulty in interpreting the finding costs per BOE ratio. Finally, there is some debate as to which reserve estimates should actually be used in the calculation. These issues are discussed below.

Two methods of accounting for exploration and development costs are currently accepted in practice: the successful efforts method and the full cost method. In general, under the successful efforts method, geological and geophysical (G&G) exploration costs are written off as incurred. The costs of dry exploratory wells are written off when the determination is made that the well is dry. The costs of successful exploratory wells and successful and dry development wells are capitalized and amortized over production. Under the full cost method, all costs incurred in exploration, drilling, and development are capitalized and amortized. Clearly, any attempt to calculate finding costs for a sample of both full cost and successful efforts companies would require a detailed evaluation of all of the reported costs to ensure that equivalent costs, whether expensed or capitalized, are used.

In computing finding costs per BOE, there should be correspondence or matching between the costs in the numerator and the reserves in the denominator. The difficulty is determining which reserves should be used to correspond to the costs in the numerator. Reserves categories include reserves added through discoveries and extensions, purchases of reserves in place, revisions in reserve estimates, and enhanced recovery. If reserves added through extensions and discoveries are to be used, then finding costs per BOE should be calculated as follows:

Formula 1 without revisions:

$$\text{Finding costs/BOE} = \frac{\text{G\&G exploration costs} + \text{All exploratory drilling costs}}{\text{Reserve extensions and discoveries (excluding revisions)}}$$

Formula 1 with revisions:

$$\text{Finding costs/BOE} = \frac{\text{G\&G exploration costs} + \text{All exploratory drilling costs}}{\text{Reserve extensions and discoveries} + \text{Revisions in previous estimates}}$$

Some analysts include current reserve revisions, while others exclude them. There is no single, consistently used method. One possible solution that is frequently used is recomputing the previous year's finding costs per BOE by adjusting the reserves for that year by revisions reported in the current year. The logic is that since the revisions relate to prior years rather than the current year, the revisions should be reflected in the ratio for the prior years. Obviously this solution assumes that finding costs per BOE would be analyzed in the context of several years' data rather than for a single year.

The next issue is whether to include reserves purchased in place. If these reserves are to be included in the calculation, proved properties must be included in the numerator so that costs and reserves correspond. The formula would be as follows:

Formula 2 without revisions:

$$\text{Finding costs/BOE} = \frac{\text{G\&G exploration costs + All exploratory drilling costs + Proved properties}}{\text{Reserve extensions and discoveries + Purchased reserves in place}}$$

Formula 2 with revisions:

$$\text{Finding costs/BOE} = \frac{\text{G\&G exploration costs + All exploratory drilling costs + Proved properties}}{\text{Reserve extensions and discoveries + Purchased reserves in place + Revisions in previous estimates}}$$

Finally, sometimes finding costs per BOE is computed by attempting to include all costs necessary to replace reserves. This ratio includes all of the costs in the above ratio plus the cost of unproved properties and development drilling expenditures, as follows:

Formula 3 with revisions:

$$\text{Finding costs/BOE} = \frac{\text{G\&G exploration costs + All exploratory and development drilling costs + Proved and unproved property acquisition costs}}{\text{All reserve additions (including revisions in previous estimates)}}$$

The purpose of a development well is not to find reserves but to produce previously discovered proved reserves. Consequently, when development drilling costs are included in the ratio, this ratio is sometimes referred to as a **finding and development cost ratio** or a **reserve replacement cost ratio**.

These different formulae can result in significantly different finding costs per BOE. Gaddis, Brock, and Boynton used the financial statements of several oil and gas producers to demonstrate the vastly different results in finding costs per BOE.[2]

- Method A: G&G costs and exploratory drilling costs only are divided by reserve additions, excluding revisions.
- Method B: G&G costs and exploratory drilling costs only are divided by reserve additions, including revisions.
- Method C: G&G costs, exploratory drilling costs, and development drilling costs are divided by all reserve additions, including revisions.

The following results were computed using a five-year period (1986–1990):

Company	Method A	Method B	Method C
Amoco	$6.59	$3.20	$6.15
ARCO	$4.19	$2.93	$4.46
Chevron	$8.75	$2.90	$5.23
Conoco	$5.58	$3.27	$6.69
Exxon	$4.53	$3.18	$9.27
Marathon	$2.68	$2.38	$5.68
Mobil	$5.97	$2.34	$3.54
Phillips	$3.53	$1.61	$2.74
Texaco	$4.71	$1.81	$4.04
Unocal	$2.04	$1.93	$4.43
Average	$4.86	$2.56	$5.22

Given the lack of correspondence between the timing of the expenditure and the addition of the reserves, Gaddis, Brock, and Boynton recommend computing finding costs using moving averages across several years.

EXAMPLE

Finding Costs per BOE

Using the reserve disclosure data presented earlier for Tyler Company, finding costs per BOE are computed below using the various formulae. Assume the following costs (in thousands):

	2010	2011	2012
Unproved property acquisition	$ 25	$ 45	$ 80
Proved property acquisition	50	50	50
G&G	300	350	325
Exploratory drilling (including dry hole)	321	400	450
Development drilling	150	90	200

Formula 1 without revisions:

2010: $$\frac{\$300 + \$321}{53} = \$11.717/\text{BOE}$$

2011: $$\frac{\$350 + \$400}{73} = \$10.274/\text{BOE}$$

2012: $$\frac{\$325 + \$450}{65} = \$11.923/\text{BOE}$$

Formula 1 with revisions:

2010: $$\frac{\$300 + \$321}{53 + 60} = \$5.496/\text{BOE}$$

2011: $$\frac{\$350 + \$400}{73 + 130} = \$3.695/\text{BOE}$$

2012: $$\frac{\$325 + \$450}{65 + 221} = \$2.710/\text{BOE}$$

Formula 2 without revisions:

2010: $$\frac{\$300 + \$321 + \$50}{53 + 50} = \$6.515/\text{BOE}$$

2011: $$\frac{\$350 + \$400 + \$50}{73 + 15} = \$9.091/\text{BOE}$$

2012: $$\frac{\$325 + \$450 + \$50}{65 + 10} = \$11.000/\text{BOE}$$

Formula 2 with revisions:

2010: $$\frac{\$300 + \$321 + \$50}{53 + 50 + 60} = \$4.117/\text{BOE}$$

2011: $$\frac{\$350 + \$400 + \$50}{73 + 15 + 130} = \$3.670/\text{BOE}$$

2012: $$\frac{\$325 + \$450 + \$50}{65 + 10 + 221} = \$2.787/\text{BOE}$$

Formula 3 with revisions:

2010: $$\frac{\$300 + \$321 + \$150 + \$50 + \$25}{53 + 50 + 60 + 210} = \$2.268/\text{BOE}$$

2011: $$\frac{\$350 + \$400 + \$90 + \$50 + \$45}{73 + 15 + 130 + 132} = \$2.671/\text{BOE}$$

2012: $$\frac{\$325 + \$450 + \$200 + \$50 + \$80}{65 + 10 + 221 + 115} = \$2.689/\text{BOE}$$

In summary, since there is no consensus regarding the most appropriate formula, it is important to consistently use the formula or formulas selected. Also, in any analysis, it is important to look at trends and averages over several years.

Lifting costs per BOE

Lifting costs per BOE is a very popular performance indicator. Lifting costs per BOE is a measure that may be used to evaluate the extent to which a company is controlling its operating costs or how efficiently the company is getting oil and gas out of the ground, or both. The basic formula is as follows:

$$\text{Lifting costs/BOE} = \frac{\text{Total annual lifting costs}}{\text{Annual production (BOE)}}$$

In order to use this ratio appropriately, it is important to understand the costs that may or may not be included and how the results should be interpreted.

Frequently the terms *lifting costs* and *production costs* are used interchangeably. The FASB defines production costs in *SFAS No. 19*, par. 24, as follows:

> *Those costs incurred to operate and maintain an enterprise's wells and related equipment and facilities, including depreciation and applicable operating costs of support equipment and facilities (paragraph 26) and other costs of operating and maintaining those facilities.*

Using this definition, examples of production costs would include labor, repairs and maintenance, supplies, materials, fuel, property taxes, and severance taxes.

If lifting cost per BOE is used to measure how effectively companies are controlling their costs, then it is important that the ratio include only those costs actually controllable at the field level. For example, depreciation, property taxes, and severance taxes are rarely within the control of field and operating personnel. Accordingly, if all production costs are included, then the ratio may actually measure overall operating efficiency rather than the control of costs at the field level.

There are several factors that should be considered when using lifting cost per BOE in benchmarking, or when comparing several companies. For example, the ratio is computed across oil and gas production; however, it typically costs more to produce oil than to produce gas. Therefore, if one company has primarily oil production, while another company has primarily gas production, the first company may appear to be much less efficient compared to the second company. In reality, that may not be the case. Careful selection of peer companies will help mitigate this issue.

Lifting costs per BOE should not be used without giving adequate consideration to revenues and net income. A company having higher lifting cost per BOE may also have higher revenues and higher net income per BOE. For example, assume that Company A's lifting cost (not including depreciation) per BOE is $24.50, and Company B's lifting cost per BOE is $21.00. It would appear that Company B is more efficient, and therefore, the more profitable operator. Assume, however, that Company A's costs are largely driven by the fact that it is operating in an offshore location where the crude oil is of much higher quality. As a result, Company A is able to sell its oil for $126/bbl, while Company B's average revenue is $120/bbl. Company A's gross profit per barrel is $101.50 (i.e., $126 − $24.50), while Company B's gross profit per barrel is $99.00 (i.e., $120 − $21.00). This example illustrates the need to utilize lifting cost per BOE in the context of other financial ratios. It also illustrates the need to choose peer companies with similar operations, e.g., mostly offshore versus onshore, etc.

When used appropriately, lifting cost per BOE is an informative performance measure. When computing lifting cost per BOE by using financial statement information, *SFAS No. 69* is commonly used as a source of lifting or production costs. The reserve report is the source of current period production information.

EXAMPLE

Lifting Costs per BOE

Assume that the production costs that appear on Tyler Company's annual report are (in thousands):

Year	Lifting (Production) Costs (excluding DD&A)	DD&A
2010	$500	$800
2011	525	775
2012	510	750

Using the reserve production information presented previously, lifting cost per BOE would be computed as follows:

$$2010: \quad \frac{\$500}{249} = \$2.008/BOE$$

$$2011: \quad \frac{\$525}{243} = \$2.160/BOE$$

$$2012: \quad \frac{\$510}{275} = \$1.855/BOE$$

DD&A per BOE

DD&A reflects the historical cost of finding and developing reserves and, as such, relates to prior periods, while lifting costs relate to the current period. Consequently, DD&A per BOE is not helpful in assessing current period efficiencies. However, since DD&A appears on the income statement, it does affect current profitability. One means of dealing with depreciation, depletion, and amortization (DD&A) is to omit DD&A on all wells, equipment, and facilities from the lifting cost per BOE calculation and separately compute a ratio of DD&A per BOE. The formula is as follows:

$$DD\&A/BOE = \frac{\text{Total annual DD\&A}}{\text{Annual production (BOE)}}$$

EXAMPLE

DD&A per BOE

Using the information presented previously for Tyler Company, DD&A per BOE is given in the following:

2010: $\dfrac{\$800}{249} = \$3.213/\text{BOE}$

2011: $\dfrac{\$775}{243} = \$3.189/\text{BOE}$

2012: $\dfrac{\$750}{275} = \$2.727/\text{BOE}$

RESERVE VALUE RATIOS

Value of proved reserve additions per BOE

One criticism often leveled at ratio analyses using reserve quantities is that the analyses do not convey information about the "quality" of the reserves added. Some reserves are of high quality, i.e., they require relatively less in production, transportation, or refining costs. In contrast, the production of other reserves requires extensive expenditures for lifting, transportation, or refining. While the disclosures mandated by *SFAS No. 69* are controversial, the use of certain value-related amounts found in the disclosures in ratio analyses has become a popular means of incorporating reserve value. One such ratio is the value of proved reserve additions per BOE. This ratio is computed using certain elements of the changes in the standardized measure disclosure and the reserve quantity information disclosure. For example:

Formula 1 without revisions:

$$\text{Value of proved reserve additions/BOE} = \dfrac{\text{Changes due to extensions, discoveries, and improved recovery}}{\text{Reserve extensions and discoveries + Improved recovery}}$$

Formula 1 with revisions:

$$\text{Value of proved reserve additions/BOE} = \frac{\text{Changes due to extensions, discoveries, and improved recovery + Changes due to revisions in estimates}}{\text{Reserve extensions and discoveries + Improved recovery + Revisions in previous estimates}}$$

Formula 2 without revisions:

$$\text{Value of proved reserve additions/BOE} = \frac{\text{Changes due to extensions, discoveries, and improved recovery + Changes due to purchases of reserves in place}}{\text{Reserve extensions and discoveries + Improved recovery + Purchases of reserves in place}}$$

Formula 2 with revisions:

$$\text{Value of proved reserve additions/BOE} = \frac{\text{Changes due to extensions, discoveries, and improved recovery + Changes due to purchases of reserves in place + Changes due to revisions in estimates}}{\text{All reserve additions (including revisions in previous estimates)}}$$

As is the case with finding costs per BOE, the calculation of value of proved reserve additions per BOE requires consistency between the value figures included in the numerator and the reserve categories included in the denominator. Again, for the best results, multiple year analyses should be used.

EXAMPLE

Value of Proved Reserve Additions per BOE

The following information is taken from *SFAS No. 69* disclosures in Tyler Company's annual report:

Changes in Standardized Measure of Discounted Future Net Cash Flows
in Thousands

Year ended Dec. 31	2010	2011	2012
Beginning of year	$20,355	$20,212	$19,792
Changes due to prices and costs	801	500	(902)
Changes due to purchases in place	530	220	139
Changes due to extensions, discoveries, and improved recovery	599	890	712
Changes due to revisions in estimates	699	1,100	989
Changes due to sales in place	(225)	(350)	(400)
Changes due to production	(2,700)	(2,980)	(3,000)
Accretion of discount	153	200	250
Balance end of year	$20,212	$19,792	$17,580

Formula 1 without revisions:

2010: $\dfrac{\$599}{53 + 210} = \$2.278/\text{BOE}$

2011: $\dfrac{\$890}{73 + 132} = \$4.341/\text{BOE}$

2012: $\dfrac{\$712}{65 + 115} = \$3.956/\text{BOE}$

Formula 1 with revisions:

2010: $\dfrac{\$599 + \$699}{53 + 210 + 60} = \$4.019/\text{BOE}$

2011: $\dfrac{\$890 + \$1,100}{73 + 132 + 130} = \$5.940/\text{BOE}$

2012: $\dfrac{\$712 + \$989}{65 + 115 + 221} = \$4.242/\text{BOE}$

Formula 2 without revisions:

2010: $\dfrac{\$599 + \$530}{53 + 210 + 50} = \$3.607/\text{BOE}$

2011: $\dfrac{\$890 + \$220}{73 + 132 + 15} = \$5.045/\text{BOE}$

2012: $\dfrac{\$712 + \$139}{65 + 115 + 10} = \$4.479/\text{BOE}$

Formula 2 with revisions:

2010: $\dfrac{\$599 + \$530 + \$699}{53 + 210 + 50 + 60} = \$4.901/\text{BOE}$

2011: $\dfrac{\$890 + \$220 + \$1{,}100}{73 + 132 + 15 + 130} = \$6.314/\text{BOE}$

2012: $\dfrac{\$712 + \$139 + \$989}{65 + 115 + 10 + 221} = \$4.477/\text{BOE}$

Value added ratio

Another popular reserve value ratio is the value added ratio. First, the value of proved reserve additions per BOE ratio as described above is computed. This ratio is then compared to finding costs per BOE. Specifically, the value added ratio is computed by dividing the value of proved reserve additions per BOE ratio by the finding costs per BOE ratio. The objective of this analysis is to compare the cost of finding reserves with the value added by those reserves. Obviously such a comparison requires consistent use of costs and reserve categories in each formula. For example, assume the following:

$$\text{Finding costs/BOE} = \dfrac{\text{G\&G exploration costs + All exploratory and development drilling costs + Proved and unproved property acquisition costs}}{\text{All reserve additions (including revisions in previous estimates)}}$$

The comparison of finding costs and value added would require that the value of proved reserve additions ratio per BOE include similar categories of values and reserves. Accordingly:

$$\text{Value of proved reserve additions/BOE} = \frac{\text{Changes due to extensions, discoveries, and improved recovery} + \text{Changes due to purchases of reserves in place} + \text{Changes due to revisions in estimates}}{\text{All reserve additions (including revisions in previous estimates)}}$$

A value added ratio may then be computed:

$$\text{Value added ratio} = \frac{\text{Value of proved reserve additions/BOE}}{\text{Finding costs/BOE}}$$

Companies should strive to maximize the value added ratio. Companies having the highest value added ratio are adding maximum reserve value at minimum finding costs.

EXAMPLE

Value Added Ratio

The value added ratios for 2010, 2011, and 2012 for Tyler Company are computed below. (The finding costs per BOE and value of proved reserve additions per BOE ratios were computed previously.)

Finding costs/BOE (formula 3 with revisions):

$$\text{Finding costs/BOE} = \frac{\text{G\&G exploration costs} + \text{All exploratory and development drilling costs} + \text{Proved and unproved property acquisition costs}}{\text{All reserve additions (including revisions in previous estimates)}}$$

Divided into:

Value of proved reserve additions/BOE (formula 2 with revisions):

$$\text{Value of proved reserve additions/BOE} = \frac{\text{Changes due to extensions, discoveries, and improved recovery} + \text{Changes due to purchases of reserves in place} + \text{Changes due to revisions in estimates}}{\text{All reserve additions (including revisions in previous estimates)}}$$

Value added ratio:

2010: $\dfrac{\$4.901}{\$2.268} = 2.161$

2011: $\dfrac{\$6.314}{\$2.671} = 2.364$

2012: $\dfrac{\$4.477}{\$2.689} = 1.665$

Financial Ratios

There are many other ratios that are frequently used in financial statement analysis. These ratios are used to evaluate companies in all industries, including oil and gas companies. These ratios include the following:

Liquidity ratios:

Current ratio $= \dfrac{\text{Current assets}}{\text{Current liabilities}}$

Quick ratio $= \dfrac{\text{Liquid current assets}}{\text{Current liabilities}}$

Working capital $=$ Current assets − Current liabilities

Financial strength ratios:

$$\text{Debt to stockholder's equity} = \frac{\text{Total long and short-term debt}}{\text{Total stockholder's equity}}$$

$$\text{Debt to assets} = \frac{\text{Total long and short-term debt}}{\text{Total assets}}$$

$$\text{Times interest earned} = \frac{\text{Net income}}{\text{Total interest expense}}$$

Profitability ratios:

$$\text{Net income to sales} = \frac{\text{Net income}}{\text{Sales}}$$

$$\text{Return on stockholder's equity} = \frac{\text{Net income}}{\text{Total stockholder's equity}}$$

$$\text{Return on assets} = \frac{\text{Net income}}{\text{Total assets}}$$

$$\text{Cash flow from operations to sales} = \frac{\text{Current year cash flow}}{\text{Current year sales}}$$

$$\text{Price/earnings ratio} = \frac{\text{Market value of common/share}}{\text{Adjusted earnings/share}}$$

$$\text{Price/cash flow ratio} = \frac{\text{Market value of common/share}}{\text{Cash flow/share}}$$

Some analysts incorporate *SFAS No. 69* reserve values into some of these formulae. For example, in computing ratios that include total assets or stockholder's equity (i.e., return on assets or return on stockholder's equity), the standardized measure of discounted future net cash flows may be substituted for proved properties as reported in the balance sheet. Additionally, a comparison between the book value of oil and gas assets and the standardized measure of discounted future net cash flows is sometimes used as a measure of the financial strength of underlying oil and gas assets.

PROBLEMS

1. What is benchmarking?

2. What is the reserve replacement ratio? What is the reserve replacement ratio attempting to measure? How would you interpret it?

3. What is the reserve life ratio? What is the reserve life ratio attempting to measure? How would you interpret it?

4. What does a high net wells to gross wells ratio indicate?

5. What does a low average reserves per well ratio indicate?

6. How do you interpret the average daily production per well ratio?

7. Discuss the difficulties with computing and applying finding costs ratios.

8. How do you interpret finding cost ratios? Why are they so popular in financial statement analysis?

9. What does a high (low) lifting cost per BOE indicate? When does lifting cost per BOE indicate that costs are effectively being controlled at the field level?

10. How do you interpret DD&A per BOE?

11. What is the value of proved reserve additions ratio attempting to measure? How do you interpret the value of proved reserve additions ratio?

12. What is the value added ratio? What does a high (low) value added ratio indicate?

13. The following reserve table appeared in the financial statements of Lomax Company:

Estimated Quantities of Net Proved Crude Oil and Natural Gas
(Worldwide Totals only)
in Thousands of Barrels and Millions of Cubic Feet

Year ended Dec. 31	2015		2016		2017	
	Oil	Gas	Oil	Gas	Oil	Gas
Beginning of year	171	779	234	783	335	724
Revisions of previous estimates	10	12	15	31	(11)	22
Improved recovery	21	30	25	23	15	50
Purchases of reserves in place	0	0	12	12	0	24
Sales of reserves in place	(12)	(12)	(20)	(99)	(70)	(24)
Extensions & discoveries	69	78	90	42	6	150
Production	(25)	(104)	(21)	(68)	(24)	(76)
End of year totals	234	783	335	724	251	870

Year	Net Wells	Gross Wells
2015	750	2010
2016	840	1910
2017	900	2050

REQUIRED: Compute the following ratios for all three years:

a. The reserve replacement ratio computed for all three methods and for oil and gas separately

b. The reserve life ratio computed for oil and gas separately

c. The net wells to gross wells ratio

d. The average reserves per well ratio computed using BOE, i.e., combining reserves based on relative energy content

e. The average daily production per well computed using BOE

14. Lomax Company reported the following costs on its financial statements (in thousands):

	2015	2016	2017
Unproved property acquisition	$405	$200	$800
Proved property acquisition	0	500	350
G&G	500	650	425
Exploratory drilling (including dry hole)	221	700	650
Development drilling	50	90	300

REQUIRED: Using the reserve disclosure for Lomax Company in problem 13 and the data presented in this problem, compute finding costs per BOE using the various formulae.

15. Lomax Company reported the following expenses in its financial statements (in thousands):

Year	Lifting Costs	DD&A
2015	$211	$500
2016	226	450
2017	183	525

REQUIRED: Using the reserve disclosure for Lomax Company given in problem 13 and the data presented in this problem:

a. Compute lifting costs per BOE.

b. Compute DD&A per BOE.

16. Lomax Company's Statement No. 69 disclosures included the following information:

Changes in Standardized Measure of Discounted Future Net Cash Flows
in Thousands

Year ended Dec. 31	2015	2016	2017
Beginning of year	$1,131	$ 643	$1,656
Changes due to prices and costs	(830)	550	200
Changes due to purchases in place	0	156	139
Changes due to extensions, discoveries, and improved recovery	400	180	712
Changes due to revisions in estimates	99	310	39
Changes due to sales in place	(22)	(35)	(40)
Changes due to production	(270)	(398)	(500)
Accretion of discount	135	250	199
Balance at end of year	$ 643	$1,656	$2,405

REQUIRED: Using the information for Lomax Company in problems 13, 14, and 15 and in this problem:

a. Compute the value of proved reserve additions per BOE using the various formulae.

b. Compute the value added ratio for each year, utilizing formula 3 for finding costs per BOE, and formula 2 with revisions for value of reserves added per BOE.

17. Discuss the ratios computed for Lomax Company in problems 13, 14, 15, and 16. What is your assessment of the performance and future potential of Lomax Company?

REFERENCES

1. U.S. Securities and Exchange Commission. 2007. "Disclosure of Oil and Gas Operations." *SEC Industry Guide 2*. Washington, D.C.: Securities and Exchange Commission.

2. Gaddis, Dwight, Horace Brock, and Charles Boynton. 1992. "Pros, Cons of Techniques Used to Calculate Oil, Gas Finding Costs." Oil & Gas Journal. June: pp. 93–95.

APPENDIX A

AUTHORIZATION FOR EXPENDITURE

When joint working interests are involved, the working interest owner designated as operator obtains approval for estimated expenditures by the use of an Authorization for Expenditure (AFE). The AFE generally includes cost estimates of the projected work in enough detail for the nonoperators to determine the reasonableness of the dollar amounts.

The following AFE may be used for drilling and completing a well. As shown, a completed AFE delineates the cost of the activities that are necessary in drilling and completing a well. The AFE breaks out these costs by IDC and equipment. The nonoperators are interested, for tax purposes, in the amounts to be incurred for IDC and thus potentially subject to an immediate write-off.

The AFE is a cost estimate; however, the authorization extends to the actual costs incurred. The authorization may apply to only dry-hole costs (i.e., drilling costs only) and not to completion costs. In most instances, the nonoperators authorizing the expenditures would authorize both dry-hole and completion costs.

AUTHORITY FOR EXPENDITURE

WELL/PROSPECT: J. Stevenson #1 **FIELD:** Austin Chalk

LOCATION: 467' FSL & 4625' FSWL of the T. Roberts A-1122

COUNTY: Nueces County **STATE** Texas

AFE DESCRIPTION: Drill & Complete 8900' Test

	DRY HOLE	PRODUCER
INTANGIBLE COST ESTIMATE:		
01 Roads, Location, Survey	30,000	30,000
02 Legal, Damages, Cleanup	1,500	1,500
03 Drilling Contractor, MI, RU, RD		
04 Drilling Contractor, Footage, Daywork	126,000	145,000
05 Completion, Workover, Swab Unit		10,000
06 Mud Chemicals	25,000	25,000
07 Cement, Cementing Services	10,000	18,000
08 Float Equipment	2,000	5,000
09 Open Hole Logging & Evaluation	20,000	20,000
10 Drillstem Testing		
11 Mud Logging		
12 Directional Drilling		
13 Fishing Tools and Services		
14 Water, Fuel	9,000	9,000
15 Bits	2,500	2,500
16 Rentals	25,000	25,000
17 Trucking/Transportation	10,000	15,000
18 Boats/Dockage		
19 Csg. Crews, Tongs, Handling Tools, LD/PU Machine	3,000	6,500
20 BHP, GOR, Potential Tests		
21 Cased Hole Logging and Perforating		5,000
22 Stimulation		
23 Inspection, Testing (BOP's, Csg.,Tbg)		1,500
24 Misc. Labor and Materials	3,000	5,000
25 Supervision and Overhead	6,500	8,000
Contingencies: 3%	8,000	9,500
SUB-TOTAL	**$281,500**	**$341,500**

TANGIBLE COST ESTIMATE:						
26 Conductor OD:	14"	Footage:	40'		3,500	3,500
27 Surface OD:	9 5/8"	Footage:	2000'		26,000	26,000
28 Intermediate OD:		Footage:				
29 Liner/Hanger OD:		Footage:				
30 Production OD:	7"	Footage:	8900'			86,000
31 Tubing OD:	2 7/8"	Footage:	8500'			28,000
32 Subsurface Equip.						6,000
33 Wellhead					2,500	10,000
34 Surface Fac. and Installation Incl. Pumping Unit						25,000
		Contingencies:				
		SUB–TOTAL			$32,000	$184,500
		TOTAL COST			$313,500	$526,000

APPROVE: _____

DISAPPROVAL: _____

COMPANY/INDIVIDUAL: _____

DATE: _____

SIGNATURE: _____ PREPARED BY: _____

NAME/TITLES: _____ DATE: _____

Appendix B

Regulation S-X 4-10

Reg. § 210.4-10. Financial Accounting and Reporting for Oil and Gas Producing Activities Pursuant to the Federal Securities Laws and the Energy Policy and Conservation Act of 1975

This section prescribes financial accounting and reporting standards for registrants with the Commission engaged in oil and gas producing activities in filings under the federal securities laws and for the preparation of accounts by persons engaged, in whole or in part, in the production of crude oil or natural gas in the United States, pursuant to Section 503 of the Energy Policy and Conservation Act of 1975 [42 U.S.C. 6383] ("EPCA") and section 11(c) of the Energy Supply and Environmental Coordination Act of 1974 [IS U.S.C. 796] ("ESECA"), as amended by section 505 of EPCA. The application of this section to those oil and gas producing operations of companies regulated for rate-making purposes on an individual-company-cost-of-service basis may, however, give appropriate recognition to differences arising because of the effect of the rate-making process.

> *Exemption.* Any person exempted by the Department of Energy from any record-keeping or reporting requirements pursuant to Section 11(c) of ESECA, as amended, is similarly exempted from the related provisions of this section in the preparation of accounts pursuant to EPCA. This exemption does not affect the applicability of this section to filings pursuant to the federal securities laws.

Definitions

(a) *Definitions*. The following definitions apply to the terms listed below as they are used in this section:

(1) *Oil and gas producing activities.*

(i) Such activities include:

(A) The search for crude oil, including condensate and natural gas liquids, or natural gas ("oil and gas") in their natural states and original locations.

(B) The acquisition of property rights or properties for the purpose of further exploration and/or for the purpose of removing the oil or gas from existing reservoirs on those properties.

(C) The construction, drilling and production activities necessary to retrieve oil and gas from its natural reservoirs, and the acquisition, construction, installation, and maintenance of field gathering and storage systems—including lifting the oil and gas to the surface and gathering, treating, field processing (as in the case of processing gas to extract liquid hydrocarbons) and field storage. For purposes of this section, the oil and gas production function shall normally be regarded as terminating at the outlet valve on the lease or field storage tank; if unusual physical or operational circumstances exist, it may be appropriate to regard the production functions as terminating at the first point at which oil, gas, or gas liquids are delivered to a main pipeline, a common carrier, a refinery, or a marine terminal.

(ii) Oil and gas producing activities do not include:

(A) The transporting, refining and marketing of oil and gas.

(B) Activities relating to the production of natural resources other than oil and gas.

(C) The production of geothermal steam or the extraction of hydrocarbons as a by-product of the production of geothermal steam or associated geothermal resources as defined in the Geothermal Steam Act of 1970.

(D) The extraction of hydrocarbons from shale, tar sands, or coal.

(2) *Proved oil and gas reserves.* Proved oil and gas reserves are the estimated quantities of crude oil, natural gas, and natural gas liquids which geological and engineering

data demonstrate with reasonable certainty to be recoverable in future years from known reservoirs under existing economic and operating conditions, i.e., prices and costs as of the date the estimate is made. Prices include consideration of changes in existing prices provided only by contractual arrangements, but not on escalations based upon future conditions.

(i) Reservoirs are considered proved if economic producibility is supported by either actual production or conclusive formation test. The area of a reservoir considered proved includes (A) that portion delineated by drilling and defined by gas-oil and/or oil-water contacts, if any; and (B) the immediately adjoining portions not yet drilled, but which can be reasonably judged as economically productive on the basis of available geological and engineering data. In the absence of information on fluid contacts, the lowest known structural occurrence of hydrocarbons controls the lower proved limit of the reservoir.

(ii) Reserves which can be produced economically through application of improved recovery techniques (such as fluid injection) are included in the "proved" classification when successful testing by a pilot project, or the operation of an installed program in the reservoir, provides support for the engineering analysis on which the project or program was based.

(iii) Estimates of proved reserves do not include the following:

(A) oil that may become available from known reservoirs but is classified separately as "indicated additional reserves";

(B) crude oil, natural gas, and natural gas liquids, the recovery of which is subject to reasonable doubt because of uncertainty as to geology, reservoir characteristics, or economic factors;

(C) crude oil, natural gas, and natural gas liquids, that may occur in undrilled prospects; and

(D) crude oil, natural gas, and natural gas liquids, that may be recovered from oil shales, coal, gilsonite and other such sources.

(3) *Proved developed oil and gas reserves.* Proved developed oil and gas reserves are reserves that can be expected to be recovered through existing wells with existing equipment and operating methods. Additional oil and gas expected to be obtained through the application of fluid injection or other improved recovery techniques for supplementing the natural forces and mechanisms of primary recovery should be included as "proved developed reserves" only after testing by a pilot project or after the operation of an installed program has confirmed through production response that increased recovery will be achieved.

(4) *Proved undeveloped reserves.* Proved undeveloped oil and gas reserves are reserves that are expected to be recovered from new wells on undrilled acreage, or from existing wells where a relatively major expenditure is required for recompletion. Reserves on undrilled acreage shall be limited to those drilling units offsetting productive units that are reasonably certain of production when drilled. Proved reserves for other undrilled units can be claimed only where it can be demonstrated with certainty that there is continuity of production from the existing productive formation. Under no circumstances should estimates, for proved undeveloped reserves, be attributable to any acreage for which an application of fluid injection or other improved recovery technique is contemplated, unless such techniques have been proved effective by actual tests in the area and in the same reservoir.

(5) *Proved properties.* Properties with proved reserves.

(6) *Unproved properties.* Properties with no proved reserves.

(7) *Proved area.* The part of a property to which proved reserves have been specifically attributed.

(8) *Field.* An area consisting of a single reservoir or multiple reservoirs all grouped on or related to the same individual geological structural feature and/or stratigraphic condition. There may be two or more reservoirs in a field which are separated vertically by intervening impervious strata, or laterally by local geologic barriers, or by both. Reservoirs that are associated by being in overlapping or adjacent fields may be treated as a single or common operational field. The geological terms "structural feature" and "stratigraphic condition" are intended to identify localized geological features as opposed to the broader terms of basins, trends, provinces, plays, areas-of-interest, etc.

(9) *Reservoir.* A porous and permeable underground formation containing a natural accumulation of producible oil and/or gas that is confined by impermeable rock or water barriers and is individual and separate from other reservoirs.

(10) *Exploratory well.* A well drilled to find and produce oil or gas in an unproved area, to find a new reservoir in a field previously found to be productive of oil or gas in another reservoir, or to extend a known reservoir. Generally, an exploratory well is any well that is not a development well, a service well, or a stratigraphic test well as those items are defined below.

(11) *Development well.* A well drilled within the proved area of an oil or gas reservoir to the depth of a stratigraphic horizon known-to be productive.

(12) *Service well.* A well drilled or completed for the purpose of supporting production in an existing field. Specific purposes of service wells include gas

injection, water injection, steam injection, air injection, salt-water disposal, water supply for injection, observation, or injection for in-situ combustion.

(13) *Stratigraphic test well.* A drilling effort, geologically directed, to obtain information pertaining to a specific geologic condition. Such wells customarily are drilled without the intention of being completed for hydrocarbon production. This classification also includes tests identified as core tests and all types of expendable holes related to hydrocarbon exploration. Stratigraphic test wells are classified as (i) "exploratory type," if not drilled in a proved area, or (ii) "development type," if drilled in a proved area.

(14) *Acquisition of properties.* Costs incurred to purchase, lease or otherwise acquire a property, including costs of lease bonuses and options to purchase or lease properties, the portion of costs applicable to minerals when land including mineral rights is purchased in fee, brokers' fees, recording fees, legal costs, and other costs incurred in acquiring properties.

(15) *Exploration costs.* Costs incurred in identifying areas that may warrant examination and in examining specific areas that are considered to have prospects of containing oil and gas reserves, including costs of drilling exploratory wells and exploratory-type stratigraphic test wells. Exploration costs may be incurred both before acquiring the related property (sometimes referred to in part as prospecting costs) and after acquiring the property. Principal types of exploration costs, which include depreciation and applicable operating costs of support equipment and facilities and other costs of exploration activities, are:

(i) Costs of topographical, geographical and geophysical studies, rights of access to properties to conduct those studies, and salaries and other expenses of geologists, geophysical crews, and others conducting those studies. Collectively, these are sometimes referred to as geological and geophysical or "G&G" costs.

(ii) Costs of carrying and retaining undeveloped properties, such as delay rentals, ad valorem taxes on properties, legal costs for title defense, and the maintenance of land and lease records.

(iii) Dry hole contributions and bottom hole contributions.

(iv) Costs of drilling and equipping exploratory wells.

(v) Costs of drilling exploratory-type stratigraphic test wells.

(16) *Development costs.* Costs incurred to obtain access to proved reserves and to provide facilities for extracting, treating, gathering and storing the oil and gas. More specifically, development costs, including depreciation and applicable operating costs of support equipment and facilities and other costs of development activities, are costs incurred to:

(i) Gain access to and prepare well locations for drilling, including surveying well locations for the purpose of determining specific development drilling sites, clearing ground, draining, road building, and relocating public roads, gas lines, and power lines, to the extent necessary in developing the proved reserves.

(ii) Drill and equip development wells, development-type stratigraphic test wells, and service wells, including the costs of platforms and of well equipment such as casing, tubing, pumping equipment, and the wellhead assembly.

(iii) Acquire, construct, and install production facilities such as lease flow lines, separators, treaters, heaters, manifolds, measuring devices, and production storage tanks, natural gas cycling and processing plants, and central utility and waste disposal systems.

(iv) Provide improved recovery systems.

(17) *Production costs.*

(i) Costs incurred to operate and maintain wells and related equipment and facilities, including depreciation and applicable operating costs of support equipment and facilities and other costs of operating and maintaining those wells and related equipment and facilities. They become part of the cost of oil and gas produced. Examples of production costs (sometimes called lifting costs) are:

(A) Costs of labor to operate the wells and related equipment and facilities.

(B) Repairs and maintenance.

(C) Materials, supplies, and fuel consumed and supplies utilized in operating the wells and related equipment and facilities.

(D) Property taxes and insurance applicable to proved properties and wells and related equipment and facilities.

(E) Severance taxes.

(ii) Some support equipment or facilities may serve two or more oil and gas producing activities and may also serve transportation, refining, and marketing activities. To the extent that the support equipment and facilities are used in oil and gas producing activities, their depreciation and applicable operating costs become exploration, development or production costs, as appropriate. Depreciation, depletion, and amortization of capitalized acquisition, exploration, and development costs are not production costs but also become part of the cost of oil and gas produced along with production (lifting) costs identified above.

Successful Efforts Method

(b) A reporting entity that follows the successful efforts method shall comply with the accounting and financial reporting disclosure requirements of Statement of Financial Accounting Standards No. 19, as amended.

Full Cost Method

(c) *Application of the full cost method of accounting.* A reporting entity that follows the full cost method shall apply that method to all of its operations and to the operations of its subsidiaries, as follows:

(1) *Determination of cost centers.* Cost centers shall be established on a country-by-country basis.

(2) *Costs to be capitalized.* All costs associated with property acquisition, exploration, and development activities (as defined in paragraph (a) of this section) shall be capitalized within the appropriate cost center. Any internal costs that are capitalized shall be limited to those costs that can be directly identified with acquisition, exploration, and development activities undertaken by the reporting entity for its own account, and shall not include any costs related to production, general corporate overhead, or similar activities.

(3) *Amortization of capitalized costs.* Capitalized costs within a cost center shall be amortized on the unit-of-production basis using proved oil and gas reserves, as follows:

(i) Costs to be amortized shall include (A) all capitalized costs, less accumulated amortization, other than the cost of properties described in paragraph (ii) below; (B) the estimated future expenditures (based on current costs) to be incurred in developing proved reserves; and (C) estimated dismantlement and abandonment costs, net of estimated salvage values.

(ii) The cost of investments in unproved properties and major development projects may be excluded from capitalized costs to be amortized, subject to the following:

(A) All costs directly associated with the acquisition and evaluation of unproved properties may be excluded from the amortization computation until it is determined whether or not proved reserves can be assigned to the properties, subject to the following conditions: (1) Until such a determination is made, the properties shall be assessed at least annually to ascertain whether impairment has occurred. Unevaluated properties whose costs are individually significant shall be assessed individually. Where it is not practicable to individually assess the amount of impairment of properties for which costs are not individually significant, such properties may be grouped for purposes of assessing impairment. Impairment may be estimated by applying factors based on historical experience and other data such as primary Lease terms of the properties, average holding periods of unproved properties, and geographic and geologic data to groupings of individually insignificant properties and projects. The amount of impairment assessed under either of these methods shall be added to the costs to be amortized. (2) The costs of drilling exploratory dry holes shall be included in the amortization base immediately upon determination that the well is dry. (3) If geological and geophysical costs cannot be directly associated with specific unevaluated properties, they shall be included in the amortization base as incurred. Upon complete evaluation of a property, the total remaining excluded cost (net of any impairment) shall be included in the full cost amortization base.

(B) Certain costs may be excluded from amortization when incurred in connection with major development projects expected to entail significant costs to ascertain the quantities of proved reserves attributable to the properties under development (e.g., the installation of an offshore drilling platform from which development wells are to be drilled, the installation of improved recovery programs, and similar major projects undertaken in the expectation of Significant additions to proved reserves). The amounts which may be excluded are applicable portions of (1) the costs that relate to the major development project and have not previously been included in the amortization base, and (2) the estimated future expenditures associated with the development project. The excluded portion of any common costs associated with the development project should be based, as is most appropriate in the circumstances, on a comparison of either (i) existing proved reserves to total proved reserves expected to be established upon completion of the project, or (ii) the number of wells to which proved reserves have been assigned and total number of wells expected to be drilled. Such costs may be excluded from costs to be amortized until the earlier determination of whether additional reserves are proved or impairment occurs.

(C) Excluded costs and the proved reserves related to such costs shall be transferred into the amortization base on an ongoing (well-by-well or property-by-property) basis as the project is evaluated and proved reserves established or impairment determined. Once proved reserves are established, there is no further justification for continued exclusion from the full cost amortization base even if other factors prevent immediate production or marketing.

(iii) Amortization shall be computed on the basis of physical units, with oil and gas converted to a common unit of measure on the basis of their approximate relative energy content, unless economic circumstances (related to the effects of regulated prices) indicate that use of units of revenue is a more appropriate basis of computing amortization. In the latter case, amortization shall be computed on the basis of current gross revenues (excluding royalty payments and net profits disbursements) from production in relation to future cross revenues, based on current prices (including consideration of changes in existing prices provided only by contractual arrangements), from estimated production of proved oil and gas reserves. The effect of a significant price increase during the year on estimated future gross revenues shall be reflected in the amortization provision only for the period after the price increase occurs.

(iv) In some cases it may be more appropriate to depreciate natural gas cycling and processing plants by a method other than the unit-of-production method.

(v) Amortization computations shall be made on a consolidated basis, including investees accounted for on a proportionate consolidation basis. Investees accounted for on the equity method shall be treated separately.

(4) *Limitation on capitalized costs*:

(i) For each cost center, capitalized costs, less accumulated amortization and related deferred income taxes, shall not exceed an amount (the cost center ceiling) equal to the sum of:

(A) the present value of estimated future net revenues computed by applying current prices of oil and gas reserves (with consideration of price changes only to the extent provided by contractual arrangements) to estimated future production of proved oil and gas reserves as of the date of the latest balance sheet presented, less estimated future expenditures (based on current costs) to be incurred in developing and producing the proved reserves computed using a discount factor of ten percent and assuming continuation of existing economic conditions; plus

(B) the cost of properties not being amortized pursuant to paragraph (i)(3)(ii) of this section; plus

(C) the lower of cost or estimated fair value of unproven properties included in the costs being amortized; less

(D) income tax effects related to differences between the book and tax basis of the properties referred to in paragraphs (i)(4)(i)(B) and (C) of this section.

(ii) If unamortized costs capitalized within a cost center, less related deferred income taxes, exceed the cost center ceiling, the excess shall be charged to expense and separately disclosed during the period in which the excess occurs. Amounts thus required to be written off shall not be reinstated for any subsequent increase in the cost center ceiling.

(5) *Production costs.* All costs relating to production activities, including workover costs incurred solely to maintain or increase levels of production from an existing completion interval, shall be charged to expense as incurred.

(6) *Other transactions.* The provisions of paragraph (h) of this section, "Mineral property conveyances and related transactions if the successful efforts method of accounting is followed," shall apply also to those reporting entities following the full cost method except as follows:

(i) *Sales and abandonments of oil and gas properties.* Sales of oil and gas properties, whether or not being amortized currently, shall be accounted for as adjustments of capitalized costs, with no gain or loss recognized, unless such adjustments would significantly alter the relationship between capitalized costs and proved reserves of oil and gas attributable to a cost center. For instance, a significant alteration would not ordinarily be expected to occur for sales involving less than 25 percent of the reserve quantities of a given cost center. If gain or loss is recognized on such a sale, total capitalization costs within the cost center shall be allocated between the reserves sold and reserves retained on the same basis used to compute amortization, unless there are substantial economic differences between the properties sold and those retained, in which case capitalized costs shall be allocated on the basis of the relative fair values of the properties. Abandonments of oil and gas properties shall be accounted for as adjustments of capitalized costs; that is, the cost of abandoned properties shall be charged to the full cost center and amortized (subject to the limitation on capitalized costs in paragraph (b) of this section).

(ii) *Purchases of reserves.* Purchases of oil and gas reserves in place ordinarily shall be accounted for as additional capitalized costs within the applicable cost center; however, significant purchases of production payments or properties with lives substantially shorter than the composite productive life of the cost center shall be accounted for separately.

(iii) Partnerships, joint ventures and drilling arrangements.

(A) Except as provided in subparagraph (i)(6)(i) of this section, all consideration received from sales or transfers of properties in connection with partnerships, joint venture operations, or various other forms of drilling arrangements involving oil and gas exploration and development activities (e.g., carried interest, turnkey wells, management fees, etc.) shall be credited to the full cost account, except to the extent of amounts that represent reimbursement of organization, offering, general and administrative expenses, etc., that are identifiable with the transaction, if such amounts are currently incurred and charged to expense.

(B) Where a registrant organizes and manages a limited partnership involved only in the purchase of proved developed properties and subsequent distribution of income from such properties, management fee income may be recognized provided the properties involved do not require aggregate development expenditures in connection with production of existing proved reserves in excess of 10% of the partnership's recorded cost of such properties. Any income not recognized as a result of this limitation would be credited to the full cost account and recognized through a lower amortization provision as reserves are produced.

(iv) *Other services.* No income shall be recognized in connection with contractual services performed (e.g. drilling, well service, or equipment supply services, etc.) in connection with properties in which the registrant or an affiliate (as defined in § 210.1-02(b)) holds an ownership or other economic interest, except as follows:

(A) Where the registrant acquires an interest in the properties in connection with the service contract, income may be recognized to the extent the cash consideration received exceeds the related contract costs plus the registrant's share of costs incurred and estimated to be incurred in connection with the properties. Ownership interests acquired within one year of the date of such a contract are considered to be acquired in connection with the service for purposes of applying this rule. The amount of any guarantees or similar arrangements undertaken as part of this contract should be considered as part of the costs related to the properties for purposes of applying this rule.

(B) Where the registrant acquired an interest in the properties at least one year before the date of the service contract through transactions unrelated to the service contract, and that interest is unaffected by the service contract, income from such contract may be recognized subject to the general provisions for elimination of intercompany profit under generally accepted accounting principles.

(C) Notwithstanding the provisions of (A) and (B) above, no income may be recognized for contractual services performed on behalf of investors in oil and gas producing activities managed by the registrant or an affiliate. Furthermore, no income may be recognized for contractual services to the extent that the consideration received for such services represents an interest in the underlying property.

(D) Any income not recognized as a result of these rules would be credited to the full cost account and recognized through a lower amortization provision as reserves are produced.

(7) *Disclosures.* Reporting entities that follow the full cost method of accounting shall disclose all of the information required by paragraph (k) of this section, with each cost center considered as a separate geographic area, except that reasonable groupings may be made of cost centers that are not significant in the aggregate. In addition:

(i) For each cost center for each year that an income statement is required, disclose the total amount of amortization expense (per equivalent physical unit of production if amortization is computed on the basis of physical units or per dollar of gross revenue from production if amortization is computed on the basis of gross revenue).

(ii) State separately on the face of the balance sheet the aggregate of the capitalized costs of unproved properties and major development projects that are excluded, in accordance with paragraph (i) (3) of this section, from the capitalized costs being amortized. Provide a description in the notes to the financial statements of the current status of the significant properties or projects involved, including the anticipated timing of the inclusion of the costs in the amortization computation. Present a table that shows, by category of cost, (A) the total costs excluded as of the most recent fiscal year; and (B) the amounts of such excluded costs, incurred (1) in each of the three most recent fiscal years and (2) in the aggregate for any earlier fiscal years in which the costs were incurred. Categories of cost to be disclosed include acquisition costs, exploration costs, development costs in the case of significant development projects and capitalized interest.

INCOME TAXES

(d) *Income taxes.* Comprehensive interperiod income tax allocation by a method which complies with generally accepted accounting principles shall be followed for intangible drilling and development costs and other costs incurred that enter into the determination of taxable income and pretax accounting income in different periods.

Index

Illustrations and diagrams are indicated with bolded page locators.

1995 COPAS Accounting Procedure, 531
2000 Current Issues and Rulemaking Projects (SEC), 623
2001 PricewaterhouseCoopers Survey of U.S. Petroleum Accounting Practices, 78
    assigning costs to inventory, 285
    recognition of inventories, 284
    seismic costs, 162
    significant properties, 108
2005 COPAS Accounting Procedure, 492, **493–515**

A

AAPL Form 610, 467, **468–489**
Abandonment
    costs of, 619
    of individually insignificant (group) properties, 113–115
    of individually significant properties, 112–113
    of portions of wells, 159–160
    of properties, 244–245
    of proved property, 245
    and reclamation, 521
    of an unproved property, 112
    of unproved property, 432
    of wells and equipment, 432
Abstract of title, 94
Accounting examples
    abandonment of individually insignificant (group) properties, 113–115
    abandonment of individually significant properties, 112–113
    abandonment of proved property, 245
    accounting rate of return, 298–299
    accumulation of joint costs in regular accounts, 525–526
    acquisition, exploration, and development costs disclosure, 632
    acquisition costs, 420
    additional development costs, 161
    allocation back to well, 399
    annual expenditures for exploration operations, 522
    capitalized IDC of integrated producer, 426–427
    carried interest created by a farm-out/farm-in, 564–568
    carried working interest, 528–530
    carved-out production payment payable in money, 597–599

carved-out production payment payable in product, 601–603
ceiling test, 259–260
ceiling test income tax effect, 262–263
changes due to the passage of time, 322–323
company-wide capitalized costs, 628–631
complex DD&A, 250–252
comprehensive field DD&A, 213–216
concession agreement, 679–680
cost allocation, 287–288
DD&A, 184–185, 186–188
DD&A for production payment interest, 201–202
DD&A future development costs, 238–239
DD&A on ORI, 200
DD&A rates, revision of, 203–205
DD&A under three accounting methods, 443–446
decimal ownership, 369
delinquent taxes and mortgage payments, 104–105
depreciation and depletion, 431
depreciation of seven-year property, 430
depreciation of support equipment and facilities, 198–199
development drilling, 153–154
distribution of joint costs as incurred, 527
exchange of working interests, 569
exclusion of costs, 195–196
exclusion of reserves, 196
expected cash flow approach, 319
expected present value technique, 335–337
exploratory drilling costs, 147–150
farm-in/farm-out, 554–555
fieldwide DD&A, 189–190
full cost DD&A, 235–237
full cost entries, 230–234
full cost vs. successful efforts, 253–256
future revenue, 639
G&G Costs, 417–418
gain on final settlement, 324
gas injection or gas lift, 384–385
gas revenue, 382–383
gas used off lease by operator, 384
geological and geophysical (G&G) costs, 78
IDC vs. equipment, 424–426
initial estimation and recording, 320–321
interest capitalization, 166, 257–258
intracompany sales, 375–376
IRR, NPV, PI comparison, 303–305
joint production DD&A, 192–194
joint working interest, 546

lease abandonment, 207–208
lessor's acquisition costs, 452
major development project exclusion, 242
minimum royalties, 402–403
multiple nonworking interests, 201, 376–378
net present value, 300
nonmonetary exchange, 552–553
oil and gas disclosures, 619–622
oil revenue, 374–375
oil used off lease by operator, 378–379
option to lease, 102–103
overriding royalty interest (ORI), 546
partial abandonment, 115
partial reclassification, 118
participation factors based on multiple elements, 372
payback method, 296–297
percentage depletion, 447
percentage depletion: 100% and 65% limitations, 448
pipeline gas imbalances, 397–398
pooling, 570
post-balance sheet events, 116, 163
pressure base, 381
production advance, 596–597
production costs, 283
production payment interest (PPI), 547–548
profitability index (PI), 302–303
proved property cost disposition under successful efforts, 209–213
PSC cost recovery, 686–688
purchase of new material, 531
reclassification of individually insignificant (group) properties, 117–118
reclassification of individually significant properties, 117
reserve report, 626
retained production payment payable in money-reasonably assured, 591–592
retained production payment payable in product, 594–595
retained production payments payable in money-not reasonably assured, 593–594
revenue and cost allocation to ownership interests, 367–368
revenue recognition at time of production, 389
revenue recognition at time of sale, 388
revision of DD&A rates, 203–205
risk service agreement, 690–691
sale, 209

sale and abandonment-unproved and proved properties, 605–606
sale and purchase of entire interest in a proved property, 582
sale and purchase of partial interest in a proved property, 584–585
sale of divided interest in a proved property, 585–586
sale of entire overriding royalty interest in proved property, 583
sale of entire unproved property, 575–576
sale of overriding royalty interest in proved property, 588
sale of overriding royalty interest in unproved property, 580
sale of portion of ORI in unproved property, 581
sale of portion of unproved property-group impairment, 578
sale of portion of unproved property-individual impairment, 577
sale of significant reserves, DD&A rate distortion, 606
sale of working interest in unproved property with retention of ORI, 579
sales of a partial interest in unproved property, 576
secondary recovery system, 290
standard measure of oil and gas (SMOG), 640–641
subleasing, 421
successful efforts vs. full cost, 253–256
support equipment and facilities, 162
take-or-pay, 387
tax accounting for both independent and integrated producers, 437–440
top leasing, 105–106
transfer from one property to another property-Condition B, 534
transfer from property to another property-Condition C, 535
transfer from Warehouse—Condition A, 532
transfer from warehouse to joint property-Condition B, 533
tubular goods, 292
two-stage allocation, 400–401
unitization, 572–573
unitization-determination of interest, 370–371
unproved property inclusion and exclusion, 240–241
upward revision in estimate, 325–328
well abandonment, 207
working interest method, 182
workovers and recompletions, 157–158
Accounting for asset retirement obligations, 313–331
  asset recognition, 317
  conditional AROs, 329–330
  funding and assurance provisions, 328
  initial measurement-fair value, 317–321
  legally enforceable obligations, 315–316
  miscellaneous expenses, 330–331
  obligating event, 316–317
  scope of *SFAS No. 143*, 314–315
  subsequent recognition and measurement, 322–328
Accounting for asset retirement obligations and asset impairment, 313–338
  accounting for asset retirement obligations, 313–331
  accounting for the impairment and disposal of long-lived assets, 331–338
Accounting for different types of nondrilling exploration costs, 84
Accounting for international petroleum operations, 677–695
  concessionary agreements with government participation, 680–681
  concessionary systems, 679–680
  contractual systems, 681–682
  financial accounting issues, 692–694
  joint operating agreements, 691–692
  petroleum fiscal systems, 678–679
  production sharing contracts, 682–683
  production sharing contracts (PSCs), 682–683
  service contracts, 689–691
Accounting for materials, 530–535
Accounting for production activities, 280–305
  accounting treatment, 282–285
  accumulation and allocation of costs, 286–288
  decision to complete a well, 293–296
  individual production costs, 288–292
  joint interest operations, 293
  production costs statements, 292
  project analysis and investment decision making, 296–305
Accounting for revenue from oil and gas sales, 348–402
  crude oil exchanges, 379
  definitions, 349–351

determination of revenue, 366–369
measurement and sale of oil and natural gas, 351–363
miscellaneous revenue accounting topics, 391–392
oil and gas revenues, 370–372
recording gas revenue, 380–382
recording oil revenue, 373–379
revenue reporting to interest owners, 391
standard division order, 363–366
timing of revenue recognition, 386–390
unitizations, 370–372
Accounting for suspended well costs, 164–165
Accounting for the impairment and disposal of long-lived assets, 331–338
asset groups, 332
impairment for full cost companies, 338
long-lived assets to be disposed of, 337–338
long-lived assets to be held and used, 332–337
scope, 331–332
Accounting guidelines (AGs), 50
Accounting methods and current status, historical development of, 48–50
Accounting procedures, 491–516
direct charges, 517–521
general provisions, 516–517
inventories, 524
overhead, 521–522
pricing of joint account material purchases, transfers, and dispositions, 523–524
Accounting rate of return, 298–299
Accounting Research Bulletin (ARB) No. 43, "Restatement and Revision of Accounting Research Bulletins", 285
Accounting Research Study No. 11, "Financial Reporting in the Extractive Industries", 48
Accounting Series Release (ASR) 253 (SEC), 49
Accounting treatment, 282–285
cost of production versus inventory, 283–284
of the four basic types of costs, **229**
lower-of-cost-or-market valuation, 285
recognition of inventories, 284
Accounts payable, 59, 69
Accounts receivables, 55, 66
Accretion expense, 195, 314
Accretion of discount, 195, 314, 650, 664–665
Accruals, 55, 66
Accrued liabilities, 59, 69
Accrued state and federal taxes, 59, 69

Accumulated DD&A and writedowns, 59, 68
Accumulation and allocation of costs, 286–288
Accumulation of joint costs in regular accounts, 525–526
Acidizing, 8
Acquisition, exploration, and development costs disclosure, 632
Acquisition, maintenance, or transfer of interest, 491
Acquisition costs, 37
accounting examples, 420
defined, 94
lessee's transactions, 419–421
lessor's transactions, 452–453
successful efforts (SE) accounting methods (*ASR 257*), **95, 96**
successful efforts (SE) accounting methods (*ASR 257*) example, 97
Acquisition costs of unproved property, successful efforts, 92–128
delinquent taxes and mortgage payments, 103–105
disposition of capitalized costs-impairment of unproved property, 106–111
disposition of capitalized costs-surrender or abandonment of property, 112–127
internal costs, 99–100
land department, 127–128
options to lease, 100–103
option to lease, 100–103
purchase in fee, 98–99
top leasing, 105–106
Acquisition of mineral interests in property, 11–16
carved-out net profits interest created from working interest, 16
carved-out production payment, 15
mineral interests, 12–14
mineral rights, 11
net profits interest created from mineral interest, 16
retained ORI, 14–15
upstream oil and gas operations, 11–16
Activities, defined, 166
Adjustments, 516
Ad valorem taxes, 419
Advance royalties, 452
Advances and payments by the parties, 516
AFES and drilling contracts, 155–156
Affiliates, 520
Allocable production costs vs. directly attributable costs, **287**

Allocation back to well, 399
Allocation of income and expenditures, 537
Allocation of oil and gas, 398–400
Allocations, 517
Allowance account, 248
Allowance account vs. impairment account, 248
American Association of Petroleum Geologists (AAPG), 622
American Institute of Certified Public Accountants (AICPA), 315
Amortization, 181
Analysis of changes in development costs, 661
Analysis of changes in price and cost and quantity revisions and sales and transfers, 664
Analysis of oil and gas companies' financial statements, 701–725
    comparing financial reports, 703–704
    financial ratios, 724–725
    reserve cost ratios, 711–719
    reserve ratios, 704–711
    reserve value ratios, 719–724
Annual average pricing-crude oil, **264**
Annual average pricing-crude oil vs. end of year pricing, **264**
Annual average pricing-natural gas, **264**
Annual average pricing-natural gas vs. end of year pricing, 264
Annual expenditures for exploration operations, 522
Anticline, 5
API gravity, 349, 352
API schematic flow of natural gas, 362
Approvals by parties, 517
Area of interest, 9
ARO. See SFAS No. 143, "Accounting for Asset Retirement Obligations (AROs)"
ASR 257. See successful efforts (SE) accounting methods (*ASR 257*)
ASR 258. See full cost method accounting (*ASR 258*)
Assessment of the ceiling test, 263
Asset groups, 332
Asset recognition, 317
Asset-related expenditures, 334
Asset retirement obligations, 260–261
Associated gas, 349
Authorization for expenditure (AFE), 155, 516
Automobile depreciation, 199
Average daily production per well, 710–711
Average reserves per well ratio, 709–710

B

"Back in" rights, 680
Barrels, 349, 352
Barrels of oil equivalent (BOE), 191
Basic oil and gas tax accounting, 414–454
    lessee's transactions, 414–452
    lessor's transactions, 452–454
Basic royalty interest (RI), 546
Basic sediment and water (BS&W), 28, 291, 350, 352
Basic working interest (WI), 546
Black's Law Dictionary, 316
Booking charges to joint accounts
    accumulation of joint costs in operator's regular accounts, 524–526
    distribution of joint costs as incurred, 526–527
Bottom-hole contributions, 82–83, 419
British thermal unit (Btu), 191
Brokers, 363
Brokers' fees, 94
BS&W (basic sediment and water). *See* basic sediment and water (BS&W)
Btu, 350

C

Capital and maintenance jobs, 537
Capital assets, 537
Capitalization rules, 77
Capitalized IDC of integrated producer, 426–427
Capital uplifts, 688
Carried interest created by a farm-out/farm-in, 564–568
Carried interests, 563
Carried interests or sole risk, 563–568
Carried party/carrying parties, 527, 563
Carried working interest, 527, 528–530
Carrying and retaining costs, 80–82
Carved-out net profits interest created from working interest, 16
Carved-out production payments, 15, 548, 595–604
    interest on, 549
    payable in money, 595, 597–599
    payable in product, 601–603
    payable in product or volumetric production payment (VPP), 600

Cash accounts, 55, 66
Cash call, 516
Cash flows, 639
Casing, 23
Casinghead gas, 350
Catastrophe overhead, 522
Catastrophic events, 208
Ceiling test, 259–260
Ceiling test income tax effect, 262–263
Changes
    analysis of reasons for, 652–666
    due to the passage of time, 322–323
    due to the revisions in estimates, 324
    in estimated future development costs, 658–659
    from extensions, discoveries, and improved recovery, 656
    in prices and costs, 653
    in the standardized measure of discounted future net cash flows, 652–666
Chart of accounts
    for a full cost company, 66–70
    for a successful efforts company, 55–63
Chart of accounts for a successful efforts company
    accounts payable, 59
    accounts receivables, 55
    accruals, 55
    accrued liabilities, 59
    accrued state and federal taxes, 59
    accumulated DD&A and writedowns, 59
    cash accounts, 55
    deferred federal income tax, 59
    general and administrative expense, 61–62
    lease operating expense (LOE), 60–61
    long-term notes payable, 59–60
    marketing and transportation expense, 61
    material & supplies, 55
    miscellaneous assets, 59
    miscellaneous expenses, 63
    nonproductive expenses, 63
    offshore development drilling, 56–57
    offshore development well plug and abandonment, 57–58
    offshore exploratory drilling, 58
    offshore facilities, 57
    offshore lease acquisition, unproved property, 58–59
    offshore recompletion, 57
    offshore workover, 57
    oil and gas accounting introduction, 55–63
    onshore development drilling, 58
    onshore development well plug and abandon, 58
    onshore exploratory drilling, 58
    onshore facilities, 58
    onshore recompletions, 58
    onshore workover, 58
    prepaids, 55
    producing leaseholds-proved property, 55
    revenues and gains, 60
    short-term notes payable, 59
    unevaluated leaseholds-unproved property, 55
    wells and equipment, 56
Chart of accounts for full cost (FC) company
    accounts payable, 69
    accounts receivables, 66
    accruals, 66
    accrued liabilities, 69
    accrued state and federal taxes, 69
    cash accounts, 66
    deferred federal income tax, 69
    general and administrative expense, 69
    lease operating expense (LOE), 69
    long-term notes payable, 69
    long-term receivables, 69
    marketing and transportation expense, 69
    material & supplies, 66
    miscellaneous assets, 68
    miscellaneous expenses, 69
    nonproductive costs, 67
    offshore development drilling, 67
    offshore development well plug and abandonment, 67
    offshore exploratory drilling, 67
    offshore facilities, 67
    offshore G&G, 68
    offshore lease acquisition, unproved property, 68
    offshore recompletion, 67
    offshore workover, 67
    oil and gas accounting introduction, 66–70
    onshore development drilling, 67
    onshore development well plug and abandon, 68
    onshore exploratory drilling, 68
    onshore facilities, 68
    onshore G&G, 68
    onshore recompletions, 68
    onshore workover, 68
    prepaids, 66
    producing leaseholds-proved property, 68
    revenues and gains, 69

Index 755

short-term notes payable, 69
stockholders equity and related accounts, 69
unevaluated leaseholds-unproved property, 67
wells and equipment, 67
Christmas tree, 25, 138
City gate, 350, 389
Claims and lawsuits, 491
Combined fixed rate, 522
Commercial substance, 551
Commingled gas, 350
Communications, 521
Company labor, 536
Company-wide capitalized costs, 628–631
Completion decision, 26, 294–295
Compliance with laws and regulations, 491
Computerized Equipment Pricing System (CEPS), 531
Concession agreements, 679–680
Concessionary agreements with government participation, 680–681
Concessionary contract/concessionary agreement, 678
Concessionary systems, 678, 679–680
Condensate, 350
Conditional AROs, 329–330
Condition D, 523
Condition E, 523
Confirmation, 396
Confirmed nominations, 396
Construction overhead, 522
Contractor, defined, 678
Contractual systems, 678, 681–682
Conversions, 191
Converting volume, 380–381
Conveyances, 544–607
  defined, 545
  exchanges and poolings, 551–573
  full cost, 604–607
  general rules, 549–-550
  mineral interests, 545–549
  production payments, 589–604
  sales, 573–588
Conveyances, exchanges and poolings, 551–573
  carried interests or sole risk, 563–568
  farm-ins/farm-outs, 553–555
  farm-ins/farm-outs with a reversionary working interest, 555–560
  free wells, 560–562
  joint venture operations, 568–569
  poolings and unitizations, 570
  unitizations, 571–573

Conveyances, full cost, 604–607
Conveyances, general rules, 549–-550
Conveyances, production payments, 589–604
  carved-out, 595–604
  retained, 589–595
Conveyances, sales, 573–588
  proved property sales, 581–588
  unproved property sales, 573–581
Cooperative unit, 372
COPAS AG-19, "Expenditure Audits in the Petroleum Industry: Protocol & Procedure Guidelines", 536
COPAS AG-22, "Producer Gas Imbalances", 392
COPAS MFI-27, "Employee Benefits Chargeable to Joint Operations and Subject to Percentage Limitation", 519
COPAS MFI-51, "2005 COPAS Accounting Procedure", 492
COPAS offshore accounting procedures, 536
Core sample, 24
Cost accounting methods, historical, 44–48
Cost allocation, 287–288
Cost centers, 228
Cost depletion, 428
Cost disposition, nonworking interests, 199–202
Cost disposition through abandonment, retirement, or sale of proved property, 202–216
Cost disposition through amortization, 179–206
  cost disposition-nonworking interests, 199–202
  DD&A calculation, 183–188
  DD&A on a fieldwide basis, 188–190
  DD&A when oil and gas reserves are produced jointly, 191–194
  depreciation of support equipment and facilities, 197–199
  estimated future dismantlement, site restoration, and abandonment costs, 194
  exclusion of costs or reserves, 195–197
  reserves owned or entitled to, 181–183
  revision of DD&A rates, 202–203
Cost of production versus inventory, 283–284
Cost oil cap, 684
Cost oil/cost gas, 684
Cost ratios, 711–716
Cost recovery, 684–685
Cost recovery order, 685

Costs
    accounting treatment of the four basic types, **229**
    four basic types of, **180**
    of materials, supplies, and fuel, 289
Council of Petroleum Accountants Societies (COPAS), 50, 491–492
Credit-adjusted risk-free discount rate, 318
Credit-adjusted risk-free rate, 318, 328
Cross-fence recovery, 688
Crude oil, 350
Crude oil exchanges, 379
Crude oil measurement, 351–354
Crude oil sales, 360–361

D

Damaged or lost equipment and materials, 158
Damages and losses, 520
Day rate, 23
DD&A (depreciation, depletion, and amortization), 44
    accounting examples, 184–188, 200, 203–205, 238–239, 443–446
    calculation of, 183–188
    complex, 250–252
    comprehensive field, 213–216
    defined, 181
    expense, 62, 69
    on a fieldwide basis, 188–190
    future development costs, 238–239
    on ORI, 200
    per BOE, 718–719
    for production payment interest, 201–202
    rates, revision, 203–205
    under successful efforts versus full cost, 253–256
    under three accounting methods, 443–446
    when oil and gas reserves are produced jointly, 191–194
Decimal ownership, 369
Decision to complete a well, 293–296
Decommissioning costs, 619
Deferred federal income tax, 59, 69
Deferred taxes, 261
Definitions, 516
Delay rental payment, 21
Delay rental payments, 418
Delay rentals, 80–81
Delineation well, 43
Delinquent taxes and mortgage payments, 103–105
Depletion, 428
Depreciable property, 428–429
Depreciation, 428–429
Depreciation, depletion, and amortization (DD&A). *See* DD&A (depreciation, depletion, and amortization)
Depreciation and depletion, 431
Depreciation and operating costs, 289
Depreciation of seven-year property, 430
Depreciation of support equipment and facilities, 197–199
Derivatives, 604
Detailed survey, 9, 77
Determination of revenue, 366–369
Deterministic reserve estimation, 622
Development costs, 37, 51
    additional, 160–161
    defined, 160–161
    development drilling costs, 151–154
    incurred during the period that reduce future development costs, 659–660
Development drilling, 153–154
Development drilling accounting, 151
Development dry holes vs. exploratory dry holes, 141–142
Development wells, 39, 41, **42**, 142, 152
Direct charges, 517–521
Direct costs, 517
Directional wells, 23–24
Directly attributable costs vs. allocable production costs, **287**
Direct sales, 363
Disclosure of reserves, 693–694
Disclosures, 77. *See also* SFAS No. 69, "Disclosures about Oil and Gas Producing Activities"
Discount rate, 299
Dismantlement, restoration, and abandonment costs, 619
Disposal groups, 338
Disposition of capitalized costs, 234–256
    abandonment of properties, 244–245
    DD&A under successful efforts versus full cost, 253–256
    exclusion of costs, 238–242
    impairment of unproved properties costs, 242–244
    impairment of unproved property, 106–111
    inclusion of estimated future decommissioning costs, 237–238

inclusion of estimated future development expenditures, 237
reclassification of properties, 245–252
support equipment and facilities, 253
surrender or abandonment of property, 112–127
Disposition of material, 524
Dissolved gas, 349, 350
Distribution of joint costs as incurred, 527
Divided working interest, 13
Division of revenue between owners, 366, 403
Division order, 364, **365–366**
Domestic market obligation, 688–689
Downstream activities, defined, 1
Downward revisions, 324
Drilling and development, 490
Drilling and development costs, successful efforts, 137–167
    accounting for suspended well costs, 164–165
    additional development costs, 160–161
    AFES and drilling contracts, 155–156
    development drilling costs, 151–154
    drilling and development seismic, 162
    exploratory drilling costs, 144–150
    financial accounting for drilling and development costs, 141–142
    income tax accounting for drilling costs, 137–140
    interest capitalization, 165–167
    offshore and international operations, 167
    post-balance sheet events, 163
    special drilling operations and problems, 156–160
    stratigraphic test wells, 154
    support equipment and facilities, 162
    well classification, 142–144
Drilling and development seismic, 162
Drilling barges and ships, 26
Drilling capitalization rules, 157
Drilling contracts, 23, 155–156
Drilling exploratory and developmental dry holes, 432
Drilling fluid, 23
Drilling operations, 22–27, 422–423
Dry-hole contribution, 82–83
Dry holes, 107

E

E&P joint ventures, 13
Ecological and environmental costs, 521
Economic interest, 11
Economic interest method, 694
Employee benefits, 288, 519
End of year pricing, 264, **624**
Energy Policy and Conservation Act of 1975, 3
Enhanced recovery methods, 31–32
Entitlement, 397
Entitlement method, 392
Entitlement reserves,, 693
"Environmental Remediation Liabilities" (AICPA) *SOP 96-1*, 315
Equipment costs, 137–138, 422, 423–427
Equipment vs. IDC, 139–140
Equivalent formula:, 183
Equivalent Mcfs (Mcfe), 191
Estimated future decommissioning costs, 237–238
Estimated future dismantlement, site restoration, and abandonment costs, 194
Estimation approach, 333
Estimation of proved reserves, 693
Estimation periods, 333–334
Exchange of working interests, 569
Exclusion of costs, 195–196, 239–242
Exclusion of costs or reserves, 195–197
Exclusion of reserves, 196
Exclusively owned equipment and facilities of the operator, 520
Exhibits, 490
Expected cash flow approach, 318, 319
Expected present value approach, 335
Expected present value technique, 335–337
Expenditure audits, 517
Expenditures and liabilities, 490
Expense of conducting periodic inventories, 524
Exploration and production (E&P) activities, defined, 1
Exploration costs, 37
    accounting for, 51
    defined in *Reg. S-X 4-10*, 75
Exploration methods and procedures, 9–10
Exploratory drilling costs, 144–150
Exploratory dry holes vs. development dry holes, 141–142
Exploratory-type stratigraphic test wells, 146
Exploratory wells, 39, 41, **42**, 142, 146

Extensions, discoveries, and improved recovery, net of future production and development costs, 657–658

F

Farm-ins/farm-outs, 553–555
Farm-ins/farm-outs with a reversionary working interest, 555–560
FASB (EITF) Issue No. 88-18, "Sales of Future Revenues", 599
FASB Interpretation No. 33 (to *SFAS No. 34*), 257
Fault trap, 5
Federal and state government regulation, 48
Federal Energy and Conservation Act, 48
Fee interest, 11
Field, 39
Field facility, 350
Fieldwide DD&A, 189–190
Fill gas, 284
Financial accounting for drilling and development costs, 141–142
Financial accounting issues
  accounting for international petroleum operations, 692–694
  disclosures of proved reserves-*SFAS No. SFAS No. 69*, "Disclosures about Oil and Gas Producing Activities", 692–694
  financial accounting versus contract accounting, 692–693
  international accounting standards, 692–694
  SFAS No. 69, "Disclosures about Oil and Gas Producing Activities", 692–694
Financial Accounting Standards Board (FASB), 48
Financial accounting versus contract accounting, 692–693
Financial ratios, 724–725
Financial reports, 703–704
Financial Statements example, 46
FIN No. 47, Interpretation No. 47, "Accounting for Conditional Asset Retirement Obligations: An Interpretation of FASB Statement No. 143", 329
First-level supervision, 518
Fiscal system, 678
Fishing and sidetracking, 159
Fishing operations, 159

Floating production/storage/offloading vessels (FPSOs), 27
Flowchart in exploring for oil and gas., 10
Footage rate, 23
Force majeure, 491
Foreign tax credit carryforwards, 262
Fracturing, 8
"free fuel" clause, 378
Free wells, 560–562
FSP 19-1 "Accounting for Suspended Well Costs", 164
Full cost (FC) accounting methods, 45–46
  entries example, 65–66
  overview, **64**
Full cost accounting, 227–265
  disposition of capitalized costs, 234–256
  interest capitalization, 257–258
  limitation on capitalized costs-ceiling, 257–258
  reserves in place-purchase, 256
  vs. successful efforts accounting, 695
Full cost accounting rules, 228
Full cost ceiling test, 263–264
  Reg. S-X 4-10, 261
  SFAS No. 144, "Accounting for the Impairment or Disposal of Long-Lived Assets", 265
Full cost DD&A, 235–237
Full cost entries, 230–234
Full cost method, 44
Full cost method accounting (*ASR 258*), 49
Full cost vs. successful efforts, 253–256
Funding and assurance provisions, 328
Future cash flows, 639
Future cash inflows, 644–648
Future decommissioning costs, 195
Future income taxes, 642–643
Future income tax payable, 639
Future of upstream oil and gas operations, 30–31
Future revenue, 639
Futures prices, 363

G

G&G (geological and geophysical) costs. *See* geological and geophysical (G&G) costs
Gain on final settlement, 324
Gain or loss recognition upon settlement, 324
Gas balancing agreement, 350

Gas balancing agreement (*AAPL Form 610-E*), 392
Gas-cap gas, 349
Gas imbalances, 391
Gas injection, 384
Gas injection or gas lift, 384–385
Gas inventory, 284
Gas-lift gas, 384
Gas revenue, 382–383
Gas revenue recognition, 390
Gas settlement statement, 28, 350, 363
Gathering systems, 291
Gauging, 350, 352
General administrative overhead costs, 290
General and administrative expense, 61–62, 69
Generalized crude oils API correction to 60° F, **358**
Generalized crude oils volume correction to 60° F, **359**
General provisions, 516–517
Geological and geophysical (G&G) costs, 417–418
    accounting examples, 78
    exchanged for interest in property, 79–80
    in international operations, 87
    nondrilling exploration costs, successful efforts, 77–80
    offshore G&G vs. onshore surveys, 87
Geological and geophysical (G&G) techniques, 8
Glossary of common terms, 38–43
Good used material (Condition B), 523
Government involvement in operations, 681–682
Government participation, 680, 684
Grantee, 420
Grantor, 420
Gravity, 352
Gross acquisition costs, 51
Gross purchase and sales method, 379
Gross revenue from oil sales, 373–374
Group properties, 110

H

Heater-treater, 350
Henry Hub, 363
Historical world oil prices, **4**
History of the U.S. oil and gas industry, 2–4
Horizontal wells, 23–24
Hurdle rate, 300

I

IDC vs. equipment, 139–140, 424–426
Impairment, 332
    defined, 107
    for full cost companies, 338
    of individually insignificant (group) properties, 109, 110
    of individually significant and insignificant properties, 243
    of individually significant unproved properties, 108–109
    for individually significant unproved properties, 243–244
    of multiple groups of properties, 111
    of unproved properties costs, 242–244
Impairment, abandonment, and reclassification of individually insignificant unproved properties, 249
    for significant unproved properties., 246–247
Impairment account, 248
Impairment account vs. allowance account, 248
Impairment allowance
    individually insignificant (group), 119, 121
    individually significant unproved leases, 119, 121
Inclusion of estimated future decommissioning costs, 237–238
Inclusion of estimated future development expenditures, 237
Income tax accounting for drilling costs, 137–140
Income tax effects, 261–263
Independent oil and gas company, defined, 1
Independent producer, 422
Independent producer abandonment, 433–434
Indications of impairment, 332–333
Indirect costs, 517
Individually insignificant (group) properties
    abandonment of, 113–115
    accounting examples, 117–118
    impairment allowance, 119, 121
    impairment of, 109, 110
    reclassification of, 117–118
Individually significant and insignificant properties, 243
Individually significant properties
    abandonment of, 112–113
    reclassification of, 117
Individually significant unproved leases, 119, 121

Individually significant unproved properties, 108–109, 243–244
Individual production costs, 288–292
    gathering systems, 291
    saltwater disposal systems, 291
    secondary and tertiary recovery, 290
    severance taxes, 292
    tubular goods, 291–292
Industry Guide for the Extractive Industries (SEC), 627
In fee, defined, 93
Initial estimation and recording, 320–321
Initial measurement-fair value, 317–321
"In-kind" balancing, 392
Insurance, 520
Insurance coverage, 289
Intangible drilling and development costs (IDC), 422–423
Intangible drilling costs (IDC), 137–138
Integrated oil and gas company, defined, 1
Integrated producer, 422
Integrated producer abandonment, 435–437
Interest capitalization, 257–258
    accounting examples, 166, 257–258
    drilling and development costs, successful efforts, 165–167
    SFAS No. 34, 257
Interest cost recovery, 684
Interest method of allocation, 322
Interests, 449
Interests, mineral types, 546–549
Interests of parties, 490
Internal costs, 99–100
Internal employee cost allocations
    SFAS No. 19, 99
    successful efforts (SE) accounting methods (*ASR 257*), 99
Internal rate of return (IRR) method, 301
Internal revenue code election, 491
International accounting standards, 692–694
International Accounting Standards Board (IASB), 695
International Financial Reporting Standards (IFRS), 695
International operations
    division of revenue between owners, 366
    early termination provisions, 112
    geological and geophysical (G&G) costs in, 87
    mineral interests acquisitions, 93
    negotiation of petroleum contracts in, 94
    obtaining contract areas in, 108
    reserves ownership in, 182–183
    tax laws of, 415–416
    turnkey contracts in, 156
Interpretation No. 47, "Accounting for Conditional Asset Retirement Obligations: An Interpretation of FASB Statement No. 143" (*FIN No. 47*), 329
Interpretations and Guidance (SEC, March 2001), 285
Intracompany sales, 375–376
Inventories, 284, 524
Inventory method, 379
Investment credit, 688
Investment tax credits, 261
IRR, NPV, PI comparison, 303–305
ITC carryforward, 262

J

Jack-up drilling platforms, 26
Joint accounts, 517
Joint interest accounting, 465–538
    accounting for materials, 530–535
    the accounting procedure, 491–516
    booking charges to the joint account: accumulation of joint costs in operator's regular accounts, 524–526
    booking charges to the joint account: distribution of joint costs as incurred, 526–527
    joint interest accounting, 524
    joint interest audits, 536–538
    the joint operating agreement, 467–491
    joint operations, 465–467
    nonconsent operations, 527–530
    offshore operations, 536
Joint interest audits, 536–538
Joint interest operations, 293
Jointly owned corporations, 466
Joint operating agreements (JOAs), 50, 293, 467–491
    recoverable versus nonrecoverable costs, 691–692
    sample, **468–489**
Joint operations
    joint interest accounting, 465–467
    joint venture contracts, 466–467
Joint production DD&A, 192–194
Joint venture contracts, 466–467
Joint venture operations, 568–569

Joint ventures of undivided interests, 466
Joint working interest, 13, 546

L

Labor, 518
Labor costs, 288
Land department, 127–128
Landman, 21
Lease & well equipment, 179
Lease abandonment, 207–208
Lease automatic custody transfer (LACT) unit, 350, 354
Lease bonus, 21
Lease bonuses, 452
Lease operating costs, 281
Lease operating expense (LOE), 60–61, 69
Lease or signature bonus, 94
Lease provisions, 16–22
Legal costs, 419
Legal expense, 520
Legal fees, 94
Legally enforceable obligations, 315–316
Legal partnerships, 466
Lessee, 21, 420
Lessee's transactions, 414–452
    acquisition costs, 419–421
    drilling operations, 422–423
    equipment costs, 423–427
    losses from unproductive property, 432–446
    nondrilling costs, 416–419
    percentage depletion, 446–448
    production operations, 428–431
    property taxes, 448–449
    recapture of IDC and depletion, 449–450
Lessor, 21
Lessor's acquisition costs, 452
Lessor's revenue, 452–453
Lessor's transactions
    acquisition costs, 452–453
    revenue, 453–454
Lifting costs, 281
Lifting costs per BOE, 716–718
Limitation on capitalized costs-ceiling, 257–266
    assessment of the ceiling test, 263
    asset retirement obligations, 260–261
    deferred taxes, 261
    income tax effects, 261–263
    post-balance sheet events and the ceiling test, 265–266
    SFAS No. 144, 265

Line fill, 284
Local distribution company (LDC), 350
Long-lived assets
    to be disposed, 337–338
    to be disposed of by sale, 337
    to be disposed of other than by sale, 337
    to be held and used, 332–337
Long-term notes payable, 59–60, 69
Long-term receivables, 69
Losses from unproductive property, 432–446
Lower-of-cost-or-market valuation, 285
Lowest Known Horizon (LKH) criteria, 623

M

Major development project exclusion, 242
Marketers, 363
Marketing and transportation expense, 61, 69
Market risk premiums, 320
Mass asset accounting, 432
Material & supplies, 55, 66
Material and supplies transferred to or from the operation, 537
Materials and services purchased externally, 537
Materials and supplies, 520
Material transfer pricing, 523
Maximum efficiency rate (MER), 30
Mcfs, 191, 350
Mcfs vs. MMBtus, 380
Measurement and sale of oil and natural gas, 351–363
    crude oil measurement, 351–354
    crude oil sales, 360–361
    natural gas measurement, 361–363
    natural gas sales, 363–364
    run ticket calculation, 354–359
Measuring devices, 363
Measuring impairment, 335
Measuring liabilities value, 318
Mid-month convention, 430
Midstream activities, defined, 1
Million British thermal units (MMBtu), 191, 380
    conversion calculations, 381
Mineral interests (MI), 11, 12–14
    conveyances, 545–549
    types of interests, 546–549
Mineral rights (MR), 11
Minimum royalties, 22, 402–403
Minutes of operator's meetings, 536
Miscellaneous assets, 59, 68

Miscellaneous expenses, 63, 69, 330–331
Miscellaneous JOA terms, 491
Miscellaneous revenue accounting topics, 391–392
    allocation of oil and gas, 398–400
    gas imbalances, 391
    minimum royalty-an advance revenue to royalty owners, 401–403
    pipeline gas imbalances, 396–398
    producer gas imbalances, 391
Mixed dollars, 650
MMBtus vs. Mcf, 380
MMBtu to Mcf conversion, 380
Model Form Interpretations (MFIs), 50
Modified Accelerated Cost Recovery System (MACRS), 429
Mousehole connection, 23
Mud, 23
Multiple completion, 25
Multiple nonworking interests, 201, 376–378

N

Natural gas, 350
Natural gas historical prices, **4**
Natural gas measurement, 361–363
Natural gas processing, 386
Natural gas processing plant, 350
Natural gas sales, 363–364
Net change in income taxes, 665–666
Net operating loss (NOL) carryforwards, 262
Net present value (NPV) method, 299–300
Net profits interest (NPI), 15, 549
Net profits interest created from mineral interest, 16
Net proved reserves, 694
Net wells to gross wells ratio, 708–709
New field wildcat well, 43
New material (Condition A), 523
Nomination, 396
Nonassociated gas, 349, 350
Nonconsent operations, 527–530
Nondrilling costs, 416–419
Nondrilling exploration costs, successful efforts, 75–87
    accounting for different types of, 84
    carrying and retaining costs, 80–81
    G&G costs, 77–80
    geological and geophysical (G&G) costs, 77–80
    offshore and international operations, 87
    support equipment and facilities, 86
    test-well contributions, 82–85
Nondrilling exploration costs and acquisition costs, 121–127
Nonmonetary exchange, 552–553
Nonoperating interest, 12
Nonoperators, 13
Nonprocessed natural gas, 385
Nonproductive costs, 67
Nonproductive expenses, 63
Nonrisk service contracts, 689
Nonworking interest, 12
North Sea Brent Blend, 360–361
Notices, 491

O

Obligating event, 316–317
Offices, camps, and miscellaneous facilities, 521
Off lease use of gas by operator, 384
Offset clause, 22
Offsetting cash inflows, 321
Offset well situation, **22**
Offshore and international operations
    drilling and development costs, 167
    nondrilling exploration costs, 87
Offshore development drilling, 56–57, 67
Offshore development well plug and abandonment, 57–58, 67
Offshore drilling, 26
Offshore exploratory drilling, 58, 67
Offshore facilities, 57, 67
Offshore G&G, 68
Offshore G&G vs. onshore surveys, 87
Offshore lease acquisition, unproved property, 58–59, 68
Offshore operations, 536
Offshore recompletion, 57, 67
Offshore workover, 57, 67
Oil, gas, and mineral lease, **17–20**
Oil and gas accounting introduction, 37–70
    chart of accounts for a full cost company, 66–70
    chart of accounts for a successful efforts company, 55–63
    glossary of common terms, 38–43
    historical cost accounting methods, 44–48
    historical development of accounting methods and current status, 48–50

successful efforts accounting, 51–54
Oil and gas accounting methodologies, 32
Oil and gas disclosures, 617–666
    accounting examples, 619–622
        capitalized costs relating to oil and gas producing activities, 627–631
        changes in the standardized measure of discounted future net cash flows relating to proved oil and gas reserve quantities, 649–666
        costs incurred for property acquisition, exploration, and development activities, 631–633
        proved reserve quantity information, 622–627
        required disclosures, 618–619
        results of operations for oil and gas producing activities, 633–637
        standardized measure of discounted future net cash flows relating to proved oil and gas reserve quantities, 637–648
Oil and gas exploration flowchart, 10
Oil and gas joint interest operations, 50
Oil and gas producing activities, 40
    operations disclosure, 635–636
Oil and gas revenues, 370–372
Oil inventory, 284
Oil revenue accounting examples, 374–375
Oil used off lease by operator, 378–379
Onshore development drilling, 58, 67
Onshore development well plug and abandon, 58, 68
Onshore exploratory drilling, 58, 68
Onshore facilities, 58, 68
Onshore G&G, 68
Onshore recompletions, 58, 68
Onshore workover, 58, 68
Operating agreement and AFE, 536
Operator designation, 293
Operators, 13, 466–467, 490
Option costs and delinquent taxes, 121–127
Option payment, 22
Options, 360
Option to lease, 100–103
Order of drilling, 47, **47**
Ordinary income, 452
Organization of Petroleum Exporting Countries (OPEC), 3
ORI (overriding royalty interest). *See* overriding royalty interest (ORI)
Orifice meter, 363
Overhead, 521–522, 537

Overriding royalty interest (ORI)
    accounting examples, 546
    DD&A on, 200
    retained, 14–15
    sale of, in unproved property, 581
    sale of working interest, in unproved property, 579

P

Paid up leases, 21
Parcel, 449
Partial abandonment, 115
Partial reclassification, 118
Participation factors, 370, 371, 571
Participation factors based on multiple elements, 372
Payback method, 296–297
Payout, 528
Percentage depletion, 452
    100% and 65% limitations, 448
    accounting examples, 447
    lessee's transactions, 446–448
Perforating, 25
Period, beginning of, 183
Period-end, 183
Permeability, 7, 7
Petroleum, origin of, 5–9
Petroleum fiscal systems, 678–679
Petroleum Management Resources System, 622–623
Pipeline gas imbalances, 391, 397–398
Pooled or unitized working interest, 16
Pooling, 465
    accounting examples, 570
    vs. unitization, 16, 30
    and unitizations, 570
Pooling provisions, 22
Porosity, 7, 7
Post-balance sheet events
    accounting examples, 116, 163
    and the ceiling test, 265–266
    drilling and development costs, successful efforts, 163
    SFAS No. 19 on, 115
Post-balance sheet period, 115
Posted field price, 351
Posted prices, 360–361
Prepaids, 55, 66
Pressure base, 381

Pricing of joint account material purchases, transfers, and dispositions, 523–524
Primary recovery, 27
Primary term, 21
Probabilistic methods, 623
Probabilistic reserve estimation, 622
Processed gas, 351
Producer gas imbalances, 391, 393–396
Producing leaseholds-proved property, 55, 68
Production advance, 596–597
Production and sales, 28
Production balancing, 392
Production bonuses, 682
Production costs, 37, 51, 281, **283**, 283
Production costs statements, 292
Production facilities, 25
Production operations, 428–431
Production payment interest (PPI), 14, 547–548
Production payment interests, **590**
Production sharing contracts (PSCs), 678
    accounting for international petroleum operations, 682–683
    cost recovery, 684–685
    government participation, 684
    other terms and fiscal incentives, 688–689
    profit oil, 685–687
    royalties, 683
    signature and production bonuses, 682
Production taxes, 292
Profitability, 295
Profitability index (PI), 301, 302–303
Profit oil, 685–687
Profit oil/profit gas, 685
Project analysis and investment decision making, 296–305
    accounting rate of return, 298–299
    internal rate of return (IRR method), 301
    net present value (NPV) method, 299–300
    payback method, 296–297
    profitability index, 301
Project area, 416
Promissory estoppel, 316
Property classification, 51
Property taxes, 289, 448–449
Proportionate consolidation, 13
Prospecting costs, 75
Proved area, 39, 41, 143
Proved developed oil and gas reserves, 143
Proved developed reserves, 38–39, 43, 181, 623–624
Proved property, 38, 40, 179

Proved property cost disposition, successful efforts, 179–216
    cost disposition through abandonment, retirement, or sale of proved property, 202–216
    cost disposition through amortization, 179–206
    impairment, 216
Proved property cost disposition under successful efforts, 209–213
Proved property sales, 581–588
Proved reserve quantity information
    reserve definitions, 623–624
    reserve quantity disclosure, 624–627
    use of end of year prices, 624
Proved reserves, 43, 181, 234, 237, 623
Proved undeveloped reserves, 39, 143, 624
PSC cost recovery, 686–688
Psia, 351
Pumping service, 289
Purchase in fee, 98–99
Purchase in fee, both FMVs known, 98
Purchase in fee, one FMV known, 99
Purchase of new material, 531
Purchases, 523
Put on test, 398

R

Reassessment, 328
Recapturable IDC and depletion given, 449–450
Recapturable IDC and depletion not given, 450–451
Recapture of IDC and depletion, 449–450
Reclassification of an unproved property, 116
Reclassification of individually significant properties, 117
Reclassification of properties, 245–252
Recognition of inventories, 284
Recompletions, 144, 156, 289
Reconnaissance survey, 9, 77
Recording gas revenue, 380–382
    natural gas processing, 386
    nonprocessed natural gas, 385
    stored natural gas, 386
    take-or-pay provision, 386
    vented or flared gas, 385
Recording oil revenue, 373–379
Recording revenue at the time of settlement, 390

Recoverable versus nonrecoverable costs, 691–692
Recovery processes, 27–28
Refiners, 422
Reg. S-X 4-10
    acquisition costs specified in, 94
    conveyance provisions, 549–550
    DD&A on a fieldwide basis, 188–189
    development costs, 160–161
    estimated future decommissioning costs, 237–238
    exclusion of costs, 239
    exploration costs defined in, 75
    exploratory and development wells, 43
    full cost accounting rules, 228
    full cost ceiling test, 261
    future decommissioning and environmental remediation costs, 313
    material distortion of amortization rate, 604–605
    production activities, 281–282
    production operations, 282
    proved area, 143
    proved developed oil and gas reserves, 143
    proved reserves, 237
    proved undeveloped reserves, 143
    reserve definitions, 623–624
    reserve valuation, 237
    reservoirs, 189
    sales, abandonments, and drilling arrangements, 607
    successful efforts vs. full cost methods, 44, 50
    well classification in, 142
Rentals and royalties, 518
Repairs and maintenance, 288–289
Reserve categories, **41**
Reserve cost ratios
    analysis of oil and gas companies' financial statements, 711–719
    DD&A per BOE, 718–719
    finding costs ratios, 711–716
    lifting costs per BOE, 716–718
Reserve definitions, 623–624
Reserve estimates, 293
Reserve life ratio, 704–708
Reserve quantity disclosures
    proved reserve quantity information, 624–627
    vs. reserve value disclosures, 666

Reserve ratios
    analysis of oil and gas companies' financial statements, 704–711
    average daily production per well, 710–711
    average reserves per well ratio, 709–710
    net wells to gross wells ratio, 708–709
    reserve life ratio, 704–708
    reserve replacement ratio, 704–707
Reserve recognition accounting (RRA), 49
Reserve replacement ratio, 704–707
Reserve report, 626
Reserves at year-end, 428
Reserves in place, 256
Reserves owned or entitled, 181–183
Reserve valuation, 237
Reserve value disclosures vs. reserve quantity disclosures, 666
Reserve value ratios
    analysis of oil and gas companies' financial statements, 719–724
    value added ratio, 722–724
    value of proved reserve additions per BOE, 719–722
Reservoirs, 7, 38
Retailer, 422
Retained ORI, 14–15
Retained production payments, 548, 589–595
    payable in money, not reasonably assured, 593–594
    payable in money, reasonably assured, 590, 591–592
    payable in product, 594–595
    payable in production, 594
Retirement, 195, 314
Retirement obligation, 330
Revenue
    and cost allocation to ownership interests, 367–368
    from crude oil, 388–389
    lessor's transactions, 453–454
    from natural gas, 389–390
Revenue audits, 538
Revenue recognition
    at time of production, 389
    at time of sale, 388
Revenue reporting to interest owners, 391
Revenues and gains, 60, 69
Revision of DD&A rates, 202–205
Revision of quantity, 661
Revisions of quantity estimates and changes in timing of production, 662

Rev. Rule. 77-188, 416–417
Rigging up, 23
Rights to free use of resources for lease operations, 22
Right to assign interest, 22
Ringfencing, 688
Risk service agreement, 690–691
Risk service contracts, 689
Royalties, 683
Royalty expense, 428
Royalty holidays, 689
Royalty interest (RI), 12
Royalty provision, 21
Run ticket, 28, 351, **353**
Run ticket calculation, 354–359
Run ticket data, **356**

S

SAB 106, 261
Sales
    and abandonment-unproved and proved properties, 605–606
    accounting examples, 209
    of divided interest in a proved property, 585–586
    of entire interest in unproved property, 574
    of entire overriding royalty interest in proved property, 583
    of entire unproved property, 575–576
    of overriding royalty interest in proved property, 588
    of overriding royalty interest in unproved property, 580
    of a partial interest in unproved property, 576
    of portion of unproved property-group impairment, 578
    of portion of unproved property-individual impairment, 577
    of property, **574**
    and purchase of entire interest in a proved property, 582
    and purchase of partial interest in a proved property, 584–585
    of significant reserves and DD&A rate distortion, 606
    of a working interest in a proved property with retention of a nonworking interest, 586
    of working interest in a proved property with retention of a nonworking interest, 586–587
Sales and purchases
    of an entire interest in a proved property, 581
    of a partial interest in a proved property, 583
    Sales and transfer prices, net of production costs, 652, 654–655
Sales method, 392
Salt dome, 5
Saltwater disposal systems, 291
Salvageable, defined, 138
Salvage value, 430
Scrubbers, 351
Secondary and tertiary recovery, 290
Secondary recovery methods, 27–28
Secondary recovery system, 290
SEC *Reg. S-X 4-10*. *See Reg. S-X 4-10*
Security and Exchange Commission (SEC), 49, 108
Sedimentary rock, 5
Seismic costs, 162
Seismic studies, 23
Seismic technology, 78
Seismology, 8
Separate tract, 449
Separators, 351
Service contracts, 678, 689–691
Services, 520
Services and facilities, 537
Service wells, 39, 142, 143, 161
Setting pipe, 23
Severance taxes, 292
SFAC No. 7, "Using Cash Flow and Present Value in Accounting Measurements", 317–318
SFAS No. 19, "Financial Accounting and Reporting by Oil and Gas Producing Companies", 44, 49, 563
    abandonment or retirement from catastrophic event, 206
    accounting for suspended well costs, 164
    advances for future production, 599
    amortization rate change, 208
    capitalization rules contained in, 77
    carried interests or sole risk, 563–564
    carved-out production payments, 596, 600
    conveyance provisions, 549–550
    cost disposition through amortization, 181
    cost of acquiring unproved properties, 94

DD&A on a fieldwide basis, 188
DD&A rate. determination, 195
development costs, 160–161
development drilling accounting, 151
exploratory dry holes vs. development dry holes, 141–142
exploratory-type stratigraphic test wells, 146
exploratory wells, 146
farm-in/farm-out, 553
on foreign concession covering a large geographical area, 118
free wells, 560–562
funds advanced, 597
future decommissioning and environmental remediation costs, 313
on impairment, 107–108
internal employee cost allocations, 99
international mineral interests acquisitions, 93
joint venture operations, 568–569
nondrilling exploration costs, 76
nonstandard situations, 159–160
post-balance sheet events, 115, 163
reclassification of an unproved property, 116
retained production payments, 589
retained production payments payable in production, 594
revision of DD&A rates, 202
royalty interests, 199
sale of property, 575
sales and purchases of an entire interest in a proved property, 581
sales and purchases of a partial interest in a proved property, 583
of sales of a partial interest in unproved property, 577
sales of a working interest in a proved property with retention of a nonworking interest, 586
seismic costs, 162
significant properties, 109
surrendered lease expense, 112
types of sharing arrangements, 551
unitizations, 571–572
SFAS No. 25, "Suspension of Certain Accounting Requirements for Oil and Gas Producing Companies", 49
SFAS No. 34, "Capitalization of Interest", 164–165
FASB Interpretation No. 33 (to *SFAS No. 34*), 257

SFAS No. 69, "Disclosures about Oil and Gas Producing Activities", 49–50, 182
  acquisition, exploration, and development costs disclosure, 631–632
  capitalized costs relating to oil and gas producing activities, 627
  changes in standard measure of oil and gas (SMOG), 649, 651–652
  company-wide capitalized costs, 627–628
  disclosure of reserves, 693–694
  disclosures mandated by, 77
  effects of changes in prices and costs, 662
  end of year prices, 624
  financial accounting issues, 692–694
  net proved reserves, 694
  oil and gas disclosures, 617
  oil and gas producing activities operations disclosure, 633–634
  required disclosures, 618
  reserve quantity disclosure, 624–625
  reserve report, 622–623
  standard measure of oil and gas (SMOG), 637–638
  substantial disclosures, 228
SFAS No. 121, "Accounting for the Impairment of Long-Lived Assets and for Long-Lived Assets to Be Disposed Of", 216, 331
SFAS No. 131, "Disclosures about Segments of an Enterprise and Related Information", 617
SFAS No. 133, "Accounting for Derivative Instruments and Hedging Activities", 604
SFAS No. 143, "Accounting for Asset Retirement Obligations (AROs)", 195, 260–261
  accounting for asset retirement obligations, 313
  applicability of, 314
  credit-adjusted risk-free rate, 328
  expected cash flow approach, 318–319
  gain or loss recognition upon settlement, 324
  initial measurement of an ARO liability, 317
  Interpretation No. 47 (*FIN No. 47*), 329
  reporting and disclosures, 330
  retirement, 314
  scope of, 314–315
SFAS No. 144, "Accounting for Impairment or Disposal of Long-Lived Assets", 111, 216, 337
  accounting for asset retirement obligations, 313

asset groups, 332
component of an entity, 338
full cost ceiling test, 265
scope of, 331–332
SFAS No. 153, "Exchanges of Nonmonetary Assets", 551
Sharing factors, 370
Sharing in-kind, 12
Shooting rights, 77, 100, **101**, 418
Short-term notes payable, 59, 69
Shrinkage, 351
Shut-in payments, 22, 289
Shut-in royalty payments, 419
Sidetracking, 159
Signature and production bonuses, 682
Significance guidelines, 108
Significant properties, 242
Signing bonus/signature bonus, 680
Sliding scale royalties, 683
Society of Petroleum Engineers (SPE), 622
Society of Petroleum Evaluation Engineers (SPEE), 622
Sole risk, 527, 563
Solution gas, 350
Sour crude, 360
Special drilling operations and problems
    abandonment of portions of wells, 159–160
    damaged or lost equipment and materials, 158
    drilling and development costs, successful efforts, 156–160
    fishing and sidetracking, 159
    workovers, 156–158
Special drilling situations, **159**
Special inventories, 524
Spot markets, 360
Spot prices, and futures prices, 363
Spot sales, 351, 363
Spud date, 23
Spudded in, 23
Stand, 23
Standard conditions, 361
Standard division order, 363–366
Standard measure of oil and gas (SMOG), 637–638, 640–641
State and federal regulations, 28–30
Statement of Financial Accounting Concepts (SFAC) No. 7, 108
Statements and billings, 516
Stockholders equity and related accounts, 69
Stored natural gas, 386
Strapped, defined, 352

Strapping table, **357**
Stratigraphic condition, 188–189
Stratigraphic test wells, 26, 39, 142
    classification of, 155
    drilling and development costs, successful efforts, 154
Structural feature, 188
Subleasing, 421
Sublessee, 420
Submersible and semisubmersible drilling platforms, 27
Subsequent recognition and measurement, 322–328
Substantial disclosures, 228
Successful efforts (SE) accounting methods (ASR 257), 44, 45–46, 49, 51–54
    acquisition costs, **95, 96**
    acquisition costs example, 97
    chart of accounts for, 55–63
    DD&A (depreciation, depletion, and amortization) under, 253–256
    exploration costs, **76**
    vs. Full Cost, **45**
    vs. Full Cost, entries for, 53–54
    vs. full cost accounting, 695
    vs. full cost accounting examples, 253–256
    internal employee cost allocations, 99
    oil and gas accounting introduction, 51–54
    overview of, **52**
    purchase in fee, both FMVs known, 98
    purchase in fee, one FMV known, 99
Successful efforts (SE) companies
    chart of accounts for, 55–63
    development costs, **152**
    drilling capitalization rules., 157
    exploration costs, **145**
    impairment, 216
    proved property cost disposition, 216
    summary, **180**
Successful efforts vs. full cost methods, 44, 50
Sufficient progress, 164–165
Summary of accounting treatment of costs, successful efforts vs. full cost vs. tax, 441–442
Support equipment and facilities, 253
    accounting examples, 162
    accounting for, 86
    drilling and development costs, successful efforts, 162
    nondrilling exploration costs, successful efforts, 86
Surface owner, 11

Surrendered lease expense, 112
Suspended well costs, 164
Sweet crude, 360

T

Take-or-pay, 387
Take-or-pay provision, 386
Tangible, 138
Tank battery, 351, 352
Tank strapping, 351
Tax accounting for both independent and integrated producers, 437–440
Taxes, licenses, permits, 520
Taxes and insurance, 537
Tax holidays, 689
Technical labor, 519
Terms and fiscal incentives, 688–689
Terms of agreement, 491
Tertiary recovery, 28
Testing for recoverability, 333
Test-well contributions, 82–85
Thief, 351, 352
Thousand cubic feet (Mcf), 380
Three-party top lease, 105
Timing of production, 293
Timing of revenue recognition, 386–390
    revenue from crude oil, 388–389
    revenue from natural gas, 389–390
Titles, 490
Top leasing
    accounting examples, 105–106
    acquisition costs of unproved property, successful efforts, 105–106
    defined, 105
Traditional approach for measuring liabilities value, 318
Traditional present value approach, 335
Transfer from one property to another property, Condition B, 533, 534
Transfer from property to another property, Condition C, 534, 535
Transfer from warehouse, Condition A, 531, 532
Transfer from warehouse to joint property, Condition B, 532, 533
Transportation, 520
Traps
    defined, 5
    types of, **6**
Trigger events, 328

Tripping in, 23
Tripping out, 23
Tubular goods, 291–292
Turnkey basis, 23
Two-stage allocation, 400–401

U

Ultra deepwater, defined, 27
Under construction costs, 257
Undivided working interest, 13
Unevaluated leaseholds-unproved property, 55, 67
Unit conversions, 191
Unitizations, 370–372, 465, 571–573
    accounting examples, 572–573
    defined, 570
    determination of interest, 370–371
    vs. pooling, 16, 30
Unit-of-production formula:, 183
Units of volume, 184
Unproved properties, 40
Unproved property inclusion and exclusion, 240–241
Unproved property sales, 573–581
Un-ringfencing, 688
Unwinding of the discount, 314
Upstream activities, defined, 1
Upstream oil and gas operations, 1–36
    acquisition of mineral interests in property, 11–16
    brief history of the U.S. oil and gas industry, 2–4
    common state and federal regulations, 28–30
    drilling operations, 22–27
    exploration methods and procedures, 9–10
    the future of, 30–31
    lease provisions, 16–22
    origin of petroleum, 5–9
    production and sales, 28
    recovery processes, 27–28
Upward revision in estimate, 325–328
Upward revisions, 324
Used material (Condition C), 523
U.S. Generally Accepted Accounting Principles (GAAP), 77
U.S. Geological Survey (USGS), 77

V

Value added ratio, 722–724
Value of proved reserve additions per BOE, 719–722
Vented or flared gas, 385
Volume units, 184

W

Warehouse depreciation, 198
Warehousing, 537
Well abandonment, 207
Well classification, 142–144
Wellhead prices, 363
Well logging, 25
Wells and equipment, 56, 67
Wells and related equipment and facilities, 146
Wells-in-progress, 146
Well spacing, 28
West Texas Intermediate (WTI), 360
Working interest method, 182, 694
Working interest or operating interest (WI), 12
Workover operations, 289
Workovers, 156–158
Workovers and recompletions, 157–158
World gas reserves, **31**
World oil reserves, **31**
World Petroleum Congress (WPC), 622
Worthlessness, 432